S0-CQP-399

ASTRONOMICAL DATA ANALYSIS SOFTWARE AND SYSTEMS III

A SERIES OF BOOKS ON RECENT DEVELOPMENTS IN ASTRONOMY AND ASTROPHYSICS

A.S.P. CONFERENCE SERIES
BOARD OF EDITORS

Dr. Sallie L. Baliunas, Chair
Dr. John P. Huchra
Dr. Roberta M. Humphreys
Dr. Catherine A. Pilachowski

© Copyright 1994 Astronomical Society of the Pacific
390 Ashton Avenue, San Francisco, California 94112

All rights reserved

Printed by BookCrafters, Inc.

First published 1994

Library of Congress Catalog Card Number: 94-72083
ISBN 0-937707-80-5

D. Harold McNamara, Managing Editor of Conference Series
408 ESC Brigham Young University
Provo, UT 84602
801-378-2298

A SERIES OF BOOKS ON RECENT DEVELOPMENTS IN ASTRONOMY AND ASTROPHYSICS

Vol. 1-Progress and Opportunities in Southern Hemisphere Optical Astronomy: The CTIO 25th Anniversary Symposium
ed. V. M. Blanco and M. M. Phillips ISBN 0-937707-18-X

Vol. 2-Proceedings of a Workshop on Optical Surveys for Quasars
ed. P. S. Osmer, A. C. Porter, R. F. Green, and C. B. Foltz ISBN 0-937707-19-8

Vol. 3-Fiber Optics in Astronomy
ed. S. C. Barden ISBN 0-937707-20-1

Vol. 4-The Extragalactic Distance Scale: Proceedings of the ASP 100th Anniversary Symposium
ed. S. van den Bergh and C. J. Pritchet ISBN 0-937707-21-X

Vol. 5-The Minnesota Lectures on Clusters of Galaxies and Large-Scale Structure
ed. J. M. Dickey ISBN 0-937707-22-8

Vol. 6-Synthesis Imaging in Radio Astronomy: A Collection of Lectures from the Third NRAO Synthesis Imaging Summer School
ed. R. A. Perley, F. R. Schwab, and A. H. Bridle ISBN 0-937707-23-6

Vol. 7-Properties of Hot Luminous Stars: Boulder-Munich Workshop
ed. C. D. Garmany ISBN 0-937707-24-4

Vol. 8-CCDs in Astronomy
ed. G. H. Jacoby ISBN 0-937707-25-2

Vol. 9-Cool Stars, Stellar Systems, and the Sun. Sixth Cambridge Workshop
ed. G. Wallerstein ISBN 0-937707-27-9

Vol. 10-The Evolution of the Universe of Galaxies. The Edwin Hubble Centennial Symposium
ed. R. G. Kron ISBN 0-937707-28-7

Vol. 11-Confrontation Between Stellar Pulsation and Evolution
ed. C. Cacciari and G. Clementini ISBN 0-937707-30-9

Vol. 12-The Evolution of the Interstellar Medium
ed. L. Blitz ISBN 0-937707-31-7

Vol. 13-The Formation and Evolution of Star Clusters
ed. K. Janes ISBN 0-937707-32-5

Vol. 14-Astrophysics with Infrared Arrays
ed. R. Elston ISBN 0-937707-33-3

Vol. 15-Large-Scale Structures and Peculiar Motions in the Universe
ed. D. W. Latham and L. A. N. da Costa ISBN 0-937707-34-1

Vol. 16-Atoms, Ions and Molecules: New Results in Spectral Line Astrophysics
ed. A. D. Haschick and P. T. P. Ho ISBN 0-937707-35-X

Vol. 17-Light Pollution, Radio Interference, and Space Debris
ed. D. L. Crawford ISBN 0-937707-36-8

Vol. 18-The Interpretation of Modern Synthesis Observations of Spiral Galaxies
ed. N. Duric and P. C. Crane ISBN 0-937707-37-6

Vol. 19-Radio Interferometry: Theory, Techniques, and Applications, IAU Colloquium 131
ed. T. J. Cornwell and R. A. Perley ISBN 0-937707-38-4

Vol. 20-Frontiers of Stellar Evolution, celebrating the 50th Anniversary of McDonald Observatory
ed. D. L. Lambert ISBN 0-937707-39-2

Vol. 21-The Space Distribution of Quasars
ed. D. Crampton ISBN 0-937707-40-6

Vol. 22-Nonisotropic and Variable Outflows from Stars
ed. L. Drissen, C. Leitherer, and A. Nota ISBN 0-937707-41-4

Vol. 23-Astronomical CCD Observing and Reduction Techniques
ed. S. B. Howell ISBN 0-937707-42-4

Vol. 24-Cosmology and Large-Scale Structure in the Universe
ed. R. R. de Carvalho ISBN 0-937707-43-0

Vol. 25-Astronomical Data Analysis Software and Systems I
ed. D. M. Worrall, C. Biemesderfer, and J. Barnes ISBN 0-937707-44-9

Vol. 26-Cool Stars, Stellar Systems, and the Sun, Seventh Cambridge Workshop
ed. M. S. Giampapa and J. A. Bookbinder ISBN 0-937707-45-7

Vol. 27-The Solar Cycle
ed. K. L. Harvey ISBN 0-937707-46-5

Vol. 28-Automated Telescopes for Photometry and Imaging
ed. S. J. Adelman, R. J. Dukes, Jr., and C. J. Adelman ISBN 0-937707-47-3

Vol. 29-Workshop on Cataclysmic Variable Stars
ed. N. Vogt ISBN 0-937707-48-1

Vol. 30-Variable Stars and Galaxies, in honor of M. S. Feast on his retirement
ed. B. Warner ISBN 0-937707-49-X

Vol. 31-Relationships Between Active Galactic Nuclei and Starburst Galaxies
ed. A. V. Filippenko ISBN 0-937707-50-3

Vol. 32-Complementary Approaches to Double and Multiple Star Research, IAU Colloquium 135
ed. H. A. McAlister and W. I. Hartkopf ISBN 0-937707-51-1

Vol. 33-Research Amateur Astronomy
ed. S. J. Edberg ISBN 0-937707-52-X

Vol. 34-Robotic Telescopes in the 1990s
ed. A. V. Filippenko ISBN 0-937707-53-8

Vol. 35-Massive Stars: Their Lives in the Interstellar Medium
ed. J. P. Cassinelli and E. B. Churchwell ISBN 0-937707-54-6

Vol. 36-Planets and Pulsars
ed. J. A. Phillips, S. E. Thorsett, and S. R. Kulkarni ISBN 0-937707-55-4

Vol. 37-Fiber Optics in Astronomy II
ed. P. M. Gray ISBN 0-937707-56-2

Vol. 38-New Frontiers in Binary Star Research
ed. K. C. Leung and I. S. Nha ISBN 0-937707-57-0

Vol. 39-The Minnesota Lectures on the Structure and Dynamics of the Milky Way
ed. Roberta M. Humphreys ISBN 0-937707-58-9

Vol. 40-Inside the Stars, IAU Colloquium 137
ed. Werner W. Weiss and Annie Baglin ISBN 0-937707-59-7

Vol. 41-Astronomical Infrared Spectroscopy: Future Observational Directions
ed. Sun Kwok ISBN 0-937707-60-0

Vol. 42-GONG 1992: Seismic Investigation of the Sun and Stars
ed. Timothy M. Brown ISBN 0-937707-61-9

Vol. 43-Sky Surveys: Protostars to Protogalaxies
ed. B. T. Soifer ISBN 0-937707-62-7

Vol. 44-Peculiar Versus Normal Phenomena in A-Type and Related Stars
ed. M. M. Dworetsky, F. Castelli, and R. Faraggiana ISBN 0-937707-63-5

Vol. 45-Luminous High-Latitude Stars
ed. D. D. Sasselov ISBN 0-937707-64-3

Vol. 46-The Magnetic and Velocity Fields of Solar Active Regions, IAU Colloquium 141
ed. H. Zirin, G. Ai, and H. Wang ISBN 0-937707-65-1

Vol. 47-Third Decinnial US-USSR Conference on SETI
ed. G. Seth Shostak ISBN 0-937707-66-X

Vol. 48-The Globular Cluster-Galaxy Connection
ed. Graeme H. Smith and Jean P. Brodie ISBN 0-937707-67-8

Vol. 49-Galaxy Evolution: The Milky Way Perspective
ed. Steven R. Majewski ISBN 0-937707-68-6

Vol. 50-Structure and Dynamics of Globular Clusters
ed. S. G. Djorgovski and G. Meylan ISBN 0-937707-69-4

Vol. 51-Observational Cosmology
ed. G. Chincarini, A. Iovino, T. Maccacaro, and D. Maccagni ISBN 0-937707-70-8

Vol. 52-Astronomical Data Analysis Software and Systems II
ed. R. J. Hanisch, J. V. Brissenden, and Jeannette Barnes ISBN 0-937707-71-6

Vol. 53-Blue Stragglers
ed. Rex A. Saffer ISBN 0-937707-72-4

Vol. 54-The First Stromlo Symposium: The Physics of Active Galaxies
ed. Geoffrey V. Bicknell, Michael A. Dopita, and Peter J. Quinn ISBN 0-937707-73-2

Vol. 55-Optical Astronomy from the Earth and Moon,
ed. Diane M. Pyper and Ronald J. Angione ISBN 0-937707-74-0

Vol. 56-Interacting Binary Stars
ed. Allen W. Shafter ISBN 0-937707-75-9

Vol. 57-Stellar and Circumstellar Astrophysics
ed. George Wallerstein and Alberto Noriega-Crespo ISBN 0-937707-76-7

Vol. 58-The First Symposium on the Infrared Cirrus and Diffuse Interstellar Clouds
ed. Roc M. Cutri and William B. Latter ISBN 0-937707-77-5

Vol. 59-Astronomy with Millimeter and Submillimeter Wave Interferometry
ed. M. Ishiguro and Wm. J. Welch ISBN 0-937707-78-3

Vol. 60-The MK Process at 50 Years: A Powerful Tool for Astrophysical Insight
ed. C. J. Corbally, R. O. Gray, and R. F. Garrison ISBN 0-937707-79-1

Vol. 61-Astronomical Data Analysis Software and Systems III
ed. Dennis R. Crabtree, R. J. Hanisch, and Jeannette Barnes ISBN 0-937707-80-5

Inquiries concerning these volumes should be directed to the:

Astronomical Society of the Pacific
CONFERENCE SERIES
390 Ashton Avenue
San Francisco, CA 94112-1722
415-337-1100

ASTRONOMICAL SOCIETY OF THE PACIFIC
CONFERENCE SERIES

Volume 61

ASTRONOMICAL DATA ANALYSIS SOFTWARE AND SYSTEMS III

Edited by
Dennis R. Crabtree, R. J. Hanisch, and Jeannette Barnes

Stack Scimon 8/2/94

Table of Contents

Part 1. Network Information Systems and On-Line Services

Part 2. Visualization and User Interfaces

Part 3. Archives, Catalogs, Surveys, and Databases

Part 4. Data Analysis

Section A. Image Analysis

Section B. Spectral Analysis

Section E. Radio Astronomy

Section F. Multi-Wavelength Analysis

Part 5. Real-Time Software, Control Systems, and Scheduling

Part 6. Systems Software, Software Development Methods, and Data Structures

Preface

This volume contains the papers presented at the third annual conference on Astronomical Data Analysis Software and Systems (ADASS III) which was held at the Victoria Conference Centre in Victoria, British Columbia, from the 13th to the 15th of October 1993. There were 217 registered participants at the meeting with 52 people representing 16 countries outside the United State and Canada. The number of participants from outside North America emphasizes the fact that ADASS is the world's premier meeting on astronomical software.

The special topics for ADASS III were Surveys and Catalogs, Multiwavelength Analysis, Networking and Distributed Computing, and Image and Spectral Analysis. This volume is organized into sections and subsections which reflect these categories as well as sections on Visualization and User Interfaces, Real-time Software, and Systems Software for which there many contributed papers.

This year's conference also included a number of BOF (Birds of a Feather) sessions on special topics including IDL, Electronic Publishing, IRAF Site Management, and Radio-astronomical Databases. In addition there was an IRAF-STSDAS-PROS-EUVE status report and IRAF panel discussion on the last day of the meeting. There were 13 computer demonstrations which were available during the meeting. The meeting had a notable start with the opening reception which was held in the First People's Gallery of the Royal British Columbia Museum, located next to the Conference Centre on Victoria's Inner Harbour.

The conference was sponsored by the National Optical Astronomy Observatories, the National Research Council of Canada, the Smithsonian Astrophysical Observatory, the Space Telescope Science Institute and the University of Victoria. Corporate sponsors for the conference included APUNIX, Barrodale Computing Services Ltd., Digidyne Inc., Digital Equipment Corporation, NCD Inc., Silcon Graphics Canada Inc., SYBASE Canada Inc., and ZED data. We thank both the sponsoring organizations and the corporate sponsors for their generous support.

The conference Program Organizing Committee was comprised of Carol Christian (Center for EUV Astronomy, U.C. Berkeley), Tim Cornwell (National Radio Astronomy Observatory), Dennis Crabtree (Canadian Astronomy Data Centre, Dominion Astrophysical Observatory), Daniel Durand (Canadian Astronomy Data Centre, Dominion Astrophysical Observatory), Robert Hanisch (Space Telescope Science Institute), F. Rick Harnden, Jr. (Smithsonian Astrophysical Observatory), Douglas Rabin (National Solar Observatory, National Optical Astronomy Observatories), Richard Shaw (Space Telescope Science Institute), Doug Tody (National Optical Astronomy Observatories), and Diana Worrall (Smithsonian Astrophysical Observatory).

The Local Organizing Committee was chaired by Dennis Crabtree and included Daniel Durand, Severin Gaudet, Norman Hill, Gerald Justice, James Hesser and Chris Pritchet. Special thanks are extended to Pat McGuire of Conference Services at the University of Victoria whose experience and hard work was a major factor in the success of the meeting. Thanks are also extended to Cynthia Ho whose co-op student term turned into more than she imagined.

The proceedings of this conference are also being made available on the World Wide Web thanks to the permission of the Board of Editors of the Astronomical Society of the Pacific. This "experiment" is being done so that the Society can prepare for the transition to the era of electronic publishing. Many of the papers at this year's meeting were on the subject of Network Information Systems, inlcuding electronic publishing, which is indicative of the extremely rapid development in this area.

The editors would like to thank Chris Biemesderfer of *ferberts associates* for preparation of the LaTeXstyle files used for the meeting abstracts and these proceedings and for the editing tools which made producing this volume much less work.

Dennis R. Crabtree
Dominion Astrophysical Observatory

Robert J. Hanisch
Space Telescope Science Institute

Jeannette Barnes
National Optical Astronomy Observatories

April 1994

Cover Illustration: Sample map generated by automatic procedures designed for the VLA's FIRST Survey, provided by Rober Becker (see his paper, page 165).

Participant List

Mark Abbott, Center for EUV Astrophysics, 2150 Kittredge St., Berkeley, CA 94720, USA ⟨ mabbott@cea.berkeley.edu ⟩

Hans-Martin Adorf, ESO/ST-ECF, Karl-Schwarzschild-Str. 2, D-85748 Garching, Germany ⟨ adorf@eso.org ⟩

Chris Aikman, Dominion Astrophysical Observatory, National Research Council, 5071 West Saanich Road, Victoria, BC V8X 4M6, Canada ⟨ aikman@dao.nrc.ca ⟩

Miguel Albrecht, European Southern Observatory, Karl-Schwarzschild-Str. 2, D-85748 Garching, Germany ⟨ malbrech@eso.org ⟩

Greg Aldering, University of Minnesota, 116 Church St. SE, Minneapolis, MN 55455, USA ⟨ alg@aps1.spa.umn.edu ⟩

David Allan, University of Birmingham, Space Research Group, Department of Physics & Space Research, Edgbaston, Birmingham B15 5TT, United Kingdom ⟨ dja@star.sr.bham.ac.uk ⟩

Marsha Allen, The Johns Hopkins University, Department of Physics & Astronomy, Baltimore, MD 21218, USA ⟨ allen@msx1.pha.jhu.edu ⟩

Bruce Altner, Applied Research Corporation, 8201 Corporate Drive, Suite 1120, Landover, MD 20785, USA ⟨ altner@fosvax.arclch.com ⟩

Heinz Andernach, Observatoire de Lyon, 9 Avenue Charles Andre, Saint Genis Laval, Cedex F-69561, France ⟨ heinz@image.univ-lyon1.fr ⟩

Salim Ansari, European Space Agency, Via Galileo Galilei, Frascati I-00044, Italy ⟨ salim@esrin.esa.it ⟩

Behram Antia, Center for EUV Astrophysics, 2150 Kittredge St., Berkeley, CA 94720, USA ⟨ behram@ssl.berkeley.edu ⟩

Alex Antunes, Institute of Space & Astronomical Sciences, 3-1-1 Yashinodai, 3-Chome, Sagamihara, Kanagawa 299, Japan ⟨ alex@astro.isas.ac.jp ⟩

Karin Babst, Computer Sciences Corporation/GSFC, 7926 Helmart Drive, Laurel, MD 20723, USA ⟨ KBabst@gsfcmail.nasa.gov ⟩

Pascal Ballester, European Southern Observatory, Karl-Schwarzschild-Str. 2, D-85748 Garching, Germany ⟨ pballest@eso.org ⟩

Raymond Bambery, 4800 Oak Grove Drive, MS-168-427, Pasadena, CA 99109, USA ⟨ Jet Propulsion Laboratory ⟩

Klaus Banse, European Southern Observatory, Karl-Schwarzschild-Str. 2, D-85748 Garching, Germany ⟨ banse@eso.org ⟩

Irene Barg, Steward Observatory, University of Arizona, Room 260, 949 N. Cherry Ave., Tucson, AZ 85718, USA ⟨ barg@as.arizona.edu ⟩

Jeannette Barnes, National Optical Astronomy Observatories/IRAF Group, PO Box 26732, 950 N. Cherry Ave., Tucson, AZ 85726, USA ⟨ jbarnes@noao.edu ⟩

Andrea Baruffolo, Osservatorio Astromomico Di Padova, Violo Osservatorio 4, Padova I-35122, Italy ⟨ baruffolo@astrpd.infn.it ⟩

John Basart, Iowa State University, 333 Durham Center, Ames, IA 50011, USA ⟨ jpbasart@iastate.edu ⟩

Robert Becker, University of California, Physics Department, Davis, CA 95616, USA ⟨ bob@wyrd.llnl.gov ⟩

Gareth Bevan, Center for EUV Astrophysics, 2150 Kittredge St., Berkeley, CA 94720, USA ⟨ gareth@cea.berkeley.EDU ⟩

Chris Biemesderfer, ferberts associates, PO Box 1180, Oracle, AZ 85623, USA ⟨ cbiemes@noao.edu ⟩

Peter Biereichel, European Southern Observatory, Karl-Schwarzschild-Str. 2, D-85748 Garching, Germany ⟨ biereichel@eso.org ⟩

Teresa Bippert-Plymate, Steward Observatory, University of Arizona, Room 276, 949 N. Cherry Ave., Tucson, AZ 85721, USA ⟨ teresa@as.arizona.edu ⟩

Mark Birkinshaw, Smithsonian Astrophysical Observatory, 60 Garden St., Cambridge, MA 02138, USA ⟨ mb1@cfa.harvard.edu ⟩

Richard Bochonko, University of Manitoba, Department of Math & Astronomy, Winnipeg, MB R3T 3M8, Canada ⟨ bochonk@ccu.umanitoba.ca ⟩

Bruce Bohannan, Kitt Peak National Observatory, PO Box 26732, 950 N. Cherry Ave., Tucson, AZ 85726, USA ⟨ bruce@noao.edu ⟩

Peter Boyce, American Astronomical Society, 1630 Connecticut Ave. NW, Suite 200, Washington, DC 20009, USA ⟨ pboyce@blackhole.aas.org ⟩

Bill Boyd, Center for EUV Astrophysics, 2150 Kittredge St., Berkeley, CA 94720, USA ⟨ bboyd@cea.berkeley.edu ⟩

Martin Bremer, Leiden Observatory, PO Box 9513, Leiden 2300 RA, The Netherlands ⟨ bremer@nfra.nl ⟩

Alan Bridger, Joint Astronomy Centre, 660 N. Aohoku Pl., University Park, Hilo, HI 96720, USA ⟨ ab@jach.hawaii.edu ⟩

Luciano Burderi, University of Palermo, Istituto di Fiscia, Via Archirafi 36, Palermo I-90133, Italy ⟨ burderi@ifcaiv.ifcai.pa.cnr.itng ⟩

Howard Bushouse, Space Telescope Science Institute, 3700 San Martin Dr., Baltimore, MD 21218, USA ⟨ bushouse@stsci.edu ⟩

Ivo Busko, Astrophysics Division, INPE, CP 515, 12201-970 Sao Jose Dos Campos, Sao Paulo, Brazil ⟨ busko@stsci.edu ⟩

Derek Buzasi, National Aeronautics & Space Administration, Code SZE, Washington, DC 20546, USA ⟨ dbuzasi@nhqvax.hq.nasa.gov ⟩

Carol Christian, C&M Science Innovations Ltd., 2550 Shattuck Ave., Berkeley, CA 94704, USA ⟨ carol@cea.berkeley.edu ⟩

Judith Cohen, California Institute of Technology, Department of Astronomy, MS 105-24, Pasadena, CA 91125, USA ⟨ jlc@deimos.caltech.edu ⟩

Betty Colhoun, Computer Sciences Corporation, NASA – Goddard Space Flight Center, Code 440.9, Greenbelt, MD 20771, USA ⟨ BColhoun@gsfcmail.nasa.gov ⟩

James Condon, National Radio Astronomy Observatory, 520 Edgemont Rd., Charlottesville, VA 22903, USA ⟨ jcondon@nrao.edu ⟩

Al Conrad, W. M. Keck Observatory, PO Box 220, 65-1120 Mamalahoa Highway, Kamuela, HI 96743, USA ⟨ aconrad@keck.hawaii.edu ⟩

Maureen Conroy, Smithsonian Astrophysical Observatory, 60 Garden St., Cambridge, MA 02138, USA ⟨ mo@cfa.harvard.edu ⟩

Kem Cook, Lawrence Livermore National Laboratory, MS L-401, PO Box 808, Livermore, CA 94550, USA ⟨ kcook@imager.llnl.gov ⟩

Dennis Crabtree, Dominion Astrophysical Observatory, 5071 West Saanich Rd., Victoria, BC V8X 4M6, Canada ⟨ crabtree@dao.nrc.ca ⟩

Richard Crutcher, Beckman Institute, Drawer 25, 405 N. Mathews Ave., Urbana, IL 61801, USA ⟨ crutcher@sirius.astro.uiuc.edu ⟩

Daniele Dal Fiume, Istituto Tesre CNR Italy, Via De' Castagnoli 1, Bologna I-40126, Italy ⟨ daniele@botes1.bo.cnr.it ⟩

Heather Dalterio, American Astronomical Society, 1630 Connecticut Ave. NW, Suite 200, Washington, DC 20009, USA ⟨ dalterio@blackhole.aas.org ⟩

Philip Daly, Joint Astronomy Centre, 660 N. Aohoku Pl., University Park, Hilo, HI 96720, USA ⟨ pnd@jach.hawaii.edu ⟩

Lindsey Davis, National Optical Astronomy Observatories/IRAF Group, PO Box 26732, 950 N. Cherry Ave., Tucson, AZ 85726, USA ⟨ ldavis@noao.edu ⟩

J.-P. de Cuyper, Royal Observatory of Belgium, Ringlaan 3, B1180 Ukkel, Belgium ⟨ jeanpierre@astro.oma.be ⟩

Michele de La Peña, Computer Sciences Corporation, IUE 0BS – 10000A Aerospace Rd., Green Tec 1 Building, Lanham-Seabrook, MD 20706, USA ⟨ delapena@iuegtc.dnet.nasa.gov ⟩

C. de Vries, Laboratory for Space Research Leiden, Niels Bohrweg 2, Leiden 2333 AL, Netherlands ⟨ devries@rulrol.LeidenUniv.nl ⟩

Susana Delgado, Instituto de Astrofisica De Canarias, Via Lactea, E-38200 La Laguna, Tenerife, Spain ⟨ sdm@iac.es ⟩

Janet DePonte, Smithsonian Astrophysical Observatory, 60 Garden St., Cambridge, MA 02138, USA ⟨ janet@cfa.harvard.edu ⟩

Erik Deul, Sterrwacht Leiden, Niels Bohrweg 2, Leiden 2300 RA, The Netherlands ⟨ Erik.Deul@strw.LeidenUniv.nl ⟩

S. G. Djorgovski, California Institute of Technology, Department of Astronomy, MS 105-24, Pasadena, CA 91125, USA ⟨ george@deimos.caltech.edu ⟩

Edmund Dombrowski, Naval Research Laboratory/BDC/SFA Inc., Inglewood Office Community, 1401 McCormick Dr., Landover, MD 20785, USA ⟨ dombrowski@bocv8.nrl.navy.mil ⟩

Bryan Dorland, Naval Research Laboratory, Backgrounds Data Center, Code 7604, 4555 Overlook Ave. SW, Washington, DC 20375, USA ⟨ dorland@bdcv8.nrl.navy.mil ⟩

Daniel Durand, Dominion Astrophysical Observatory, 5071 West Saanich Rd., Victoria, BC V8X 4M6, Canada ⟨ durand@dao.nrc.ca ⟩

Rick Ebert, Infrared Processing and Analysis Center, California Institute of Technology, MS 100-22, Pasadena, CA 91125, USA ⟨ rick@ipac.caltech.edu ⟩

Daniel Egret, Infrared Processing and Analysis Center, California Institute of Technology, MS 100-22, Pasadena, CA 91125, USA ⟨egret@ipac.caltech.edu⟩

Guenther Eichhorn, Smithsonian Astrophysical Observatory, 60 Garden St., Cambridge, MA 02138, USA ⟨gei@cfa.harvard.edu⟩

Keith Feggans, ACC, Inc., 300 69th Pl., Seat Pleasant, MD 20743, USA ⟨hrsfeggans@hrs.gsfc.nasa.gov⟩

Wes Fisher, Dominion Astrophysical Observatory, 5071 West Saanich Rd., Victoria, BC V8X 4M6, Canada ⟨fisher@dao.nrc.ca⟩

Michael Fitzpatrick, National Optical Astronomy Observatories/IRAF Group, PO Box 26732, 950 N. Cherry Ave., Tucson, AZ 85726, USA ⟨fitz@noao.edu⟩

Murray Fletcher, 5071 W. Saanich Rd., Victoria, BC V8X 4M6, Canada ⟨fletcher@dao.nrc.ca⟩

Tom Fuller, University of California, Physics Department, Santa Barbara, CA 93106-9530, USA ⟨fuller@voodoo.physics.ucsb.edu⟩

Jim Fullton, CNIDR, 3021 Cornwallis Road, Research Triangle Park, NC 27709, USA ⟨Jim.Fullton@cnidr.org⟩

Laura Fullton, University of North Carolina, Department of Physics & Astronomy, Phillips Hall, CB 3255, Chapel Hill, NC 27599, USA ⟨laura@gluttony.astro.unc.edu⟩

Tom Garrard, California Institute of Technology, 220-47 Downs, Pasadena, CA 91125, USA ⟨garrard@ipac.caltech.edu⟩

Severin Gaudet, Dominion Astrophysical Observatory, 5071 West Saanich Rd., Victoria, BC V8X 4M6, Canada ⟨gaudet@dao.nrc.ca⟩

Pedro Gigoux, National Optical Astronomy Observatories, Cerro Tololo Inter-American Observatory, Casilla 603, La Serena, Chile ⟨pgigoux@noao.edu⟩

Paolo Giommi, European Space Agency/ESRIM, Via G Galilei, Frascati I-00044, Italy ⟨giommi@esis.esrin.esa.it⟩

Brian Glendenning, National Radio Astronomy Observatory, 520 Edgemont Rd., Charlottesville, VA 22903, USA ⟨bglenden@nrao.edu⟩

Ann Gower, 1615 McTavish Rd. RR#2, Sidney, BC V8L 3S1, Canada ⟨agower@otter.phys.uvic.ca⟩

Perry Greenfield, Space Telescope Science Institute, 3700 San Martin Dr., Baltimore, MD 21218, USA ⟨greenfield@stsci.edu⟩

Philip Gregory, University of British Columbia, Department of Physics, 6224 Agricultural Rd., Vancouver, BC V6T 1Z1 Canada ⟨p_greg@physics.ubc.ca⟩

Gerald Grieve, University of British Columbia, 2219 Main Mall, Vancouver, BC V6T 1Z4, Canada ⟨grieve@astro.ubc.ca⟩

Ted Groner, 2251 W. Calle Comodo, Tucson, AZ 85705, USA ⟨ted@as.arizona.edu⟩

Robert Hanisch, Space Telescope Science Institute, 3700 San Martin Dr., Baltimore, MD 21218, USA ⟨ hanisch@stsci.edu ⟩

Joseph Hardin, National Center for Supercomputing Applications, University of Illinois, 405 N. Mathews Ave., Urbana, IL 61801, USA ⟨ hardin@ncsa.uiuc.edu ⟩

Frank R. Harnden, Smithsonian Astrophysical Observatory, 60 Garden St., Cambridge, MA 02138, USA ⟨ frh@cfa.harvard.edu ⟩

Dan Harris, Smithsonian Astrophysical Observatory, 60 Garden St., Cambridge, MA 02138, USA ⟨ harris@cfa.harvard.edu ⟩

Kimberly Hawkins, Computer Sciences Corporation, 7635 Green Dell Lane, Highland, MD 20777, USA ⟨ KHawkins@gsfcmail.nasa.gov ⟩

Rommie Hawkins, Wellesley College, Department of Astronomy, KNAC/Whitin Observatory, Wellesley, MA 02181, USA ⟨ lhawkins@annie.wellesley.edu ⟩

Jeffrey Hayes, Space Telescope Science Institute, 3700 San Martin Dr., Baltimore, MD 21218, USA ⟨ hayes@stsci.edu ⟩

Andre Heck, Strasbourg Astronomical Observatory, 5 Rue Des Mesanges, F-67120 Duttleheim, France ⟨ heck@cdsxb6.u-strasbg.fr ⟩

John Heise, Space Research Organization Netherlands, Sorbonnelaan 2, Utrecht 3548 CA, Netherlands ⟨ johnh@sron.ruu.nl ⟩

Arne Henden, USRA/NOFS, PO Box 1149, Flagstaff, AZ 86002, USA ⟨ aah@nofs.navy.mil ⟩

James Hesser, Dominion Astrophysical Observatory, 5071 West Saanich Rd., Victoria, BC V8X 4M6, Canada ⟨ hesser@dao.nrc.ca ⟩

Lloyd Higgs, Dominion Radio Astrophysical Observatory, PO Box 248, Penticton, BC V2A 6K3, Canada ⟨ lah@drao.nrc.ca ⟩

Graham Hill, Dominion Astrophysical Observatory, 5071 West Saanich Rd., Victoria, BC V8X 4M6, Canada ⟨ hill@dao.nrc.ca ⟩

Norman Hill, Dominion Astrophysical Observatory, 5071 West Saanich Rd., Victoria, BC V8X 4M6, Canada ⟨ nhill@dao.nrc.ca ⟩

Phil Hodge, Space Telescope Science Institute, 3700 San Martin Dr., Baltimore, MD 21218, USA ⟨ hodge@stsci.edu ⟩

Mark Holdaway, National Radio Astronomy Observatory, PO Box 0, Socorro, NM 87801, USA ⟨ mholdawa@aztec.vla.nrao.edu ⟩

Allan Hornstrup, Danish Space Research Institute, Gl. Lundtoftevej 7, DK-2800 Lyngby, Denmark ⟨ allan@dsri.dk ⟩

J.-C. Hsu, Space Telescope Science Institute, 3700 San Martin Dr., Baltimore, MD 21218, USA ⟨ hsu@stsci.edu ⟩

Stephen Hulbert, Space Telescope Science Institute, 3700 San Martin Dr., Baltimore, MD 21218, USA ⟨ hulbert@stsci.edu ⟩

Gareth Hunt, National Radio Astronomy Observatory, 520 Edgemont Rd., Charlottesville, VA 22903, USA ⟨ ghunt@nrao.edu ⟩

Alan Irwin, 552 Broadway, Victoria, BC V8Z 2G2, Canada ⟨ irwin@otter.phys.uvic.ca ⟩

Masayuki Itoh, Institute of Space & Astronautical Science, 3-1-1 Yashinodai, 3-Chome, Sagamihara, Kanagawa 299, Japan ⟨ itoh@astro.isas.ac.jp ⟩

Robert Jackson, Space Telescope Science Institute, 3700 San Martin Dr., Baltimore, MD 21218, USA ⟨ jackson@stsci.edu ⟩

George Jacoby, Kitt Peak National Observatory, PO Box 27632, 950 N. Cherry Ave., Tucson, AZ 85726, USA ⟨ gjacoby@noao.edu ⟩

Don Jennings, Computer Sciences Corporation, NASA – Goddard Space Flight Center, Code 668.1, Greenbelt, MD USA ⟨ jennings@enemy.gsfc.nasa.gov ⟩

Mark Johnston, Space Telescope Science Institute, 3700 San Martin Dr., Baltimore, MD 21218, USA ⟨ johnston@stsci.edu ⟩

James Jordan, Computer Sciences Corporation, NASA – Goddard Space Flight Center, Code 668.1, Greenbelt, MD 20771, USA ⟨ jmj@enemy.gsfc.nasa.gov ⟩

Gerald Justice, Dominion Astrophysical Observatory, 5071 W. Saanich Rd., Victoria, BC V8X 4M6, Canada ⟨ justice@dao.nrc.ca ⟩

Marin Kalinkov, Institute of Astronomy, Bulgarian Academy of Sciences, 72 Lenin Blvd., Sofia 1784, Bulgaria ⟨ markal@bgearn.bitnet ⟩

Michael Kesteven, ATNF/National Radio Astronomy Observatory, PO Box 0, Socorro, NM 87801, USA ⟨ mkesteve@atnf.csiro.au ⟩

Stephen Kent, Fermilab, MS 127, PO Box 500, Batavia, IL 60510, USA ⟨ skent@fnal.fnal.gov ⟩

Joao Kohl-Moreira, Observatorio Nacional CNPQ, Rua Gal Jose Cristino 77, Rio de Janeiro RJ 20921-400, Brazil

Gerard Kriss, The Johns Hopkins University, Department of Physics & Astronomy, Baltimore, MD 21218, USA ⟨ gak@perseus.pha.jhu.edu ⟩

Reinhold Kroll, Instituto de Astrofisica De Canarias, Via Lactea, E-38200 La Laguna, Tenerife, Spain ⟨ kroll@iac.es ⟩

Iliana Kuneva, Institute of Astronomy, Bulgarian Academy of Sciences, 72 Lenin Blvd., Sofia 1784, Bulgaria ⟨ markal@bgearn.bitnet ⟩

Wayne Landsman, Hughes STX, NASA – Goddard Space Flight Center, Code 681, Greenbelt, MD 20771, USA ⟨ landsman@uit.dnet.nasa.gov ⟩

Zoltan Levay, Space Telescope Science Institute, 3700 San Martin Dr., Baltimore, MD 21218, USA ⟨ levay@stsci.edu ⟩

Jim Lewis, Center for EUV Astrophysics, 2150 Kittredge St., Berkeley, CA 94720, USA ⟨ lewis@cfa.harvard.edu ⟩

Jim Lewis, Royal Greenwich Observatory, Madingly Rd., Cambridge CB3 0EZ, United Kingdom ⟨ jrl@ast-star.cam.ac.uk ⟩

Peter Linde, Lund Observatory, Box 43, S-22100 Lund, Sweden ⟨ peter@astro.lu.se ⟩

Don Lindler, ACC, Inc., 11518 Gainsborough Rd., Potomac, MD 20854, USA ⟨ lindler@hrs.gsfc.nasa.gov ⟩

Bob Link, Canada-France-Hawaii Telescope Corporation, PO Box 1597, Kamuela, HI 96743, USA ⟨ link@cfht.hawaii.edu ⟩

Knox Long, Space Telescope Science Institute, 3700 San Martin Dr., Baltimore, MD, USA 21218 ⟨ long@stsci.edu ⟩

William Lupton, W. M. Keck Observatory, PO Box 220, 65-1120 Mamalahoa Highway, Kamuela, HI 96743, USA ⟨ wlupton@keck.hawaii.edu ⟩

Dyer Lytle, National Optical Astronomy Observatories/IRAF Group, PO Box 26732, 950 N. Cherry Ave., Tucson, AZ 85719, USA ⟨ dlytle@noao.edu ⟩

Maria Maccarone, IFCAI/CNR, Piazza G Verdi 6, Palermo I-90138, Italy ⟨ maccarone@ipacuc.cuc.unipa.it ⟩

Oleg Malkov, Institute of Astronomy, Russian Academy of Sciences, 48 Pyattnitskaya St., Moscow 109017, Russia ⟨ omalkov@airas.msk.su ⟩

Eric Mandel, Smithsonian Astrophysical Observatory, 60 Garden St., Cambridge, MA 02138, USA ⟨ eric@cfa.harvard.edu ⟩

Herman Marshall, Massachusetts Institute of Technology, Room 37-667A, 70 Vassar St., Cambridge, MA 02139, USA ⟨ hermanm@space.mit.edu ⟩

Peregrine McGehee, Gemini Project, PO Box 26732, 950 N. Cherry Ave., Tucson, AZ 85726, USA ⟨ mcgehee@noao.edu ⟩

Thomas McGlynn, Computer Sciences Corporation, NASA – Goddard Space Flight Center, Code 668.1, Greenbelt, MD 20771, USA ⟨ mcglynn@grossc.gsfc.nasa.gov ⟩

Brian McIlwrath, Rutherford Appleton Laboratory, Starlink, Chilton, Didcot, Oxfordshire 0X11 OQX, United Kingdom ⟨ bkm@star.rl.ac.uk ⟩

Stephen Meatheringham, Mount Stromlo & Siding Spring Observatories, Private Bag, Weston Creek Post Office, ACT 2611, Australia ⟨ sjm@mso.anu.edu.au ⟩

Glenn Miller, Space Telescope Science Institute, 3700 San Martin Dr., Baltimore, MD 21218, USA ⟨ miller@stsci.edu ⟩

Doug Mink, Smithsonian Astrophysical Observatory, 60 Garden St., Cambridge, MA 02138, USA ⟨ mink@cfa.harvard.edu ⟩

John Moran, Smithsonian Astrophysical Observatory, 60 Garden St., Cambridge, MA 02138, USA ⟨ moran@cfa.harvard.edu ⟩

Jeffrey Morgan, University of Washington, Department of Astronomy, FM-20, Seattle, WA 98115, USA ⟨ morgan@phast.phys.washington.edu ⟩

Stephen Morris, Dominion Astrophysical Observatory, 5071 West Saanich Rd., Victoria, BC V8X 4M6, Canada ⟨ morris@dao.nrc.ca ⟩

Simon Morris, Dominion Astrophysical Observatory, 5071 W. Saanich Rd., Victoria, BC V8X 4M6, Canada ⟨ simon@dao.nrc.ca ⟩

Rick Murowinski, Dominion Astrophysical Observatory, 5071 W. Saanich Rd., Victoria, BC V8X 4M6, Canada ⟨ mski@dao.nrc.ca ⟩

Stephen Murray, C & M Science Innovations/Smithsonian Astrophysical Observatory, 60 Garden St., Cambridge, MA 02138, USA ⟨ ssm@cfa.harvard.edu ⟩

Maria Negri, Italian Space Agency, Viale Regina Margherita 202, Roma I-00198, Italy ⟨ negri@irmias.ias.fra.cnr.IT ⟩

James Nemec, University of British Columbia, Department of Geophysics & Astronomy, Vancouver, BC V6T 1W5, Canada ⟨ nemec@dao.nrc.ca ⟩

Tom Nicinski, Fermilab, PO Box 500, Batavia, IL 60510, USA ⟨ nicinski@fnal.fnal.gov ⟩

Jan Noordam, NFRA, PO Box 2, 7990 AA Dwingeloo, The Netherlands ⟨ noordam@nfra.nl ⟩

Ray Norris, ATNF, PO Box 76, Epping NSW 2121, Australia ⟨ rnorris@atnf.csiro.au ⟩

Earl O'Neil, Jr., National Optical Astronomy Observatories/KPNO, PO Box 26732, 950 N. Cherry Ave., Tucson, AZ 85726, USA ⟨ eoneil@noao.edu ⟩

Eric Olson, Center for EUV Astrophysics, 2150 Kittredge St., Berkeley, CA 94720, USA ⟨ ericco@cea.berkeley.edu ⟩

Fabio Pasian, Osservatorio Astronomico Di Trieste, Via BG Tiepolo 11, Trieste I-34131, Italy ⟨ pasian@oat.ts.astro.it ⟩

Laszlo Pasztor, Research Institute of Soil Science & Agriculture, Hungarian Academy of Sciences, Herman Otto ut 15, Budapest H-1022, Hungary ⟨ h2295pas@huella.bitnet ⟩

Harry Payne, Space Telescope Science Institute, 3700 San Martin Dr., Baltimore, MD 21218, USA ⟨ payne@stsci.edu ⟩

Jeffrey Pedelty, NASA – Goddard Space Flight Center, Code 934, Greenbelt, MD 20771, USA ⟨ pedelty@jansky.gsfc.nasa.gov ⟩

William Pence, NASA – Goddard Space Flight Center, HEASARC, Code 668, Greenbelt, MD 20771, USA ⟨ pence@tetra.gsfc.nasa.gov ⟩

Jeffrey Percival, University of Wisconsin, Space Astronomy Laboratory, Chamberlin Hall, 1150 University Ave., Madison, WI 53706, USA ⟨ jwp@sal.wisc.edu ⟩

Jeff Pier, US Naval Observatory, PO Box 1149, Flagstaff, AZ 86002, USA ⟨ jrp@nofs.navy.mil ⟩

Benoit Pirenne, ESO/ST-ECF, Karl-Schwarzschild-Str. 2, D-85748 Garching, Germany ⟨ bpirenne@eso.org ⟩

Joe Pollizzi, Space Telescope Science Institute, 3700 San Martin Dr., Baltimore, MD 21218, USA ⟨ pollizzi@stsci.edu ⟩

Chris Pritchet, 3627 Iona Rd., Victoria, BC V8P 4S6, Canada ⟨ pritchet@otter.phys.uvic.ca ⟩

Mauro Pucillo, Osservatorio Astronomico Di Trieste, Via GB Tiepolo 11, Trieste I-34131, Italy ⟨ pucillo@oat.ts.astro.it ⟩

Bo Rasmussen, ESO/ST-ECF, Karl-Schwarzschild-Str. 2, D-85748 Garching, Germany ⟨ bfrasmus@eso.org ⟩

Katherine Rhode, Smithsonian Astrophysical Observatory, 60 Garden St., Cambridge, MA 02138, USA ⟨ rhode@cfa.harvard.edu ⟩

Alan Richmond, Hughes-STX, NASA – Goddard Space Flight Center, Code 664, T2/76, Greenbelt, MD 20771, USA ⟨ richmond@guinan.gsfc.nasa.gov ⟩

Hermann-Josef Roeser, Max Planck Institut Für Astronomie, Koenigstuhl 17, D-69117 Heidelberg 1, Germany ⟨ roeser@mpia-hd.mpg.de ⟩

Chris Rogers, Dominion Radio Astrophysical Observatory, Box 248, Penticton, BC V2A 6K3, Canada ⟨ crogers@drao.nrc.ca ⟩

Mary Alice Rose, Space Telescope Science Institute, 3700 San Martin Dr, Baltimore, MD 21218, USA ⟨ marose@stsci.edu ⟩

Arnold Rots, USRA, NASA – Goddard Space Flight Center, Code 668, Greenbelt, MD 20771, USA ⟨ arots@xebec.gsfc.nasa.gov ⟩

Lee Rottler, UCO Lick Observatory, Natural Sciences II, Santa Cruz, CA 95060, USA ⟨ rottler@lick.ucsc.edu ⟩

Krista Rudloff, Space Telescope Science Institute, 3700 San Martin Dr., Baltimore, MD 21218, USA ⟨ rudloff@stsci.edu ⟩

Raymond Rusk, 1784 Kisber Pl., Victoria, BC V8P 5H7, Canada ⟨ rusk@orca.drep.dnd.ca ⟩

Leslie Saddlemyer, 11304 Piers Rd. RR #3, Sidney, BC V8L 5J5, Canada ⟨ saddlmyr@dao.nrc.ca ⟩

Skip Schaller, Steward Observatory, University of Arizona, 949 N. Cherry Ave., Tucson, AZ 85721, USA ⟨ schaller@as.arizona.edu ⟩

Marion Schmitz, Infrared Processing and Analysis Center, California Institute of Technology, MS 100-22, Pasadena, CA 91125, USA ⟨ mschmitz@ipac.caltech.edu ⟩

Ethan Schreier, Space Telescope Science Institute, 3700 San Martin Dr., Baltimore, MD 21218, USA ⟨ schreier@stsci.edu ⟩

Joseph Schwarz, European Southern Observatory, Via Dell'Orso 7A, Milano I-20121, Italy ⟨ schwarz@hq.eso.org ⟩

Rob Seaman, National Optical Astronomy Observatories/IRAF Group, PO Box 26732, 950 N. Cherry Ave., Tucson, AZ 85726, USA ⟨ seaman@noao.edu ⟩

Dick Shaw, Space Telescope Science Institute, 3700 San Martin Dr., Baltimore, MD 21218, USA ⟨ shaw@stsci.edu ⟩

Amotz Shemi, Tel Aviv University, School of Physics & Astronomy, Tel Aviv 69978, Israel ⟨ amotz@wise.tau.ac.il ⟩

Eliot Shepard, Smithsonian Astrophysical Observatory, 60 Garden St., Cambridge, MA 02138, USA ⟨ shepard@cfa.harvard.edu ⟩

David Silberberg, Space Telescope Science Institute, 3700 San Martin Dr., Baltimore, MD 21218, USA ⟨ silberberg@stsci.edu ⟩

Richard Simon, National Radio Astronomy Observatory, 520 Edgemont Rd., Charlottesville, VA 22903, USA ⟨ rsimon@nrao.edu ⟩

Riccardo Smareglia, Osservatorio Astronomico Di Trieste, Via GB Tiepolo 11, Trieste I-34131, Italy ⟨ smareglia@oat.ts.astro.it ⟩

Oleg Smirnov, Institute of Astronomy, Russian Academy of Sciences, 48 Pyatnitskaya St., Moscow 109017, Russia ⟨oms@airas.msk.su⟩

Eric Smith, NASA – Goddard Space Flight Center, Code 681, Greenbelt, MD 20771, USA ⟨esmith@hubble.gsfc.nasa.gov⟩

William Snyder, Naval Research Laboratory, 4555 Overlook Ave. SW, Building 209, Code 7604, Washington, DC 20375-5352, USA ⟨snyder@bdcv8.nrl.navy.mil⟩

David Stern, Research Systems Inc., 2995 Wilderness Pl., Suite 203, Boulder, CO 80301, USA ⟨stern@rsinc.com⟩

Peter Stetson, Dominion Astrophysical Observatory, 5071 W. Saanich Rd., Victoria, BC V8X 4M6, Canada ⟨stetson@dao.nrc.ca⟩

Betty Stobie, Space Telescope Science Institute, 3700 San Martin Dr., Baltimore, MD 21218, USA ⟨stobie@stsci.edu⟩

Karen Strom, University of Massachusetts, Five College Astronomy Department, GRC-517, Amherst, MA 01003, USA ⟨kstrom@donald.phast.umass.edu⟩

Sandra Terranova, Smithsonian Astrophysical Observatory, 60 Garden St., Cambridge, MA 02138, USA ⟨nova@cfa.harvard.edu⟩

Doug Tody, National Optical Astronomy Observatories/IRAF Group, PO Box 26732, 950 N. Cherry Ave.,Tucson, AZ 85726, USA ⟨tody@noao.edu⟩

Francesco Tribioli, Osservatoria Astrofisico Di ARcetri, Largo Enrico Fermi 5, Firenze I-50127, Italy ⟨ftribioli@arcetri.astro.it⟩

Ginevra Trinchieri, Osservatorio Astronomico Di Brera, Via Brera 28, Milano I-20121, Italy

Benoit Turgeon, Institute for Space & Terrestrial Science, Space Astrophysics Laboratory, Floor 2, 4850 Keele St., North York, ON M3J 3K1, Canada ⟨turgeon@nereid.sal.ists.ca⟩

Francisco Valdes, National Optical Astronomy Observatories/IRAF Group, PO Box 26732, 950 N. Cherry Ave., Tucson, AZ 85726, USA ⟨fvaldes@noao.edu⟩

Ger Van Diepen, NFRA, PO Box 2, 7790 AA Dwingeloo, The Netherlands ⟨gvandiep@nrao.edu⟩

Michael van Hilst, University of Washington, 12300 Roosevelt Way NE, #204, Seattle, WA 98125, USA

Jean-Pierre Veran, Canada-France-Hawaii Telescope Corporation, PO Box 1597, Kamuela, HI 96743, USA ⟨veran@cfht.hawaii.edu⟩

Stephen Voels, NASA – Goddard Space Flight Center, Astrophysics Data Facility, Code 631, Greenbelt, MD 20771, USA ⟨voels@nssdca.gsfc.nasa.gov⟩

Patrick Wallace, Rutherford Appleton Laboratory, Starlink, Chilton, Didcot, Oxfordshire 0X11 0QX, United Kingdom ⟨ptw@star.rl.ac.uk⟩

Steve Wampler, Gemini Project, PO Box 26732, 950 N. Cherry Ave., Tucson, AZ 85719, USA ⟨swampler@noao.edu⟩

Archibald Warnock, Hughes STX, NASA – Goddard Space Flight Center, Greenbelt, MD 20771, USA ⟨ warnock@hypatia.gsfc.nasa.gov ⟩

Barbara Weibel-Mihalas, National Center for Supercomputing Applications, Beckman Institute, Drawer 25, 405 N. Mathews Ave., Urbana, IL 61801, USA ⟨ bmihalas@ncsa.uiuc.edu ⟩

Rick Wenk, AT&T Bell Labs, 600 Mountain Ave., Murray Hill, NJ 07974, USA ⟨ raw@physics.att.com ⟩

Richard White, Space Telescope Science Institute, 3700 San Martin Dr., Baltimore, MD 21218, USA ⟨ rlw@stsci.edu ⟩

A. Wicenec, AIT, Waldhaeuserstr. 64, Tuebingen 72076, Germany ⟨ wicenec@ait.physik.uni-tuebingen.de ⟩

Belinda Wilkes, Smithsonion Astrophysical Observatory, 60 Garden St., Cambridge, MA 02138, USA ⟨ belinda@cfa.harvard.edu ⟩

Winifred Williams, National Optical Astronomy Observatories/NSO Gong Project, PO Box 26732, 950 N. Cherry Ave., Tucson, AZ 85726, USA ⟨ wwilliams@noao.edu ⟩

John Williams, Space Telescope Science Institute, 3700 San Martin Dr., Baltimore, MD 21218, USA ⟨ williams@stsci.edu ⟩

Ramon Williamson II, Space Telescope Science Institute, 3700 San Martin Dr., Baltimore, MD 21218, USA ⟨ ramon@stsci.edu ⟩

George Wolf, SW Missouri State University, Department of Physics & Astronomy, Springfield, MO 65807, USA ⟨ gww836f@smsvma.bitnet ⟩

Andy Woodsworth, Dominion Astrophysical Observatory, Gemini Project, 5071 W. Saanich Rd., Victoria, BC V8X 4M6, Canada ⟨ wdswrth@dao.nrc.ca ⟩

Diana Worrall, Smithsonian Astrophysical Observatory, 60 Garden St., Cambridge, MA 02138, USA ⟨ dmw@cfa.harvard.edu ⟩

Nailong Wu, Space Telescope Science Institute, 3700 San Martin Dr., Baltimore, MD 21218, USA ⟨ nailong@stsci.edu ⟩

Carlos Wuensche, University of California, Physics Department, Santa Barbara, CA 93106-9530, USA ⟨ alex@cfi.ucsb.edu ⟩

Nelson Zarate, Space Telescope Science Institute, 3700 San Martin Dr., Baltimore, MD 21218, USA ⟨ zarate@stsci.edu ⟩

Karl Zimmerman, Max Planck Institut Für Astronomie, Koenigstuhl 17, D-69117 Heidelberg 1, Germany ⟨ zimmerman@mpia-hd.mpg.de ⟩

Third Annual Conference on Astronomical Data Analysis Software and Systems
October 13-15, 1993 Victoria, B.C.

Astronomical Data Analysis Software and Systems III
ASP Conference Series, Vol. 61, 1994
D. R. Crabtree, R. J. Hanisch, and J. Barnes, eds.

Distributed Astronomical Data Archives

Jim Fullton

CNIDR, 3021 Cornwallis Road Research Triangle Park, NC 27709

Abstract. The prevalence of high speed network connectivity and powerful computing platforms has brought the concept of distributed data and image archives to the forefront of network communications research. This paper discusses the applicability of a distributed information model for national and international astronomical data archives based on evolving communications standards. A prototype archive node at the National Optical Astronomy Observatories which utilizes prevalent network information protocols is described, and the extension of this archive into the proposed U.S. National Research and Education Network (NREN) is examined.

1. Issues in the Distributed Data Model

From the viewpoint of the network systems designer, the astronomical community's data location requirements present an ideal testbed for several new network resource management technologies. The community is relatively small and well connected in terms of network access. Modern astronomical research is as nearly dependent upon the use of computers as it is observational instruments. While the value of archival data for research has been recognized, some of the problems inherent in mass storage and retrieval of large amounts of information have discouraged further work.

1.1. Standardized Communications

The adoption of one or more standardized communications interfaces into a distributed archive is of critical importance. In the past, publishers of large information archives have had little recourse but to adopt proprietary communications and authentication protocols. This required information providers and consumers to purchase software, hardware and services from single sources rather than reaping the technological benefits of an open and diverse market based on widely accepted standards. These proprietary systems were developed to overcome the poor performance of earlier computing platforms, which inhibited the development of generalized standards due to the overhead imposed by systems capable of operation over a broad range of communications and computing hardware.

The adoption of standard, non-proprietary communications protocols provides archive designers with tremendous flexibility, both in terms of design and budget. The use of these standards allows the parallel development of both components of the client/server environment by many different parties. Furthermore, the choice of such protocols is usually quite straightforward. Implementations of even the most complex search and retrieval protocols make few demands on system resources, thus allowing multiple protocol access into data collections with minimal overhead.

1.2. Ease of Data Access

An issue of great concern to consumers of astronomical data is ease of access. The term "ease of access" has many implications ranging from the style and functionality of the user interface to security and authentication issues.

For the purposes of this paper, we will totally ignore user interface requirements and concentrate on the system characteristics required for efficient, low maintenance production operation.

1.3. Volume of Data

While the volume of information of interest to astronomers is vast, consisting of many terabytes of image data, the quantity of searchable information is proportionally small. Typically, information that defines the content of images may be found within image headers. These headers can easily be processed and stored in relational databases. Other information such as that found within object catalogs may be stored in the same manner.

Observation notes and logs can be stored as free text, which may then be searched and ranked using standard relevance ranking techniques (Salton 1991). Free text databases typically occupy between 30% to 75% of the space of the original data.

1.4. Storage and Location of Data

We must distinguish between the physical and financial problems involved in the storage and retrieval of astronomical data and the problems inherent in the location of it. By storing data centrally, most resource location problems can be disposed of. Unfortunately, the expense and maintenance involved in central storage have posed significant obstacles to its implementation. If a location-independent model rather than a central-storage model is considered, many of these financial and support obstacles fade. Functionally, this allows the expense of data storage to be spread across many network sites.

This method of data access and management presents several advantages. Prohibitive central storage costs are avoided since the cost of storage is borne by the owner of the data—typically, this cost must be met regardless of the availability of the data. Additionally, the owner of the data has complete control over it—access permission may be revoked at any time by the owner. The owner may mandate any retrieval mechanism or mechanisms, and these mechanisms may be "mixed and matched" by the client access system with no user intervention. Because of its distributed nature, the loss of any data repository or repositories will not affect the entire system. Finally, the central data requirements of the system are quite small and consist solely of the mapping system which links resources to host machines. This system may be duplicated at many sites to provide a level of redundancy adequate for production use.

This distributed system allows relatively small sites such as individual researchers with personal collections of data to exist on an equal footing with large, agency maintained archives which might contain many gigabytes or terabytes of professionally maintained data.

1.5. Ownership of Information and Security

Many network data collections are restricted to access by a particular group, or at least are not available without the permission or notification of the owner. This arrangement necessitates a mechanism for providing access control at the level of the individual data object in the collection. While the best mechanism for securing data is to not make it available on the network, it is commonly desired that data be searchable but not retrievable without explicit permission from the owner. This is an "unrestricted search, restricted retrieve" model, and can be easily implemented with most network information systems. Security issues are controlled in the access protocol, which can be defined on an individual basis by the data owner. In the case of the preferred low level access protocols for the prototype system described below, Z39.50(NISO 1992) and the well-known File Transfer protocol (FTP), access can be restricted on a per user/per image basis using security challenges.

1.6. Diversity of Data

The term "astronomical data" covers a variety of diverse data sets, ranging from image headers (relational), log entries and observers notes (free text), journal articles (formatted text), and image data itself (large volume, essentially unsearchable pixels). This entire range of information needs to be coordinated and combined into a single, reasonably seamless archive system.

2. Network Information Protocols and Applications

A brief discussion of several popular information dissemination technologies is useful before moving into the details of the prototype. More detailed discussions of information systems protocol integration are available.

2.1. Z39.50

Z39.50 is a US National Information Standards Organization (NISO) standard for Information Search and Retrieval. The interoperable European counterpart of Z39.50 is called ISO 10162/10163 Search and Retrieval (SR). Z39.50 is a general information search and retrieval protocol. Unlike the other systems described in this paper, it is a pure protocol and not a stand-alone application. Z39.50 provides mechanisms for controlling access to individual databases, retrieving arbitrarily formatted data and variants of given data objects, customizing search parameters, and other features discussed more fully in the standard and in supporting papers(Lynch 1992).

2.2. WAIS

WAIS(Kahle 1991) is an acronym for Wide Area Information Servers. Based on Z39.50-1988, sometimes called Version 1, WAIS enjoys great popularity as a search and retrieval application and is currently being used for several astronomical data applications (e.g., STELAR—Study of Electronc Literature in Astronomy (Van Steenberg 1993)). WAIS as an application is being upgraded to implement the newer release of Z39.50 — Version 2/Version 3.

2.3. Gopher

Gopher(McCahill 1992)is a resource browsing tool developed at the University of Minnesota. Gopher is designed to present information as a series of menus which can be navigated by the user. Data objects themselves are quite simple — they are either menu items with associated location information or data objects which may or may not be of interest to the user. The Gopher client either displays the menu item or retrieves the data object for perusal by the user.

2.4. World Wide Web

World Wide Web (WWW)(Berners-Lee 1992a) is a hypertext publishing tool developed by Tim Berners-Lee at CERN. WWW allows navigation through hypertext links defined as Uniform Resource Locators (URLs). The user simply specifies the link to be followed and a new data object is retrieved and presented. WWW provides gateway access to several of the other systems mentioned in this paper, including Z39.50.

2.5. NCSA Mosaic

NCSA Mosaic is an integrated information access environment for a variety of network information systems and protocols. Mosaic is primarily a WWW client in that data is presented and accessed in a manner closely related to the hypertext environment presented by WWW. Mosaic allows the use of graphics and sound embedded within text and provides access to many different systems through its own native interface and stand-alone gateways. Mosaic has become quite popular as the client of choice for most information access systems.

3. Uniform Resource Location Protocols

At the most abstract level, astronomical resource location is very similar to other data location problems addressed within the Internet Engineering Task Force (IETF). A data item must be assigned a unique permanent identifier that is completely disassociated from its location information, as well as a description bound to that identifier. Once an identifier is obtained, it must be resolved into this location information, which may then be used to retrieve the data object for local analysis or examination. These permanent identifiers, known as Uniform Resource Names (URNs), their associated transient locators, called Uniform Resource Locators, and ther content descriptors, Uniform Resource Citations (URCs) have been described in a series of Internet Drafts(Berners-Lee 1993)(Emtage Weider, 1992). These drafts have been accepted by the Internet community as the standards for the description and location of Internet information resources.

By binding a permanent identifier to any astronomical data object, its current location on the network and preferred access mechanism can be established with a simple query to a central resource locator server. Given this information, properly constructed software can immediately retrieve the data object from its current storage location using any suitable data retrieval mechanism.

4. A Prototype Distributed Data Archive

A prototype data archive has been created which contains solar magnetograms collected by the National Solar Observatory (NSO). NSO has made daily observations of the solar magnetic field since 1977. These 2048 x 2048 x 2 byte magnetograms provide an interesting test bed for the use of various network information dissemination standards and systems in astronomical archives. The prototype makes use of the well known information protocol HTTP (Hypertext Transfer Protocol)(Berners-Lee 1992b) as implemented in World Wide Web and NCSA Mosaic. The archive itself is implemented on a Digital Equipment Corporation Alpha AXP and DECStation 5000/200.

4.1. Searchable Data Components

An integrated HTTP server provides access to the searchable component of the archive. The exact search systems used are changeable, but these changes can occur without the knowledge of the system users. The relatively small searchable component of this archive is ideally maintained at a single site, although it may be spread over multiple sites with only a small loss of search efficiency. We plan to replace the direct HTTP link with a gateway to a Z39.50 Version 3 server. This will greatly extend the search capabilities of the server while providing an additional network communications back end.

Each searchable data component must be entered into the master search directory. At entry time, a unique URN is assigned to the data object. This URN is entered, along with the current URL, into a master resource location server. At this point, the components of a basic URC are also created and stored in the resource location database.

4.2. Relationship Between Functional Components

The relationship between the different components of the system can seem quite complex. It is important to understand the location and function of each of the components already mentioned. URLs are stored in the Resource Location Server, which may be located anywhere on the network. URCs are also stored in the resource location server. A URC in this case consists of usage information, and a brief descriptive "headline". The URC provides enough information to the user to allow a decision to be made regarding the relevance and availability of the information, and an attempt is made to briefly summarize the data as the owner chooses.

Searchable data is stored in the Master Archive Search Server. Although searches are currently performed through a HTTP server linked to a relational database, this will be replaced with a Z39.50 server and gateway. After this enhancement is complete, the client will have no direct connection to the master search server.

4.3. Search Results

When a search is complete, the master search server automatically creates a page of information in Hypertext Markup Language (HTML)(Berners-Lee 1991) format. This page of information consists of: URLs pointing to "thumbnail" and original images, derived from the URNs stored in the master archive mapped against the resource location server; descriptive information from the archives and the URC; derived links to other interesting data objects, such as journal articles referring to the data in question or other observations of the same or similar objects.

The HTML page is not cached—rather, it is immediately passed to the server and then deleted. This feature forces the URLs to be re-resolved after each search, allowing archive data to be arbitrarily moved and re-registered without interrupting usage.

4.4. Data Location and Retrieval

Public data is served through very simple, low overhead HTTP servers. It is very important to note that images are never retrieved after a search and are not moved through any sort of gateway — only their pointers are transferred, embedded in hypertext. The client interacts directly with the server at the data publishing site to retrieve images and relevant textual data. No database management software is required at the image server side.

Secure data can be published through any protocol (Z39.50 and FTP are good examples) that provides access control. Secure data is registered as such through its URL. The URL maintains protocol information, and when a connection through a secure protocol is initiated, authentication procedures can occur. A description of these procedures is outside the scope of this document, but are described in the Z39.50 standard. As a trivial but illustrative example, the URL for a given object can specify FTP as the desired access protocol. The FTP server at the publisher side can require a username and password for access to the data object, which the user must provide through the integrated Mosaic client.

5. Conclusion

Many benefits may be derived from publishing astronomical data through a searchable, distributed archive system. By using the model described in this paper, entire image archives can be seamlessly distributed across the Internet, while providing the semblance of centralized archive management to the user. Security can be fully controlled by the owner of the data, with security options ranging from none to user and password control for specific images within a collection.

A mechanism such as this for distributed image and data access blends well and interoperates fully with other traditional information systems on the Internet, and can be easily upgraded as new communications technologies are developed. Furthermore, the entire system is based on open, non-proprietary technologies which can be integrated into various astronomical data processing applications with ease and consistency.

Acknowledgments. The author wishes to thank several individuals and organizations for their support, encouragement, and critical comments during the preparation of this paper. On the astronomy side, Doug Rabin of the National Solar Observatory for fostering my initial interest in the problem, and Bruce Carney at UNC for encouraging my continued work. On the network side, my colleagues at CNIDR and UNC for indulging my interest in astronomical information dissemination, and Clifford Lynch of the University of California Office of the President and Cecilia Preston at the UC Berkeley School of Information Sciences for the use of their fine facilities while this paper was being written. Digital Equipment Corporation kindly furnished the equipment used for the development and operation of the prototype archive. Last but not least, I thank my wife Laura for taking time during the writing of her observing proposals to review and comment upon this paper, not to mention her constant support and patience.

References

Alberti, B., Ankelesaria, F., Lindner, P., McCahill, M. & Torrey, D. (1992). The Internet Gopher protocol: a distributed document search and retrieval protocol. University of Minnesota

Berners-Lee, T. Caiilliau, R., Groff, J. F. & Pollermann, B. (1992). World Wide Web: The Information Universe. Electronic Networking: Research, Application and Policy, 2(1), 75-77

Berners-Lee, T. (1993) Uniform Resource Locators, RFC 1446

Berners-Lee, T. (1993) Hypertext Markup Language, Proposed Internet RFC

Deutsch, P., Emtage, A. & Weider, C. Uniform Resource Names. (1993), Proposed Internet RFC

Fullton, J. (1993) Intelligent Information Retrieval: The Case of Astronomy and Related Space Sciences, 113-117. Kluwer Academic Publishers

Lynch, C. A., Hinnenbusch, M.. Peters, P. E. & McCallum, S. (1990).

Information Retrieval as a Network Application. Library Hi Tech. 8,(4), 57-72

Kahle, B., Morris, H., Davis, F. & Tiene, K. (1992) Wide Area Information Servers: An Executive Information System for Unstructured Files. Electronic Networking: Research, Applications and Policy, 2(1), 59-68

National Information Standards Organization. (1992). Z39.50 Version 2

Van Steenberg, M., Warnock, A., Brotzman, L. & Gass, J. (1992). STELAR: A Study of Electronic Literature in Astronomy. National Aeronautics and Space Administration

Astronomical Data Analysis Software and Systems III
ASP Conference Series, Vol. 61, 1994
D. R. Crabtree, R. J. Hanisch, and J. Barnes, eds.

Gopher and World Wide Web - Successors to FTP

Robert E. Jackson

Astronomy Programs, Computer Sciences Corporation, Space Telescope Science Institute, 3700 San Martin Drive, Baltimore, MD 21218

Abstract. Gopher and World Wide Web are easier to use and more powerful than Anonymous FTP. Information providers should build their information delivery systems around Gopher or World Wide Web instead of Anonymous FTP.

1. Introduction

Gopher and World Wide Web (WWW) were created with the goal of providing easy access to distributed information. Gopher users navigate a network of menus, while WWW users navigate a network of hypertext documents.

Like FTP, Gopher and WWW allow the information provider to serve a file system directory tree. Unlike FTP, which can only access information on one server at a time, a Gopher menu or WWW hypertext document can point to any file or directory located on any FTP or Gopher or WWW server. A menu item can also start a Telnet session, query a Wide Area Information System (WAIS) index, or run a program on the server. Enhancements to Gopher and WWW systems are providing multi-field forms and per-user security. From the perspective of an information provider, Gopher and WWW provide capabilities far beyond those of Anonymous FTP.

2. Better User Interface

Gopher and WWW can provide more meaningful information about a file or directory than FTP. With FTP, the user sees the names used by the filesystem for the directories and files and often those names are cryptic or misleading. If a README file providing more information is present, it can be difficult to view both the README file and the FTP directory listing at the same time.

With Gopher, the mapping of the filesystem onto the menu is completely under the control of the information provider. File and directory names can be replaced by a line of text and the order of menu items can be changed to a logical order instead of an alphabetical order. Files outside the directory can be listed or files can be hidden from view.

WWW provides even more control over the display presented to the user. A directory can be displayed as a structured document with any text in any order and with extensive control over the font size, indentation, etc. While Gopher allows only one line of text to describe a directory or file, WWW can provide several pages for the description or it can put links to several files or directories in the same line.

3. Information Discovery Tools

There is little point in making information available when it is difficult for the user to discover the existence of or location of the information. Information served by Gopher or WWW can be easily "discovered" globally or within a single server. With FTP, it is much harder to find which server contains the desired information or where the information is in the server.

If the Gopher server is registered with the central registry at the University of Minnesota, then all the menu entries for that Gopher server are indexed and can be searched by the "Veronica" tool. The user does not have to know the address of any server or the directory path of any file. They simply ask Veronica for all menu items containing a certain pattern. Veronica is the Gopher equivalent of FTP's "Archie".

Gopher and WWW provide a convenient framework to create catalogs and indices of links to related information. The Usenet University Network Academy has indexed a large number of Internet accessible resources. At ST ScI, "Astronomical Internet Resources" is a catalog of WWW servers, Gopher servers, WAIS indices, Telnet sessions, and FTP servers of potential use for astronomers. Such indices and catalogs will be incomplete as long as information providers do not publicize their resources and rely on the catalog maintainers to discover the existence of the resources.

Gopher and WWW also have tools which make it easy to find information within a single server. All the Gopher menu items in a server can be indexed and made searchable, thus eliminating the need to wander through the directory tree searching for a specific file. The WAIS tools can be used to index the contents of a filesystem, thus allowing the user to find the file with specific contents. These search tools are automated and do not require manually created indices or README files.

4. Link Distributed Information

Gopher and World Wide Web make it very easy to connect information residing on different machines. A single Gopher menu or WWW document can point to resources located anywhere on the Internet. Astronomers in Europe, Canada, and Australia can user Gopher or WWW to build a single information system rather than presenting the user with three separate systems. FTP provides no means to automaticly jump from one server to another.

5. Easier To Set Up Server

It is easy to set up a Gopher server. With the Unix software from the University of Minnesota, the information provider executes the following commmand, where the "config-file" contains server parameters, such as file types to ignore and access permissions.

```
gopherd -c -o <config-file> <information-directory> <portnum>
```

It is equally easy to set up a WWW server. With the Unix software from the National Center for Supercomputing Applications, the information provider executes the following command, where the "config-file-directory" contains a few files which determine the port number, information directory, etc.

```
httpd -d <config-file-directory>
```

Because both Gopher and WWW can use non-protected ports, i.e., above 1024, any user connected to the Internet can make their information accessible, without the assistance of a System Manager.

Setting up an information server using Anonymous FTP is much more complicated and requires several directories, special files, and specific permissions. Because FTP uses a protected port number, the installation procedure also requires access to a root or superuser account.

6. Serve More Users

Gopher and WWW use stateless protocols, i.e., they serve the client request and shut down. The only time a user has a process running on the server is when they are getting information. When a user is staring at their client screen, the server is free to handle other users. A Gopher or WWW server can serve many more "simultaneous" users than FTP, which creates two processes for each session. Alternately, a smaller machine can be used for the Gopher or WWW server.

7. Server Availability

Gopher and WWW server software is available for the most of the platforms. currently used by astronomers. Gopher servers are available for Unix, VMS, Macintosh, VM/CMS, MVS, and DOS PC systems. WWW servers are available for Unix, VMS, Macintosh, and VM/CMS systems and systems which can run Perl. WAIS servers are available for Unix and VMS systems. Gopher and WWW clients are available for even more systems.

8. Alternative to Telnet or Custom Server

Gopher or WWW servers can provide almost as much functionality as a Telnet or custom server at a fraction of the software and manpower costs.

Various reasons contribute to the reduced software and manpower costs. Since existing clients are used, no effort is needed to develop or distribute client software. Because Gopher and WWW are widely installed, public domain systems, a great many people are writing tools for them and making those tools available. It is easy for a Gopher or WWW server to run existing local programs. They can also access tools written in interpreted scripting languages, e.g., Perl or Tcl, which speeds the creation of small tools. WAIS provides a powerful search tool, and WAIS indices are easy to create and easy to access from a Gopher or WWW server.

An information system built around Gopher or WWW would also support more "simultaneous" users than a Telnet or custom server. Each Gopher or WWW server user creates a relatively small process only when they are obtaining information. Each Telnet server user creates a relatively large process for the duration of their entire session.

The NCSA Mosaic client can be used as a graphical user interface and its widgets include: Button, Text Entry Box, Check Box, Radio Button, Selection List, and Select Region of a Bitmap. Several of the available WWW servers can process the information from these widgets and send it to tools and programs residing on the server. Any new information systems should seriously consider using the WWW/Mosaic tools instead of writing large amounts of custom server or client code.

Astronomical Data Analysis Software and Systems III
ASP Conference Series, Vol. 61, 1994
D. R. Crabtree, R. J. Hanisch, and J. Barnes, eds.

In the Jungle of Astronomical On–line Data Services

Daniel Egret[1]

IPAC, Caltech, Pasadena, CA 91125, U.S.A.

Abstract. The author tried to survive in the jungle of astronomical on–line data services. In order to find answers to common scientific data retrieval requests, he had to collect many pieces of information, about a number of different data bases, catalog services, or information systems. And, before long, he knew he would have to navigate through several systems as no one was offering a general answer to his questions.

1. Introduction

I want to propose here some elements of classification which should help the user to evaluate how adequate the different information services are for providing satisfying answers to specific queries. For that, many aspects of the user interaction have to be considered: documentation, access, query formulation, functionalities, qualification of the data, overall efficiency, etc.

2. The Jungle of Astronomical On-Line Services

The number of on-line astronomical services (data bases, datasets, catalogs, archives, `ftp`-services, gophers, WAIS indexes, information systems, directories of services, World Wide Web, etc.) is such that we now need not simply directories, but directories of directories, and that we need homogeneous interfaces to underlying individual heterogeneous interfaces... The paradox is then that simple things tend to lose visibility, in the same time that sophisticated functions are made available as buttons on an X-window.

Very often the user has no adequate access to the information about these services: what is the actual data contents ? how frequently is it updated ? what are the data retrieval functionalities, and limits ?

Obviously, important efforts are made, here and there, to help organizing the information: ADASS meetings, ALDB meetings (Astronomy from Large Databases), or the books by Albrecht and Egret (1991), Heck and Murtagh (1993) are recent examples.

But astronomers are not necessarily interested to attend these meetings or read these books... until they face data retrieval problems, in their scientific work.

---

[1]Currently on leave from CDS, Strasbourg, France

3. Exploring Different Ways to Retrieve a Set of Data

I will take here an example from the common astronomical activity:

> "I want to list all the stars from the SAO Catalog lying within $\alpha = 12$ to 13 hours, and $\delta = -20$ to $-10°$ ".

This is a simple standard request, involving a reasonably large area of the sky ("I can't do that by hand"), and that can be solved typically in the following different ways: through a local customized implementation, which in generally implies a preliminary data import; through catalog handling systems; through integrated databases; through general purpose information systems.

I will show in more details the characteristics of these public services, as they are available now.

But, before that, I want to note that all these approaches are equally essential in the landscape of astronomical data handling, all contributing to different aspects of it. And, furthermore, all rely on the work of those (from individual scientists and technicians, to data centers) who contribute in: collecting data, homogenizing formats, cross-identifying lists and catalogs, providing bibliographical references, writing documentation, etc.

3.1. Local Answers

The SAO catalog is one of the most widely distributed catalogs: many institutes have installed a local computer version, and provide their local users with simple pieces of software allowing to select SAO stars in a given area of the sky (and to make overlays for Palomar plates, for instance).

This approach may be efficient in specialized domains, when there is local expertise on the data, but, in all other cases, we expect databases and information systems to better handle these aspects for us, in order to avoid redundant work: we know that observatories are in general eager to optimize the use of their resources (storage space) and manpower (maintenance and documentation)!

3.2. Data Browsers and Catalog Handling Systems

Examples of such publicly available services are STARCAT, and XCATSCAN (Gaudet et al., Ebert et al. 1994). Other examples are HEASARC/BROWSE which serves the High-energy community, or DIRA2 which serves the Astronet community in Italy.

These systems include functionalities allowing the user to: select a subset from one or several catalogs, select fields from a catalog, build result files, import them to his/her home computer. They also frequently propose additional data analysis tools.

With the increasing number of computerized catalogs made available each year, the user will have to know in advance which catalog he/she wants to use in order to find his/her way through the jungle (see, e.g., Andernach 1994).

Once the catalog is selected, like in our SAOC example, the user will have to submit the query in a SQL or pseudo-SQL mode: as this language may not be familiar to all users, it may be interfaced by a "query by form" approach (STARCAT) or a Graphical User Interface (XCATSCAN). While the processing time is in general not critical (in our example it will not exceed a few minutes), the learning curve may be too long for unfrequent users when only command-line interfaces are available.

3.3. Databases

Integrated databases such as the NASA/IPAC Extragalactic Database (NED), and the SIMBAD astronomical database, offer a new specificity: the organization of the information is based on the careful cross-identification of the astronomical objects; this identification can then be used as a pointer for retrieving compiled data and bibliographical references regardless of the different names and aliases of the object used therein.

As explained by Schmitz (1994) in this Conference, adding a new catalog is not just a question of format and data handling, but addresses the scientific problem of actually identifying the sources before folding the catalog into the database. For this reason, there is generally a significant delay between the publication of the data through the data centers, and its integration in the databases.

Both SIMBAD and NED are optimized for query by names and by positions, and are less efficient when dealing with full-sky queries.

NED allows to search for all the extragalactic objects (of course our example of the SAOC is not relevant here, but a similar query would be listing all extragalactic IRAS sources in the field), within a search radius of 5 degrees centered on any given position. Constraints on Object type and Redshift can be added. In the recent X Window version, NED also offers SAO Overlay plots as a help for locating and identifying sources.

SIMBAD allows to specify constraints on coordinates, magnitudes, and a series of other parameters. A Graphical User Interface, is here also, in preparation, and should help the user to overcome the current difficult syntax needed for constraints specification.

A number of specialized databases and archives dealing with specific instruments, or space missions, are not considered here: they play a critical role in the astronomers' activity, but as they propose an exclusive access to the data, there is no question of alternate choice, and the related user community is in general well kept informed of the functionalities of the archives.

3.4. Information Systems

General Information Systems such as ADS or ESIS are able to answer a wider range of queries (a large part of them being forwarded to specialized information servers such as those considered earlier).

Ideally, the user should first refer to these more general systems, especially when he/she needs additional information before formulating and refining the query. But we are not in an ideal world...

Also, information services currently found on the networks, such as WAIS servers, gophers, and World Wide Web are excellent for pointing to existing resources. However they do not in general address the question of data selection.

In our example, the ADS would help the user, through a Graphical User Interface, to select the SAO Catalog, among a wide variety of catalogs, and then to formulate a SQL query, which would produce the requested table.

ESIS also offers a convenient approach for selecting a catalog (through keywords; the SAOC being absent from the list, I tried the PPM), and for formulating 'search by cone' queries. It is even possible to specify an object name rather than coordinates for defining the center of the field (SIMBAD is used as a name resolver).

4. Conclusions

As a conclusion to this very quick survey, here are some of the parameters to be considered when comparing available services:

> friendliness and ease of use; learning curve; functionalities and modern information retrieval tools; performance (response time); data quality, and data integrity; scientific documentation; network access and file transfer.

Clearly enough these criteria and their weighting are subjective (personal backgrounds, etc.), time-dependent (the landscape is changing very rapidly), place-dependent (for response time and network performances), while the scientific environment and expertise are also important aspects in the choice.

Additional difficulties related to the different levels of heterogeneity are also to be overcome (see Albrecht and Egret 1993).

More details about the different systems mentioned here are to be found in Albrecht and Egret (1991), and in the second volume, in preparation.

Let me add two final comments:

The existing tools (the information systems, the World Wide Web, and the client/server model in general) help the user to navigate from one service to the other until he finds the answer to his/her scientific problem. This implies more than ever a proper documentation of all the scientific aspects of the stored data.

We need to optimize the resources: competition is useful, but an increased collaboration between data managers will also benefit the whole community.

Acknowledgments. Enlightening discussions with Barry Madore, Dave Van Buren, and Miguel Albrecht, are gratefully acknowledged, as well as the kind support of the NED and SIMBAD teams.

References

Albrecht, M.A., & Egret., D., Eds. 1991, "Databases & on-line Data in Astronomy", Kluwer Acad. Publ.

Albrecht, M.A., & Egret, D. 1993, in "Intelligent Information Retrieval", A. Heck & F. Murtagh (eds.), Kluwer Acad. Publ., 135

Andernach, H., et al., 1994, this volume

Ebert, R., et al., 1994, this volume

Gaudet, S., Hill, N., & Pirenne, B. 1994, this volume

Heck, A., & Murtagh, F. 1993, "Intelligent Information Retrieval: The Case of Astronomy and Related Space Sciences", Kluwer Acad. Publ.

Schmitz. M., et al., 1994, this volume

Astronomical Data Analysis Software and Systems III
ASP Conference Series, Vol. 61, 1994
D. R. Crabtree, R. J. Hanisch, and J. Barnes, eds.

The Astrophysics Data System: Distributed Data and Services for the Astronomical Community

Guenther Eichhorn

Smithsonian Astrophysical Observatory, Cambridge, MA

Abstract. The Astrophysics Data System (ADS) is a NASA sponsored effort to make astronomical data and services available to the astronomical community. It is based on a client/server architecture. The data are maintained and stored by the ADS data centers. The users selectively retrieve only data necessary for their research. Services currently available include catalog access to over 250 catalogs, access to data archives at HEASARC (ROSAT), IPAC (IRAS all sky survey), NSSDC (NDADS archive), SAO (Einstein archive), access to over 150,000 astronomy and astrophysics abstracts, the SIMBAD astronomical object database in Strasbourg, France and the NASA Extragalactic Database at IPAC. Local services provided by ADS include a local database system and table editor, as well as plot, image display, mapping, and coordinate conversion tools. The ADS is available to all interested users free of charge.

1. Introduction

The Astrophysics Data System (ADS) is a NASA sponsored project to provide access to astronomical and astrophysical data for the astronomical community. The project started in 1988 and released the first operating version in 1991. This release as well as the second release in April 1992 was based on a character user interface. The third release in November 1992 was based on a much easier to use graphical user interface (GUI). The last version, released in April 1993, provides access to 200 astronomical catalogs and 150,000 astronomy and astrophysics abstracts. The next release, scheduled for January, 1994 will increase the number of catalogs to about 250 and in addition will provide access to several data archives (IRAS archive at IPAC, Einstein archive at SAO, NDADS archive at NSSDC).

The ADS's goal is to enable science, especially multi-mission, multi-spectral interdisciplinary science. It was clear very early on, that it was not feasible to concentrate all the data in one location. The maintenance of such a data repository would overwhelm any reasonable budget. It was therefore decided to base the ADS on a heterogeneous distributed client/server architecture.

The ADS is not a data analysis system. There are several analysis systems available that have been specifically developed to handle astronomical data. The ADS is a bridge between the data at the data centers (nodes) and the data analysis system. As such it provides the user easy access to a wide variety of data and retrieves and stores them in a form that can be imported into the

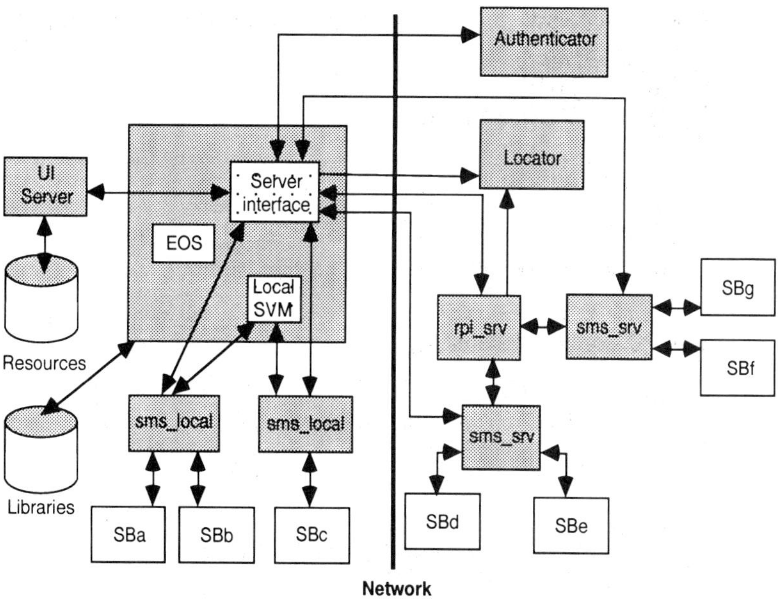

Figure 1. ADS System Architecture.

data analysis systems. ADS does provide capabilities for the user to display and preview the data and do some simple data manipulation.

2. ADS Architecture

The ADS system is a distributed data access system. The data are stored at various nodes. These nodes store and maintain the data that they want to make available through ADS. The user is provided with the ADS client software that can contact the different nodes and select and retrieve data from these nodes over the network in real time. This allows the nodes to update their data at any time. The users will always have the latest version of the data available. Since the system allows the user to select data in real time, only the data needed for a particular research project need to be stored on the user machine.

Figure 1 shows a schematic overview of the ADS architecture. The gray boxes are parts of the ADS supplied software. The white boxes are the server bodies that perform the services. They are written by the service provider. On the left side is the software that resides on the client machine, on the right the node software.

The Locator is software that stores and makes available the knowledge about the location of the available services. Whenever a service is started up by running the Remote Process Invocation Service (rpi-srv) it broadcasts its existence to the Locators along with a property list. This allows subsequently the client to find this service through the locators.

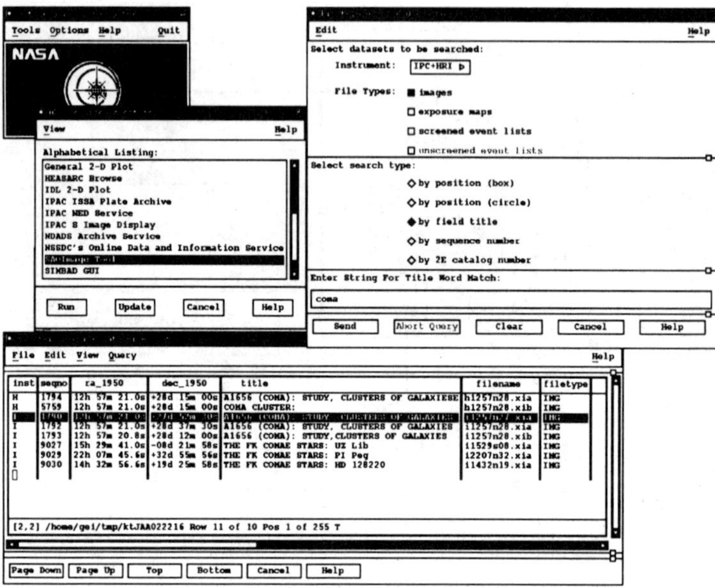

Figure 2. Einstein Archive Query.

When a users starts the ADS on her/his workstation, the software on the left half of Figure 1 is started. The EOS kernel requests a password from the user and authenticates it through one of the nodes with a Kerberos Authentication Server. The user interacts with the system through a graphical user interface (GUI). Whenever the user requests service, the system checks if the service is available locally or remotely. In the case of a local service, a local Session Manager Service (sms-local) is started on the client machine. For a remote service, the client contacts one of the locators to find the address of the requested service. It then contacts the server node and requests a session from the rpi-srv. After the session (sms-srv) is established the client communicates directly with the sms-srv. The session manager (either local or remote) in turn starts the server body that performs the requested service. The communication between the sms and the server body is through Standard Input/Standard Output. This makes it very easy to build and debug a server body or to incorporate an existing service into ADS.

The user interfaces are developed within ADS. They use X/Motif as the underlying windowing software. The software is written in an interpreted language called C-Lite which closely resembles C. This makes it easy to develop and debug GUI code.

3. Example

Figure 2 shows an example of an archive query to the Einstein archive. It shows

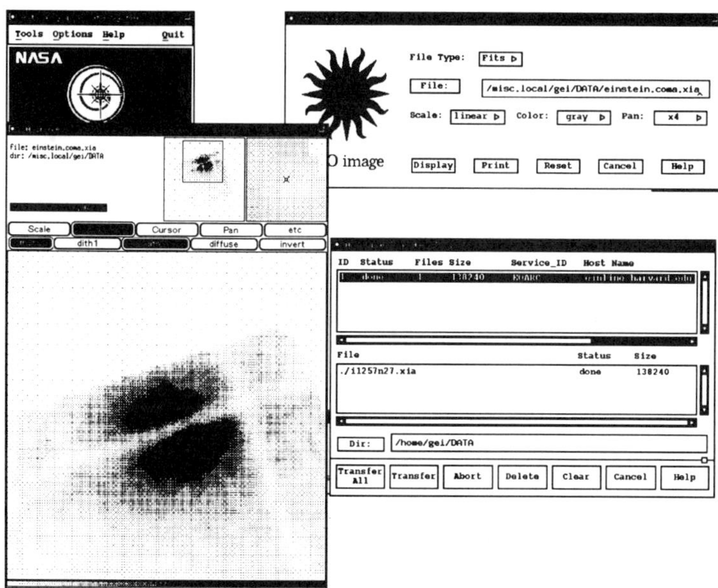

Figure 3. Einstein Image Display.

the service selection windows on the top left, the Einstein archive query window on the top right and the results window on the bottom. A query for "coma" in the title resulted in the selection of eight archive files. One of these files was selected and its retrieval requested. Figure 3 shows the file transfer window with the "done" status for the file transfer (lower right). The file was then displayed with SAOimage (Figure 3 bottom left). The ADS has a integrated interface to SAOimage (top right of Figure 3). This allows the user to call and control SAOimage from within ADS. This capability allows the user to quickly determine whether the retrieved images are indeed what is necessary for the intended research and to perform some preliminary data analysis.

Acknowledgments. The ADS is a cooperative effort of groups at SAO in Cambridge, MA, IPAC in Pasadena, CA, CASA in Boulder, CO, Pennsylvania State University, Space Telescope Science Institute in Baltimore, MD, CEA in Berkeley, CA, and HEASARC, NSSDC, GRO, and IUE at Goddard Space Flight Center. Also cooperating is the SIMBAD project in Strasbourg, France. The project is funded by NASA grant NCCW-0024.

Astronomical Data Analysis Software and Systems III
ASP Conference Series, Vol. 61, 1994
D. R. Crabtree, R. J. Hanisch, and J. Barnes, eds.

The European Space Information System (ESIS)

P. Giommi and S. G. Ansari

ESIS, Information Systems Division of ESA, ESRIN, Via G. Galilei, 00044 Frascati, Italy

Abstract. The European Space Information System (ESIS) is an ESA service to access, retrieve and compare data from a number of remote archives of astronomical data. By means of a graphical user interface, ESIS provides simple and uniform access to data from the IUE, EXOSAT, HST, *Einstein* databases and to a large amount of bibliographical information. ESIS also gives access to SIMBAD and more than 50 astronomical catalogues and mission logs including GINGA, ROSAT, EUVE and GRO. From the user's point of view all the local and remote data archives and catalogues appear as a single large database. ESIS services are available via remote access through computer lines or through the ESIS client software that can be installed on the user's machine.

1. Introduction

The European Space Information System (ESIS) is an ESA service to browse, retrieve and manipulate data from a number of remote and heterogeneous databases. ESIS supports the fields of Astronomy and Space Physics. The ESIS software is a second generation database system that provides uniform access to astronomical catalogues and data products such as images, spectra and lightcurves, independently of the site where the data are actually stored. After the retrieval the data can be manipulated and compared in a variety of ways and can be exported to specialized packages for detailed analysis. A bibliographic service that makes use of SIMBAD and of a database including references and abstracts of more than 500,000 scientific publications is also provided. Information about astronomical sources and data sets can easily be accessed through a *single* user-friendly graphical user interface.

ESIS currently gives access to data from the following archives of astronomical data :

EXOSAT, Noordwijk, The Netherlands
IUE, Villafranca, Spain
ST-ECF, Garching, Germany
SIMBAD, Strasbourg, France

Data from some historical archives, i.e., the EXOSAT FOTs archive, the complete set of *Einstein* images and part of the COS-B data, are stored in a central ESIS computer. In addition, many popular catalogues of astronomical objects and the mission logs of ROSAT, GINGA, COS-B, EUVE and other satellites are also available within ESIS.

ESIS provides four main applications (corresponding to the top icons of the ESIS session manager shown in Figure 1) where different kinds of data access and manipulation can be done. The first application handles data from catalogues and mission logs, while the other three concentrate on astronomical images, spectra, and lightcurves respectively. Some of these applications are described in some detail elsewhere in these proceedings (Ansari et al. 1993a, Ansari et al. 1993b)

2. Search Capabilities

Data can be searched in one or a selection of catalogues using several methods:

- Search by cone: with this method searches can be made around any position in the sky with any radius. Coordinates can be input directly or by giving a the name of an object, in which case ESIS first retrieves the coordinates from SIMBAD then performs the search.
- Search by name: Any object name may be given. If the object is known to SIMBAD, the system retrieves all other known names and then searches on the selected catalogues.
- Search by parameters: The names of parameters in all catalogues have been homogenized, so that a search on a given field found in more than one catalogue would retrieve all records in a pre-selected set of catalogues. The parametric search is also used in conjunction with a cone search by putting several conditions or ranges, on which a search is to be made.
- Overview of observation dates: This method provides a way to display an overview on all products accessible through ESIS for a given object. After specifying a name of a source an interactive plot of observation dates vs frequency is automatically built where detailed information and data sets can be retrieved by the click of a mouse.

3. Data Manipulation and Correlation Tools

The ESIS data manipulation applications fall into three different categories:

Imaging This application is activated by clicking on the second icon of the ESIS session manager (Figure 1.). The image application currently gives access to EXOSAT, *Einstein* and COS-B data. In the near future data from HST, IRAS and ROSAT will be made available. Based on a Graphical User Interface, the application links the ESIS catalogues with XIMAGE, a multi-wavelength imaging package (Giommi et al. 1992) commands. By use of one of the ESIS search capabilities images are identified in several ways and are remotely accessed and displayed. From here on, the user has basic analysis tools to combine images (summing or building a mosaic), to

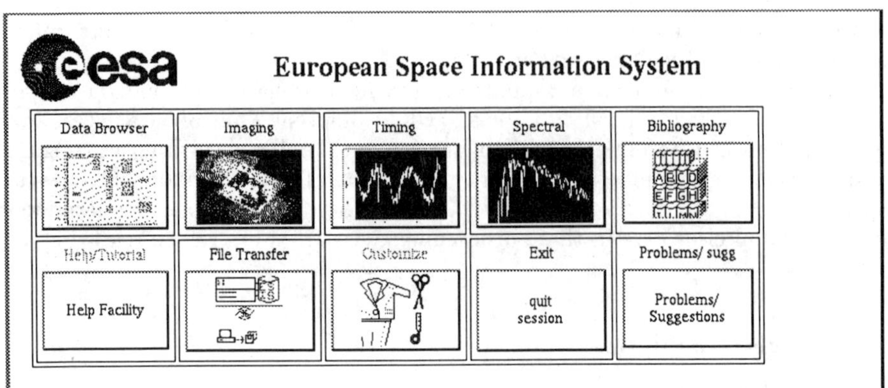

Figure 1. The ESIS session manager.

convert pixel positions to coordinates, to overlay a grid, or overlay known sources from one or several ESIS catalogues.

Since the package also provides a standard FITS reader, the application supports a wide range of images available from CD-ROMs and other sources, including ROSAT events files. The user also has tools within ESIS to import his/her own data.

Native mode of the command line XIMAGE and SAOIMAGE can also be accessed.

Spectral The spectral application (fourth icon on the ESIS session manager (see Figure 1) provides on-line access to IUE Low Dispersion spectra, EXOSAT ME and GSPC data, and IRAS Low Resolution spectra. Searches are made only on those catalogues or missions' logs that contain spectral information. Several spectra can be simultaneously compared. The manipulation features provided include : Radio to X-ray energy distribution automatically constructed from catalogue data, spectral line identification, etc.

In a future release, we intend to provide automatic redshift calculation from catalogues providing redshift values, co-addition and multi-panel plot capabilities as well as XSPEC routines to fit models to X-ray spectra.

Timing The Timing Application (third icon in Figure 1.) provides multiple display and basic manipulation of light curves, available from all the EXOSAT instruments. The list of functions provided include basic statistical analysis, data rebinning, Fast Fourier Transform, Auto-correlation function, folding etc. Native mode to XRONOS (Stella & Angelini, 1992), a widespread analysis package is provided.

4. Bibliography

The bibliographic service (fifth icon of Figure 1) provided within ESIS is based on the NASA Abstracts File and the set of references included in the SIMBAD database. Search capabilities include:

- An object identifier, where ESIS retrieves reference codes from SIMBAD for a given time range and displays the record and the abstract.
- Author(s), references, keywords, or abstracts may all be searched. In the case of abstracts, for example, several words may be provided by the user, and the system retrieves accordingly all the references containing these phrases or words.
- Multiple fields may be involved in one search.

References

Ansari, S.G., Giommi P., Micol A., & Natile P. 1993a "Homogeneous Access to Data: The ESIS Reference Directory", this volume

Ansari, S.G., Giommi P., Ulla A. & Berle H. 1993b "The Spectral Package", this volume

Giommi, P., Angelini, L., Jacobs, P. & Tagliaferri, G. 1992, "XIMAGE: A Multi-Mission X-Ray Image Analysis Package", Astronomical Data Analysis Software and Systems I, A.S.P. Conf. Ser., Vol. 25, eds. D.M. Worrall, C. Biemesderfer & J. Barnes, 100

Stella L. & Angelini L. 1993, "XRONOS: A Timing Analysis Package", Astronomical Data Analysis Software and Systems I,A.S.P. Conf. Ser., Vol. 25, eds. D.M. Worrall, C. Biemesderfer & J. Barnes, 103

Astronomical Data Analysis Software and Systems III
ASP Conference Series, Vol. 61, 1994
D. R. Crabtree, R. J. Hanisch, and J. Barnes, eds.

European Networked Information Resources for Space and Earth Science

S. G. Ansari

ESIS - Information Systems Division of ESA, ESRIN, Italy

R. Albrecht

ST-ECF, Space Telescope - European Coordinating Facility, ESA, Germany

G. Triebnig

Environmental Data Network Section of ESA, ESRIN, Italy

Abstract. Impressive developments can be observed of the so-called *Networked Information Resource Discovery and Retrieval* systems like Archie, Veronica, Gopher, World Wide Web, WAIS, Hyper-G etc. These systems have in common the utilization of the Internet and its client-server architectures. Most of the software of these systems has been put into the public domain and there is a permanent improvement of server technology, graphical clients and gateways to interconnect the systems. Standardization of client server protocols and of exchange data formats is evolving. This paper addresses the emerging Internet information service technologies by asking how they can be of use to the space and earth science communities. A number of European services - all still in demonstration status - are described. The paper considers both the information provider's and the user's point of view.

We summarize the available and planned services, their inter connectivity and the network protocols they use, giving examples in Astronomy, Earth Observation and Space Sciences. In particular, we try to extrapolate into the future on possibilities of providing comprehensive services using common network resources.

1. Introduction

During the initial days of the Internet, two basic network functions were possible: a connection to a tty-based terminal (*telnet*) and a file transfer protocol (*ftp*). The simplicity of these protocols and the widespread use of the Internet in the academic world gave rise to a need to share and exchange software. Software archives were established around the world, giving access to file systems, where users were allowed to download software. The concept of *anonymous ftp* was created. Soon, the number of network sites offering an anonymous ftp grew beyond control. The necessity to provide an overview of what is available became crucial. Many sites were copying files from other sources and providing their local services, because TCP/IP network connections were not reliable enough. The concept of the *Archie* archives was established. File lists were duplicated by several main sites across the North American, the European, the Asian and the Australian continents.

The file naming convention to identify software was, however, not necessarily very efficient. Many users had to download README files from anonymous ftp sites to search for the pieces of information they seek. With the dawn of X-based terminals, this became easy, but cumbersome: one window to connect to the site and the other to read the README files and find the information searched. Another, more coordinated service had to be established. WAIS provided the link between the files and their READMEs. Anybody could join in and set up a server, providing information on their services, or filesystems. It became extremely easy for the user to review the contents of directories and files, by accessing a summary description of each entry in the filesystem. WAIS was and is still used for various purposes, one of which is to give access to bibliographic references. It had one disadvantage, however. After identifying a file on an anonymous ftp site, the user had to still connect to the ftp site, either via an *Xarchie* GUI or from a tty window to download a file. WAIS was a YAG (Yet another GUI) to clutter the Xterminal or workstation.

2. The Philosophy: World Wide Web

Originally aimed for the High Energy Physics community, the World Wide Web was created to meet the needs of the academic world to distribute and share information (Berners-Lee, 1993). Quoting from the on-line WWW presentation:

> *The W3 principle of universal readership is that once information is available, it should be accessible from any type of computer, in any country, and an (authorized) person should only have to use one simple program to access it. This is now the case. In practice the web hangs on a number of essential concepts. Though not the most important, the most famous is that of hypertext.*

Adding on top of this concept the NCSA Mosaic tool, supporting all major hardware platforms and all major Internet protocols in a single Graphical User Interface, provides a powerful networked information discovery, retrieval and collaboration tool.

The World Wide Web along with the NCSA Mosaic has caught on very rapidly in all major institutions and plays a very important rôle in providing access to distributed information.

3. The Astronomical World

Being at the forefront of technology, the astronomical world has caught on. All the required tools are already available. What lacks is coordination among the various groups. It is, however, only a matter of time, when this powerful tool is recognized and used in a cohesive manner.

Many NCSA Mosaic Homepages exist today in Europe, each providing their own services. Most of them are experimental, since user feedback and usage statistics must first be evaluated before services are recognized by the astronomical community at large. In Europe, ESIS and the Space Telescope - European Coordinating Facility in ESO, for example, collaborated very closely on the setting up of their homepages, each providing project-specific information and a more general page of access points across the community. The astronomical resource page (H.M. Adorf, private communications) is currently being widely used by other institutions to give access to other groups. The activity of setting up a homepage can be very stimulating and can bring a number of different archive groups together.

The ESIS homepages may be found in:

http://mesis.esrin.esa.it/html/esis.html

The ST-ECF homepage may be found in:

ftp://ecf.hq.eso.org/pub/WWW/ST-ECF-homepage.html

The astronomical community wants to go further than providing information. The next step is providing on-line newsletters (like the AAS initiative), or bibliographical services (like STELAR or ESIS and other various groups.). But what about the actual data? Will it one day be possible that we share the same interface to give access to distributed catalogues, all having a common protocol? What about the visualization of data products? The community already has the FITS standard format. Couldn't the same be done for the interface to visualize and manipulate the data?

4. The Experiment of the GDS

The ESA Earth-observation Guide and Directory Service (GDS), developed by the European Space Agency (ESA), is a new freely accessible on-line database system providing information about Earth observation data products, systems, applications, etc. The design of the GDS information base – and the GDS user interface metaphor– is targeted towards efficient browsing of the knowledge, based on hypertext principles, in a way predetermined by authors versus finding facts through keywords, full-text or parametric searches. It can, however, also do full-text searches of course. Some of this information includes:

- Landsat images stored in ESA archives.
- Remote sensing experiments.
- Earth observation information, including: public domain digital terrain model data from the US Geological Survey.
- Institution and organization information

The system behind it is called *Hyper G*, developed by an Austrian-Italian university industry team. The system is designed as a general-purpose, large-scale, multi-user, distributed hypermedia system information system and has a seamless integration to the World Wide Web, Gopher or WAIS.

More details on the GDS services may be found in:

http://tracy.esrin.esa.it/ROOT

5. Where to Go from Here ?

Prototypes already exist, where datasets may be identified by a parametric search and extraction preview images (quick-looks) (e.g., quicklooks on the Advanced Very High Resolution Radar (AVHRR) experiment of Earth images. ESIS provides an experimental bibliographic search capability within its homepage, based on an application which accesses a FUL/Text reference collection.

With the advent of Version 2.0 of NCSA Mosaic and the capability of defining forms, there are no limits of imagination as to what can be done. With the simple HTTP protocol, data providers can easily make their data available to the rest of the community. As scientific requirements on such a distributed system begins to unfold, scientific packages may one day be *integrated* within this this interface and tailored to everyone's needs. But a clear goal must be addressed and groups must begin to concentrate on the *standards* in interfacing and accessing data.

References

Adorf, H.M. 1993, ST-ECF Portal,
ftp://ecf.hq.eso.org/pub/WWW/ST-ECF-homepage.html

Ansari, S.G., Micol, A., & Walker, S. 1993, The European Space Information System,
http://mesis.esrin.esa.it/html/esis.html

Berners-Lee, Tim, & Cailliau, R. 1993, Giving this seminar,
http://info.cern.ch/hypertext/WWW/Talks/General/About.html

Bina, E., & Andriessen, M. 1993, About NCSA Mosaic for the X Window System,
http://www.ncsa.uiuc.edu/SDG/Software/Mosaic/Docs/help-about.html

Astronomical Data Analysis Software and Systems III
ASP Conference Series, Vol. 61, 1994
D. R. Crabtree, R. J. Hanisch, and J. Barnes, eds.

Online Access to IPAC Datasets and Services

R. Ebert for the IPAC Tools Group

Infrared Processing and Analysis Center, Caltech, Pasadena, CA 91125

Abstract. In the past, IPAC has developed its tools and maintained its data archives to be maximally useful to visiting scientists working with IRAS data. IPAC is now making the transition to an online archive in order to support not only the continuing demand for access to IRAS catalogs and images, but also in preparation for support of future infrared missions. We discuss the implementation of the new services, Xcatscan and IRSKY, and other systems planned for the future.

1. Introduction

NASA created its data centers to co-locate satellite mission data with the scientific and technical expertise needed to understand and use the data and derived data products. This was a great improvement over the past when data were taken, processed, and given to the investigating science team for analysis. Once that team had moved on to other work, use of the data depended on one's ability to understand the contents of magnetic tapes, printouts, and published literature. The Infrared Processing and Analysis Center (IPAC) was created as a data center to process and analyze data from the Infrared Astronomy Satellite (IRAS) in 1983. IPAC has three main goals: develop data products; maintain expertise in the mission and data products; and provide that expertise and data to the scientific community. This paper is about the last of these goals and our plans for the near future.

For the past ten years, a scientist wanting to use the IRAS mission data products would typically travel to IPAC to consult with experts on the various data products, and to learn and use specific analysis tools. In a time when computer networks were *slow* (1/1000th or less of current bandwidth), this was the only practical method of accessing the *large* volume of data archived.

Networks are now faster, computers are faster, software is more complex, and users are more sophisticated. IPAC is now in a transition period, focusing more of its resources on reaching its users over the network. IPAC will continue to support visitors who travel to IPAC with projects too large or involved to conduct over the Internet, or who do not have access to the Network. The challenge is to provide the online user with the same access to data and expertise as the traveler.

2. The Data & The Missions

IRAS was a joint mission of the US, the Netherlands, and the UK that surveyed 96% of the sky in four wavelength bands (12, 25, 60, & 100 μm) in 1982–83. IRAS returned about 24GB of raw data which produced an archive of 132 GB when combined with time, pointing, and calibration information. From this archive IPAC has produced the IRAS Sky Survey Atlas (ISSA), an all–sky dataset containing over 4GB of flux and position calibrated images with 4 arc-minute resolution, and 10 IRAS specific catalogs containing nearly 1 GB of information including source associations with other major astronomical catalogs. IPAC supports the continuing demand for access to IRAS datasets and 36 other major astronomical catalogs.

As these data products approach their tenth anniversary, IPAC is turning its attention to supporting large new missions. The largest of these will be the Two Micron All-Sky Survey (2MASS) which as currently planned, will survey the sky from the ground in 3 near–infrared wavelength bands (J, H, K) with 2 arc-second resolution. Beginning in 1996 and lasting 2 – 3 years, this project will return nearly 10 TeraBytes of data, at the rate of 10 GB per night, or the equivalent of an IRAS mission every 3 days. The processing is expected to produce over 4 TB of images, and 2 GB of catalog information containing 10^8 stars and 10^6 galaxies. Preparation for this project is already under way. The data processing requirements can be met with currently available processors. However, it is clear that there is much to gained from improvements in data handling and storage technology for providing access to terabyte archives.

IPAC is also the US science support center for the European Space Agency's (ESA) Infrared Space Observatory (ISO). The ISO mission is expected to return 350 GB of data, 30 – 60 GB of which will be for US Observers. In the coming year, IPAC will provide ISO science and instrument expertise, and software to assist the US community with proposal preparation, and to help the US Guaranteed Time Observers (GTO) and Guest Observers (GO) with detailed observation planning. During and after the mission, IPAC will support the community with detailed analyses of the data.

3. The Tools: Interface Design

We have chosen to provide online services reached via the Internet (**telnet**) as opposed to exporting large software applications for several reasons. Foremost is the large size of the datasets which precludes most users from installing them online at their workstations. Second is the great effort required to prepare and support software portable to multiple platforms, operating systems, and environments. This approach also avoids the delays of distributing and installing software upgrades. IPAC services run on IPAC computers and rely on the remote user's MIT X Window System server for display.

Users of online services (so–called *collaborative* systems excluded) have a disadvantage: they cannot ask questions. To take this into account, an online interface must present the information needed to use the service. This has some implications on the systems we have created.

Interfaces must be "easy to learn" and "easy to use." We opted to model in software the type of interaction that a traveling visitor would have with a staff scientist or analyst when learning to use a system at IPAC. The procedural or checklist approach is often effective. For example, first select the catalog of interest, then define the search constraints, third, specify what should be included in the report, and finally submit the search. This method is functional, and familiar to most users, but does not rely on familiarity with a specific interface *style*. Emphasis is on convenient access to functionality rather than on modeling a particular presentation style or convention (look & feel).

Of course the point of these interfaces is to provide quick, convenient, easy access to a large archive of data and information (*meta*–data). However, to "really work" an interface must *codify expertise*. Most interfaces do this to some extent. For example, a warning to the user of input errors such as "you've just entered a DEC greater than 90 degrees...," might be used. We have tried to extend this by choosing default values that guide new users into certain selections and by showing key pieces of information in context.

4. The Applications: Xcatscan & IRSKY

Xcatscan (X–window system Catalog Scanning) is a general search and retrieval tool for accessing IPAC's 46 online astronomical catalogs. Searches can be limited on position or field value constraints, by IRAS associations to major catalogs and IRAS source name (in the case of IRAS catalogs), or by positional associations with a list of sources uploaded by the user via FTP. These search constraints may be defined using a *query–by–example* style input, by user written SQL, or a combination of the two.

The scientist can select fields to be included in the search output individually, or in predefined sets. The user may start with a predefined set and modify it at will. Search results and session log are written to a user specific subdirectory of an anonymous ftp archive for retrieval.

The Infrared Sky (IRSKY) is the centerpiece application to IPAC's support of ISO proposers and observers. It provides, for the first time, interactive online access to the all–sky ISSA. IRSKY displays catalog sources from the IRAS PSC, FSC and Hubble GSC overlaid on ISSA images. Included are calculators to estimate the total sky brightness at any wavelength between 5 and 200 μm, and confusion noise due to faint galaxies and infrared cirrus.

IRSKY provides a summary guide to the ISO focal plane, as a list of instruments and modes available at a given wavelength and spectral resolution. The system provides for graphic overlays of raster scan maps, the ISO focal plane with instrument apertures showing *chopping directions*, and the Sun vector.

Comments on these tools are welcome. For information on these applications contact the author or send e-mail to `tools-comments@ipac.caltech.edu`.

5. The Future: Xscanpi & WWW

An interactive, X-based version of the IRAS Scan Processing and Integration tool (SCANPI) is currently under development and is expected to be available on the Net in early spring 1994. Xscanpi provides access to the nearly 133 GB of IRAS 1-D scans which is kept *near-line* in compressed form. The goal is to reduce special request processing that now takes 1-2 weeks from receipt of request to a 10-20 minute job online.

In addition, the interactive tool will provide the user with the functionality to preview and (de)select individual scans before doing the coaddition. Xscanpi also provides for point source fitting and removal (SCLEAN), and extended source fitting for determining accurate fluxes for galaxies larger than the IRAS beam.

IPAC is also considering use of the World-Wide Web (WWW) for distribution and online access to IPAC documentation and services. Distributed information systems like the WWW and tools like National Center for Supercomputing Applications' (NCSA) Mosaic are promising to change the way we use the network. For those who are already familiar with them, they are helping users stem the tide of information overload on the Net. The WWW and Mosaic in particular seem to be well on their way to establishing themselves as the way to *surf on the Net.* IPAC will continue to look at the emerging standards and tools in this arena to offer its users the benefit of new powerful software.

6. Conclusion

IPAC's vitality derives from the synergy of its three main goals: produce high quality data products, develop expertise in those data products, and provide the astronomical science community with access to those products and expertise. The success of our efforts to use the Network to provide online access depends on our ability to make our services functional and friendly to novice and expert user alike, even as the volume of data and information increases 1000-fold in the next five years.

7. Acknowledgement

All members of the IPAC Tools Group are contributing to the software and systems discussed here. They are J. Bennett, B. Hartley, G. Helou, L. Hermans, I. Khan, G. Laughlin, S. Lord, J. Mazzarella, R. Narron, D. Van Buren, and S. Wheelock. Additional software was provided by S. Woods, and R. M. Melnyk.

IPAC is operated by the Jet Propulsion Laboratory, California Institute of Technology for the Astrophysics division of NASA.

Astronomical Data Analysis Software and Systems III
ASP Conference Series, Vol. 61, 1994
D. R. Crabtree, R. J. Hanisch, and J. Barnes, eds.

SkyView: An All-Sky Data Service

Thomas A. McGlynn[1], Nicholas E. White, Keith Scollick[1]
Goddard Space Flight Center, Greenbelt, MD 20771

Conrad Sturch[1]
Space Telescope Science Institute, Baltimore, MD 21218

Abstract. *SkyView* is a new network service which allows users to get digital images of selected regions from all-sky and large area surveys. In recent years digital information from such surveys has become much more widely distributed within the astronomical community, but it remains difficult to use. Data come in different formats, projections, coordinate systems, equinoxes and scales. This makes it tedious for users to get at the data. *SkyView* addresses these essentially geometric issues — providing the data to users in a convenient form, and allowing astronomers to concentrate on research rather than geometry. *SkyView* is also a single source for many of the most useful astronomical surveys.

1. Introduction

In the past decade digital versions of several all-sky surveys have become available. The explosion in the information processing capability of computer systems has meant that the actual images from these surveys can be delivered to the community rather than simply catalogs of interesting objects derived from the images. Some of these surveys have been distributed to the astronomical community on CD-ROM's. However, in practice it can still be very difficult to use these surveys. The coordinate system, scale, orientation, epoch, ...used to distribute the survey rarely matches the users precise requirements. We describe a new facility, *SkyView*, which addresses these issues.

SkyView is an on-line service which allows users to pick a region of the sky and then obtain images of it at a number of wavelengths. The user chooses the projection, coordinate system, scale, orientation, and size of the desired image. The user selects these parameters to match exactly existing data from ground- or space-based observations or to meet any other research needs. The data is transformed from the available all-sky and wide-area surveys according to the users specifications and the users are given the resulting FITS files or PostScript images in formats that they can use right away.

---

[1]Computer Sciences Corporation

SkyView can also show coordinate and contour overlays, and locations of objects from astronomical catalogs. Future versions will provide internal capabilities to compare images including contouring one image over another and generating multicolor images from several inputs.

SkyView is available over the network to all astronomers.

2. Discussion

What are the current capabilities of SkyView*?*

Current capabilities include:

- GUI and CLI interfaces.
- Selection of data from surveys centered on any point in the sky.
- Data sampling in any of gnomonic, rectangular, orthoscopic, and Aitoff projections.
- Data sampling in any of Equatorial, Galactic or Ecliptic coordinates.
- Automatic mosaicing of data when the survey data is split into multiple regions.
- Data resampling to the desired output pixel size.
- Coordinate system overlays.
- Overlays of objects from catalogs.
- Generation FITS or PostScript files.

What enhancements are planned?

- Further integration of *SkyView* into the ADS.
- Data overlays of two or more images including user supplied images.
- Automated region selection where a user supplies a standard data product from some processing system (e.g., ROSAT, ST ScI) and *SkyView* provides appropriate comparison images.
- Additional projections and coordinate systems.
- Many more catalogs and surveys.

What can I use SkyView *for?*

Multiwavelength astronomy, finding charts, searches for objects, looking for temporal variations, proper motions, observation planning, large scale structure, source identification, ...

How do I get access to SkyView*?*

Since November 1993 *SkyView* has been available through the HEASARC `xray` account. You may login as `xray` on `legacy.gsfc.nasa.gov`. No password is required. *SkyView* is also available through the Astronomical Data System. We encourage any interested astronomer to try *SkyView* and we welcome any comments you may have.

Figure 1. A sample *SkyView* display. A user can generate get this image of the Andromeda galaxy in a few moments. *SkyView* automatically takes care of formatting the data properly. In this case *SkyView* has determined the sub-images of the IRAS 100 micron survey to use, transformed to J2000 coordinates and resampled the image.

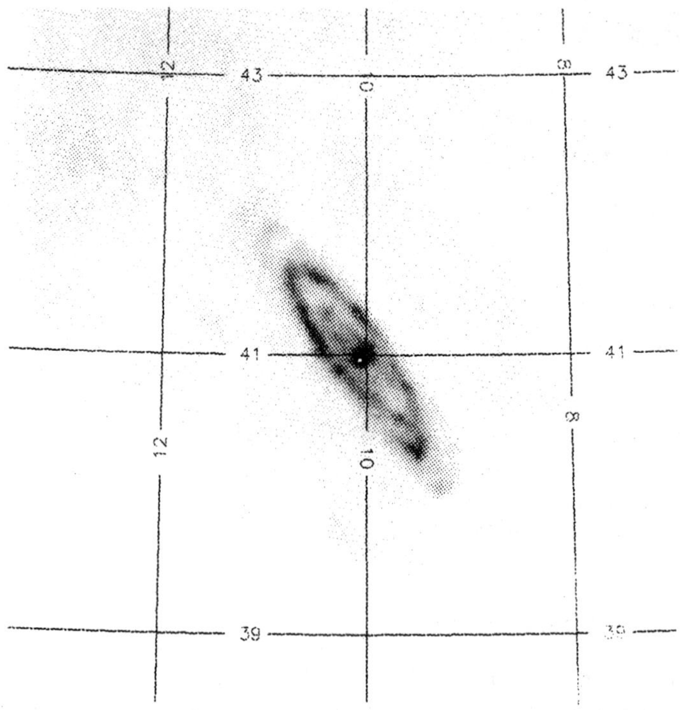

What are the current and planned holdings for SkyView*?*

Surveys currently available or scheduled for ingest in *SkyView* are shown in Table 1. We welcome any suggestions for additional surveys, and we anticipate adding many small surveys not shown in the next few months.

Current catalogs, shown in Table 2, are intended to be representative of the kinds of catalogs that will be available. Adding catalogs to *SkyView* is very easy and as *SkyView* is further integrated into the ADS and HEASARC the vast catalog holdings of those systems will also be available.

How does SkyView *work?*

SkyView currently has both a simple GUI interface and a CLI both of which work within IDL. Using the GUI you are presented with the form shown in Figure 1 when you login. Fill out the form to specify the region and presentation of data that you wish, press the `GO` button and your data will be retrieved. *SkyView* surveys are stored near-line in an optical disk jukebox. It typically takes a minute or two to generate an output image. You can display the images on your home machine or generate FITS and PostScript files which can be retrieved to

Table 1. Surveys included in *SkyView*.

Name	Regime	Resolution	Available	Notes
Bonn 21cm	Radio	.5°	Now	$\delta > -19°$
IRAS	Infrared	90"	Now	100, 60, 25, 12 μm
HEAO A-2	X-Ray	1°	Now	Contains glitches
Preliminary EUVE	UV	1°	Now	555, 405, 171, 83 A
Palomar SS	Opt	1.4"	Starting 1994	Compressed data
Southern SS	Opt	1.4"	Starting 1994	Compressed data
COS-B	γ-ray	1.5°	1994	Galactic plane.
COMPTEL	γ-ray	2°	1994	
EGRET	γ-ray	1°	Early 1994	Several energies
ROSAT WFC	UV		1995	
Final EUVE	UV		When available	

your home system using anonymous FTP. With the CLI the same information is provided using either command line arguments or a simple dialog.

Table 2. Some typical catalogs included in *SkyView*.

Catalog	Regime
4C	Radio
Veron-Cetty	Radio
IRAS point source	IR
Dark Nebulae(Lynds)	Optical-nebulae
Bright Nebulae(Lynds)	Optical-nebulae
HII (Sharpless)	Optical-nebulae
FK5	Optical-stars
Bright Star	Optical-stars
GSSS	Optical
POSS Plate centers	Optical-observations
Seyfert (Weedman)	Optical-galaxies
RNGC	Optical-galaxies
Markarian	Optical-galaxies
Uppsala	Optical-galaxies
Abell	Optical-clusters
4th Uhuru	X-ray

Where can I get more information?

Preliminary versions of the *SkyView* User's Guide and Design Document are available via anonymous FTP on `skview.gsfc.nasa.gov`. Once you have started *SkyView* there is extensive on-line help. If you wish to talk to someone about the system contact:

Thomas McGlynn 301-286-7743 `mcglynn@grossc.gsfc.nasa.gov`

or

Keith Scollick 301-286-8143 `scollick@skview.gsfc.nasa.gov`

Acknowledgments. *SkyView* has been funded by NASA's Astrophysics Data Program, grant NAS5-32068.

Astronomical Data Analysis Software and Systems III
ASP Conference Series, Vol. 61, 1994
D. R. Crabtree, R. J. Hanisch, and J. Barnes, eds.

An IRAF Solar Data Pipeline into the World Wide Web

Dyer Lytle

National Optical Astronomy Observatories, Tucson

Abstract. The recent advent of Internet filing and search tools such as Gopher, WAIS, and the World Wide Web make easy global access to public domain astronomical data a reality. The IRAF project at NOAO is working to both build archives of such data and to make that data available to researchers world wide. As part of this project, daily synoptic observations from the solar spectromagnetograph on Kitt Peak are being made available to users of the World Wide Web via HTML documents. Data is reduced automatically by an IRAF script which is initiated by the Unix task "cron". HTML documents are built that include data from the resulting image headers and pointers to the image data for remote display. As part of the pipeline, data is also sent to specific sites via "rdist". The intent of this automated pipeline is to make this data available to the scientific community as soon as possible after the observations are made.

1. Introduction

The system described above is up and running in Tucson on a prototype basis. Additions and modifications are ongoing at this time. This paper describes what has been done and future additions to the project.

2. Data Acquisition

The data described here are full disk images of the sun taken at the Kitt Peak Vacuum Telescope. The images as taken are three dimensional (1788x1788xN where N is different for the two types of images described) and contain various data products. The two types of images, one taken in a magnetically sensitive line of iron (868.8 nanometers) and the other in a line of helium at 1083 nanometers, are taken on a daily basis when weather allows.

Previously, this data would accumulate on computer disks sometimes for as long as 2 weeks before the data was processed. Then, when the data was processed by hand the results would be made available on an anonymous FTP disk for anyone who knew the proper address. The aim of the current project is to make this data available as soon as possible to as many people as possible.

To do this, various tasks have been written in IRAF, perl, and csh to check a specific directory for new data and, when it is found, to process it through a reduction pipeline.

3. Automatic Data Manipulation with IRAF

Using an experimental version of the IRAF command language that allows for scripts which load the CL automatically upon execution, a pipeline for data analysis has been constructed. This pipeline consists of a perl script, a self loading IRAF script, and a normal IRAF script which calls various tasks in the vacuum telescope package.

The root perl script is run on the data analysis computer (SPARC 2) by the Unix "cron" facility every minute from 8 AM to 6 PM every day. When this script finds new data, it first writes a lock file to stop parallel execution of the script and then calls the IRAF scripts to process the data. These scripts produce various data products including a scaled down version of the full disk solar images destined for distribution. When done, the script renames the original data file and removes the lock file. Both the magnetogram images and the helium 10830 images are processed in this way.

When these final distribution images are complete they are copied to the archive computer which is located at the NOAO offices in Tucson (also a SPARC 2).

4. The 'rdist' and FTP Archives

On the downtown archive computer, various perl scripts are run by cron every hour. These scripts look for the arrival of the distribution images from the mountain. When these images appear, various mirror archives are built for distribution to some sites and data is also FTPed to other sites.

Currently mirror archive directories are kept for the Yohkoh X-ray satellite people in Japan and for the NOAA/SEL people in Boulder, Colorado. The Yohkoh archive contains the seven most recent magnetograms and the seven most recent helium 10830 grams. The NOAA archive contains the most recent 15 days of each. For these "mirror" archives, the Unix task "rdist" is run every hour to keep the directories at the local and remote sites identical. This provides a more robust archive facility than the FTP archive described below.

Some sites run the VMS operating system and cannot receive archive distributions via rdist. For these sites FTP is used for data transfer. When the new data is available, it is sent to the remote site by FTP. Currently there is no automatic checking for success or failure of the FTP transfer. This means that if the network is down when the FTP transfer is to occur, the data never gets to the remote site. We have plans to remedy this situation in the near future in order to make these data transfers more reliable. FTP distributions are running to Solar Dispatch in Canada, Big Bear in California and NASA/Goddard.

5. The Connection to the World Wide Web

The downtown archive computer also has a perl script that is run once an hour by cron that checks for new data and, when it is found, pipes it through an IRAF script which produces GIF images that are made available to the Internet community on the World Wide Web.

Currently, the way the images are made available is to give them standard names which are already included in existing HTML documents. The housekeeping data for these images, date, wavelength, sky conditions, etc., are extracted from the headers and written in to special blank images which also have standard names which are embedded in the afore mentioned HTML documents. This way the HTML documents remain static and just the images pointed to by those documents change.

Currently seven days worth of magnetograms and helium 10830 grams are kept online in HTML documents. There is a page for each day as well a page showing all seven magnetograms and a page showing all seven helium grams. The HTML documents also contain pointers to the magnetogram archive project which will soon allow the user to search and retrieve spectromagnetograph data spanning the entire 18 year life of the daily synoptic program. This author is a co-investigator on that project also.

Work is in progress to flesh out this work including the addition of tutorials on the sun and the data available, other data products like plots of the history of the intensity of the solar magnetic field, and links to other, related, archives.

6. Future of the Project

In the coming year we hope to make more and better data products available both to our archive sites and to World Wide Web users. A "gopher" server will be set up to access this data as well. We hope to soon have the entire magnetogram and helium 10830 archives online via the Web. If you are interested in Web access, keep an eye on http://argo.tuc.noao.edu/ and http://argo.tuc.noao.edu:2001/synoptic.html

If you would like to have an online data archive set up for you please contact me at lytle@noao.edu.

Astronomical Data Analysis Software and Systems III
ASP Conference Series, Vol. 61, 1994
D. R. Crabtree, R. J. Hanisch, and J. Barnes, eds.

An Astronomical Software Directory Service

Robert J. Hanisch, Harry Payne, and Jeffrey J. E. Hayes

Space Telescope Science Institute, 3700 San Martin Drive, Baltimore, MD 21218

Abstract. We will be developing an on-line Astronomical Software Directory Service (ASDS) with funding from NASA's Astrophysics Data Program. Our primary objective is to allow astronomers and astronomical software developers to easily locate existing programs for their use and re-use, providing a uniform level of high-level documentation (package capabilities, system dependencies, installation requirements, etc.). The ASDS will catalog a comprehensive set of astronomical data reduction and analysis software. The ASDS is intended as a directory service — a query for a certain type of software will result in the user being given information about what software is available as well as instructions for retrieving relevant packages from anonymous FTP servers. Software retrieval may also be automated. We will not archive software per se, given that problems of revision control, currency, and algorithm validation are untenable. Users will interact with the system using a public domain distributed database interface such as WAIS, Gopher, and/or WWW.

1. Introduction

Astronomers are fortunate to have a wide variety of software available for processing and analyzing their data, and funding opportunities such as NASA's Astrophysics Data Program help to both broaden the scope of the available software and improve its quality. What is missing, however, is a simple, widely available tool that would allow astronomers and astronomical software developers to find out what has already been developed and is available for use. What is needed is an on-line searchable database with a simple user interface that provides pointers to the available programs and associated documentation. Much code is rewritten these days, not because anyone has found a fundamentally better way to solve the problem, but because they simply don't know who has already done it, whether the code runs on the system the user has available, and where to get it if it does. More efficient use of software development resources will allow the astronomical community to make better use of its limited research dollars.

NASA's COSMIC project is a repository for software developed under NASA contracts. As such, it encompasses a broader spectrum of code than likely to be of interest to astronomers. The facility is not well known to astronomers, and fees are charged for the distribution of software. In this world of widely available networking and anonymous FTP archives, this approach is not very useful.

The ARCHIE service on the Internet provides a better model for a software directory service. The ARCHIE database, which was originally developed at McGill University in Montreal, indexes much of the software that is available in anonymous FTP archives worldwide. ARCHIE is most useful for users who know what particular piece of software they are searching for, by name, and mainly need to find an anonymous FTP archive that contains it. The database is huge, as it covers virtually all disciplines, but only rather rudimentary facilities are provided for locating software by topic. No information is provided about system requirements or dependencies.

Our objective is to provide an *Astronomical Software Directory Service* that improves upon the COSMIC and ARCHIE approaches and has the following characteristics:

- A simple public domain user interface that can be distributed to user sites at no charge and is easy to install.
- A uniform standard of high-level, on-line, searchable documentation for each package.
- Validated installation procedures (but not algorithms) on the systems supported by the software developer.
- A database of system requirements (hardware, operating system, disk space, memory, etc.).

The Directory will include the software developed under NASA's Astrophysics Data Program (ADP) and all other astronomical software that is publicly available to the community (and whose developers agree to have their software indexed). The development of the ASDS is itself being funded through the ADP.

2. The Current Situation and Project Objectives

The astronomical community is currently dealing with the issue of software distribution in an uneven and uncoordinated manner. There are major astronomical institutions who distribute their software to the community at no charge (e.g., AIPS from NRAO, IRAF from NOAO, STSDAS from ST ScI, PROS from SAO, Starlink from RAL, and MIDAS from ESO). In addition, there is an extremely informal 'network', i.e., just an astronomical software grapevine, by which users hear about applications programs, math routines, graphics libraries, etc., that have been developed by other astronomers. Programs such as DAOPHOT (Peter Stetson's package for crowded field photometry) have tended to be distributed in this way, i.e., passed along from one researcher to another, often having been modified in ways unknown to the original developer.

There are two general approaches one can take in establishing a software locating service:

1. An astronomical software *library*, in which an anonymous FTP server would be set up that actually contained copies of source code, object libraries, etc.

2. An astronomical software *directory*, where information about astronomical software would be stored, but where users would be directed to other sites to actually retrieve the software. A more elaborate system would initiate retrievals on behalf of the user without need for a separate FTP session.

There are many problems in trying to set up a central software library, foremost of which is revision control. Keeping the library current given the scores of packages available would be difficult, if not impossible. A library also implies a certain level of algorithm validation. This is a worthy goal but virtually impossible to meet, even for a single system such as AIPS, MIDAS, or IRAF. Thus, the preferred and more practical approach is to develop a directory that points to sources of software (e.g., anonymous FTP archives) but leaves the responsibility for keeping those archives up to date with the developer.

Several individuals have tried to assemble catalogs of astronomical software, but these projects have been only partly successful. Rhodes et al. (1989) compiled a library of astronomical software documentation several years ago, but owing to the rapid rate of change in software systems the collection was largely obsolete by the time the project was completed. Chris Benn (RGO) compiled a small software directory as a result of sending out a questionnaire and published the results via e-mail — the result is very incomplete both in scope and detail.

Several objectives must be met in order to develop a useful astronomical software directory service:

- The directory must be as complete as possible.
- The directory must be kept current, preferably automatically.
- The software for searching the directory must be easily distributed and installed (and itself must be indexed in the directory).

3. Project Description

The development of the Astronomical Software Directory Service (ASDS) requires work in several areas in order to meet the objectives listed above.

3.1. Software Collection and Validation of Build Procedures

In order to construct the directory, information must be collected concerning the astronomical data analysis packages to be indexed. Several approaches will be utilized, including following contacts identified in previous software catalogs, contacting PIs of ADP software development projects, reviewing COSMIC and ARCHIE for useful astronomical software, and exercising network information facilities already in existence. We will ask software developers to provide a standard, high-level description of their code and its dependencies but will not rely solely on this input to build the Directory.

3.2. Documentation Database

The completeness and readability of the documentation for software packages differs tremendously, and it is beyond scope to try to rationalize all astronomical software documentation to some uniform standard. We will follow two approaches to documentation:

1. Write our own high-level descriptions of each package. If the developers themselves provide such documentation, it will form the basis for our standardized descriptions. This information will be augmented by technical details such as disk and memory requirements, compiler dependencies, operating system dependencies, and other hardware requirements.

2. Utilize developer-provided documentation to form the inputs to a context-sensitive search engine. For example, using the standard WAIS server, one could query the Directory for software related to a series of keywords and return a list of the packages whose descriptions include such terms.

3.3. User Interface and Search Engine

There are now at least three major utilities available (as free netware) for accessing and maintaining distributed databases: WAIS, Gopher, and WWW. The astronomical community has already been introduced to WAIS via NASA's Project STELAR (STudy of Electronic Literature for Astronomical Research) and the AAS Executive Office on-line abstract service. Users interact with one or more remote databases either via a simple terminal-based user interface or with a generic X-windows interface. The Gopher software was developed at the University of Minnesota, and again provides access to distributed databases via a simple, generic user interface. The most versatile and powerful interface is the NCSA Mosaic package for interacting with the World-Wide Web. Given the rapid rise in use of this facility we will probably base the initial ASDS system on this user interface. A number of software development groups are already developing HTML home pages for use with NCSA Mosaic, and we can capitalize on this work by providing a central directory of all such home pages.

4. Summary

Using currently available public domain technologies we plan to design and implement an on-line directory service for astronomical data analysis software. We intend to make the scope of the directory as broad as possible, and encourage developers to contact us if they would like their software to be indexed in the system (send e-mail to asds@stsci.edu). We will not build a software archive or library but rather will construct a directory that points to software available on developer-maintained publicly accessible systems. NCSA Mosaic is likely to be the primary user interface for the system. Documentation search and retrieval facilities will probably be implemented with WAIS.

Acknowledgments. This project is to be funded by NASA's Astrophysics Data Program (ADP) under NRA 92-OSSA-15.

References

Rhodes, C., Kurtz, M., & Rey-Watson, J. 1989, in Library and Information Services in Astronomy, IAU Colloquium No. 110, G. A. Wilkins and S. Stevens-Rayburn, Washington: United States Naval Observatory, 215

Astronomical Data Analysis Software and Systems III
ASP Conference Series, Vol. 61, 1994
D. R. Crabtree, R. J. Hanisch, and J. Barnes, eds.

The Earth Data System and The National Information Infrastructure Testbed

Carol A. Christian and Stephen S. Murray

C&M Science Innovations, Ltd., 2550 Shattuck Avenue # 105, Berkeley, CA 94704

Abstract. Earth science data is presently stored in several separate archives with different catalog and access methods. These range from large, government archives of satellite imagery with full data management and access services to small, field data sets under the control of an individual scientist. The prototype Earth Data System (EarthDS) is designed to provide access of such heterogeneous data researchers interested in global weather changes, ocean current movements, and coastal drainage. In particular, the EarthDS includes a collaborative data browser and analysis capability as an example of a truly distributed computing capability that allows people at geographically separate institutions to search, view, and analyze data together.

The EarthDS is a testbed application being developed by the National Information Infrastructure Testbed organization; a consortium of industry, government and academic institutions. The application is constructed using an implementation of OSF's Distributed Computing Environment (DCE) software, wrapped in a software toolkit reminiscent of the Astrophysics Data System toolkit. It uses a communications network of high-speed cross-country DS3 (45Mbps) lines coupled with ATM switches, and routers with SMDS, FDDI, and Ethernet interfaces. The initial tests in May, 1993, showed outstanding performance with network nodes distributed between California and New Mexico locations linked by a T3 network. The functional system will be demonstrated in November 1993 at several forums including Supercomputing 93. Future developments will incorporate distributed real time device control and data retrieval as well as high speed computing, data retrieval and network management. In this paper we will discuss the EarthDS reference application as well as several of the technical issues regarding the full deployment of large scale distributed computing infrastructure. Full text with figures can be accessed through *Xmosaic* at that *URL* http://niit1.harvard.edu.

1. The National Information Infrastructure Testbed

1.1. Purpose

The National Information Infrastructure Testbed (NIIT) is an industry-led consortium of commercial, academic and government institutions formed expressly to develop a prototype infrastructure spanning the nation to demonstrate distributed computing applications. The applications are developed by concentrating efforts on integrating existing technologies into working systems. By including users in the development of the NIIT infrastructure, the feasibility of creating and marketing high performance networking and distributed computing resources can be evaluated.

1.2. Organization

The activities of the NIIT non-profit organization are directed by a Board of Directors while a Management Committee coordinates the activities of a number of working groups within NIIT. Working groups are organized to achieve the practical and technical objectives of NIIT. Application working groups benefit specifically from the synergy between the cooperating NIIT members in a way that ordinarily would not be practical or possible. The application itself must offer technological challenges that relate to the advancement of information infrastructure. It is anticipated that successful applications will evolve to production systems supported by their interested user communities. The existing application working groups include Earth Science and Environment, Health Care, Natural Resources, and Aeronautics and Space Sciences.

Permanent infrastructure working groups are address technical and practical challenges related to high speed networking, computing resources, middleware and distributed computing. These groups exchange technical information and in turn provide support to the application working groups.

2. The Earth Data System

The first reference application identified for a demonstration of the information infrastructure testbed addresses the problem of global environmental change. Environmental researchers must consider the interplay between various complex systems that comprise the environment. Earth data research is specifically founded on the location of, acquisition of and analysis of large, heterogeneous data sets. These data tend to be geographically distributed, such as Landsat imagery at the University of New Hampshire, ocean data (color and temperature measurements) at Oregon State University and microwave measurements at the University of California Santa Barbara. Due to the disparate nature of the multi-media data, the question of data description, storage and transfer standards is relevant. Therefore conversion of information into and out of local display and visualization tools is a challenge. Adding to the complexity of the problem are the proprietary nature and security protections associated with much of the data.

In a broad discipline such as Earth Science, no one researcher can hope to accumulate all the expertise and knowledge associated with global environmental analysis. The researcher is additionally challenged because the visualization, analysis and modeling tools required are usually tuned to the local data sets and local hardware. Often the computing environments vary from institution to institution.

Today, data for environmental studies can be exceedingly difficult to locate and access. Often data must be ordered, purchased and then painstakingly evaluated before actual analysis can be accomplished. Once accumulated at an individual institution, with qualitative information (such as completeness of coverage, cloud cover, etc.) and quantitative processing added, the data and ancillary information have the potential to be invaluable resources for all researchers. However, the challenges that must be met by an Earth Data System are generic to information infrastructure.

3. EarthDS Infrastructure

The Earth Data System (EarthDS) has been designed to meet the technical hurdles discussed above. NIIT members and contributors have concentrated on addressing requirements in networking, computing power, software and data availability.

3.1. Network

The EarthDS system is built on a backbone ATM or DS3 (45 Mega bit) network linking the key data sites in the testbed. Technical support sites are linked to the ATM network through T1 (1 Mega bit) connections. Management of the system includes monitoring and optimization of connectivity, and will expand to process control, congestion control and process optimization. Interoperability within this network topology has never before been addressed in such a widely distributed configuration, and the goal is to deploy the network as well as demonstrate a reliable configuration that can be sustained.

3.2. Computing Power

For the testbed, hundreds of Gigabytes of information will be accessed over the high-speed network connecting optical disk jukeboxes, workstations and servers. Eventually, analytic and predictive modeling services will be added.

3.3. Software

Software at the local sites generally include well structured functions integrated into a distributed metadata and data search and access environment. Services local to each institution (display, visualization, analysis and modeling) should be linked into the distributed user environment. Middleware supports the integration and linking of the software at distributed sites similar to that used by the ADS, providing a truly distributed environment. Service development consists of two components: a graphical user interface that passes user requests through the DCE, and a server that responds to these requests.

Coupled with the high speed network and high performance computing environment, a user can find, retrieve and use data that is widely distributed without being required to know the physical location of any of the data or services. The ultimate goal is to present to the distributed computing user an environment that has a performance similar to the workstation, but makes use of facilities and information that is not locally offered or maintained.

3.4. Data

Data included in the EarthDS are taken from several sources. They include Landsat MSS and Thematic Mapper (TM) Imagery scenes from the Landsat

Pathfinder Project at the University of New Hampshire (Dr. D. Skole and Dr. B. Moore); Coastal Zone Color Scanner (CZCS) ocean color and temperature data from Oregon State University (Dr. M. Abbott); and Advanced Very High Resolution Radiometer (AVHRR) microwave measurement data from the University of California Santa Barbara (Dr. J. Star). Each site has a searchable metadata database that can be used to identify data of interest and supporting ancillary data associated with the primary data sets. We have applied the Hierarchical Data Format Standard (HDF) to the heterogeneous data by implementing conversion services that transform the data as they are retrieved and passed across the NIIT/EarthDS network.

4. Collaboration Environment

The NIIT Earth Data System environment goes beyond providing a distributed computing environment to individual users. Researchers can (and should) be resources for each other. A collaborative tool, which allows researchers to collectively pose problems, conduct studies and perform analysis in real time, would greatly assist in sharing of expertise and knowledge among those researchers. The EarthDS system includes such services allowing individuals at geographically disparate locations to share files, searches, data displays, and analysis as well as their expertise through software based on Shared X.

The effectiveness of the collaborative environment is significantly dependent on the bandwidth of the network connection among collaborators. In addition to sharing computer screens, effective collaboration benefits from voice and video links, and other communication channels. One of the study areas for the EarthDS will be to provide feedback on the use of collaboration tools and how they can be improved.

5. Demonstration

The NIIT is focused on building infrastructure that addresses real applications. Our long term goal is to demonstrate a working system and establish the EarthDS as a self sustaining project, analogous to the NASA Astrophysics Data System, which is supported by the Earth Science community. In addition, it is expected that commercial applications based on the EarthDS will be developed which will provide income generation for the EarthDS NIIT industry partners.

6. Acknowledgements

We would like to gratefully thank the many individuals who have consistently offered encouragement and support to us during our participation in this project. In particular SSM thanks Dr. Irwin Shapiro (SAO) for allowing him to spend time on this project, and CAC thanks Dr. Roger Malina (CEA) for his enthusiasm and advice.

Part 2. Visualization and User Interfaces

Astronomical Data Analysis Software and Systems III
ASP Conference Series, Vol. 61, 1994
D. R. Crabtree, R. J. Hanisch, and J. Barnes, eds.

The Challenge of Visualizing Astronomical Data

Ray P. Norris

Australia Telescope National Facility, CSIRO Radiophysics Laboratory, PO Box 76, Epping, NSW 2121, Australia

Abstract. Current astronomical instruments now routinely produce such large volumes of data that it is difficult for astronomers to obtain the information they want from the data. While the machines get faster and cheaper, and are certainly able to handle the load, and the astronomer's brain is certainly capable of interpreting the data, the interface between machine and brain prevents the astronomer from assembling the data in a coherent fashion in the mind, so that not all the useful information can be extracted from the data. This bottleneck between human and machine poses problems both for those users who want to obtain an intuitive, qualitative, understanding of their data, and for those who want to obtain quantitative results from their data. Here I discuss ways of representing data which attempt to overcome this bottleneck.

1. Introduction

Our computers are certainly capable of dealing with the enormous volume of data produced by current astronomical instruments, and we know that the human brain is capable of absorbing it, and of producing new scientific knowledge from it. However, there exists a bottleneck between human and machine which prevents the user from assembling this information in the mind in a coherent and integrated way, and so some of this information is inadvertently discarded.

The visualization of data cubes has four distinct functions.

- To obtain an intuitive understanding of the data
- To see features of the data which would otherwise remain unnoticed
- To obtain quantitative results from the data
- To communicate the results, both qualitative and quantitative, to others.

The challenge, then, is this. How can we break down this conceptual barrier between user and data, so that the user can get full scientific value from this hard-earned data?

Intrinsic to the nature of this subject is the difficulty of describing in two dimensions some of the three-dimensional examples. At the ADASS meeting this was overcome by projecting movies demonstrating some of the techniques described here, but, unfortunately, I am unable to reproduce these in this paper!

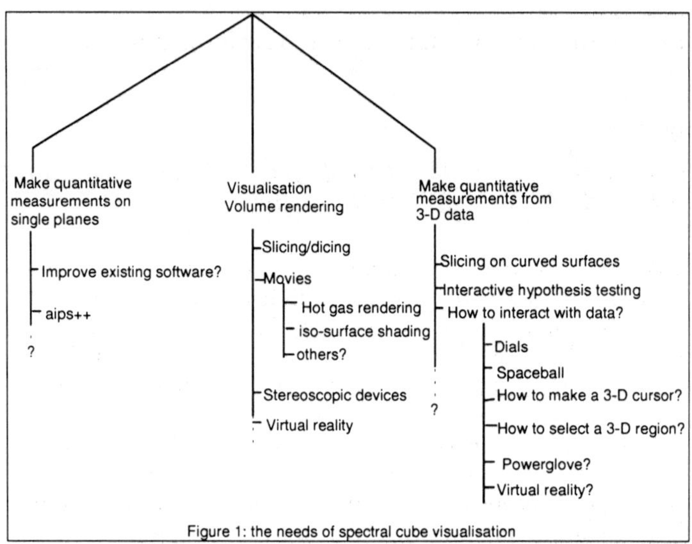

Figure 1. Aspects of visualization.

2. Spectral-line Data Cubes

Figure 1 shows the three main ways in which a user might want to interact with a spectral-line cube. First there are simple quantitative measurements to be made which can, in principle, be done with existing packages of software (AIPS, GIPSY, etc.). However, many users find that current tools are inadequate even for this simple task. While some work is being done to improve the capabilities of current software, it is likely that these demands will be met only by the release of new software, such as AIPS++ , over the next few years.

A second need, with which this paper is mainly concerned, is to visualize the cube. A third is to make quantitative measurements on the volume visualization, but as this depends on a union of techniques from the first two branches of this diagram, this must be postponed until a satisfactory solution is found for the first two.

The image of a spectral-line source will typically consist of 256 images, each of which represents a different wavelength, and each of which contains 256 x 256 pixels. These separate images can be conveniently regarded as being planes of a cube containing 256 x 256 x 256 three-dimensional pixels, or voxels (volume pixels). This cube, with one byte assigned to each voxel, occupies just 16 Mbyte, and so fits conveniently in the memory of a desktop workstation. A traditional technique of displaying this data consists of showing a two-dimensional (2-D) representation of each plane of the cube in sequence, so that the user sees a movie of one plane after another. If this is done sufficiently rapidly, then this technique can go some way towards giving the user an intuitive impression of the data.

Typically, this data may be of a cloud of interstellar gas with a velocity gradient across it. The position of the maximum emission in each plane will shift steadily as different planes are viewed, so that the movie gives the appearance of an object moving across the screen. However, this technique does not reveal subtle effects to the user, and, in the case of a large number of planes, there may be a considerable time lag between viewing different parts of the movie, and so a user will not be able to relate frames at the start of the movie to frames at the end. At the Australia Telescope National Facility (ATNF) we are exploring a number of alternative techniques which present the data cube as a single three-dimensional (3-D) object.

The algorithm which appears most successful in representing an astronomical data cube to users is the "hot gas" algorithm, which was first implemented at the ATNF by Eric Greisen. In this algorithm, a look-up table is constructed which assigns colour, brightness, and opacity to each voxel, depending on the value of the data in that voxel. An approximate radiative transfer equation is then solved through the cube for each ray of light which will strike the user's eye. The resulting image appears as a cube of glowing gas, and a galaxy will appear as a cloud of swirling mist within the cube.

Conventional movie techniques (such as AIPS TVMOVIE) take too long for the brain to associate information in different parts of the spectrum, and so do not allow the user to get a intuitive impression of the data as a whole. Volume-rendering techniques, on the other hand, show all the data at once, thus overcoming these problems, and so allow the brain to make full use of its powerful pattern-recognition capabilities. For example, a feature may be present in each plane of the cube but is indiscernible in the individual images because it is barely above the noise level, but when seen as a cube the feature may show up easily to the eye as a bar running through the cube. Similarly, subtle features such as non-circular motions in the neutral hydrogen emission from a galaxy will be far more obvious in a volume-rendered cube than in a TVMOVIE.

Although presenting the data as a static 3-D object improves on earlier visualization methods, we have found that motion is an essential cue to the brain in interpreting an image as a 3-D object. We therefore need to provide tools for rotating the image of the cube. However, the volume rendering is too slow for rotating large cubes in real time, and so most users make a movie to be displayed on the workstation screen. Such a movie then shows the cube rotating smoothly in space, with the object of interest appearing as a region of glowing gas sitting within the cube. It is likely that an even better intuitive impression of the data would be gained if the user could rotate the cube interactively. For example, a spaceball could be used to turn and twist the object, peering at it from all angles. Unfortunately, the computing power required for this is beyond the available resources.

An alternative to full interactive visualization is a technique that I call pseudo-interactive visualization, in which a movie is pre-calculated, not just of the frames needed for rotation through one turn, but of every possible view of the cube from any angle over the sphere. If images are calculated at two-degree intervals over the sphere, then the resulting 160 Mbyte can be stored in memory, and the images roamed using a spaceball or mouse, giving the user the impression that the cube's motion is fully interactive.

3. Practical Experience: the ATNF Visualization Project

The ATNF visualization project is a research project which aims to explore the use of visualization on astronomical data cubes. Because 3-D visualization is compute-intensive, it currently requires specialized machines (typically hundreds of Mflop/s), and the required performance is obtained from these machines only by writing specialized code. However, the continuing increase in the power of affordable computers means that within a few years these techniques can be transferred to standard desktop workstations. At that stage we intend to port our techniques to applications in AIPS++ and other imaging packages, but for the present we have abandoned portability and use specialized prototype code on specialized machines (such as the 400-Mflop MVX visualization accelerator).

In addition to astronomical visualization, the group has strong links into medical and geophysical imaging projects at the CSIRO Division of Radiophysics, so that a significant degree of cross-fertilization takes place.

At present, we are tackling some of the problems discussed above. In the future, we hope to extend this investigation to virtual reality techniques such as using a powerglove with tactile feedback. For example, a user could try to push down a feature in the data if she thought that it was an imaging artifact. If it was poorly constrained by the data, it would be soft and easily pushed down. If, on the other hand, it was required to be consistent with the input data, then it would feel hard and difficult to push away.

This last example is also an example of the more general technique of hypothesis testing. There are cases both in astronomy and in geophysics where the end result is constrained relatively poorly by the data, so that a researcher typically wants to ask the question: "How well is this feature constrained by the input data?". We hope to explore other techniques to allow the user to interact with the data in this way.

4. Conclusion

I have shown here some of the problems and the successes in visualizing 3-D data. The subject is probably in its infancy, and future work will encounter challenges not envisaged at present. We are currently limited by the available computing hardware, and this will probably change over the next few years. Even virtual reality techniques are likely to become much more widespread as the technology becomes more common and affordable. Our challenge then will be to implement these techniques in such a way that astronomers find them easy and intuitive to use, and, where possible, make them available within standard packages (e.g., AIPS++) so that they will run on standard equipment that sits on the desk of the average astronomer.

Astronomical Data Analysis Software and Systems III
ASP Conference Series, Vol. 61, 1994
D. R. Crabtree, R. J. Hanisch, and J. Barnes, eds.

Towards an Astrophysical Cyberspace: The Evolution of User Interfaces

Alan Richmond

Hughes STX, NASA/Goddard Space Flight Center—HEASARC, Greenbelt, MD 20771

Abstract. Our thesis is that trends in graphical user interfaces (e.g., NCSA's Mosaic; HEASARC's forthcoming **StarTrax**), and networks and distributed computing (e.g., information discovery systems like Gopher, WAIS, and hypermedia networks like the World-Wide Web), and virtual reality technology (e.g., the Electronic Visualization Laboratory's CAVE), will synergize a powerful astrophysics environment.

In sum, then, it would be unwise to ignore the design of cyberspace itself while we are engaged in the myriad considerations of particular GUI and VR implementations. The design of cyberspace is, after all, the design of another life-world, a parallel universe, offering the intoxicating prospect of actually fulfilling — with a technology very nearly achieved — a dream thousands of years old: the dream of transcending the physical world, fully alive, at will, to dwell in some Beyond — to be empowered or enlightened there, alone or with others, and to return. (Benedikt 1992).

1. The Evolution of User Interfaces

Although the 'desktop' metaphor has served tolerably well for working with the 2d world of workstation/PC screens, it is certainly not very appropriate for interacting with 3d worlds. Recent research in human-computer interaction has spurred the development of new peripherals for interacting with remote or computer-synthesized worlds. Instead of keyboard and mouse input, use is made of voice, and hand gesture and manipulation; output is to displays aimed at creating the illusion of 3d, simulating operator presence inside these worlds.

1.1. The Screen and Keyboard Paradigm

We are accustomed to thinking of user interfaces in terms of screens, keyboards, and relatively local applications. Our user conducts a dialog with our application through the mechanisms we provide; at their simplest, these are just reads and writes built into our chosen programming language. With the advent of bit-mapped graphical workstations, seamless network integration, and several other exciting new technologies, we are going to witness a profound change in this viewpoint. The user interface mediates in the processes of data acquisition, manipulation, navigation, analysis, archiving, visualization, and in the collaboration with others.

The trend is to increasing transparency, so that the user interface will become less and less of an obvious intermediary, and more of a 'looking glass' into astrophysical datascapes and cyberspaces. The user interface must place the scientist in direct contact with these processes, without superimposing its own idiosyncracies. To this end, our traditional view of the user interface as screen, keyboard, and mouse intervening between the user and application will evolve to accomodate emerging technologies and concepts, especially GUIs and GUI builders, hypertext and multimedia, virtual reality, and cyberspace.

1.2. Graphical User Interfaces

Graphical user interfaces (GUIs) are an improvement over command languages, but the next generation of user interfaces is on the way. The future will be dynamic, spatial, 3-dimensional, virtual, pervasive, gestural, colorful, frequently auditory, and sometimes immersive. GUI toolkits offer a large repertoire of user interaction devices, allowing the designer to present information and accept user input, with structures that match application and ergonomic needs most appropriately; e.g., many applications have a hierarchical command structure — with pull-down menus, you can readily see the overall program structure. These devices are well-illustrated in HEASARC's **StarTrax—NGB** (Richmond 1993).

2. Cyberspace is the Evolution of Networking.

It is a shared information universe. It is the network of interacting computers, their data and their users. It is ftp, e-mail, the World-Wide-Web, Gopher, WAIS, Mosaic, hypermedia, the Internet. VR and cyberspace are not synonyms. VR has to do with visualization and manipulation; cyberspace with navigation and collaboration. One day, VR may provide the most effective portal to cyberspace.

2.1. NCSA Mosaic and the World-Wide Web

The World-Wide Web, a distributed hypertext-based information system developed at CERN, is a globally interconnected network of hypermedia information comprising: the Internet, + a protocol for transmission of *hypermedia* documents, + a set of *servers* that respond to requests from *browsers* (or *clients*) for those documents. Hypermedia (or, more loosely, hypertext) documents contain *hyperlinks* to other documents, anywhere on the Web. A **hyperlink** is a segment of text, or an inline image that refers to another document (text, sound, image, movie) elsewhere on the Web. When a hyperlink is selected, the referenced document is fetched from the Internet, and is displayed appropriately. NCSA Mosaic is a Multi-Platform GUI Hypermedia Browser that helps you to find, fetch, and display documents and data from the Internet. Although there have been browsers for the Web since 1991, Mosaic shot rapidly to prominence because it seamlessly integrates a great deal of useful functionality in a very pleasant and easy-to-use user interface. Incorporating Web, Gopher, WAIS, NNTP, and FTP protocols, NCSA Mosaic talks natively to these servers, and can gateway to others (Hardin 1993).

2.2. Navigation in Hypermedia

As the size and sophistication of hypermedia expands the navigation problems will become more evident. The desktop metaphor and the associated tools do not well represent the spatial and structural connections which may exist inside hypermedia. New virtual space metaphors, e.g., 'cities', will provide users with a uniform visual portrayal of such structures. The assemblage of technologies that have become known as 'virtual reality' will include text either directly as documents or indirectly as menus or organizational structures. These text objects will be presented in stereoscopic 3D, due to the nature of the display technologies used.

The Navigating in Information Space project (NCSA), is mapping information — textual data — in three dimensions for navigation using virtual reality technologies. The information is processed statistically to extract relations among terms or documents in a database. Items which are more closely related are placed closer to each other in the 'information space'. The user employs the DataGlove or mouse and the NCSA VR lab's projection screen to display and move through the 3D information space. Users can reach out and 'grab' the documents of interest.

3. The Basic Properties of VR are Immersion and Mobility.

Immersion refers to the illusion of being inside a computer-generated world. The user must be persuaded to suspend disbelief in the reality of what is being observed — until he/she can ignore the interface, and concentrate on the application, VR will remain a novelty rather than a genuine scientific visualization tool. Mobility refers to the ability to 'move around' the virtual world, e.g., to examine large data sets containing spatial and temporal information. The user can employ experience to comprehend the data, by exploring it in the same way as physical space. This enhances the researcher's perception of its characteristics.

Scientific Visualization often studies multi-dimensional physical phenomena; in these systems, a computer model generates data representing the behaviour of the model. VR can help scientists explore the multi-dimensional graphical representation of their data at various levels of scale. For example, spectral-line data cubes can be viewed interactively as three-dimensional cubes, or even walked through in a virtual reality system (Norris 1993). You can use virtual reality technology to fly through 3 dimensional environments that represent extremely complex data, and reach out and manipulate these representations with your hands (Aukstakalnis 1992).

3.1. The Evolution of Novel Peripherals

These peripherals engage visual, auditory, tactile, and haptic (muscular) senses, to support user interaction with computer applications and resources. Commercially available systems support hand-tracking, head-mounted stereo displays, 3d audio, speech recognition and synthesis. Research is ongoing on force-feedback, tactile gloves, and eye-tracking. Head-mounted display (HMDs) typically weighing 4lbs, are perhaps the most popular, or well-known VR interface, comprising a pair of small displays covering the eyes (e.g., VPL's EyePhone, or 'facesucker').

Tracking Devices, like the Polhemus Isotrack, and Ascension Technology's Flock of Birds allows the computer to determine the location and orientation of a person. A head-tracking device provides location and orientation of the viewer to simulate the user's viewpoint. VPL's DataGlove tracks hand and finger movements. The Binocular Omni Orientation Monitor (Fake Space Labs) places small CRT displays (2.5" x 2.5"; 1280x500 color pixels) before the eyes. The BOOM is suspended from an articulated arm, which measures precisely its position and orientation in space, and counterbalances its mass — when you let go it remains there. The user looks into it and moves it by handles.

3.2. Examples of VR in Astrophysics

The Harvard-Smithsonian Center for Astrophysics and NCSA have used virtual reality to explore astronomical data in the Redshift Survey. Their principal objective was to map the positions of all the galaxies within about 500 ly. Seeing all the data on a 2d medium is almost meaningless. But looked at in VR, patterns start to become clear. The researchers (Margaret Geller and John Huchra) used a BOOM connected to a high-power graphics workstation. This enabled them to view the galaxy structures from any angle or perspective, giving insight into the structural layout and placement of the galaxies. This method of visualization significantly enhances the traditional methods of statistical analysis.

The Electronic Visualization Laboratory (EVL) at the University of Illinois at Chicago, also collaborates with the NCSA on a number of scientific visualization projects, e.g., in cosmology. EVL's CAVE is a room built from large displays on which the graphics are projected on three walls and the floor, to be viewed with stereo glasses. As a viewer wearing a head and hand tracking system moves inside its display area, the stereo perspectives of the environment are updated, and the image moves with and encloses the viewer. The Cosmic Explorer, a CAVE application, is a research tool for exploring the stages of the evolution of the Universe. It visualizes the result of numerical simulations to allow the exploration of the formation of the universe or colliding galaxies.

Acknowledgments. Thanks to Nick White for encouragement and the opportunity to speculate a little !

References

Aukstakalnis, S. & Blatner, D, 1992, "Silicon Mirage: The Art and Science of Virtual Reality", Peachpit Press Inc

Benedikt, M., ed., 1992, "Cyberspace: First Steps", MIT Press

Hardin, J., "Human Collaboration Technologies for the Internet — NCSA Mosaic and NCSA Collage", this volume

Norris, R. P., "The Challenge of Astronomical Visualisation", this volume

Richmond, A., et al., "StarTrax — The Next Generation User Interface", this volume

Astronomical Data Analysis Software and Systems III
ASP Conference Series, Vol. 61, 1994
D. R. Crabtree, R. J. Hanisch, and J. Barnes, eds.

StarTrax — The Next Generation Browse

Alan Richmond, Song Yom, Paul Jacobs, Margo Duesterhaus, Phil Brisco

Hughes STX, NASA/Goddard Space Flight Center—HEASARC, Greenbelt, MD 20771

Nick E. White

NASA/Goddard Space Flight Center—HEASARC, Greenbelt, MD 20771

Thomas A. McGlynn

Computer Sciences Corporation, NASA/Goddard Space Flight Center—COSSC, Greenbelt, MD 20771

Abstract. We are developing a multi-platform user interface, **StarTrax—NGB** to provide access to many services of the HEASARC, e.g., bulletins, catalogs, proposal and analysis tools. **StarTrax—NGB** will present a uniform view of HEASARC through a portable graphical user interface based on the XVT Portability Toolkit. A wide range of platforms will be supported: OSF/Motif (Unix or VMS), OPEN LOOK (Unix), Macintosh, MS—Windows (DOS), character systems. It uses the *point-and-click* metaphor of modern GUI technology, for ease of use; instead of classical command-line interfaces (CLI).

1. User Interfaces

StarTrax—NGB is a distributable user interace for HEASARC on-line services (White 1993). People use the term *user interface*, with a broad spectrum of meanings, the endpoints of this spectrum being:

- the **application functionality** provided for the user;
- the software handling **user interaction with hardware** devices, e.g., screen and keyboard.

We will understand the user interface to lie between those endpoints, encompassing both, but biased towards the technical aspects of user interaction (e.g., 'look and feel'). The user interface *mediates between the user and an application*, providing the user with mechanisms to make best use of the application (Richmond 1993). It is *not* the application itself.

1.1. Classical User Interfaces

Until recently, users of astronomical software had to memorize artificial languages (e.g., command names and syntax); make several keystrokes per function; and interpret cryptic error messages (Pasian 1991). Systems like STARCAT/Proteus (Richmond 1987) reduced these burdens with techniques such as menus and forms displayed with cursor addressing, so that relevant information could be rapidly displayed and assimilated, and the user could indicate choices with minimum effort.

1.2. Graphical User Interfaces (GUI's)

GUI toolkits offer a large repertoire of user interaction devices, allowing the designer to present information and accept user input, with structures that match application and ergonomic needs most appropriately; e.g., many applications have a hierarchical command structure — with pull-down menus, you can readily see the overall program structure. We have tried to make best use of point-and-click GUI techniques and published guidelines, prototyping continuously, soliciting feedback, to maximize **StarTrax—NGB**'s user-friendliness.

On-line help will be provided locally, allowing for a quicker response and avoiding unnecessary network load. There will be two Help systems. One will be hypertext based, using NCSA's Mosaic, for those platforms supported (we are also thinking of building **StarTrax—NGB** in Mosaic and other World-Wide Web browsers.). For the others, or for user preference, XVT's built in indexed Help system will also be provided.

1.3. XVT

XVT-Design **StarTrax—NGB** layout is generated with XVT-Design into C code and window system-specific resource files, e.g., UIL for Motif. The C code links to the XVT API (Application Programming Interface), and to API's for the client side of client-server connections over a distributed heterogeneous network. Application-specific code is written manually for the dialog layer connecting the presentation layer to the API's for the HEASARC servers.

XVT's Portability Toolkit is a thin API that allows you to build a C or C++ application that's portable to 7 popular GUI's and many OS's. XVT is at a slightly higher level than the underlying native GUI. Its easier to use than native toolkits, but less powerful (may still access the native GUI system). An executable will be created for each supported platform by linking the compiled code to an XVT library and native window system library (e.g., Motif, Xt, and Xlib) for the platform. The URL file is translated with XVT CURL to a platform-specific resource file.

The IEEE P1201.1 working group chose XVT as the base document from which to develop a standard for a uniform API for GUI development. NIST recommends XVT to government developers needing application portability. Several products are available that work with XVT, e.g., for database support.

2. Databases

The *Databases* menu on **StarTrax—NGB**'s Task (main) Menu bar, will present an organized view of all available catalogs. This will pop-up a dialog with a list of catalogs and their descriptions, and some 'search' buttons. Under that pull-down menu we will have something like: HEASARC, Compton SSC, MIPS, Leicester, ... This pull-down menu is 'dynamic', so we can add new sites very easily, e.g., without having to re-release the code. We might need a table that lists all the available sites, which is loaded each time **StarTrax—NGB** is started. (The bulletin pull-down menu is dynamic for the same reason). It also would allow us to have test or local databases that can be accessed. We may have a 'setup' under the Options menu which allows sites to be added or removed. Clicking on, say, HEASARC or Compton, one then gets a list of the available catalogs in that database.

We will allow searches on multiple catalogs. You click on a catalog and it becomes highlighted; multiple catalogs may be selected. Clicking on the Hide/Show button removes/replaces all the non-selected catalogs; when they're removed, the user sees all selected catalogs together. There will be an 'all' button to choose all the catalogs. There will be a 'zoom' pull down, which lists all the unique observatories from the observatory parameter in the metabase e.g., ROSAT, ASCA, Compton. This allows a user to zoom in on a mission.

2.1. Searching

The operation of a search will be something like this:

1. To do the search one hits a button in the search panel. This brings up a panel that lists two columns 'hit' and 'miss', plus the current search buttons currently on the 'search' panel.

2. You then can hit Cone, target, parameter or whatever and it asks for the relevant info. Then it searches through each catalog you have chosen and returns the results into a separate XVT PowerTool Spreadsheet/Table (widget).

3. But these are kept hidden, rather the number of hits or misses per catalogs are listed. You click on the hits to see the query results.

4. Subsequent queries to that catalog go to the same XVT PowerTool Spreadsheet/Table (widget), unless it is cleared somehow.

The advantage to this is that a user can churn through all the database catalogs to search for their favorite object. There will be a way to save the results of the search to an ASCII file. This will have an option to produce a file that can be input again to the search function.

3. HEASARC Data Server Internals

The GUI resides on the user's host computer and communicates with HEASARC database servers (e.g., via sockets). Application dependencies are confined (as far as possible) to the servers. The database can be Ingres, Sybase, or another SQL database. Applications will be targeted to a variety of machines, from which they can access the database.

Client/server The client/server performs non-transparent task-to-task communications via TCP/IP network protocol. The server is implemented as a daemon process capable of handling multiple connections. The client sends ASCII commands to the server and receives either ASCII or binary data from the server. The server performs necessary data alignment. The client performs necessary data conversion.

XObserver Provides user login/registration functionality, and maintains a database of user sessions. The database is implemented using the *Unix ndbm* database subroutine package.

Bulletin Provides access to bulletin board news items. The database is implemented using the *Unix ndbm* database subroutine package.

NGBrowse Provides access to data maintained in Ingres and Exosat DBMSs. The interface to DBMSs goes via STDB API and SQL CLI. It provides both internal and SQL type command syntax.

References

Pasian, F., & Richmond, A. 1991, "User Interfaces in Astronomy", in: Databases and On-line Data in Astronomy, ed: Albrecht, M., & Egret, D., Kluwer Academic Publishers

Richmond, A., McGlynn, T., Ochsenbein, F., Romelfanger, F., & Russo, G. 1987, "The Design of a Large Astronomical Database System", in: Astronomy from Large Databases: Scientific Objectives and Methodological Approaches, Garching FRG, F. Murtagh & A. Heck, eds

Richmond, A., "Towards an Astrophysical Cyberspace: The Evolution of User Interfaces", this volume

White, N., Barrett, P., Jacobs, P., & O'Neel, B., 1993, "The HEASARC Graphical User Interface", Astronomical Data Analysis Software and Systems II, A.S.P. Conf. Ser., Vol. 52, eds R. J. Hanisch, R. J. V. Brissenden, and J. Barnes

Astronomical Data Analysis Software and Systems III
ASP Conference Series, Vol. 61, 1994
D. R. Crabtree, R. J. Hanisch, and J. Barnes, eds.

IDL Widget Libraries at the Space Astrophysics Laboratory

Benoit Turgeon[1]

Space Astrophysics Laboratory, Institute for Space and Terrestrial Science, 4850 Keele Street, North York, ON, M3J 3K1, CANADA

Abstract. The Space Astrophysics Laboratory (SAL) at the Institute for Space and Terrestrial Science has compiled various libraries of procedures and functions written with the Interactive Data Language (IDL). The SAL Compound Widget Library (SALCWL) contains Graphical-User Interfaces (GUI) which can be inserted into existing applications to simplify programming of widget software.

1. Introduction

A growing number of astronomy-related applications include the use of GUIs to facilitate the input and output of information between the software and its users. Software programmers attempt to simplify their software while trying to keep computer codes relatively simple. In general, these two goals are mutually exclusive, as user-friendliness comes at the cost of programming length.

Creating widgets in the X Window System (the *X Window System* is a trademark of the Massachusetts Institute of Technology) is not an overly difficult task but most astronomers use third-party software packages to carry out their work such as IDL (*IDL* is a registered trademark of the Research Systems Inc.) or the Image Reduction and Analysis Facility (IRAF). Those software packages provide astronomers with a programming environment, but low-level interactions with the windowing system are difficult to perform.

IDL has its own version of some of the basic widget types, thus allowing programmers to build customized GUIs. As programmers gain experience with widget programming, they often find that they have to reinvent a widget cluster simply because, for instance, they need to change the value of one label widget in the entire cluster. The introduction of compound widgets in IDL is the solution to this problem.

Astronomers at the Space Astrophysics Laboratory have compiled a series of compound widgets that are as general as possible so that software applications can be built more effectively without duplicity or multiplicity by calling those compound widgets with different arguments.

This paper will show one such compound widget and explain the argument passing mechanism to generate a custom-built widget cluster. In addition, the

[1]Also with the Department of Physics & Astronomy, York University, Toronto, CANADA.

December 1992 version of the IDL Astronomy User's Library Documentation is now available from the SAL computer server. Details will be given on how to retrieve it.

2. Compound Widgets

A compound widget is a widget that holds an arbitrary number of subwidgets of any types. To the programmer, they simply are described by a widget *id* and associated embedded structures containing such information as the *state* of the widgets and the *resources* used to generate the widget. This state and resource *caching* enables the programmer to access each element of the compound widget from the event loop — to change the value of a label widget for instance.

Compound widgets can have a user-value similar to other widgets' user-values and programmers decide which information the widget will return if an event were generated within the event-loop. To ensure that compound widgets are useful, they need to be general enough to allow the creation of different applications from the same computer code. This can be problematic since *generality*, in this context, implies that the call to the widget will contain a substantial number of arguments.

3. Calling Mechanisms

To avoid unnecessary repetition, we have produced compound widgets that can be called in the following ways:

```
result=SALCW_SOMEWIDGET(parent,[UVALUE=uvalue])
```

or:

```
result=SALCW_SOMEWIDGET(parent,RESOURCES=res,[UVALUE=uvalue])
```

where `parent` is the parent of the compound widget, `uvalue` is the standard user-value for the compound widget, and `res` is a structure containing the requirements of the widget. This last component is the *raison d'être* of the SAL-CWL.

The resource structure passed to the compound widget can contain one or more elements defined in the default resources file. The contents of the structure passed to the compound widget will override the common elements found in the resources file. The compound widget will then read the resources file to fill in the missing keywords. Undefined keywords will be ignored by the compound widget.

4. SALCW_OPTIONS.PRO

This compound widget contains a series of label widgets at the top of its container, as well as an array of label and text widgets complemented by a series of button widgets at the bottom. Two examples of this compound widget are illustrated in Figure 1.

SALCW_OPTIONS is intended to display information in a tabular format which can be edited by the user. Once the various fields are filled by the user, an action can be taken by selecting and clicking on one of the buttons appearing at

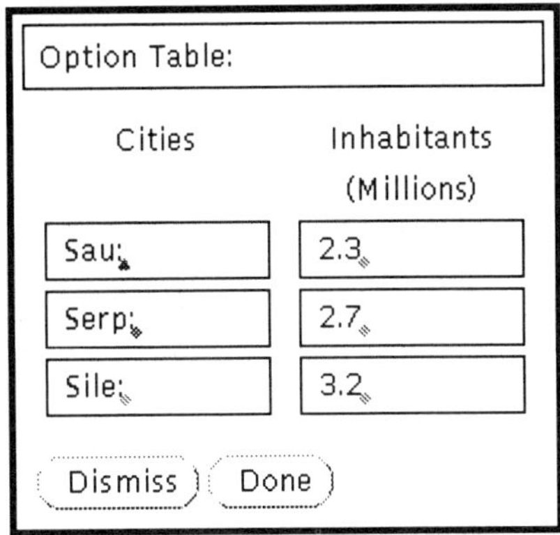

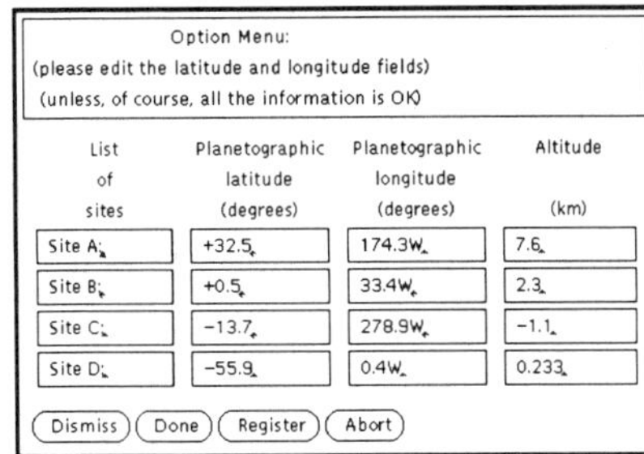

Figure 1. Left: Example of the default appearance of SALCW_ OPTIONS with a series of buttons at the bottom of the compound widget. Right: The same compound widget called with a different resource structure.

the bottom of the compound widget. In such a scenario, an event is generated by the event loop which includes information about the *state* of the compound widget such as which button was selected. The *state* of the application was stored in the first available component of the compound widget so that it can be retrieved at any moment. This allows, for instance, the programmer to access any component of the compound widget based on its user-value.

The example on the left of Figure 1 corresponds to the *default* appearance of SALCW_ OPTIONS.PRO. It was created by issuing the call
`result=SALCW_ OPTIONS(parent,UVALUE='Info').`
To obtain this result, the compound widget consulted a default resources file, in this case SALCW_OPT.RESS, to obtain the layout of its various component.

A different call,
`result=SALCW_OPTIONS(parent,RESOURCES=res,UVALUE= 'Info')`
resulted in the example shown on the right of Figure 1. The variable `res` is defined in an extended but straightforward anonymous structure as follows:

```
res={top_str:['Option Menu:',                                      $
              '(please edit the latitude and longitude fields)', $
              '(unless, of course, all the information is OK)'], $
  array:[['List',   'Planetographic','Planetographic','Altitude'],$
        ['of',     'latitude',      'longitude',      ' '      ], $
        ['sites', '(degrees)',      '(degrees)',      '(km)'   ], $
        ['Site A:',  '+32.5',         '174.3W',       '7.6'    ], $
        ['Site B:',  '+0.5',          '33.4W',        '2.3'    ], $
        ['Site C:',  '-13.7',         '278.9W',       '-1.1'   ], $
        ['Site D:',  '-55.9',         '0.4W',       '0.233'    ]], $
  type:            [[2,2,2,2],[2,2,2,2],[2,2,2,2],                  $
                    [0,1,1,0],[0,1,1,0],[0,1,1,0],[0,1,1,0]],      $
  textwidth:       10,    nbells:          2,    wait:          0.1,$
  bypass_buttons: 0,     frame:           1,    textframe:       1, $
```

```
labelframe:    0,    mainframe:     3,   xsize:          0, $
buttvalues:    ['Dismiss','Done','Register','Abort'],       $
buttuvalues:   [0,1,2,3]}
```

One important point to underline is that the above structure is the most complete and complicated resource structure that the user can use with SALCW_OPTIONS. It is perfectly valid, for instance, to only include the *bypass_buttons* tag name into a structure and set its value to 1 to remove the buttons appearing at the bottom of the compound widget. This capability is called *partial resource passing*.

5. Availability and Documentation

Software programmers are encouraged to use the compound widgets developed at SAL for their own purposes and evaluation. The author would encourage feedback of any sort to be sent by electronic mail.

SALCW is available through anonymous FTP at
`nereid.sal.ists.ca; pub/idl_widgets/aaareadme.txt`
and through the World Wide Web at
`http: //nereid.sal.ists.ca/turgeon/salcw/Intro.html.`

Each compound widget in the SAL Compound Widget Library contains an in-depth description of its components along with an example of its use so that programmers can quickly start to use the compound widgets. In addition, a hyper-media-based programmer's manual and associated documentation is available through the World Wide Web address mentioned above. This documentation includes multi-media representations of the SALCWL as well as graphical examples of its use.

6. Documentation for the IDL Astronomy User's Library

The latest release of the PostScript and DVI releases of the documentation for the IDL Astronomy User's Library (Landsman 1993) is available from SAL though anonymous FTP at
`nereid.sal.ists.ca; pub/idl_astro/aaareadme .txt`
and through the World Wide Web at
`ftp://nereid.sal.ists.ca/pub/ idl_astro.`

Acknowledgments. This paper would not have been successful without the help and generosity of Michael De Robertis from York University, Toronto, currently on leave at the Dominion Astrophysical Observatory in Victoria, BC. Thanks Mike !

References

Landsman, W.B. 1993, in Astronomical Data Analysis Software and Systems II, A.S.P. Conf. Ser., Vol. 52, eds. R.J. Hanish, R.J.V. Brissenden, & J. Barnes, 246

Astronomical Data Analysis Software and Systems III
ASP Conference Series, Vol. 61, 1994
D. R. Crabtree, R. J. Hanisch, and J. Barnes, eds.

A Prototype User Interface for *ASpect*

S. J. Hulbert, J. D. Eisenhamer, Z. G. Levay, R. A. Shaw

Space Telescope Science Institute, 3700 San Martin Drive, Baltimore, Maryland 21218

Abstract. *ASpect* is a new line and spectrum analysis package being developed at ST ScI that will incorporate a variety of analysis techniques for astronomical spectra. Work has progressed on the development of the *ASpect* interface design. We have adopted *Tcl* (and the companion toolkit *Tk*) in which to develop our graphical user interface (GUI.) *Tcl* is a public domain software environment for X11 windowing systems that provides and manages a variety of widgets. The *Tcl* environment is a mature system; it is well-supported and extensively used, even in the commercial sector. We have created a *Tcl* program that "wraps" around IRAF similar to the *xcl* prototype that has been developed at ST ScI. The *Tcl* shell provides the windowing capabilities that are required to fully support the *ASpect* analysis tasks. The shell provides access to the IRAF *cl* for task initiation as well as general data processing in the IRAF system. We have developed a prototype GUI for *ASpect* implemented in *Tcl* and using the *Tk* toolkit.

1. Overview

ASpect is a developing line and spectrum analysis package. It will provide facilities for analyzing multiple spectra from different wavelength domains with nonlinear dispersions. It will fit continuum and features with a variety of functional forms, and will permit constraints between features with line parameters. It will permit masking invalid pixels and perform robust error estimation. All of this will be available through a complete user interface, tightly integrated with the graphic display. In addition, most functions will be available in "batch" for processing large data sets more automatically. For more details on the functional capabilities planned for *ASpect* refer to Hulbert, et al. (1993).

2. The *ASpect* Package Server

We have developed a "package server" that will moderate communications between the user, the GUI, and the IRAF *cl*. This package server manages: I/O interactions with the user, I/O interactions with the *cl* and all aspects of the GUI. In monitoring the communications with the *cl*, the package server discriminates between *cl*-specific data and *Tcl*-related data that is to be used to modify the windows, widgets, etc. In fact, the prescription for the GUI is contained in a *cl* script which in turn contains a *Tcl* script for setting up the *ASpect* GUI.

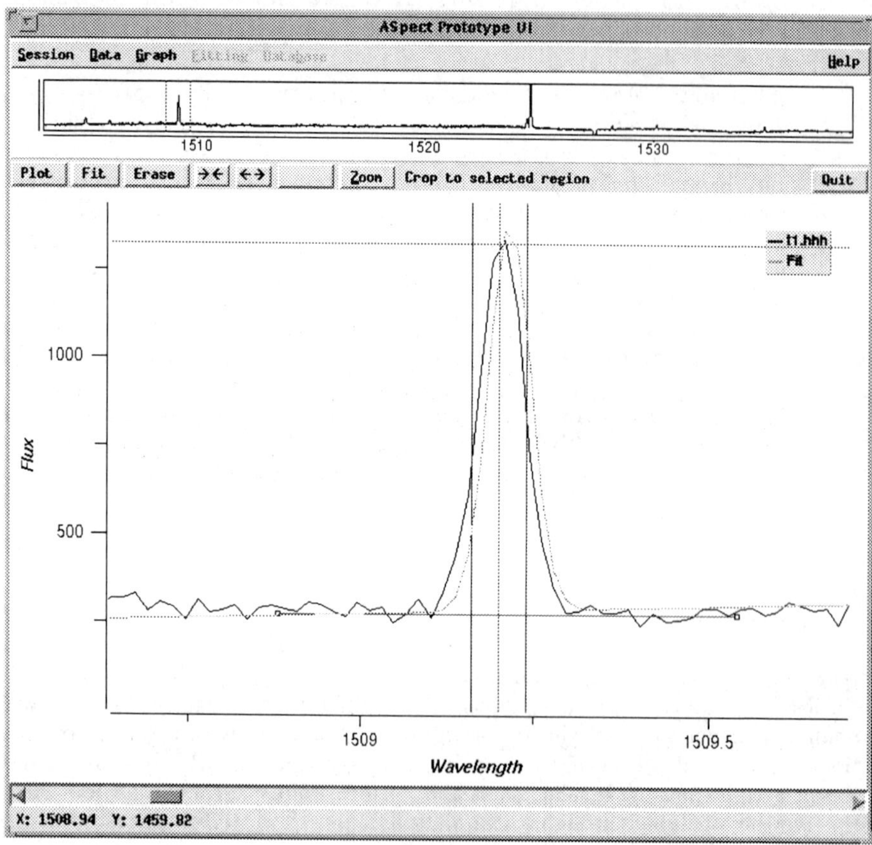

Figure 1. The *ASpect* workspace, showing elements including the menu bar, button bar, overview plot and detail plot canvas with data and fit curves, interactive feature specification cursors, plot scroll bar and status message areas with continuously updated coordinate display and context-sensitive help.

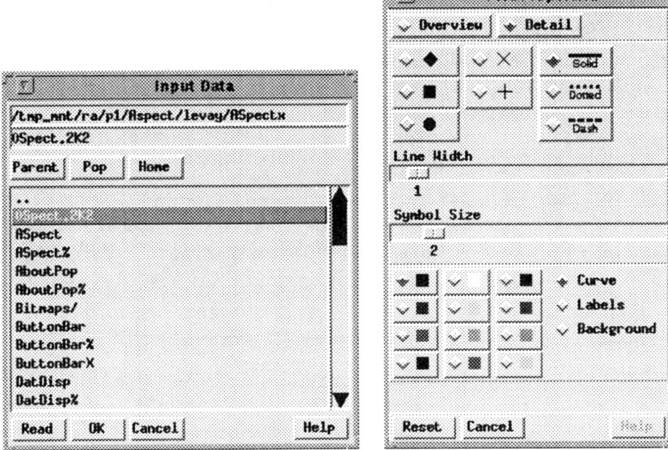

Figure 2. The *ASpect* file requestor (left) for traversing directories and specifying input data and plot preferences popup (right) for changing plot properties.

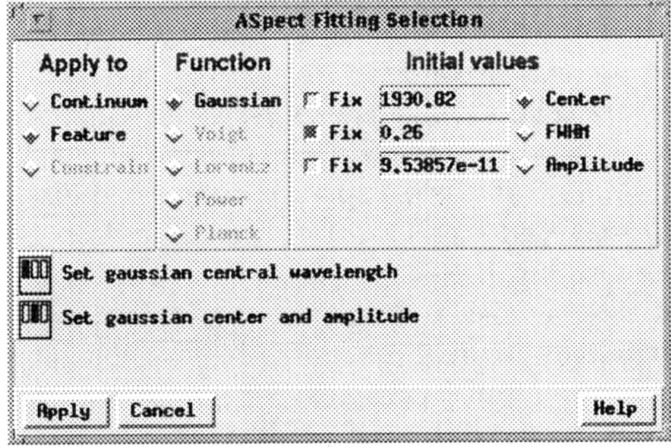

Figure 3. The *ASpect* fitting dialog popup providing facilities for specifying continuum and features to fit.

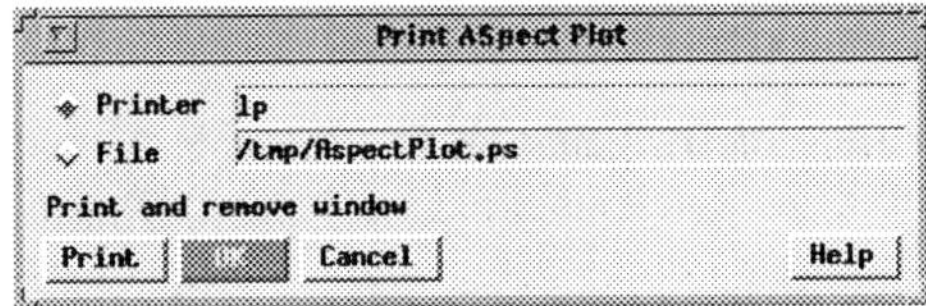

Figure 4. The *ASpect* plot printing dialog popup to print the graph to a PostScript file or printer.

The package server is able to handle the *Tcl* script since it actually contains the *Tcl* interpreter. The package server can wait for the IRAF tasks as well as send IRAF commands to the *cl*; the launching of analysis tasks is bound to buttons and initiated by user actions. The communication between *ASpect* and the IRAF *cl* flows through STDIN and STDOUT channels of the *cl*.

3. The *ASpect* GUI

The *ASpect* user interface consists of scripts written in the Tool Command Language (*Tcl*) and uses the associated X-windows toolkit *Tk*. *Tcl* is a full-featured, interpreted programming language. It is mature, freely available and redistributable, portable, robust, and has a large and growing user base.

Tk provides access to a full set of X widgets through the interpreter. *Tcl* and *Tk* together permit fast prototyping since there is no need to compile and link source code. This architecture parallels the development of the IRAF "widget server" which also uses the *Tcl* interpreter for building the GUI, but has greater integration into IRAF tasks.

We have developed the framework for the full user interface for *ASpect*. Features implemented include: interactive function and feature selection, data scrolling and zooming via scrollbars, buttons and menus, interactive file requestor, flexible plot attribute specification, help including interactive popups and context-sensitive status display and printing the plot via PostScript.

This prototype will be expanded to include all of the required features. In particular, a great deal of attention must be paid to feature selection and fitting constraints and a fully integrated help documentation. The framework developed so far incorporates acknowledged GUI design principles such as consistency of widget and window layout, attention to color selection, embedded context-sensitive help, user-configurable workspace, progressive disclosure, etc. A goal is to develop the interface, analysis, and database tools together for maximum integration as a true graphical user interface.

Additional features we wish to implement include quick keys (keyboard accelerators) and a facility for logging and restoring sessions and batch processing. In addition, there must be a full-featured facilities for managing and manipulating the databases required for data and analysis parameters.

Acknowledgments. The *ASpect* project is funded under a contract from the National Aeronautics and Space Administration Astrophysics Data Program.

References

Hulbert, S.J., Eisenhamer, J.D., Levay, Z.G., Shaw, R.A. 1993, "ASpect: A New Spectrum and Line Analysis Package" in Astronomical Data Analysis Software and Systems II, A.S.P. Conf. Ser., Vol 52, eds R.J. Hanisch, R.J.V. Brissenden & J. Barnes, 437

Astronomical Data Analysis Software and Systems III
ASP Conference Series, Vol. 61, 1994
D. R. Crabtree, R. J. Hanisch, and J. Barnes, eds.

An Xwindows/Motif Graphical User Interface for Xspec

J. M. Jordan, D. G. Jennings, T. A. McGlynn, J. T. Bonnell, G. W. Gliba, N. G. Ruggiero, T. A. Serlemitsos

CSC/COSSC/GSFC-NASA, Greenbelt, MD 20771

Abstract. The XspecGUI is an Xwindows/Motif point and click interface logically arranged around the tasks involved in Xspec analysis. Xspec functions are divided into four categories — data specification, model building, fitting, and plotting — which are all accessable via the XspecGUI control window. The functional divisions of the XspecGUI are designed to guide the novice user of Xspec through the analysis sequence without encumbering expert users. Other important attributes of the XspecGUI are run saving and restoration along with context sensitive point and click help.

Development of the XspecGUI has been performed in a rapid prototyping style and is completely separate from Xspec development. Communication with Xspec is handled via Xterm logfiles and the xkib program from the SAO ASSIST project. Xspec has received no modifications to interface with the XspecGUI program.

1. Introduction

The XspecGUI is an independent, windowed front end to the Xspec spectral analysis program. Xspec is an interactive program with a command line interface. For the novice Xspec user, the commands may tend to be obscure and easy to forget, as well as there being no indication what order the commands should be used in. The XspecGUI provides a format which leads the user through the steps in Xspec while providing added value with display and run saving features.

2. Features of the XspecGUI

Guiding the novice Xspec user is one of the goals of the XspecGUI. Xspec analysis typically consists of four tasks, selecting data files, building a model, fitting to the model, and plotting the fit. Each of Xspec's four tasks has a button associated with it on the GUI control window. The buttons are arranged from left to right in the order that the Xspec tasks are typically performed.

Clicking on each of the control window task buttons summons a window specific to that task. There are data, model, fit and plot windows. Each window is arranged to indicate the information required to finish the task.

As the steps associated with each task are performed, the running of the parent Xspec session may be observed in the GUI control window. Xspec novices have the chance to associate window actions with Xspec commands for their own future use.

While the XspecGUI is designed to lead novices through the intricacies of Xspec, it is also designed to provide added value for the expert Xspec user. The XspecGUI provides a command line entry window where Xspec can be used in the standard command line fashion, although with the addition of a visible point and click history mechanism for re-executing previous commands. Also, there are no restrictions on the order in which the Xspec task windows may be brought up and all windows may remain open.

Other features have been layered over Xspec with the XspecGUI. First, data-model-fit-plot runs may be saved for later review and comparison with other runs. Each saved run has access to an inactive data, model, fit, and plot window which displays the appropriate data for that run. Similar windows for each run may be displayed at the same time. Thus, the plots for runs one through four might be brought up and displayed for a visual comparison with the current fit. Saved runs are represented on the screen by a small window with data-model-fit-plot buttons to bring up the appropriate windows.

Second, window displays of chi-squared and model parameters are updated during fits for easy reference. There is no need to watch the central Xspec display on the control form, instead, the changing parameters and fit can be observed in the fit and parameter windows. This is an especially useful feature when a model has a large number of parameters and Xspec is displaying the numbers as multiple unlabeled rows.

Last, file selection and model building are both performed through point and click interfaces, with file selection displaying directory entries interactively. A user's current working directory is initially displayed, so, by starting the XspecGUI where one's current data of interest is, then the data files will immediately be displayed for mouse selection. Otherwise, clicking on directories will move the user up and down through a directory hierarchy to the appropriate location.

3. The XspecGUI Windows

The Control Form is the first window displayed by the XspecGUI. Functioning as the central menu and display of the GUI, the Control Form contains the direct command line interface to Xspec and the buttons which call up the subsidiary Xspec windows. Xspec may be interfaced directly to the Control Form via an editable command entry text area which is supplemented by a history list of commands which can be mouse selected for rerunning. Output from Xspec is shown in a large display in the center of the Control Form. The chattiness of the Xspec output may be adjusted by a slider located above the Xspec display. Buttons on the Control Form provide default fit and plot capabilities as well as a button to start new analysis runs. Access to the further windows in the GUI is provided via buttons arranged left to right across the top of the control form in the order that the tasks are normally performed in: data, model, fit, and plot.

Files for analysis are selected with the data selection window. Data selection is done at the bottom of the data window using a file selector which opens displaying files in the user's current directory. Directory changes may be made and files selected using the mouse. Four types of data files may be specified, PHA, background, correction, and response. Multiple PHA files may be selected and are displayed in a list window. These multiple PHA files are asociated with background, correction, and response files are all selected singly.

Generally, after the data selection is done, the model window will be brought up for model building. Additive and multiplicative models are displayed in two point and click scrolling menus which include one line descriptions of the models. Clicking on a model puts it into the Chosen Model window. Parentheses and multiplication symbols are automatically added by the model parser as required. Buttons are provided to add further parentheses and for "+" signs. Once the entire model has been chosen, the model button is clicked on and windows are displayed which collect the parameters required by the model. Each model in the chosen model has a separate parameter window displayed to collect and/or display its parameters.

Each of the model parameter windows contains the description of the model and the equation for the model. Additionally, each model parameter is named and described. Features of the model parameter windows are buttons indicating the frozen/thawed state of model parameters which will change the state of that parameter when clicked on. Additional parameters can also be entered for models with the "..." button. These additional parameters modify the default delta and the hard and soft limits of individual parameters.

During fits, parameter values can be watched by locking model parameter windows. The lock button on the parameter window will lock the window to the screen so that during a fit the values taken on by each parameter will be displayed in the parameter window. When a parameter window is locked to the screen, it can also be used to enter new values for the model parameters with the "newpar" button.

Facilities for modifying the fit iterations, the critical delta, and the fit statistic are all provided on the fit window. Additionally, the current value of the fit is displayed along with access to a number of previous fit values. During a fit, the current value of the fit is continually updated in the display on the fit window.

After a fit is performed, the plot window can be brought up to select what to plot. If the default plot is selected, then a plot of the data will be shown. Any of the other plot types available from Xspec may be chosen by a button click from a menu in the plot window. Eventually, a selection of Setplot commands will be available from the plot window. Currently, plot display is handled by the free Gnu utility ghostview. A version of the ghostview plotting package is included in the XspecGUI release. Although ghostview handles display very nicely, it will probably eventually need to be replaced when interactive plotting is added into the XspecGUI package.

4. XspecGUI Development

Independence from Xspec was a primary requirement in the development of the XspecGUI. None of the Xspec code could be touched because of the necessity of operating the GUI as a completely separate project. Running Xspec in an Xterm with its output being duplicated in a logfile and sending commands to the Xterm running Xspec with SAO's public domain xkib routine provide the required independence. A version of the xkib routine is also provided in the XspecGUI release. Although operating within the XspecGUI, Xspec can see no difference between normal and GUI operation.

Development of the GUI followed a rapid prototyping scheme. An initial stab at the project was made and presented to users, followed by successive rounds of development and presentation. Although no major modifications were initiated by user contact, many small improvements to the feel and flow of the XspecGUI were made which much improved the user appeal.

Due to limited resources, the XspecGUI was developed using only cheap, easily available tools, C, Xwindows, and Motif. The GUI was developed with a small fraction of the time of two programmers.

5. Plans for Enhancement

We haven't really had time to think beyond the beta release, but a few plans for enhancement include: mouse interaction with and channel specification for plots, an interface for user defined models, and real PLT support. Full windowing of Xspec help has been started on and is not far from implementation. Other possible changes might include the modification of some of the displays as experience develops better formats.

6. Availability

The XspecGUI is available via anonymous ftp from the gopher server or anonymous ftp on legacy.gsfc.nasa.gov.

References

ASSIST from SAO, available at
ftp://sao-ftp.harvard.edu/pub/asc/README.FIRST

Ghostview from gnu, available at
ftp://prep.ai.mit.edu/pub/gnu/README

XSPEC available at
ftp://legacy.gsfc.nasa.gov/software/xanadu/README.FIRST

Astronomical Data Analysis Software and Systems III
ASP Conference Series, Vol. 61, 1994
D. R. Crabtree, R. J. Hanisch, and J. Barnes, eds.

A GUI for an IRAF Aperture Photometry Task

L. E. Davis

NOAO/IRAF Group, Tucson, AZ 85719

Abstract. A prototype graphical user interface for an interactive IRAF aperture photometry task is presented. The GUI uses the new IRAF widget server technology, which is both window and toolkit independent, to provide a generic set of widgets for communication with the client IRAF task, and custom IRAF widgets for image display and graphics.

1. Introduction

A prototype graphical user interface for a new interactive IRAF task XPHOT, which does interactive multi-aperture photometry through circular, elliptical, rectangular, or polygonal apertures is presented. The XPHOT GUI uses the new IRAF widget server technology, which is both toolkit and window system independent, to provide a generic set of widgets for communication with the IRAF photometry task, and custom widgets for image display and graphics. Together the GUI and widget server technology: 1) simplify the management of the XPHOT input image lists and photometry algorithm parameters and the presentation of results, 2) provide the XPHOT task with a fully interactive image display capability, image cursor mode, and image graphics overlay capability, 3) encourage the development of new scientific capabilities for XPHOT by providing system support, at the application level, for common functions, e.g., interactive region definition.

2. Defining and Initializing the XPHOT GUI

The XPHOT GUI is fully specified by a GUI definition file. The XPHOT GUI definition file is a text file which defines: 1) the XPHOT GUI widget tree hierarchy and specifies its layout and appearance on the screen, 2) the action procedures to be executed when the user interacts with the GUI (e.g., pressing the space bar in the image display widget sends a message to the XPHOT task which tells it to do photometry on the object under the cursor), and 3) the GUI parameters and associated callbacks through which the XPHOT task sends messages to the the GUI (e.g., create an image list on XPHOT task startup and store it in the XPHOT GUI image list widget).

On task startup the XPHOT task requests an image list, values for all the algorithm parameters, and the name of the GUI definition file. XPHOT reads the image list and the current algorithm parameter values into memory, downloads the GUI definition file to the widget server which creates the GUI,

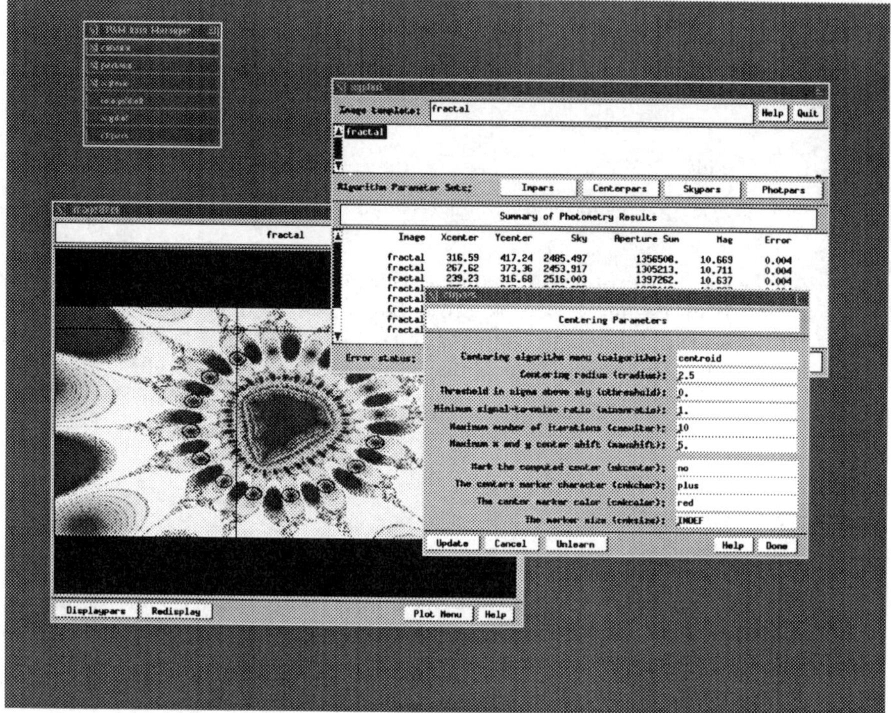

Figure 1. The XPHOT GUI command and image display windows and a sample parameter editing window.

sends the image list to the image list widget, the current parameter values to the parameter editing widgets, displays the first image, and waits for instructions from the user.

Since the GUI definition file is a text file which is read at runtime and the action procedures are written in the publicly available TCL language, an individual user or site can easily customize the default file supplied with the task to suit local tastes.

3. Image Selection and Display

The principal components of the XPHOT GUI, illustrated in Figure 1, are: 1) the command window, consisting of the image list widget, the algorithm parameter sets menu bar, and the results table, 2) the image display window, and 3) the parameter set viewing / editing window(s).

By default the first image in the image list is loaded into the image display window at task startup. New images may be selected and displayed by clicking the mouse on the appropriate image name. The image display parameters can be edited and the image redisplayed by editing the display parameters set and clicking the redisplay button.

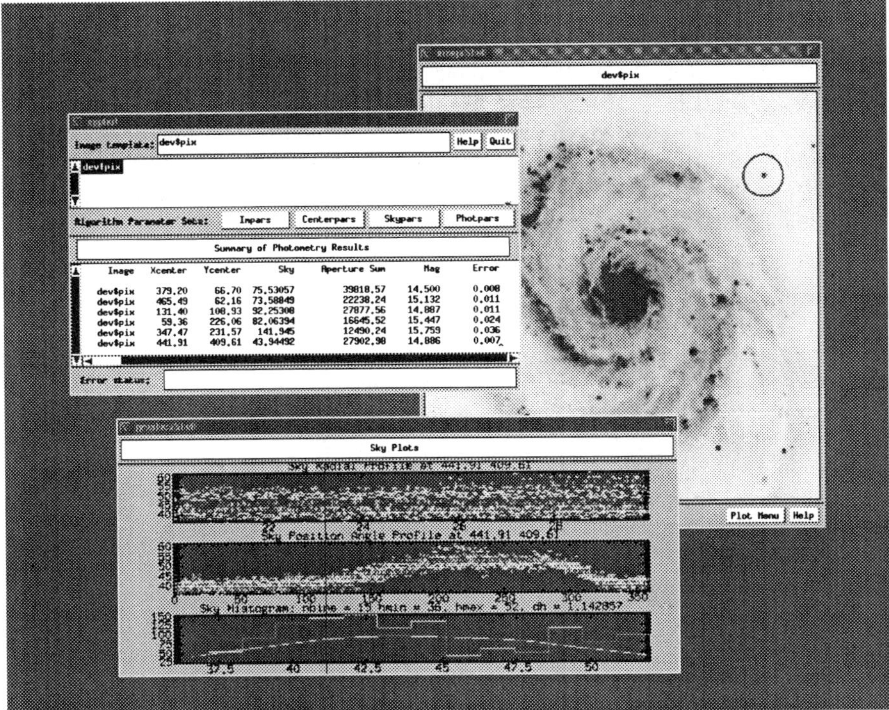

Figure 2. The XPHOT GUI command and image display windows and the optional graphical results display window.

4. Parameter Editing

The algorithm parameter sets may be viewed and / or edited by clicking on the appropriate button in parameter set menu bar and editing the parameters in the resulting form. The XPHOT GUI currently recognizes three types of parameter editing: 1) text entry where the user types in a value e.g., the radius of the photometry aperture, 2) boolean, where the user clicks on the button to toggle a value from yes to no and vice versa 3) menu item selection, where the user clicks on a button to bring up a menu and selects an item with the mouse. After parameter editing is complete the user can update the parameter set in memory, cancel all changes since the last update, or reset all the parameters to their default values, by clicking the update, cancel or unlearn buttons respectively. The parameter editing form can be left visible or closed by clicking the done button. All the parameters can be edited individually from the interactive image display window without popping up the parameter editing window, by issuing ":parameter value" commands in the usual IRAF manner. These commands will also update the appropriate parameter editing forms.

5. Markers and Region Definition

The XPHOT GUI uses the IRAF widget server marker facility to define the object and sky apertures. Markers are circular, elliptical, rectangular, or polygonal geometry widgets whose geometry can be downloaded to the client IRAF task. Markers are child widgets of the image display or graphics widget and may be dynamically created, positioned, resized, rotated, or destroyed by the user. An example of a circular marker is shown in Figure 2.

6. Doing Photometry and Viewing the Results

After selecting and displaying an image, editing the algorithm parameters, and defining the object and sky regions via markers and/or parameter values, the user can make aperture photometry measurements by moving the mouse to the object of interest, adjusting the size and shape of the apertures as necessary, and tapping the space bar. At any point the image display can be windowed, zoomed and panned in the usual manner in order to get a better view of the region of interest.

The XPHOT GUI currently displays its results in the form of: 1) numerical output, which may be scrolled vertically and horizontally by the user, as shown in Figures 1 and 2, 2) an error status message as shown Figure 2, 3) optional plots of the object and sky regions as shown in Figure 2, and 4) optional graphics overlay of the object and/or sky regions as shown in Figure 1.

7. New Scientific Capabilities

Development of XPHOT's scientific analysis capabilities has gone hand-in-hand with XPHOT GUI development. Full support, not available in existing IRAF aperture photometry tasks, for: 1) multi-aperture elliptical, rectangular, and polygonal object apertures, 2) elliptical, rectangular, and polygonal sky regions, and 3) offset sky regions, has been implemented in the XPHOT task.

8. Current Status and Future Plans

The XPHOT GUI is still in the early stages of development and can be expected to change as new system facilities are added to IRAF and feedback from users is received. Future work will concentrate on 1) increasing the image display options available to the user, 2) refining the parameter display and editing facilities, 3) adding new graphical results display options, and, 4) adding support for coordinate lists.

Astronomical Data Analysis Software and Systems III
ASP Conference Series, Vol. 61, 1994
D. R. Crabtree, R. J. Hanisch, and J. Barnes, eds.

A GUI for the IRAF Radial Velocity Task *FXCOR*

Michael Fitzpatrick

IRAF Group, NOAO[1], *PO Box 26732, Tucson, AZ 85726*

Abstract. A graphical user interface to the IRAF Radial Velocity task, *FXCOR*, is presented. The GUI allows users to make plot selections literally at the touch of a button. Analysis of data is greatly simplified since the user can now interact with the task in an intuitive manner rather than through the 172 possible commands that can be issued from the keyboard. The various sub-modes of the task each have a unique GUI optimized for the use of that particular mode. The addition of a GUI means that the user can not only control the algorithm parameters more efficiently, but the results can be presented in a way that gives the user a better understanding of the correlation.

The interface uses the prototype IRAF widget server which provides not only the familiar menus and buttons, but also custom widgets for scientific graphics and image display. Since the GUI is driven by a text file that is downloaded to the widget server and interpreted, it can be customized by the user to meet a particular preference or to streamline it for a particular application of the task. A new GUI can be rewritten quickly as new widgets are made available and layered on the task simply by changing the file which is downloaded.

1. Introduction

With a relatively modest investment in time and only a few changes to existing code, a fully interactive GUI for the IRAF[2] radial velocity task *FXCOR* has been implemented using the prototype IRAF widget server technology. The GUI provides the user with a more intuitive means of controlling task operation (managing parameters, image lists, viewing results, etc) than the traditional keystroke commands. Since the interface is written as an interpreted script, it is easily extensible by the user to add new features not provided in the distributed version. Through the use of X resources, the GUI can be additionally configured for personal preferences such as background colors and default values.

---

[1]National Optical Astronomy Observatories, operated by the Association of Universities for Research in Astronomy, Inc. (AURA) under cooperative agreement with the National Science Foundation

[2]Image Reduction and Analysis Facility, distributed by the National Optical Astronomy Observatories

Because of the rapid prototyping possible with the interpreted file, the GUI presented here uses several different approaches to common operations like parameter editing so that comparisons of which approach is best can be made. This interface should not be considered the final version, but an experimental prototype. It is expected that as new widgets become available in the server, and experience is gained in designing interfaces, this (and other) GUIs will change in the near future. We hope to demonstrate with this example that the widget server technology being used means that sophisticated and powerful GUIs can be developed for existing tasks faster and easier than with conventional programming methods.

2. Features

Aside from demonstrating how GUIs can quickly be developed for IRAF tasks, the other (obvious) goal of this project was to make the task easier to use. With the traditional interface there are over 170 possible keystroke and colon commands. Although every effort was made to make the commands easy to remember, the beginning user can easily get confused by the sheer number of possibilities. The GUI lays out for the user the commands to be used at a particular stage, and perhaps presents options that otherwise would have been missed. The commands which are sensitive to the cursor position cannot easily be replaced with a command button so it is still up to the user to remember a few keystroke commands. These keystrokes are used frequently by even novice users so this shouldn't present a problem.

The interface has several popup windows, graphics screens for each of the command modes as well as various help screens. The graphics windows consist of a *XGterm* graphics screen (a custom 2-D graphics widget similar to the Xterm graphics screen) with a command bar at the top. As each new mode is entered (either through keystrokes or with a menu selection), the current graphics input and command bar are disabled before opening a new window to avoid the chance of issuing an inappropriate command. Windows are destroyed automatically based on the context of the task. Unlike the traditional task, the separate mode windows permit the user to view the plots from previous modes. While only one window is active for input, all windows can remain open.

The presentation of the widgets is also geared towards conveying as much information as possible. For example, menu buttons are a different color than command buttons to distinguish the two, and a highlighted background color is used for the plotting options buttons to show which is the active plot. Similarly, menu panels use a check mark to show the state of boolean parameters, and a button bitmap to show which option in a group (e.g., the peak fitting function) has been selected.

3. How it Works

The prototype widget server is currently implemented as the *XGterm* terminal emulator. Non-GUI tasks running in an IRAF session in this client look much as they would running in a standard Xterm window, with a few notable extensions such as a full screen graphics cursor and color IRAF graphics. GUI tasks, on

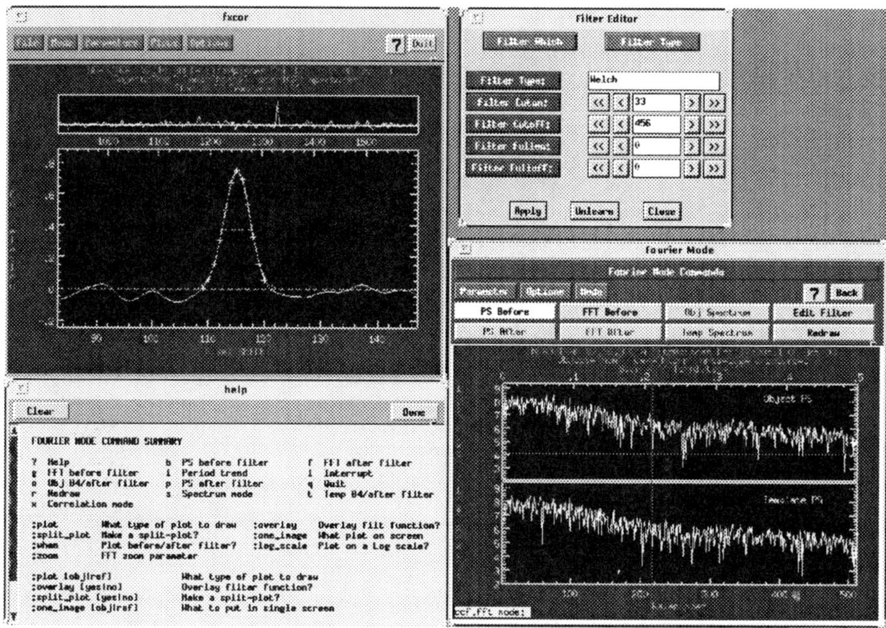

Figure 1. A sample screen layout of the FXCOR GUI showing the Fourier Mode window and some of the other popups available.

the other hand, can use any of the widgets currently available in the Athena or Xt widget sets, the XGterm client is actually controlling these widgets but to the user the task appears to be a separate X client.

The GUI is written as a *Tcl* script file which is downloaded by the client application (*FXCOR* in this case) to the XGterm Tcl interpreter and *object manager*, who is responsible for actually operating the GUI. This interpreted file contains a description of the GUI (button/menu layouts, the various popup shells, etc) and procedures to support the operation of the interface. The client generally has no knowledge that it is being run from a GUI file and so it remains unchanged with the exception of a few new hidden commands to communicate with the interface.

The GUI controls the client by means of callback procedures attached to each of the command buttons and menu options. These callbacks send keystrokes and/or colon commands to the client through an IPC message channel. Since the callback procedures send the same keystrokes that would normally be typed, the keystrokes are still available and can be thought of as accelerator commands.

Similarly, the client can send messages back to the GUI to tell it that some task value has changed, pass up a help menu or some other text output that enhances the GUI but would be too lengthy for a status line. The client does not control the UI directly, instead it sends messages to the GUI through the message channel to set the value of a GUI *parameter object*. A callback attached to the object can then be executed to update the GUI with the new status of the task and take some action such as changing a menu option or initiating a popup. These UI parameters and their callbacks allow the user to control the task either from the GUI or through the familiar keystrokes, and also help keep the client and GUI in sync.

4. Future Work

The GUI presented here is fully functional and fulfills the goals of this project. It is expected that the look of the GUI will change once users have had a chance to give their feedback on it. Some of the design choices (button layouts, menu setup, etc) so far are proving cumbersome and can probably be dealt with in a more elegant fashion. This is in part personal preference, the different ways the task can be used, and familiarity with the commands, but a truly useful GUI should be easier to use than the keyboard interface.

Aside from some small improvements to be made, it is hoped that the GUI can be expanded to provide more functionality to the task. By adding some more code to support the GUI, features can be added such as:

- a standard task parameter and pset editor
- the use of Gterm markers to specify the FFT filter function or correlation width
- enhancements to output such as a scrolling text widget of the logfile or a popup option for the verbose output page.

There will be other changes of course as the widget server facility evolves. Because no windows programming is required, even radical changes to the GUI can be done quickly so the possibilities for future work are limited only by the user's needs.

References

Fitzpatrick, M. J. 1993, "The IRAF Radial Velocity Analysis Package" in Astronomical Data Analysis Software and Systems II, A.S.P. Conf. Ser., Vol. 52, eds. R.J. Hanisch, R.J.V. Brissenden & J. Barnes, 113

Tody, D. 1993, "IRAF in the Ninetie's" in Astronomical Data Analysis Software and Systems II, A.S.P. Conf. Ser., Vol. 52, eds. R.J. Hanisch, R.J.V. Brissenden & J. Barnes, 113

Astronomical Data Analysis Software and Systems III
ASP Conference Series, Vol. 61, 1994
D. R. Crabtree, R. J. Hanisch, and J. Barnes, eds.

Graphical Interfaces for Spectral Analysis in the EUV IRAF package

Mark J. Abbott, Alex Keith, and Tom Kilsdonk

Center for EUV Astrophysics, 2150 Kittredge St., Berkeley, CA 94530

Abstract. We present a prototype of a task under development in the EUV layered package for IRAF to aid in the selection of time-tagged events based on measurements of data quality. It is based on a previous, simpler task and is being reengineered with graphical interfaces to enhance the user's ability to interact with the data in a rapid, intuitive manner. The task aids the user in constructing filters which can be applied to their data to include or remove time intervals of interest. The intervals can be chosen interactively from displays of the time behavior of one or more variables over the span of the observation. These variables can be simple values reported by the instrument or more complex quantities derived from those values or from the data itself; for example, the brightness of the background, the orientation of the spacecraft, or the fraction of data lost due to instrument deadtime. Users will be able to quickly see the effect of changing the limits on these variables in the resulting time intervals.

1. Discussion

The release of IRAF 2.10.3 will provide the first support for windowed graphical user interfaces (GUIs) to tasks within the IRAF. In anticipation of this, the EUVE Guest Observer (GO) Center has been studying how this capability can best be taken advantage of in the EUV layered package for IRAF. This package is used by the GO Center to reduce EUVE data before delivery to GOs, and is also distributed to the GOs for use at their home institutions. The initial analysis revealed an existing task which would clearly benefit from a GUI.

The task *dqselect* is designed to work with tables of time variable quantities (called monitors) to allow users to view the behavior of the monitors over time and to allow simultaneous constraints (or limits) on multiple monitors to be translated into sets of time intervals. The current implementation of this task uses a simple interface of line graphics and a command line to display monitors and place limits on them. The fundamental algorithms of the task are correct, but there are a number of enhancements which would greatly improve the ease with which the user interacts with the data and achieves the desired result.

The addition of a new window based GUI to *dqselect* is a straightforward project. The primary interactions the user has with the task are in choosing monitors to be displayed and setting the limits. Currently, the user must choose monitors for display by name; the monitor names are not intuitive and there are

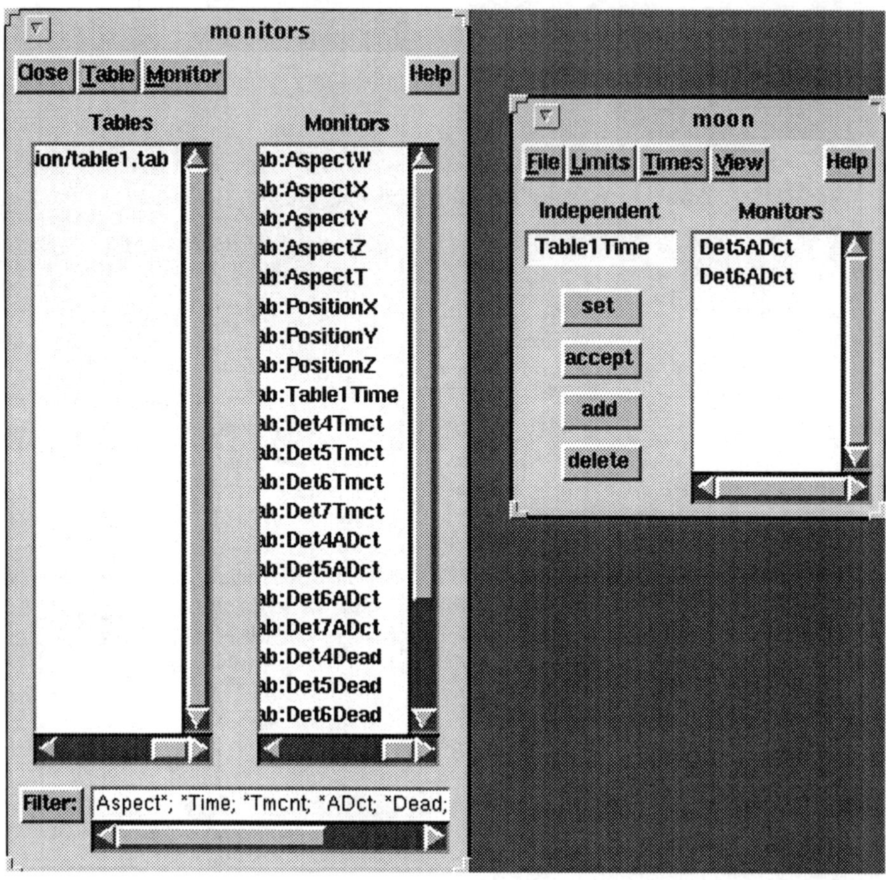

Figure 1. Example windows from the task *dqselect.* On the left is the pool of monitors available for display. On the right is a display control panel with three chosen monitors.

several hundred possibilities (an example of a name is "det4q0Sf"). For the new task, we have chosen the concept of a "pool" of monitors to select from. The user loads tables (disk files) and all of the monitors in each table go into a single pool. Through the use of user-chosen filters on monitor names and selective hiding of certain tables, the pool can be made into a managable list of only the most interesting monitors. In Figure 1, the window labeled "monitors" shows a example of a pool.

In the original task, users could display at most 4 monitors at a time. The new version will support any number of independent "displays", each of which may have one or more monitors plotted on it. Monitors in the pool are associated with each display using the familiar and efficient point-and-click operation. The window labeled "moon" in Figure 1 is the control panel for an example display. Three monitors have been selected from the pool for the display: "Table1Time" has been chosen as the independent variable and "Det5ADct" and "Det6ADct"

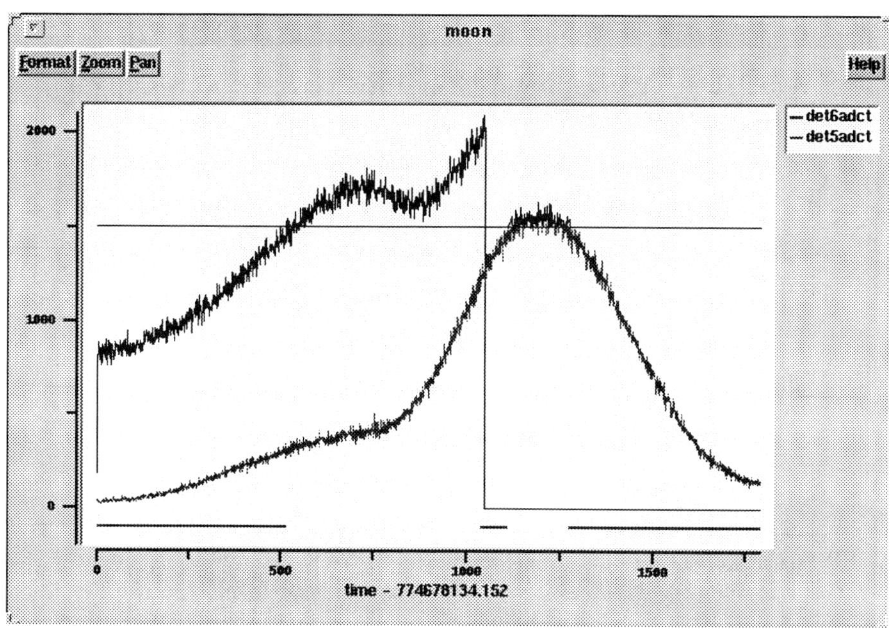

Figure 2. An example window from the task *dqselect.* This is the plot window for the display for which the control panel is shown in Figure 1. Two monitors are plotted along with an upper limit and the time intervals when the monitors are within the limits.

have been chosen as the dependent variables. For any display, there are two possible "views" of the variables, as lists of data points or as a plot.

Each display has a plot window, in which the monitors are graphed for the plot view. The plot window for our example is shown in Figure 2. The plot window supports a large number of user-configurable options for the details of the plot format. One of the key improvements over the old plots is the ability to use color to distinguish the plotted monitors. This allows us to overplot rapidly varying monitors directly, without resulting in confusion. It also allows us to visually emphasize the portions of each monitors range which are outside any limits set on it.

Figure 2 shows two overplotted monitors. Though it is difficult to see in the black and white version of this figure, the monitors are in different colors. Also, above the horizontal line showing the upper limit set at 1500, the color of each monitor changes to black. Across the bottom of the plot are horizontal line segments showing at which times both monitors are within the limit. In one glance the user can easily discern the most important pieces of information dqselect can provide.

A large number of other enhancements and improvements to *dqselect* are planned; here we have mentioned only the most important changes which would have been difficult or impossible without the availability of the new GUI support in IRAF. We anticipate a much more user-friendly and powerful task as a result.

Astronomical Data Analysis Software and Systems III
ASP Conference Series, Vol. 61, 1994
D. R. Crabtree, R. J. Hanisch, and J. Barnes, eds.

WiSPR — A Graphical User Interface for Accessing a Sybase Database

Ramon L. Williamson II

Space Telescope Science Institute, 3700 San Martin Drive, Baltimore, MD 21218

Abstract. This paper describes a graphical user interface for accessing a Sybase database written in the Tcl/Tk environment. Features of the interface are discussed, as well as future enhancement plans.

1. Introduction

WiSPR is a Tcl/Tk script in the X environment that builds a display of the fields of an arbitrary database table in an easy-to-read format on the fly. Each field of the database table has a text widget into which search strings for that field may be entered. When all desired search fields have been filled, an SQL query is constructed and values are returned into the text widgets one row at a time. Subsequent matches to the query are displayed until all matching rows have been retrieved.

Sybase database access is accomplished by means of the sybtcl library, written as an extension to Tcl/Tk by Tom Poindexter of Denver Colorado.

2. Features

Figure 1 shows a typical retrieval using WiSPR. Retrievals are done by filling the text widgets with ranges of values for each of the database columns.WiSPR takes entries from those widgets that are not blank and builds an SQL query that is sent to the database. Results of the retrieval are displayed one entry at a time in these same widgets.

One of the more important features of WiSPR is that the user needs only a minimal knowledge of SQL in order to use the task. Knowledge about the datatypes used by Sybase and the wildcards available in SQL is helpful in fully utilizing WiSPR's capabilities. WiSPR recognizes all SQL wildcards when doing queries.

Once values are returned, the results may be saved to a file or printed on a printer. In addition, if write access to the database in question is available, the user can add new rows to the database or update entries in a row already retrieved.

On-line help is available at anytime using a menu-driven help system.

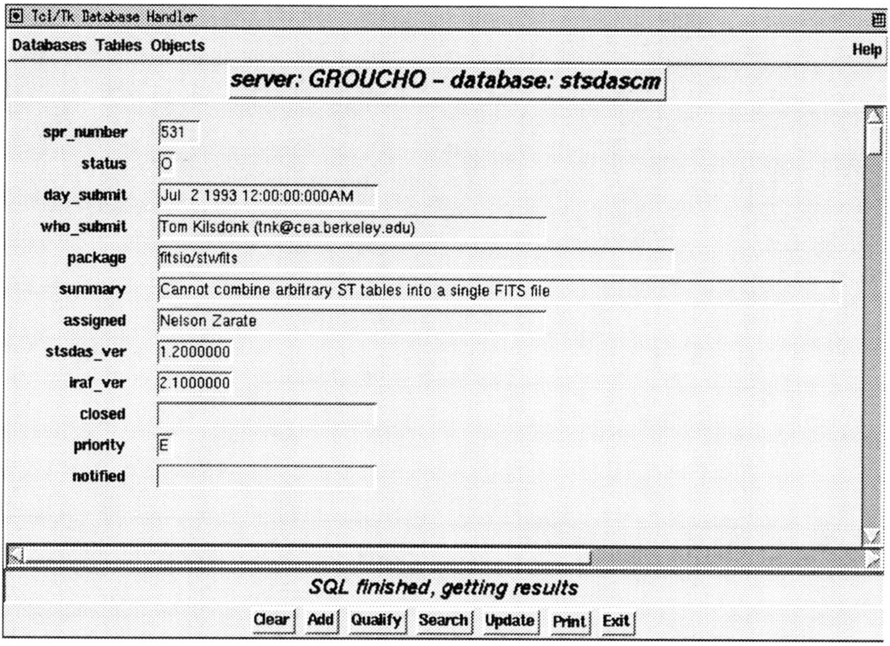

Figure 1. WiSPR's User Interface.

3. Future Enhancements

WiSPR was written as an experiment in using Tcl/Tk to write User Interfaces and to learn the Tcl/Tk language. WiSPR as it stands meets most of the needs I had in mind when writing it. Some enhancements considered for implementation as time permits include:

- WiSPR can currently only do simple queries on columns, i.e., ranges of values cannot be done. With the correct implementation, a syntax could be included to tell WiSPR to search a range of values.
- An alternate output where instead of filling the fields of the interface with the returned values one at a time all of the results of the query are output in a new window where it can be viewed or printed to a printer or saved in a file.

4. Availability

WiSPR may be obtained by contacting me via Internet at ramon@stsci.edu.

Astronomical Data Analysis Software and Systems III
ASP Conference Series, Vol. 61, 1994
D. R. Crabtree, R. J. Hanisch, and J. Barnes, eds.

Separating Form from Function: The StarView Experience

J. Pollizzi

Space Telescope Science Institute, Baltimore, Md 21218

Abstract. Modern day GUI building tools provide both a power and a flaw to the interface developer. This paper contrasts the typical GUI tool approach with that espoused by object oriented methodologies, the model—view—controller. A means of reconciling the use of both when working with an X11 application is then presented. The paper is summarized with recommendations to consider when developing a user interface.

1. Background

The adoption of standardized interfaces libraries, such as X and Motif, have allowed application developers to access the power of modern day workstations. These libraries are, however, complex to use and with subtle nuances that can be difficult to master.

The advent of GUI (Graphical User Interface) building tools over the last few years has made the use of windowing libraries more accessible to the developer. There is little question of the impact of these tools. Over the past year of conferences, more and more presentations have been made claiming the ease and rapidity with which useful applications have been made using these tools.

XVT, JAM, VUIT, are examples of such tools. In general the tools provide interactive "screen painters" to visually create displays. Using the screen painter, the developer iterates in choosing display objects (buttons, fields, menus, ...) and positioning them on the display. For the most part, the complexities associated with the chosen collection of objects is hidden from the user. Many of the tools also provide a "prototype" capability that allow the developer to test their displays interactively while the displays are still under construction. These techniques give the developer many options to quickly review and refine their work. Ultimately the tool exports code modules (usually in C or C++) that the developer then combines with their own modules to complete the application.

To explore the capabilities and implications of these tools, a sample application was developed using the ICS BuilderXcessory tool.

Using BuilderXcessory(also called BX), we quickly produced a simple, yet useful database lookup application. We then considered the implications of making derivations of the application (i.e., to meet new needs, but based on the previous work).

At minimum, any change would require using the BX tool to modify the screen display. This would be followed by the subsequent step for code generation and building. Depending on the changes to be done, either the original application would have to be cloned (i.e., one for each variation), or made significantly more complicated to support the different needs as a combined application. Cloning might be rather fast, but any future changes could require a serious maintenance effort to keep the clones all working. Similarly, making the application support all the variations significantly increases the complexity (in shear numbers) of callbacks that have to be maintained.

Although the use of the screen building tool (BX in this case) aided the developer in setting up the application's presentation format, it did little to isolate the presentation aspects of the application, from the underlying database access functions.

2. Model—View—Controller

Some of the earliest work in the development of object technology focused on the issue of isolating presentation mechanisms from the underlying data model. The approach became known as the "Model—View— Controller".

Within the model—view—controller approach, emphasis is placed equally on the three aspects of the application: the underlying data model being manipulated (called the model); the presentation component that displays the relevant parts of the model or accepts user input (the view); and the determination of when the model needs updating, or further interaction is required from the user (the controller).

The Model—View—Controller (MVC) isolates the understanding of each component there-by allowing for individual evolution:

- Additions/changes in how the data is arrived at or how defined is independent of how it's presented.
- Over the course of the interface's lifetime, changes will occur to the presentation mechanisms — but the underlying data model remains essentially intact.
- User interaction can change without necessarily impacting either the data's derivation or display.

The MVC approach was used as the basis for the StarView application. StarView is being developed by the Space Telescope Science Institute as the user interface to the Hubble Data Archives. The use of the MVC in StarView has given us a significant amount of latitude in keeping StarView flexible enough to accommodate changes to the database or forms requirements, without having to re-visit the code. This approach has also allowed the development team to quickly adapt StarView from a CRT display to the use of X Windows. Yet both versions of the application continue to share common database and form definitions.

3. Using Features of X or Motif to Help

StarView was designed from its outset to be a highly generalized interface that matches a large class of databases to a form display. Not all applications require the sophistication and complexity of StarView. However, the fundamental aspect of the Model—View—Controller should be considered in every user interface application.

The challenge is to take the best advantage of the interface building tools for creating the displays, and yet structure the application so that it is relatively resilent to most screen changes.

One generic solution would be to somehow utilize the information about the window/form setup in a way that could be dynamically mapped to the database. The X and Motif libraries have a rich class of functions that can aid the user in this process. Two steps are needed to associate a display with the data in an application:

First, the initial problem is to find a dynamic way of establishing a connection from a display element of the screen, to its data source (i.e., something that knows what the data value is and when the value changes). Assume that, through the interface tool, we named each display widget to be identical to its database field. Then the following code segment could be used to create this connection:

```
void  Associate_Widget( w )
  Widget      w;
  { void *LookupAttributeFunc();
    void  (*attributefunc)();

    /* Look-up the function to call based on the Widget's name */
    attributefunc = LookupAttributeFunc( XtName(w) );

    /* Notify the Attribute (if one) of the widget that is
       associated with it */
    if ( attributefunc ) { (*attributefunc)( w ); }
    else { printf( "\nNo Attribute found for widget: %s", XtName(w) );}
  }
```

Based on the display widget's name, we locate the proper data handling function. The function is then called with the proper widget id. This id would be used whenever the function determined that the associated data value was to be displayed (or read back as the case may be). This approach allows a generic way to connect a display element up with its source of data.

Next, we need a mechanism to locate all the display elements that are to be dealt with. Fortunately, X requires that all primitive widgets (i.e., widgets that display some item) be collected together under a container widget. The following code segment, when invoked as the "mapped" callback for a container widget, satisfies this requirement:

```
void  WhenMapped_Callback( w, client_data, callback_struct)
  Widget w;
  XtPointer  client_data;
  XmAnyCallbackStruct  *callback_struct;
{
  WidgetList all_my_children;
  Cardinal     num_children;
  void  Associate_Widget();
  int   i;

  /* First, get the list of my children */
```

```
  XtVaGetValues( w, XmNchildren,  &all_my_children,
                    XmNnumChildren,   &num_children, NULL);

  /* For each child, have it associate itself with its data attribute */
  for (i = 0; i < num_children; i++)
    { /* But only for Text or TextFields, i.e., skip buttons,
         labels, other widgets */
      if ( XmIsTextField(all_my_children[i])
           || XmIsText(all_my_children[i]) )
                          Associate_Widget( all_my_children[i] );
    };
}
```

This procedure uses X primitives to locate the children widgets associated with the current form. For each text or textfield type child, it then calls the above routine to establish the link from field to data. In this way, we have provided a generic approach in allowing the application to dynamically map to a screen that was built by a GUI tool. We have achieved, in large part, the desired isolation between the form and function of the application.

4. Summary

This paper has discussed the promise and power of the various interface building tools that exist today. The paper also points out their shortcomings. As a fair warning, remember that any marketable product must address a broad spectrum of use. Invariably that means no tool will completely satisfy any one user's specific requirements. With that in mind, the following guidelines should be followed when selecting a tool:

- In selecting tools, be sure to understand what the tool can't do as well as you come to understand what it can do.
- Look for tools that allow you work outside of their environments:
  - Tools that provide code that you can tailor or modify as you need to.
  - Tools that are driven by "editable" control/data files.

When constructing your application, especially if tools are being used, be aware of the constraints you impose on any future evolution of your application. During the design phase, ask yourself what changes might be encountered in the future. Then determine how your design will accommodate these changes.

To the extent possible, maintain an architecual abstraction between the presentation and the data being presented. Remember that a key goal of the X approach was to isolate user interaction from application design. Also, recognize how components of your application fit into the Model— View—Controller paradigm.

Finally, be open minded in employing all the tools at your disposal to create the isolation between data and display.

Astronomical Data Analysis Software and Systems III
ASP Conference Series, Vol. 61, 1994
D. R. Crabtree, R. J. Hanisch, and J. Barnes, eds.

The StarView Flexible Query Mechanism

D. P. Silberberg

Space Telescope Science Institute, 3700 San Martin Drive, Baltimore, MD 21218

R. D. Semmel

Johns Hopkins University Applied Physics Lab, Johns Hopkins Road, Laurel, MD 20723-6099

Abstract. We are using a software system known as QUICK to generate SQL for the StarView User Interface. Users query the database to which StarView is connected via predefined screens or an interface for ad hoc access. The internal software constructs a high-level request consisting only of columns to be queried and user-specified constraints, and submits it to QUICK. In turn, QUICK uses an Extended Entity-Relationship (EER) model of the database to calculate column tables and the table joins to translate the request into SQL. The SQL query is submitted to the database server and the results are displayed by StarView.

The advantage of using QUICK is that users wanting to formulate queries need not know the database topology. They need only specify the column names and request qualifications. Similarly, screen designers need only specify database columns to create screens.

1. Introduction

StarView is a distributed user interface to the Space Telescope Data Archive and Distribution Service (DADS). The DADS architecture includes the data archive as well as a relational database catalog describing the archived data. Users query the DADS catalog via StarView screens. Based on the results, users decide which data sets to retrieve from the DADS archive. Archive requests are packaged by StarView and sent to DADS, which returns the requested data sets to the users.

The DADS catalog is comprised of more than 1,500 columns distributed among 50 tables. StarView has several predefined screens serving as the interface for the most common catalog requests. Each screen field corresponds to a database column. However, the predefined screens represent only a small subset of the column combinations that users need for querying the catalog. As StarView services a wide and diverse group of users, the interface must provide the flexibility to query the catalog by the combination of columns that best suits a user's needs. Moreover, users should not be required to understand the topology of the database when formulating queries requiring joins.

StarView provides the needed flexibility by enabling users to view the columns of the DADS catalog as if they were defined in a single table known as a *Universal Relation* (UR) (Ullman 1989). In the UR approach, users need only specify columns of interest, and the interface software infers the underlying joins for the corresponding query.

This paper describes the justification for providing a UR interface within StarView and then describes QUICK (Semmel 1992), which is used to translate UR queries into relational database queries.

2. Providing an Interface to Multiple Tables

Most astronomical databases are single-table catalogs describing sky objects and their properties. Each catalog row corresponds to an object and each column refers to a property of the object. Many user interfaces are designed to display the data of a single astronomical table. The query mechanism for such interfaces is straightforward because there are no table joins to calculate.

Some applications require that more than one table be queried at a time. For instance, a user interface that cross-correlates more than one table must be able to query multiple tables. Also, a user interface requiring more than one table to describe catalog objects must be able to query multiple tables. In either case, the mechanism that produces the query is complicated. Most solutions require that predefined queries be embedded in the screen definitions or in the interface code. Unfortunately, this approach requires that screen designers be database domain experts. Often, however, designers are not aware of the subtleties of how data is related in the database. Therefore, there is much room for error when designers define screens or write user-interface code. Furthermore, if the database design changes, many screens may need to be updated and code may need to be modified. This is a tedious and error-prone process.

In addition to interface design problems, there are problems associated with ad hoc access. Specifically, ad hoc requests may correspond to complex SQL queries. However, most users are neither database domain experts nor query language experts. Consequently, all but an elite group of users is precluded from querying the database in an ad hoc manner.

Ideally, developers and users want to view the database as a single table. Furthermore, users want to query the database by specifying only the column names and qualifications, as exemplified in the following Universal SQL (USQL) request:

$$\textbf{select}\ col_1, ..., col_n where col_m > 93 and col_p = \textbf{'JONES'}$$

Note that a USQL request requires neither a FROM clause nor *join expressions* in the WHERE clause.

A relational database schema does not contain enough information about the data being modeled to enable the translation of a USQL request into an SQL statement. To perform this translation, it is necessary to have domain knowledge of the table relationships (including primary and foreign key information) as well as the contexts in which tables are related. While such knowledge is captured in an initial analysis, it is often discarded when the database is created. Yet, by exploiting initial domain knowledge, it is possible to automatically analyze the database design and calculate the correct join and table list given a list of attributes.

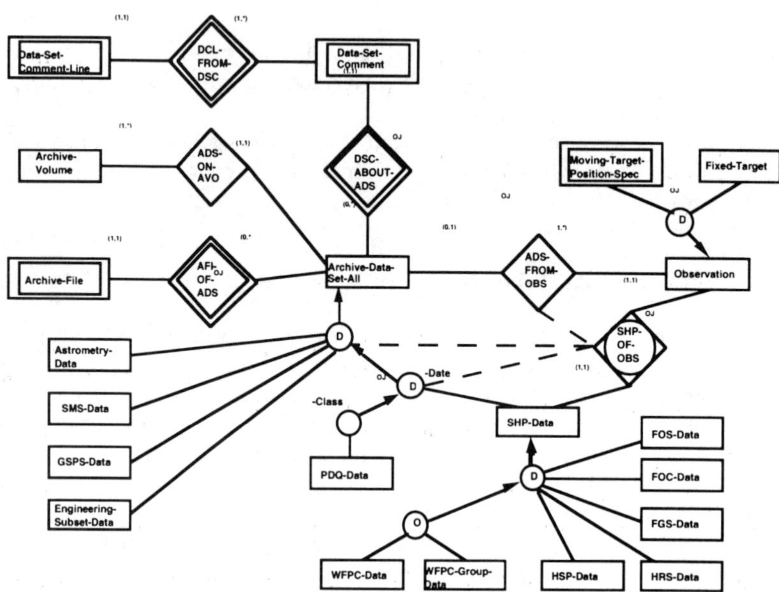

Figure 1. Sample EER Diagram

3. The StarView Approach

QUICK is a system that uses conceptual knowledge of an application domain to translate USQL requests into SQL statements (Semmel 1992, and Semmel and Silberberg 1993). QUICK is based on an EER definition of a database. From the EER specification, QUICK calculates all the maximal subgraphs, or *contexts*, of the database. A context is a composition of all the EER objects corresponding to tables that can be natural joined in a single query. When a list of attributes is presented, QUICK determines all contexts that contain the attributes. For each applicable context, QUICK calculates the joins and the table list, and generates the SQL statement. If the attributes participate in more than one context, the returned SQL statement is a union of the generated SQL statements. If the attribute set does not appear in any context, then QUICK indicates that no join is possible. Note that while it is possible to write an SQL statement joining tables that do not simultaneously participate in any context, such a join would not make sense given the design described by the EER diagram.

StarView employs the QUICK software to translate USQL requests into SQL statements. Figure 1 shows an example of an EER diagram describing approximately one-half of the DADS database.

The boxes and double boxes denote strong and weak entity types in the EER and tables in the actual database. Each EER entity has uniquely named attributes (not shown) which map to renamed (and possibly identically named) table columns. Each entity also has a group of attributes that serves as the primary key of the entity and the table.

The diamonds and double diamonds denote entity and weak-entity relationships. The pairs of numbers associated with each diamond describe the cardinality of the relationship. The **OJ** symbol indicates the possibility of an outer join. The circles denote generalization types, indicating an inheritance hierarchy among entities. A circle with a **D** denotes a disjoint generalization type, indicating that an instance of a parent entity can be a member of at most one of its child entities. A circle with an **O** denotes an overlapping generalization type, indicating that an instance of a parent entity can be a member of more than one of its child entities. A double-stemmed arrow denotes total participation among all the child entities, indicating that all instances of a parent entity must also be members of at least one of its child entities. A single-stemmed arrow denotes partial participation among the child entities, indicating that an instance of the parent entity might not be a member of any of the child entities. The relationship and generalization types are crucial for determining semantically correct join possibilities. This information is precisely what is missing from the actual database schema, and it allows QUICK to translate USQL requests into SQL statements.

4. Summary

StarView currently runs when connected to the Data Management Facility (DMF) database catalog (Silberberg et al. 1990) and the Data Archive Distribution System (DADS) database catalog (Loral Aerosys 1992). The integration of QUICK into StarView makes the query generator easily adaptable to any database with which StarView may be used. The EER of a prospective database need be designed by a domain expert of that database only once and integrated into the StarView run-time environment. Then screen designers can write screen definitions and users can create ad hoc queries without understanding the subtleties of the database domain or the complexities of the SQL language. Moreover, integrating a new database into StarView is a straightforward process, requiring little time and effort.

References

Loral Aerosys 1992, "ST-DADS Database Design Specification", Build 2, Revision B (Loral Aerosys internal document, USA)

Semmel, R. D. 1992, "Discovering Contexts in a Conceptual Schema" (Proceedings of the First International Conference on Information and Knowledge Management, Yesha Y., ed., International Society of Mini and Microcomputers - ISMM, pp. 222-230)

Semmel, R. D., & Silberberg, D. P. 1993, "The StarView Intelligent Query Mechanism" (1993 Goddard Conference on Space Applications of Artificial Intelligence, NASA, Greenbelt, MD)

Silberberg, D. P., McGlynn, T., Lubow, S. 1990, "Design of the Data Management Facility Catalog" (Space Telescope Science Institute internal document, Baltimore, MD)

Ullman, J. D. 1989, "Principles of Database and Knowledge-Based Systems", Vol. 2 (Computer Science Press, Rockville, MD)

Astronomical Data Analysis Software and Systems III
ASP Conference Series, Vol. 61, 1994
D. R. Crabtree, R. J. Hanisch, and J. Barnes, eds.

Querying Multiple Databases with StarView

J. Williams

Space Telescope Science Institute, Baltimore, Md 21218

Abstract. StarView was created to be the user interface to the Space Telescope Science Institute's Hubble Data Archive (HDA). It offers unique capabilities in exploring large scientific databases. These include the ability to create custom queries without requiring the user to understand the structure of the database and the ability to work with multiple databases. These capabilities can be combined to allow for limited cross correlation between different databases in a single query. The purpose of this paper is to explain what is required to make this happen.

1. StarView's Design

StarView's unique querying abilities are grounded in its core design. StarView was designed to work with any relational database. Its architecture differentiates between the underlying model of the data and the viewing of that data. The user's view is defined by forms or what is available through the custom query facility. The structure of the model is defined in the StarView and Quick Data Definition Languages (DDLs). StarView uses a Model-View-Controller (MVC) architecture to provide the interaction mechanism that separates the user's view and interaction from the underlying model. The MVC architecture has its origins in the Smalltalk environment. It has proven to be a powerful framework for developing user interface applications. This framework allows the user interface to change without affecting the underlying model. In StarView, this capability has been used to provide both a CRT and X Windows interface, while leaving the model unchanged. Currently, StarView has been demonstrated with different Sybase databases that are accessed through the Space Telescope Database (STDB) interface.

StarView's flexibility is accomplished through the use of its Data Definition Language. The DDL is used to describe each attribute in the database. The attribute description contains such information as data type, constraints and textual information. It may also specify associated attributes. The textual information additionally supports context sensitive help on the attributes. Figure 1 provides an example of StarView's DDL. This figure illustrates part of the DDL used to define a table of ROSAT observations.

StarView DDL - Figure 1

```
SCHEMA
    DATABASE: dmfbld
        TABLE: rosat
            ATTR: pi
             definition: Rosat principle investigator
             datatype: char
             size: 12
            END_ATTR
            ATTR: targ1
             definition: target name
             descriptive_label: target name
             datatype: char
             size: 16
            END_ATTR
            ATTR: ra
             datatype: float // int
             definition: right ascension of Rosat target
             logicaltype: ra
             units: milli_arcsecs
             associated_field: dec
             size: 4
             constraints: @ra 0..24h
             duplicate_name: 1
            END_ATTR
            ATTR: dec
             datatype: float // int
             definition: declination of Rosat target
             logicaltype: dec
             units: milli_arcsecs
             associated_field: ra
             size: 4
             constraints: @dec -90d..90d
             duplicate_name: 1
            END_ATTR
        END_TABLE
    END_DATABASE
END_SCHEMA
```

2. StarView Custom Queries

The ability to formulate custom queries is a fundamental part of StarView. It is a foolish exercise to try and create all possible screens with all possible combinations of attributes for catalogs as large as those found in the HDA. Instead, it is better to give the user the ability to easily create his own custom query. In Starview, the user can select the attributes he wishes to query on without needing to understand the structure and joins of the database tables. The custom query facility has its own DDL (known as Quick DDL) which describes relationships within a database. It can also describe relations between databases. This is how StarView supports a single query across multiple databases. As long as a relationship is defined in the DDL, a query can be generated. The user can create a custom query on the fly, or create a new form for querying using StarView's Form Definition Language (FDL).

It is possible for users to add their own databases to StarView. The changes they must make are relatively straight forward. To add a database, you must create new DDL for both StarView and the custom query facility (QUICK). This is a simple process for single table databases. You do not have to make any changes to the user interface unless you want to create a new form. Once the DDL is added, the custom query facility can automatically be used to generate a query. If you do want a custom form, one can be created in FDL.

3. Adding A Database To StarView

Let's look at an example of how this is done. We added a database of accepted ROSAT observations to StarView. We needed to take the following steps:

- Add ROSAT information to StarView DDL (shown earlier)
- Add ROSAT information to Quick DDL
- Add ROSAT to Quick contexts.

We could then use the Build Custom Query facility, SQL form or our own custom form to create a query across multiple databases. We chose to use the custom query facility to generate the SQL. This was quick and just as effective as creating a form for the query. Let's look at the details of this process.

We began by creating the DDL shown in Figure 1. We next added the Quick DDL shown in Figure 2. Its Lisp like flavor stems from the fact that Quick was developed in CLOS. It is a format that is easy for Quick to process. This DDL was supplemented by adding a new context to Quick that let it query across DMF and the ROSAT database simultaneously. This is shown in Figure 3. We then created a query on target NGC1068. There were 120 possible records in DMF with NGC1068 in some form (not all identical to what is in ROSAT). There was 1 record for this target in the ROSAT database. Quick generated a query based on this target name. From this query, 63 records were returned as an exact match.

Quick DDL - Figure 2

```
(define-database dmfbld
  :relation-schema-prefix "dmfbld..")
(set-database dmfbld)
(define-relationship-type rosat_to_hst
  :participating-entity-type-specifications
    ((exp 0 *)
```

```
    (rosat 0 1))
  :natural-join-specification
   ((rosat_to_hst targ1)
    (exp exp.targname))
  :infer-outer-joins nil)

(process-er-model)
```

Quick Context - Figure 3

```
(define-context
  :members
   (
    COSHP_OF_OBS
    OBS
    COSHP
    EXP_OF_OBS
    EXP
    DSNAME
    CFMAST
    ROSAT_TO_HST
    ROSAT
    FLS_OF_DSN
    FILES
    OBS_OF_PROP
    PROP
    GTO_GO_OF_PROP
    GTO_GO
    PEP_OF_PROP
    PEPSI
    ABS_OF_PROP
    ABSTRACT-ET
       )
)

(process-loaded-contexts)
```

4. Summary

The ability to query multiple databases in a single query is a powerful tool. StarView was designed to allow users to add and integrate their own databases into the system. Existing capabilities allow the user to correlate databases on exact matches in fields. The next logical step in this work is to allow a fuzzy join on position. We are already investigating this possibility.

Astronomical Data Analysis Software and Systems III
ASP Conference Series, Vol. 61, 1994
D. R. Crabtree, R. J. Hanisch, and J. Barnes, eds.

DRACO: An Expert Assistant for Data Reduction and Analysis

Glenn Miller and Felix Yen

Space Telescope Science Institute, 3700 San Martin Dr., Baltimore MD 21218

Abstract. The use of large format detectors, increased access to very large astronomical databases, and other developments in observational astronomy have led to the situation where many astronomers are overwhelmed by the reduction and analysis process. This paper reports a novel approach to data reduction and analysis which works in conjunction with existing analysis systems such as STSDAS/IRAF. This system, called DRACO, takes on much of the mechanics of the process, allowing the astronomer to spend more time understanding the physical nature of the data. In developing DRACO we encountered a number of shortcomings of current data analysis systems which hinder the ability to effectively automate routine tasks. We maintain that these difficulties are fundamental (not specific to our approach) and must be addressed by conventional and next-generation analysis systems.

1. Introduction

The task of data reduction presents severe obstacles to an astronomer: The volume of data may require tedious work that is susceptible to errors (e.g., the flat-fielding and bias correction of a few dozen digital images can take several day's time and it is easy to accidentally apply the wrong calibrations to some of the images). Management of the data reduction process may require tracking tens or hundreds of files through many different steps. The quality of each reduction step should be evaluated (e.g., stability of internal calibrations, or abnormally large number of bad pixels). Often the entire reduction process must be repeated several times with improved calibration data or improved reduction algorithms.

These are significant problems that inhibit progress by forcing the scientist to expend time and effort on the mechanics of reduction rather than understanding the physical nature of the data. We have developed DRACO, which is a tool for the management of data reduction and analysis. DRACO builds on the foundation of existing data analysis systems such as STSDAS/IRAF, IDL, MIDAS, etc. DRACO gathers information about the available data, develops a plan for data reduction based on a template supplied by the astronomer, and translates the plan into explicit reduction commands. An important feature of DRACO is its generality and extensibility — new types of data analysis tasks or additional data analysis systems can easily be added without modifying existing software.

This work is an extension of a successful prototype system for the calibration of CCD images developed by Johnston (1987).

2. The Draco System

To use DRACO, the astronomer first describes the reduction process by defining three key entities:

- *Procedure* — This is an abstract user program for the reduction, e.g., bias correction, dark removal, field flattening, and extraction of a spectrum
- *Primitive* — These are the abstract data analysis operations used to build a procedure (e.g., bias correction or flat fielding).
- *Implementation* — These implement primitives, usually by invoking an underlying analysis system, e.g., IRAF, STSDAS, etc.

Primitives are the basic building blocks which are used to construct the reduction procedures and insulate the astronomer from many of the details of the underlying analysis systems. Primitives form a library of routines from which new reduction and analyses can be adapted from existing ones. Adding a new analysis package to DRACO consists of creating the appropriate implementations for a set of primitives. Refer to Miller (1992), Yen (1993) for more details on Draco.

Once these entities are defined, the astronomer invokes DRACO's `start` function, specifying the directory containing the data. DRACO gathers information about the data at hand (usually by reading header information in the files), expands the procedure into a reduction plan based on the actual data, creates a command language script in the target analysis system and finally executes the script. The time-consuming reduction takes place without further attention from the astronomer. DRACO logs all steps for later review and calls attention to problems such as missing data or calibration files.

By producing a command script in the language of a data analysis system, DRACO builds on the foundation of these systems, rather than creating yet another analysis system. It is common for astronomers to write operating system command language scripts (e.g., Unix Shell or VMS DCL) to reduce data. The advantages of the DRACO-generated scripts are clear: DRACO provides a higher level of abstraction and handles many lower level details for the user. It is usually very difficult to modify custom command language scripts for different reduction tasks whereas DRACO facilitates reuse of its component data structures.

2.1. Experience

DRACO has been applied to two separate astronomical projects with successful results. The first version of DRACO was used to manage the removal of cosmic ray artifacts from a sample of HST WF/PC data for the Medium-Deep Survey Key Project (provided by Griffiths and Ratnatunga). A revised version of DRACO was used to extract spectroscopic data from ground-based telescope data (provided by MacConnell and Roberts), including bias and dark removal, flattening and extraction of the spectra. In this latter case, DRACO managed a significant amount of data: 325Mb in 650 separate files.

2.2. Availability

DRACO is an initial system which addresses the issues of automating data reduction and analysis. In order to encourage development of similar ideas and systems, DRACO (including source code and documentation) is available to the community from the Space Telescope Science Institute server (`stsci.edu`) via anonymous ftp, Gopher and World Wide Web.

3. Discussion

In the course of this project we had extensive discussions with many research groups. Contrary to the prevailing view that the lack of visualization tools or graphical user interfaces is a major impediment to research, we found that a more serious problem for all groups was the difficulty of managing the data reduction process. DRACO demonstrates a simple and effective way to perform data reduction with less human interaction and fewer errors. Other approaches are being explored such as the Khoros system (Rasure 1991) and commercial systems such as Silicon Graphics Explorer or BBN Cornerstone. We encourage developers of astronomical data analysis systems to incorporate these ideas into future work. From our experience, we can identify a number of general capabilities which would greatly enable astronomical researchers. Users need tools to:

- Describe the data at hand — Directory listings of files are usually the only tool available to describe the data. Users need more powerful tools (e.g., graphical) which understand and display the types of data (flat, bias, comparison spectrum, etc.) and the relationships between the data (e.g., identifying multiple images of the same target).
- Describe reduction process and algorithms at a high level — Users generally have to work at quite a low level, dealing with the specifics of the analysis system, operating system, file types, etc.
- Facilitate experimentation with reduction parameters and different algorithms — Data reduction is an integral part of the scientific process and it is therefore vital to experiment with relevant parameters of the reduction (number of iterations in a restoration algorithm, cosmic ray removal parameters, etc.). If reducing the data just once requires too much time and effort then experimentation becomes impractical and an important aspect of the scientific method is sacrificed.
- Make it possible to resume data reduction/analysis after interruptions.
- Perform data quality checks.
- Provide traceability: Too often data analysis systems do not to adequately document what operations were performed on the data. As a minimum, systems must document in the headers what steps were performed, including input data, algorithm, parameters and software version.

A change in the basic philosophy of scientific data analysis systems is needed. Rather than being designed as monolithic systems which are the complete environment for all data reduction and analysis, systems should begin a migration towards "interoperability" where they can be invoked by other systems and software. A client-server technology (communicating via Unix sockets or some other protocol) seen in many other computer applications seems a likely architecture to fulfill this goal.

4. Summary

Management of the data reduction process is an important problem facing astronomers who deal with observational data. The lack of effective data management tools can often be overwhelming. DRACO demonstrates one approach to providing automated assistance in order to free the astronomer to concentrate on scientific issues. DRACO works in concert with existing data analysis systems. We assert that current and future data analysis systems must provide tools for effective automation of procedures and data management.

Acknowledgments. We thank Mark Johnston, Bob Hanisch, Ron Gilliland, Richard Griffiths, Keith Horne, Jack MacConnell, Jim Roberts, Kavan Ratnatunga and Phil Martel for discussions about data reduction and their thoughtful comments on the design and development of DRACO. This work is supported by NASA's Astrophysics Information Systems Research Program through CESDIS by a contract with the Space Telescope Science Institute which is operated by AURA for NASA.

References

Johnston, M. 1987, "An Expert System Approach to Astronomical Data Analysis", Proceedings of the Goddard Conference on Space Applications of Artificial Intelligence, NASA

Miller, G. 1992, "The Data Reduction Expert Assistant", Astronomy from Large Databases II, Hagenau, France, ed. A. Heck and F. Murtagh, ESO

Rasure, J. & M. Williams 1991, "An Integrated Visual Language and Software Development Environment", Journal of Visual Languages and Computing, 2: 217-246

Yen, F. 1993 "Draco Design Document", ST ScI APSB Technical Report 1992-07, available from `stsci.edu`

Astronomical Data Analysis Software and Systems III
ASP Conference Series, Vol. 61, 1994
D. R. Crabtree, R. J. Hanisch, and J. Barnes, eds.

The Virtues of Functional CLUIs

Hans-Martin Adorf

Space Telescope – European Coordinating Facility, European Southern Observatory, Karl-Schwarzschild-Straße 2, D-85748 Garching, Germany, Phone: +49-89-32006-261 – Internet: adorf@eso.org

Abstract. The current emphasis on graphical user interfaces (GUIs) has led to a regrettable stagnation in the development of command language user interfaces (CLUIs) for applications such as image processing, database access, and information retrieval. Desiderata for a good CLUI are abstracted from the experience with various modern computer languages. In a case study involving interfaces to an SQL-based DBMS and to WAIS, the virtues of functional CLUIs have been demonstrated.

Every application [...] should have
a powerful and flexible command language
that can be used to control and extend the application.
— John K. Ousterhout, 1993

1. GUIs Have Strong and Weak Points

This is the season of graphical user interfaces (GUIs) to computer applications. GUIs certainly have merits: they are particularly easy to learn, and they suit occasional users (Adorf and di Serego Alighieri 1989). Therefore, with the advent of the Apple Macintosh and, somewhat later, of personal computers and workstations equipped with window systems, GUIs have become fashionable.

However, GUIs are not a panacea for user interface (UIF) problems: (1) Human interactions with a GUI can often not be recorded for later re-use. When they can, the statements usually form a poor programming language. (2) For communication between two GUI-driven programs, an intermediate disk file is often required, causing program execution inefficiencies. (3) GUI-driven applications are more resource demanding and cannot easily be used across a serial line. (4) It is difficult to impossible to invoke a GUI-driven application from another program; GUI-driven applications thus appear like stand-alone edifices that cannot easily serve as building blocks, thus effectively inhibiting the composition of code on a higher abstraction level.

Therefore there is reason to be concerned that much of the astronomical UIF development effort is presently absorbed by GUIs, whereas some (perhaps more fundamental) issues are left virtually unaddressed, such as the evolution of good programming languages supporting astronomical computing. Remarkably, Apple Computer has recently announced "AppleScript", a full-fledged programming language (including a GUI-action recorder and inter-application communication) for the MacOS — the once GUI-only operating system.

2. Some Widely Used CLUIs are not Ideal Either

Any computer/programming language can be considered a command language user interface (CLUI) when it permits interactive access to the functionality of some application. While well designed programming languages do not suffer from the drawbacks of GUIs outlined above, CLUIs have unfortunately somewhat fallen into dispute.

Let us briefly inspect two CLUIs used for image processing systems in optical astronomy: The MIDAS CL, modeled after DCL without achieving DCL's level of versatility and user-friendliness, is an imperative language lacking recursion. In practice this means e.g., that it is impossible to use a function at places where a value (e.g., an image name) is acceptable. The IRAF CL, modeled after the Unix shell language(s), is also imperative and non-recursive; however, a special language mechanism allows at least image sub-indexing everywhere.

The most frustrating aspect of CLs is that there are so many of them. Despite the fact that most syntax differences are superficial, they effectively prevent porting high-level code developed for one system into another one.

3. What Would a Good Command Language Look Like?

The usefulness of a programming language is generally greatly enhanced if it is "functional" (as opposed to imperative). A functional language (Hudak 1989) is, broadly speaking, a language "that emphasises the evaluation of expressions, rather than the execution of commands. The expressions in these languages are formed by using functions to combine the basic [data] values" (Jones 1994). Functional languages support and encourage programming in a functional style. They are usually very expressive and often allow a direct translation of algorithms in mathematical notation into executable code.

IDL, for instance, is equipped with a fully recursive, functional language featuring a homogeneous syntax. An FFT-based convolution of two images A and B can be written as a one-liner: `C = ffti(fft(A)*fft(B))`.

From experiments with a Lisp-based functional CLUI to SQL and WAIS (see Appendix) and from inspection of various computer languages, the following 12 desiderata for a good CL have been abstracted (roughly in decreasing order of importance): (1) simple, elegant, homogeneous and expressive syntax enhancing the code readability; (2) functional, recursive definition of the language syntax, permitting $g(f(x))$ and $h(f(x), g(y))$ constructs; (3) clear evaluation rules with variable evaluation as the default behaviour and non-evaluation as the exception; (4) positional, optional and keyword parameters; (5) concise default value mechanism minimising code clobbering; (6) lexical scoping for clear semantics; (7) collection orientation (Sipelstein and Blelloch 1990), permitting operation on vectors and images as entities; (8) macro definition capability allowing the construction of "mini-languages"; (9) interpreter for interactive code execution and compiler for efficiency; (10) programmable parser ("reader") allowing e.g., to locally modify the CL syntax; (11) dynamic variable and function type assignment permitting omission of unnecessary type declarations; (12) block comments facilitating the task of commenting out several lines of code.

Most of the above desiderata have been realised in some advanced languages such as Common Lisp (Steele 1990) and also, with some limitations, in modern scientific computing systems such as Mathematica (Wolfram 1991). Fortran-90 (Metcalf and Reid 1990) introduces collection-orientation into the mainstream of scientific, numerical computing.

Other functional languages worth inspecting are LISP-STAT (Tierney 1988) and S (Becker et al. 1988). Dylan (Shalit et al. 1992), currently under development at Apple Computer, combines a functional approach with object-orientation throughout. A promising recent development is the public domain Tcl/Tk package (Ousterhout 1993). Tk allows one to construct GUIs that issue Tcl commands for immediate execution. Tcl also includes an inter-application communication mechanism.

4. A Command Language Benefits from an Environment

CLUIs are particularly attractive when they are embedded in a suitable software development and execution environment. The synergistic effects of the tools that make up the environment greatly help to overcome those problems usually associated with CLUIs.

Among the most important environment features one finds: (1) a language sensitive editor (LSE) with automatic code indentation; (2) optional documentation strings following the definition of each function, automatically stored in a database; (3) a database of function definitions and calling sequences, automatically updated during function definition; (4) a "meta-point" mechanism which takes one to the location of a function definition in a file; (5) a symbolic debugger and a tracer; (6) command execution from an editor via a select-and-execute operation; (7) an auto-documenter extracting and collecting function definitions, documentation strings and comments from the code; (8) a cross-referencer permitting to view the call-dependencies of the functions; and potentially (9) a notebook facility (à la Mathematica) receiving results.

With the exception of notebooks, all these environment features have been present in off-the-shelf Common Lisp environments since the mid 1980s or earlier. Astronomical CLUIs for scientific computing and data analysis are thus lagging behind the state-of-the-art by about a decade, also when compared with the developments in statistics (Chambers 1989; Gaffney and Houstis 1992; McDonald and Pedersen 1988; Tierney 1991).

5. Conclusion

The question of user interfaces for astronomical data analysis, information retrieval and other tasks has been discussed. In addition to graphical user interfaces (GUIs), command line user interfaces (CLUIs) are required. However, many present-day astronomical CLUIs lack important language features which make them unnecessarily difficult to use. Attention and development efforts should therefore be devoted not only to enhancing GUIs, but also to *functional* (rather than imperative) CLUIs. IDL, Common Lisp or Mathematica may serve as examples from which astronomical software developers may borrow ideas for the future.

6. Appendix

Two case studies have been carried out using a functional programming language (Common Lisp) to construct a CLUI. In one case an interface was implemented to a WAIS client. The WAIS-search function call invoked a Unix shell command, which in turn carried out the actual search. The results were delivered directly into the Lisp process for further use, e.g., sorting or filtering. In another case an interface was constructed to the HST STARCAT relational database. By using this tool it was easy to construct a list of objects observed by HST; also various inconsistencies in the HST database have been discovered. These experiments demonstrated that such a dynamic, functional language provides a powerful and flexible environment to integrate stand-alone services.

Acknowledgments. I appreciate clarifying comments from and discussions with Alan Richmond (HEASARC/HSTX), Bob Hanisch (ST ScI), Brian Glendenning (NRAO), Herman Marshall (MIT), Jan Noordam (NFRA), Mark Johnston (ST ScI), and Skip Schaller (Steward Observatory).

References

Adorf, H.-M., & di Serego Alighieri, S., ST-ECF Newsletter, 10, 6–10, 1989

Becker, R.A., Chambers, J.M., & Wilks, A.R., "The New S Language", Wadsworth & Brooks/Cole Advanced Books & Software, Pacific Grove, California, 702 pp., 1988

Chambers, J.M., Statistical Software Newsletter, 15, 81–84, 1989

Gaffney, P.W., & Houstis, E.N., Proc. of the IFIP YC2/WG2.5 Working Conference, Karlsruhe, FRG, Elsevier Science Publishers, 406, 1992

Hudak, P., ACM Computing Surveys, bf 21, 359–411, 1989

Jones, M.P., "A frequently asked question list (FAQ) for comp.lang.functional", Yale University, Dept. of Computer Science, New Haven, CT, Jan 1994

McDonald, J.A., & Pedersen, J., SIAM J. Sci. Stat. Comput., 9, 380–400, 1988

Metcalf, M., & Reid, J., "Fortran 90 Explained", Oxford University Press, Oxford, 294 pp., 1990

Ousterhout, J.K., "An introduction to Tcl and Tk", Addison-Wesley Publishing Company, Inc., 390 pp., 1993

Shalit, A.L.M., Piazza, J., &Moon, D.A., "Dylan – an object-oriented dynamic language", Apple Computer, 1 Main Street, Cambridge, MA, 167 pp., 1992

Sipelstein, J.M., & Blelloch, G.E., Proceedings of the IEEE, 79, 504–523, 1990

Steele, G.L., "Common Lisp – The Language", 2nd edition, Digital Press, 1029 pp., 1990

Tierney, L., Lisp-Stat: An object-oriented environment for statistical computing and dynamic graphics, J. Wiley & Sons, 1991

Wolfram, S., "Mathematica – A System for Doing Mathematics by Computer", (2nd edition), Addison-Wesley Publishing Co., 1991

Part 3. Archives, Catalogs, Surveys and Databases

Astronomical Data Analysis Software and Systems III
ASP Conference Series, Vol. 61, 1994
D. R. Crabtree, R. J. Hanisch, and J. Barnes, eds.

Astronomy and Databases: A Symbiotic Relationship

M. Schmitz, G. Helou, B. F. Madore, H. G. Corwin, Jr., J. Bennett, and X. Wu

Jet Propulsion Laboratory, California Institute of Technology

Abstract. The NASA/IPAC Extragalactic Database currently represents the merger of some 40 major catalogs covering all wavelengths, and includes scores of shorter lists culled from published journal articles. Catalogs and lists are being folded into the database on a continuing basis, following a detailed cross-identification process for each object. New features recently added to the interface (freely accessible over the networks) include X Window compatibility, skyplots of NED objects and SAO stars, and detailed photometric data. Functions which will soon be available include the ability to display Spectral Energy Distributions and images. If these tools (and others now under development) are to be used properly for astronomical research, clear and complete descriptions of all published data are necessary. By providing the kind of information used within NED (such as uncertainties on all measured quantities, unique object identifiers, and detailed observational parameters), both NED users and NED developers will benefit.

1. Introduction

For the past several years, the NASA/IPAC Extragalactic Database has been used as an aid to scientific research. This paper will describe the service provided to the astronomical community and how the community can provide a service to NED and themselves.

2. Present NED Status

At present NED contains positions, cross-identifications, redshifts (or velocities), and fluxes (or magnitudes) for more than 270,000 extragalactic objects, over 600,000 detailed photometric data points from radio to X-ray, and half a million pointers to over 22,000 bibliographic references. All this, and more, are available free-of-charge over Internet provided you have a VT100 terminal, or VT100 emulation software, e.g., an 'X-term' window on a workstation. A connection to NED can be obtained with the command: "**telnet ned.ipac.caltech.edu**". Once connected to the NED platform and prompted for a "**login:**", please respond with **NED** ; no password is needed.

The system is self-documenting, especially through the HELP options which provide detailed information on the functions of that window. First-time users may want to read the tutorial in the NEWS AND INFO screen. In addition, within nearly every window, you have an opportunity to leave comments, suggestions, or questions for the NED team by using the "Comments" option. You may also send correspondence via e-mail to: "**ned@ipac.caltech.edu**".

3. Data in the Database

One important difference between NED and other databases is that NED is **OBJECT ORIENTED** and not **CATALOG ORIENTED**. This means that NED has taken separate entries from different sources and matched the objects so that each object is listed once in the database.

Since 1988, NED has been growing with more data and new functionalities. In addition to the 30-40 catalogs which have already been folded into NED, and the thousands of journal articles scanned each year which pertain to extragalactic astronomy, we have recently loaded the *Flat Galaxy Catalog* (Karachentsev et al. 1993) and the Lyon *Groups of Galaxies* (Garcia 1993).

Catalogs which are currently being folded into the data base include objects from the Arp-Madore *Catalogue of Southern Peculiar Galaxies and Associations*, (Arp & Madore 1987) and photometry from the *Catalog of Infrared Observations* (Gezari et al. 1993).

4. Database Functions and Interface

It is possible to search NED for objects selected by redshift, in addition to the selection by name, by position, or by type (e.g., QSO or infrared source). NED offers browsing capabilities for 7,400 abstracts of articles of extragalactic interest that have appeared since 1988 in A&A, AJ, ApJ, MNRAS, and PASP. Recent abstracts from PAS Japan, Astronomy Reports, and Astronomy Letters (formerly Soviet Ast. and Sov.Ast. Letters) have also been added.

For users with X Window capabilities, NED offers an easier "point and click" interface which also allows for an on-screen graphical display of SAO stars and NED objects.

5. Photometric Data

An improved feature which has been installed is a more detailed explanation of the photometric data in NED. For objects which have been detected at many frequencies, the present data in NED may be displayed in summary form with the frequency targeted, the measured value and its uncertainty and reference. Clicking on a particular entry will display not only the published flux density, uncertainty, and position, but also a numeric value for the waveband, qualifiers about how the data were obtained, and equivalent **mks** values for the flux and frequency. This much detail and more are retained by NED for each photometric data point, of which NED presently has over 600,000.

6. Help from the Community

Since NED is a tool of and for the astronomical community which is frequently accessed (over 4500 logins per month), NED needs to contain the best possible data. These come from catalogs and referreed journals, but the referees can't check every number in a paper.

Some points which would make NED's job easier (and ultimately the researchers, too), is that the authors provide clear and complete observational descriptions. How, where, and when were the data obtained. Give explicit explanations of all uncertainties. What is the significance level of an upper limit? Use correct and complete object identifications/nomenclature. One of the most time consuming aspects for us is determining what object is discussed in a paper. An IAU guideline is to supply at least two names for an object, or a name and a position. (Paturel et al. (1993) give a good discussion of problems in identifications and designations as applied to galaxies.) New discoveries should give a proposed name and the position. And finally, if authors can supply NED with machine-readable copies of the data **AS PUBLISHED**, an enormous amount of time will be saved from optically scanning the tables and proofreading. A similar set of difficulties are experienced by the SIMBAD project (see Laloé et al . 1993).

7. Summary

Fundamental research seems to be depending more and more on electronic tools such as NED and SIMBAD, but the creators of these tools depend more and more on the researchers to clearly present their results.

NED is a tool created for astronomy research and allows a user to view the sky and perform some basic analysis. This can only be truly useful in an accurate and timely way with the help of the users themselves.

The X windows and graphics capabilities of NED offer a new dimension of functionality. These new functions, however, are only as powerful as the data in NED, and are thus made possible by the solid groundwork of data collection and verification which will continue to be NED's main activity. Now, more than ever, we need the user's help in feeding more **accurate** data **faster** into NED.

Acknowledgments. The NASA/IPAC Extragalactic Database (NED) is operated by the Jet Propulsion Laboratory, California Institute of Technology, under contract with the National Aeronautics and Space Administration.

References

Arp, H.C., & Madore, B.F. 1987, "A Catalogue of Southern Peculiar Galaxies and Associations", Cambridge University Press

Garcia, A.M. 1993, A&A, 100, 47

Gezari, D.Y., Schmitz, M., Pitts, P.S., & Mead, J.M. 1993, "Catalog of Infrared Observations", Third Edition, NASA RP-1294

Karachentsev, I.D., Karachentseva, V.E., & Parnovsky, S.L. 1993, Ast. Nach., 314, 97

Laloé, S., Beyneix, A., Borde, S., Chagnard-Carpuat, C. Dubois, P., Dulou, M.R., Ochsenbein, F., Ralite, N., & Wagner, M.J. 1993, Bull. Inform. Centre de Données Astronomiques de Strasbourg 43, 57

Paturel, G., Petit, C., Bottinelli, L., & Fouque, P., Gouguenheim, L. 1993, Bull. Inform. Centre de Données Astronomiques de Strasbourg 43, 43

Astronomical Data Analysis Software and Systems III
ASP Conference Series, Vol. 61, 1994
D. R. Crabtree, R. J. Hanisch, and J. Barnes, eds.

Generation and Display of On-line Preview Data for Astronomy Data Archives

N. Hill, D. R. Crabtree, S. Gaudet, D. Durand, A. Irwin

Dominion Astrophysical Observatory/Canadian Astronomy Data Centre, 5071 W. Saanich Road, Victoria B.C. V8X 4M6

B. Pirenne

Space Telescope European-Coordinating Facility, Karl-Schwarzchild-Straße 2, D-8046 Garching bei München Germany

Abstract. This paper describes the processing procedures developed to generate, store and display the preview data for the HST and CFHT data archives. It will also describe how the preview system has been extended to other astronomy catalogs.

1. Introduction

It became apparent during the development of the Hubble Space Telescope (HST) archive at the Canadian Astronomy Data Center (CADC) and the Space Telescope-European Coordinating Facility (ST-ECF), that the archive would be more useful and efficient if archive users could view the archive data interactively. If archive users could view the image and spectral data before making requests for data, requests for unusable data would be reduced, and requests for useful data would be increased. Since the HST archive consists of approximately 320 Gigabytes of public data, of which approximately 50 Gigabytes is scientifically useful, it is not practical to store the original data on line or transmit it over the Internet fast enough to be usable interactively.

The CADC and ST-ECF have designed a system which compresses data for selected scientifically useful datasets and stores the compressed preview data in a relational database system. We also have extended the STARCAT archive interface to allow users to retrieve and preview any dataset in the preview database. Since the successful creation of the HST preview database, the CADC and ECF have generated preview data for various other archives and catalogs.

2. Generation of Preview Data for the HST and CFHT Archives

Since the CFHT and HST data archives are ongoing archives, it was necessary to develop a highly automated system for generating preview data from the archived data stored on optical disk. Figure 1 is the block diagram of the preview creation pipeline used for the CFHT and HST archives.

The HST/CFHT preview pipeline consists of the program `odpreview` which coordinates the pipeline, a set of data compression routines (Gnu zip for spectra,

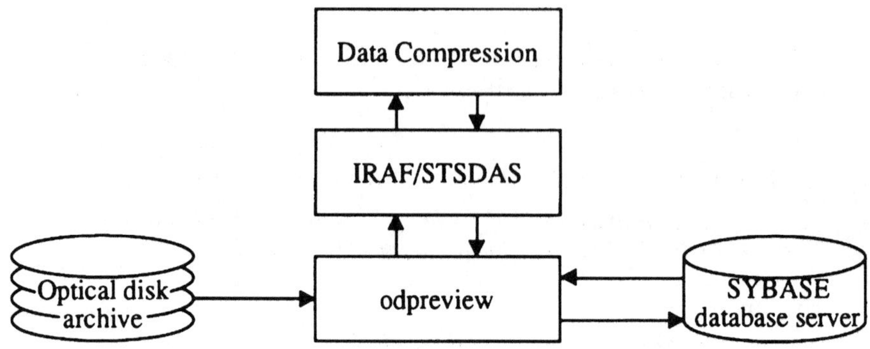

Figure 1. HST/CFHT preview creation pipeline.

and Hcompress for images), and a set of IRAF scripts. The `odpreview` program coordinates the mounting of optical disks, reading files from optical disk, processing the files with the IRAF scripts, and writing the data into the preview database. The IRAF scripts remove any extraneous information from the data files, do any data reduction necessary for an instrument and compress the data. The data are then inserted into the preview database by the `odpreview` program. Table 1 shows the average compression achieved for the various instruments in the HST archive. Figure 2 shows a compressed preview image and the original

Table 1. Average HST data compression (sizes in Kilo bytes).

Instrument	Number of datasets	Compressed preview size	Uncompressed preview size	Calibrated data size
WFPC	5246	51	1147	9043
FOC	2212	35	802	1603
FOS	2193	20	27	402
HRS	2521	14	24	624

image for a HST Faint Object Camera image of the Einstein cross (HST dataset x0cc0102t). The original image file size is approximately 1 Megabyte and the preview data size is approximately 22.5 Kilobytes.

The CFHT archive does not yet contain significant amounts of public data so only demonstration preview data have been produced. Automatic production of preview data for the CFHT archive will be more of a challenge than the HST archive was since the data reduction processes for many of the instruments are not as clearly defined. However we hope to implement a preview production pipeline for several of the CFHT instruments in 1994.

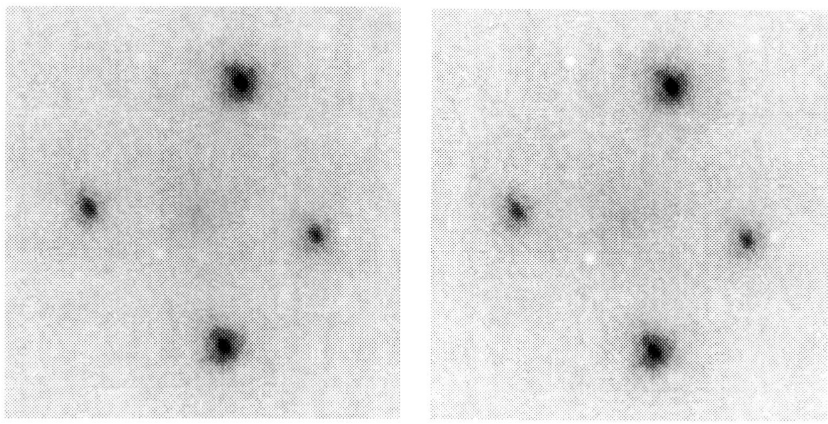

Figure 2. Preview (left) and actual (right) FOC images of the Einstein cross, HST dataset x0cc0102t.

3. Other Preview Datasets

Many of the catalogs available through STARCAT have associated *previewable* data. To help make these data available to STARCAT users, it was decided to create the program `mdpreview` to ingest FITS data files into the preview database. `Mdpreview` is a simple program which inserts compressed FITS files into the preview database and is intended to be used as one component of a processing pipeline custom designed for each data source. `Mdpreview` has allowed preview data to be created for the catalogs listed in Table 2.

Table 2. Catalogs with available preview data.

Catalog	Available datasets	Source of preview data
IUE	54247	The ULDA V4 data provided by VILSPA.
BPGS	175	The Brunzual Perrson Gunn Stryker spectral library.
IRAS PSC	11227	The University of Calgary provided the data.
Jacoby	162	The Brunzual Jacoby spectral library.

In addition to the currently available data, preview for several other catalogs should be available in the near future.

4. Preview Display

STARCAT users initiate the display of preview data by selecting the desired record in a catalog and using the preview command. The preview command causes a request to be made to the STARCAT display program `stdisplay`.

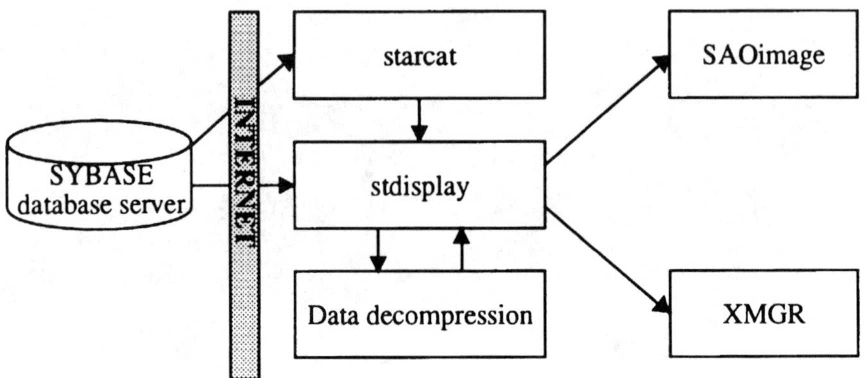

Figure 3. Preview data display.

`Stdisplay` retrieves the preview data from the database, decompresses the data, selects and activates the appropriate display tool (SAOimage for image data, and xmgr for spectral data) and sends the data to the display tool. Figure 3 shows the block diagram for the display of preview data. Since only the compressed data is transferred over the network, this design provides reasonable performance over typical network connections.

5. Conclusions

The systems we have developed have been effective for the generation and display of preview data. The HST preview pipeline is now fully operational and preview data is available for all HST WFPC, FOC, FOS, and GHRS *science* data archived at the CADC. In 1994 we hope to begin production of preview data for the CFHT archive and for the new HST WFPC2 camera.

More information on the STARCAT interface and information on how to obtain a copy can be obtain through NCSA Mosaic at
- "http://cadc.dao.nrc.ca/CADC-homepage.html" in North America or
- "ftp://ecf.hq.eso.org/pub/WWW/ST-ECF-homepage.html" in Europe.

References

Crabtree, D., Irwin, A. Blaber, R., Gaudet, S., & Durand, D. 1993 "Pipeline Processing, Automatic Image Quality Estimates & Preview Images of CFHT CCD Images", Proceedings ESO-OAT Workshop on Handling and Archiving Data from Ground-based Telesopes

White, R.L. 1992, "High-Performance Compression of Astronomical Images", Proceedings of the NASA Space and Earth Science Data Compression Workshop

Turner, P. J. 1993, "ACE/gr User's Manual"

VanHilst, M. 1991, "User Manual for SAOimage"

Astronomical Data Analysis Software and Systems III
ASP Conference Series, Vol. 61, 1994
D. R. Crabtree, R. J. Hanisch, and J. Barnes, eds.

NOAO/IRAF's *Save the Bits*, A Pragmatic Data Archive

Rob Seaman

IRAF Group, National Optical Astronomy Observatories, P.O. Box 26732, Tucson, AZ 85726

Abstract. Archival data have been pivotal to astronomy throughout history. Without Tycho's carefully recorded observations, Kepler's insight into the elliptical nature of the orbits of the planets would not have been possible. In our lifetime, the Palomar Sky Survey has been scrutinized daily for four decades by thousands of pairs of eyes. However, for a variety of reasons ground-based optical observatories have been late-comers, by and large, to the archiving of digital images. The era of the CCD has been one of data diaspora.

Many of the impediments to creating an archive are peripheral to any archive's principal mission of saving the information for future generations. There is a moral imperative to do at least this much. Modern computer networks now provide the tools to build an image archive that is pragmatic both in terms of programming effort and cost.

The NOAO *Save the Bits* archive described is nightly (and daily) automatically archiving the raw data from a half dozen Kitt Peak telescopes via the ICE (IRAF Control Environment) data acquisition software.

As images are acquired at each telescope using IRAF/ICE, a unique identifying string constructed from the telescope name, the date, and the UT of the observation is edited into each image header. The IRAF `wfits` task translates the images to FITS from within the ICE postprocessing script. A print spooler (e.g., Unix `lpr/lpd`) provides a secure mechanism for transferring the FITS files across the network to a central archive server where they are queued for processing.

The data from several telescopes are multiplexed onto a single archive medium (e.g., exabyte). As the FITS files are processed by the queue software, each is stamped with a running sequence number and the resulting FITS header is appended to a catalog file, cross-referenced to the archive tape index. To promote efficient I/O, the individual images are concatenated into FITS Image Extension tape files several tens of Megabytes in size. A mini-language tape monitor program provides a screen oriented status display that may be run from any of the domes, or from a privileged account when swapping full archive tapes.

1. Introduction

With some notable exceptions the automatic archiving of ground based astronomical imaging data has yet to become the rule in the optical and infrared communities. Many of the last decade's gigabytes of lovingly acquired CCD exposures will never become accessible to anyone other than the original observers. Continuing this data diaspora becomes more and more unacceptable as CCDs become bigger and better, and as a new generation of large high performance IR detectors arrives.

The plate vaults of the last century contributed immensely to the field. The large areal coverage of photographic plates leverages the scientific cross section of each exposure for objects imaged by happenstance along with an observer's own program. Solid state detectors and mosaics of such detectors are now beginning to rival a photographic field of view, proportionally increasing the scientific value of archiving each image. An image archived last night at the telescope may benefit data scavengers past the retirement age of the graduate student who observed it.

In this era of sometimes vicious zero-sum budget competition, observatories must strive to extract the maximum science from level or dwindling resources. A re-emphasis on the fundamentals of the field is inevitable (or should be). NOAO's *Save the Bits* archiving initiative is a natural companion to other observatory initiatives to retune and improve telescope optics, to address dome seeing concerns, and to upgrade telescope control and data acquisition software and hardware. Each of these programs directly multiplies the scientific value of the community's many clever new instruments and detectors.

2. Design Decisions

The breadth of NOAO's stable of telescopes on Kitt Peak leads naturally to a network based archiving design. A network archive also easily allows data to be merged in from outside sources, for instance *Save the Bits* also archives visible, infrared, and water vapor GOES weather satellite images. The basic idea is to multiplex the data from all of the telescopes into a single archive queue. NOAO's Central Computer Services has used Unix print spooler techniques with great success for several years to queue jobs for two generations of photographic hardcopy units. *Save the Bits* uses this scheme with a `lpd` filter program called `bitf`, described below.

This queue writes the data to a recording medium, currently 8mm Exabyte tapes. An Exabyte 8505 drive, even in uncompressed mode, provides 5 gigabytes of streaming storage at about ten dollars a tape.

The natural format for an astronomical archive is FITS, but FITS has grown in the last few years. What kind of FITS? Given the residency of the archive on a streaming medium, I/O concerns become important. Tapemarks are much smaller for the Exabyte 850X than the 820X (where they rival the size of all but the largest image files), but it remains impractical to consider writing (and more importantly, reading) perhaps thousands of files on a given tape. Tape operations would slow to a crawl. We solve this problem by concatenating the individual images from the multiple domes into large FITS Image Extension files several tens of megabytes in size. Rather than writing many hundreds or thousands of images onto each tape through the course of a night, *Save the Bits* reduces this to less than a hundred.

The *Save the Bits* catalog consists of the concatenated FITS headers from each observation. We may choose to recast the catalog into more tractable data structures in the future, but simply separating the metadata from the data is the most important first step. Recasting a catalog should not require reading through the actual archive data.

The key function of a catalog is cross-referencing the actual data. This is provided via a separate tape index consisting of a single record for each observation. The index records are tied to the catalog using a unique sequence ID assigned to each image header by `bitf`. Recasting an archive onto a different medium should not require recasting (and possibly reinterpreting) its catalog, but only a separately maintained index.

3. The Front End

Given NOAO's heterogeneous collection of observing software, the ingest end of the archive assumes that data will arrive as FITS objects. This also simplifies the back end of the online archive system, since FITS is a reliable data format to parse. In fact, *Save the Bits* accepts not only garden variety FITS "classic" files, but also any conforming FITS extensions as input. All input FITS objects will be passed through or turned into FITS extensions and concatenated into a FITS extension file with an informative dataless primary HDU. Multiple FITS objects may be concatenated in the input, as for instance when specified on a single `lpr` command line. Non-FITS files are discarded (but saved for perusal) if they are encountered.

Network queuing is provided by the host print spooler, currently the Berkeley `lpr/lpd` spooler, but the System V `lpsched` queue scheduler would do as well. All that is necessary to hook a data acquisition environment up to the archive is to arrange for converting the data into some FITS format, and to install the host level `bits` queue on the acquisition computer. One other useful step is to stamp each FITS header with a unique telescope observation identifier. We have chosen a string consisting of the standard observatory name for the telescope (e.g., `kp4m` for the Mayall 4 meter telescope) concatenated with the UT date (`YYMMDD`) and the UT time (`HHMMSS`). The three fields are separated by periods for clarity. This telescope identifier should first be edited into the copy of the data that the observers take home with them, before sending the data to the archive queue.

4. The Back End

FITS files can arrive from any location on the mountain and will be queued reliably. (Note that by design data can be archived at any point after it is acquired, even manually by the observer after it has been reduced. We are not currently making use of this feature.)

The `lpd` daemon that is spawned when a file arrives on the archive computer, in turn spawns the `bitf` input filter to dispose of the file. Concurrency protection is provided by the spooler itself ensuring that only one daemon is actively reading files from the spool directory at any one time, and consequently that only one `bitf` is ever processing these files.

`Bitf` operates in two stages. First the input FITS files are translated into individual FITS Image extensions. Files that arrive already in a conforming FITS extension format are passed through. Multiple FITS objects (files or extensions) may be concatenated in the input stream for a single instance of the `bitf` filter, although normally each FITS object is processed by a successive instance of `bitf`. All arriving files are stamped with a header keyword specifying a unique running sequence number. This number will later serve to cross reference the header catalog and the tape index. At this point, it serves to provide a name for each FITS extension file which is staged back to a disk file.

The second stage is actually reading the staged FITS extension files back off of the disk and writing them to the exabyte tape in a single large FITS file consisting of a dataless primary HDU and a number of appended (Image) extensions. The data structure used by the program at this point is a linked list of FITS header objects.

After a complete tape file is written (consisting of several individual input files), the catalog is updated by simply traversing the linked list a second time and appending each FITS header to the catalog in turn. After the catalog is updated, the index is updated in a similar fashion. Only after the tape, the catalog, and the index are updated are the staged FITS extensions deleted from the disk. Each of these four steps (tape, catalog, index, deletion) depends on the success of the preceding steps. If taping fails, the files remain on the disk and will automatically be retaped after the offending tape is swapped out.

The archive automatically transfers taping duties between an arbitrary number of exabyte drives — although we have chosen to this point to use a single drive, replacing the tape around lunchtime every day whether or not it has filled up.

In between particular instances of `bitf`, the archive status is stored as simple `keyword=value` pairs in a disk status file. The status is read by each `bitf` when it starts up and is updated (and the file system buffers synced) periodically as the status changes.

The status file also serves as half of the communications link between `bitf` and the program that the archive managers actually interact with, `bitmon`. The other half of this link is the `bitmon` request file. The request file has the same `keyword=value` format as the status file. The keywords (representing structure variables in each program) are identical. `Bitmon` appends status change requests to the request file and `bitf` reads those requests, overlaying fields in a copy of the internal status data structure. After a successful request is made (careful checks are made for each individual variable to ensure that a request is rational), the archive status is updated on disk and read back again.

`Bitmon` must be able to gain control of the archive to provide for changing tapes. The communication between the two programs is very loosely coupled. `Bitf` ignores garbled requests — garbled, for instance, if `bitmon` is still writing a request when `bitf` tries to read it. The request will normally be granted the next time `bitf` checks. On the other hand, `bitmon` is required to stop the queue, thus halting the status updates, before it may perform chores such as rewinding the current tape and mounting the next one. In addition to provisions for swapping tapes and starting and stopping the queuing and taping functions, `bitmon` also provides a tape drive status loop display and a dome status report.

Astronomical Data Analysis Software and Systems III
ASP Conference Series, Vol. 61, 1994
D. R. Crabtree, R. J. Hanisch, and J. Barnes, eds.

The Archives of the Canadian Astronomy Data Centre

Dennis R. Crabtree, Daniel Durand, Severin Gaudet, Norman Hill, Stephen C. Morris

Canadian Astronomy Data Centre, Dominion Astrophysical Observatory, 5071 W. Saanich Rd., Victoria, B.C. Canada V8X 4M6

Abstract. The Canadian Astronomy Data Centre (CADC) was created in 1986 as the group responsible for the Canadian copy of the Hubble Space Telescope (HST) Archive. The CADC mandate has since expanded and we are now also archiving data from the Canada-France-Hawaii Telescope (CFHT). In this paper we will describe the operation and maintenance of these two archives.

These archives are accessed using the STARCAT software which was developed originally by the Space Telescope European Coordinating Facility (ST-ECF) for access to the HST Archive. We are now also heavily involved with STARCAT development and will describe a data preview mechanism which we have developed within STARCAT for use with both the CFHT and HST archives.

1. Introduction

In 1986, with the launch of the Hubble Space Telescope imminent, the Dominion Astrophysical Observatory (DAO) established an agreement with the Space Telescope Science Institute (ST ScI) whereby the DAO would receive a copy of HST data and serve as the Canadian archive centre. The CADC was the group which was formed to manage the Canadian HST archive.

Besides serving as the Canadian HST archive centre, the CADC is also responsible for archiving data from the CFHT. This is one of the few archives of data from large, ground-based telescopes which currently exists. The design of the CFHT archive was based upon the experience we had gained from establishing the HST archive in Canada.

The CFHT and HST archives are accessed via STARCAT which was developed by the ST-ECF and ST ScI to access the interim HST archive known as the Data Management Facility (DMF). At ST ScI the DMF and STARCAT will soon be replaced by the Data Archive and Distribution System (DADS). However, as we needed an interface for both the CFHT and HST archives, the ST-ECF planned on maintaining STARCAT and ESO needed an interface for the planned VLT archive, we all decided to continue development of STARCAT and to expand its capabilities to handle multiple archives.

One of the major capabilities which we have recently added to STARCAT is the *data preview* capability which allows users to examine the archival data before requesting a copy.

We will describe the basic design and operation of the HST and CFHT archives at the CADC as well as the new preview capability of STARCAT, including a description of the processing steps used to produce the HST preview data.

2. HST Archive

The HST archive can be divided into two parts, the archive catalog and the actual science data. The archive catalog is a set of tables in a relational database system while the science data is stored on 12-inch optical WORM disk. The ST ScI maintains the master database and writes data to optical disk as soon as it comes out of the data-processing pipeline.

We maintain a copy of the HST database on a Sparcstation 2 running the commercial database system Sybase, which is the same database system used at ST ScI. In order to keep our copy of the database current with respect to the master copy we have a developed a software tool named DBsync. For each table in the database DBsync compares the structure in the master database to that of our copy. Any rows which have been added since the last update are copied over the Internet and used to update our database. Any changes in the schema of the table, i.e., columns added or deleted, are also detected. In this case, our copy of that table is deleted and the entire table is copied from the master database. For more details on DBsync see the paper by Norman Hill in this volume.

Unlike the ST-ECF which receives an exact duplicate of the data as it is archived at the ST ScI, we receive a copy of the data once it becomes public, which is usually one year after the observation is taken. We also do not get all of the data but only the *CAL* class and a few other assorted files.

We have developed software, named CADCOD, to perform this copy of public data. CADCOD runs on the DMF VAX at the Institute and works by querying the archive database to produce a list of files which meet certain criteria:

- The dataset is public
- This is a dataset which we do not yet have
- The dataset is wanted!

Once a list of required files have been produced, CADCOD copys the files from the optical disks at the Institute to optical disks which are then sent to the CADC.

The above procedures have worked very well for the past 3 years. We have a current copy of the archive database available and receive public data in a timely manner, except when exceptional operational circumstances arise at the ST ScI.

3. CFHT Archive

Data taken at most ground-based observatories is not currently archived. We decided to use our expertise and experience gained by working with the HST archive and apply it to producing an archive of CFHT data. Most of the work on the CFHT archive has been done at the CADC but this project would not have been possible without the support of the staff at CFHT and the CFHT Board of Directors.

The *data handling pipeline* at the CFHT has been configured so that every data frame taken at the summit is automatically and transparently copied to a staging area on magnetic disk at the headquarters building in Waimea. The observer cannot interfere with this procedure and thus has no say in which frames are saved. All of the data is stored in FITS format.

There is a process on the archive computer in Waimea which periodically looks for new files in this staging area. When it finds new files it first makes a copy of the FITS header in a separate file. The process then copies the file to optical disk (Sony 12-inch WORM, capacity 6.5 GB), verifies the copy on the optical disk and then deletes the file from magnetic disk. The next day all of the FITS headers are sent electronically to the CADC while the optical disk is sent when it is full.

When the FITS headers from the previous night's observations are received they are automatically parsed and used to update a series of tables in the database. Unless there is a communications problem the CFHT observation catalog is never more than 24 hours out of date. It is worth noting that the information in the database is potentially useful for investigating the performance and efficiency of the telescope and instruments. We are currently planning to automatically estimate the seeing on each image taken with the CCD cameras and load these seeing estimates into the database. This will then allow one to investigate the seeing as a function of other information stored in the database such as date, hour angle, air mass, etc.

When the full optical disks are received their contents are scanned and the table in the database which acts as a master directory is updated. A consistency check that a file exists for each entry in the database is also performed. Thus far we have not lost any data files.

We are planning on passing the direct imaging data through an automatic processing pipeline which will produce bias subtracted and flat-fielded images (Crabtree et al. 1994). While the processing may not be the best that is possible, the data will be of high enough quality to be useful for certain applications. These processed images will also be used to generate *preview* images, in a similar manner as we are doing for the HST images (see the following section).

Data in the CFHT archive will start becoming public in September 1994 and will be available to the world astronomical community.

4. Accessing the Archives: STARCAT

Access to the HST and CFHT archives is provided through the STARCAT software. STARCAT was originally developed by the ST-ECF and ST ScI as the interface to the interim HST archive system. While ST ScI decided to develop StarView for access to the permanent archive system DADS, the ST-ECF, ourselves and ESO decided to continue developing STARCAT. Currently, the development is being done primarily by ourselves and the ST-ECF. It is worth noting that we have developed tools which allow effective joint development of this software, even though the two groups are on different continents separated by nine time zones. This process has proved to be very effective when combined with face-to-face meetings once or twice a year.

While the original STARCAT was essentially an ASCII terminal-based interface, the current STARCAT has evolved into much more of an X Windows application. The core part of STARCAT is still terminal based but many of the new features are *spawned* off as X applications. STARCAT now works in client-server mode. Users run STARCAT on their local workstation and the connection to the database, either at the CADC or the ST-ECF is made over the Internet. This makes very efficient use of the network as only a minimum of information is transferred.

A major innovation which we have developed within STARCAT is the ability to preview data in the archives before requesting the data. Since HST and CFHT images are about 10 MB in size, it is impractical to keep them all online. However, archival researchers would benefit from being able to visualize the data before requesting it from the archive. Ideally, they would be able to confirm that their object of interest was present, the exposure was long enough, and generally assess the quality of the data. This preview capability allows archival researchers to do all of the above.

The data for the *Preview* mechanism is stored within the relational database. This data is in a highly compressed format so it takes up much less space than the original data. In the case of HST WF/PC images a typical image is stored in $\approx$ 60 KB rather than original size of 10 MB. Briefly, this compression factor is achieved by converting the data to 2-byte integers, spatially sampling the data by a factor of two in each dimension and applying a lossy compression algorithm to the data.

When an archival researcher using STARCAT somewhere on the Internet requests a *preview* the compressed image is retrieved out of the database, transferred over the Internet to the users machine, decompressed and displayed. Since the *compressed* image is sent over the network and the decompression is quite fast, the previewing process is fairly efficient.

References

Crabtree, D.R, Irwin, A., Blaber, R., Gaudet, S., & Durand, D., in Proceedings of the ESO-OAT Workshop on Handling and Archiving Data from Ground-based Telescopes, eds. M. Albrecht & F. Pasian (ESO Conference Proceedings), in press

Astronomical Data Analysis Software and Systems III
ASP Conference Series, Vol. 61, 1994
D. R. Crabtree, R. J. Hanisch, and J. Barnes, eds.

The IUE Final Archive — NEWSIPS Algorithms and Results

M. D. De La Peña, J. S. Nichols and K. L. Levay

Computer Sciences Corporation, IUE Observatory, 10000-A Aerospace Road, Lanham, MD 20706

A. Michalitsianos

NASA/GSFC/IUE Observatory

Abstract. The culmination of the IUE Project will be the creation of a Final Archive which will contain all scientific and calibration data acquired during the mission, processed in a consistent and homogeneous fashion, using an enhanced image processing system designed specifically for this purpose. Fundamental observational and image processing parameters will be incorporated into the headers of the image data files, and compiled into an IUE Final Archive Database, further enhancing the utility of the archival information.

1. Introduction

In order to make IUE data most useful to the astronomical community beyond the lifetime of the IUE Observatory, the IUE Project has undertaken the task of creating a Final Archive. The IUE Final Archive data can be characterized by three major goals: (1) process all data in a consistent fashion, making the dataset fully intercomparable; (2) utilize sophisticated image processing techniques to process the data properly, and thereby, enhance the signal-to-noise ratio (S/N) of the output data; and (3) verify the observational parameters for each exposure, correcting these data as necessary, and incorporate this self-documenting information into the FITS headers of the output data products as well as into the IUE Final Archive Database (Nichols et al. 1993).

2. Overview of the Image Processing Algorithms

The Final Archive data reduction system, NEWSIPS, utilizes innovative image processing techniques in order to achieve an improved S/N in the extracted spectrum. In the NEWSIPS system, the trifold combination of: the Intensity Transfer Functions (ITFs) created in their own "raw space" — eliminating the need for resampling (and consequent degradation) of these fundamental calibration data; an explicit image registration between the raw science image and the ITF via a cross correlation algorithm — exploiting the "fixed pattern noise" inherent in all IUE raw images as a successful fiducial; and an improved application of the ITF to the science images — requiring only one resampling of the calibration data, yield a more accurate photometric correction of the raw data (De La Peña et al. 1992 and references therein).

A flux- and line shape-preserving resampling of the photometrically corrected image incorporates multiple spatial "corrections" in a single resampling which maps the data to a geometric space where the echelle orders are parallel to an image axis and the wavelength dispersion is linear within each order (Bushouse et al. 1992 and references therein).

2.1. Specifics of the Low Dispersion Algorithms and Results

The low dispersion data are extracted from the geometrically corrected image with a signal-weighted technique based on an empirically determined profile (Kinney et al. 1991) which: conserves flux; removes most cosmic ray hits; provides an error estimate for each wavelength bin; and yields an improved S/N ratio of the extracted spectrum. The absolute flux calibration is based upon white dwarf models which have been used to determine the relative shapes of the inverse sensitivity functions, while previous UV satellite and rocket observations of η UMa and other standard stars have been used to set the absolute flux scale. Finally, sensitivity degradation corrections as a function of wavelength are implemented to account for changes in the detectors over time.

Figure 1 illustrates a comparison of NEWSIPS and IUESIPS (the current image processing system) data. This image is a 608 minute exposure characterized by a high background and marred by cosmic ray hits. The top panel illustrates the overall S/N characteristics of the two spectra; the middle panel shows a subset of each spectrum. To aid in the evaluation of the quality of the data, NEWSIPS has *both* an error estimator (sigma value) for each wavelength bin (vertical lines overplotted on the NEWSIPS data where length is indicative of the magnitude of the error) and ν-flags which are uniquely (bit-) encoded values (bottom panel) denoting various pathologies in the data. The ν-flags indicate the presence of saturation and extrapolation in the region of Lyα. The IUESIPS ϵ-flags are represented by the horizontal line overplotted on the data in the middle panel where negative dropouts denote flagged bins. Only saturation is indicated in the IUESIPS data in the region of Lyα. The NEWSIPS data has two wavelength bins in the Lyα region which have both a net flux and an associated sigma value of zero. These wavelength bins did not contain any valid data (all pixels flagged); in order to indicate the presence of extremely bad data, the flux and sigma have been set to zero.

IUESIPS ϵ-flags indicate the presence of reseaux at 1195 Å and 1325 Å. In this case, NEWSIPS neither flags this problem nor shows an indication of the reseaux in the spectrum; the signal-weighted extraction of NEWSIPS ignores the reseaux because they affect only pixels located out on the wings of the spectral order. Since the reseaux did not affect the integrated flux to an appreciable degree, the condition is not flagged.

In general, the increase in the S/N has been shown to range from 10 – 50%, with factors of 2 – 4 improvement in some cases. The greatest improvements have been seen in under-exposed, high radiation, and high sky background images. The S/N for NEWSIPS data is often better than that of the IUESIPS data for a single spectrum, and multiple NEWSIPS images can be co-added to attain further increases in the S/N.

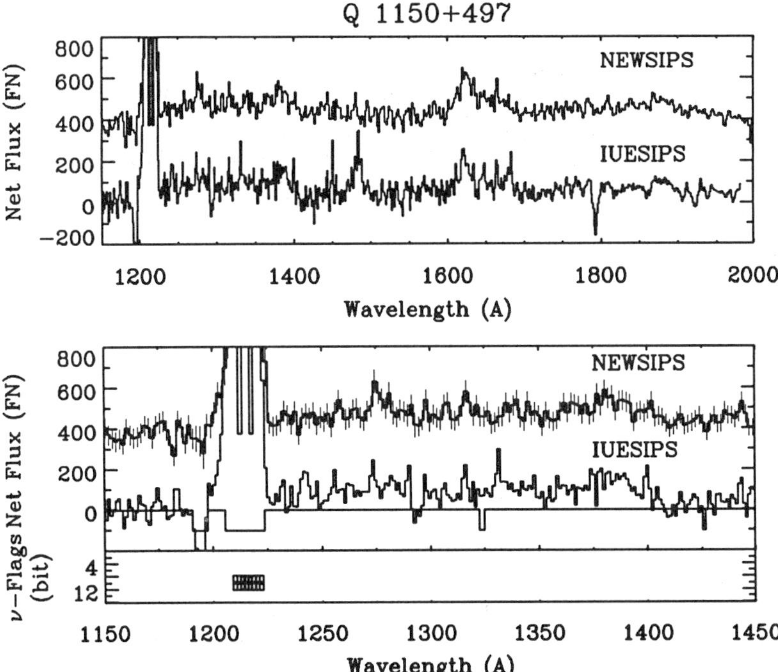

Figure 1. Comparison of NEWSIPS and IUESIPS SWP low dispersion, high sky background data for Q 1150 +497. NEWSIPS is a signal-weighted extraction; IUESIPS is a boxcar extraction. Spectra have been offset for clarity.

2.2. Specifics of the High Dispersion Algorithms

For the Final Archive there is a high dispersion geometrically rectified and rotated image with all orders parallel to the horizontal image axis, and the wavelength dispersion linear within each order, analogous to the low dispersion data. A global background is determined via the application of multiple one-dimensional Chebyshev interpolations in the spatial and then the dispersion direction (Smith 1990). In addition, corrections for the inter-order overlap problem are determined by modeling of the point spread function; this is important where the echelle orders are most closely spaced and overlap. The signal-weighted extraction method, based on an assumed profile, incorporates the background determination and applies the inter-order overlap corrections as necessary. Figure 2 illustrates a comparison of both a NEWSIPS and IUESIPS average of six spectra, and medium resolution GHRS data for μ Col. This plot is only for the illustration of the S/N characteristics of the datasets, as there are no calibrations for the NEWSIPS data in high dispersion at this time.

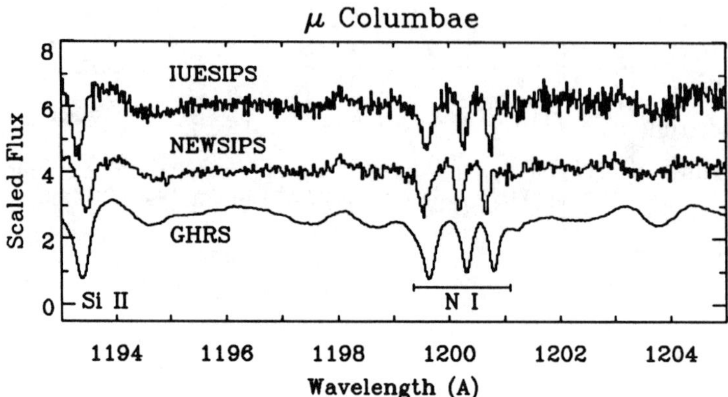

Figure 2. High dispersion comparison of data for μ Columbae.

3. Final Archive Database

The IUE Final Archive Database will contain the basic observational parameters for each exposure (i.e., accurate coordinates and exposure times, homogeneous object names) verified to ensure accuracy and clarity of the information. A substantial amount of effort has gone into accumulating the observational documentation (hand-written observing scripts, image label data, guest observer's personal log) and developing a system whereby a relational database can be populated with this critical information for use by the image processing system. A database forms-based user interface was designed as a front-end for the database which aids in the streamlining of the verification procedure and, more importantly, allows for a consistent decision-making mechanism, particularly in the event of conflicting data. Exposure-dependent parameters previously not easily available electronically (i.e., temperature, radiation counts, focus), as well as improved background and continuum measurements are now accessible.

References

Bushouse, H., De La Peña, M., Nichols-Bohlin, J. & Shaw, R. 1992, in Astronomical Data Analysis and Software Systems I, A.S.P. Conf. Ser., Vol. 25, eds. D.M. Worrall, C. Biemesderfer, & J. Barnes, 460

De La Peña, M.D., Shaw, R.A., Bushouse, H.A., & Nichols-Bohlin, J.S. 1992, in Astronomical Data Analysis and Software Systems I, A.S.P. Conf. Ser., Vol. 25, eds. D.M. Worrall, C. Biemesderfer, & J. Barnes, 457

Kinney, A., Bohlin, R., & Neill, D. 1991, PASP, 103, 694

Nichols, J.S., Garhart, M.P., De La Peña, M.D., Bushouse, H.A., & Levay, K.L. 1993, CSC/SD-93/6062, 2

Smith, M.A. 1990, in Proc. Int. Symp. "Evolution in Astrophysics", ESA SP-310, 631

Astronomical Data Analysis Software and Systems III
ASP Conference Series, Vol. 61, 1994
D. R. Crabtree, R. J. Hanisch, and J. Barnes, eds.

DENIS: Source Extractions

Erik R. Deul and András Holl[1]

Leiden Observatory, P.O. Box 9513, 2300 RA Leiden, The Netherlands

N. Epchtein

Observatoire de Paris-Meudon, DESPA, 92195 - Meudon Principal CEDEX, France

Abstract. A description of the source extraction method as employed in the Deep Near Infrared Survey of the Southern Sky is given. Some first insights, obtain by analyzing artificial images, of the performance and quality of the extraction technique are presented.

1. Introduction

The DEep Near Infrared Survey of the Southern Sky (DENIS) (Epchtein et.al. 1992) has the objective to provide full sky coverage in 2 near infrared bands (J at $1.25 mum$ and K at 2.1μm) and one optical band (I at 0.8μm), using a ground-based telescope and digital array detectors. DENIS is a joint project of 18 European and South American Institutes, aiming to provide digitized maps of and source lists for objects in the southern sky. The products of this survey will be databases of calibrated images, extended sources, and small objects. In addition catalogs of small and extended sources will be produced. With the anticipated 3σ limiting magnitudes of 18, 16 and 14 at I,J, and $K\prime$, respectively we expect to detect several 10^7 sources. With a datarate of $\sim$ 5Gb per observing night an efficient and reliable source extraction algorithm is required. Two datacenters at Institut d'Astrophysique de Paris and at Leiden Observatory provide a full off-line data reduction pipeline. Paris is responsible for the standard detector array reduction steps such as flat-fielding, bias corrections, etc., while Leiden Observatory is responsible for the actual source extraction. The observations will be performed in step-and-stare mode, using "strips" as basic units. Each strip is $12'$(one frame) wide in RA, and $30°$ long in declination, consisting of 180 overlapping frames. For each IR channel the sky is micro-scanned at 9 different matrix positions to mimic a 768×768 detector with a 256×256 array.

The extracted objects, being either survey object, photometric calibration sources, or position calibration sources are all stored in a tailor made database management system. This database management system allows for storing all the information in raw parameter format. Upon extraction or the creation of a final Small Sources Catalog the conversion to astronomically relevant units

[1]Konkoly Observatory, H-1525 P.O. Box 67, Budapest, Hungary

will take place. This allows for the use of the most up to date version of the conversion algorithms and their associated parameters.

2. Source Detection

Before we describe the results on quality assessment of the source extraction method, we will first describe the actual algorithm in some detail. For a more thorough description see Epchtein et.al. (1992).

2.1. Background Determination

To allow extraction of objects from the basic images, we need to subtract any "large" scale structure, which may be due to detector imperfections, remaining sky emission or true diffuse emission. The background determination algorithm divides a basic image (768x768 pixels) into equal squared sub-matrixes, producing a detection cutoff at a given spatial resolution twice that of the sub-matrix size ($1'$). For each submatrix a single background level and an rms estimate are computed by fitting a gaussian curve to the low intensity end of the matrix histogram. The background pixels have a mean position usually not coincident with the center of the sub-matrix. Background maps are interpolated using one of the well established procedures for unequal spaced grids.

2.2. Object Detection

Using the background determination algorithm's rms estimate for each submatrix the background subtracted input images will be thresholded at a positive fraction of this rms noise level. Pixels with values above the threshold are denoted object pixels. A software pattern analyser using an 8-fold neighbour connectivity algorithm determines the pixels belonging to a coherent structure (object). The shape of the object is irrelevant in this scheme. Rejection of objects can be done on the basis of the number of pixels per threshold level, effectively removing "noise" objects and/or using interband correlations.

2.3. Interband Connectivity

All three wavelength bands will be processed simultaneously. Objects are searched for in each band separately, meaning that there will be no bias to any of the three passbands. Having found an object in one band, this bands object geometry parameters are used to derive photometric properties of the same area in the other passbands, provided no objects were detected there, effectively producing lower limits for this object in the other passbands. During the extraction when objects are positively detected at the other passbands, this information is used to combine the multi-color information of the single object.

The result will be a catalog that is not biased to any of the three passbands and contains, for passbands at which the object has not been positively detected, a lower limit photometric value.

3. Deblending

Once an object has been detected above the lowest threshold level it may consist of a number of blended point sources (or extended sources). The purpose of the deblending algorithm is to separate the individual components both in position and intensity.

3.1. Blending Detection

Using the above mentioned thresholding/pattern-analyser technique, objects are detected at a number of equidistant intensity levels. The effect of this thresholding is to generate contour information in log(intensity) space. At higher levels the object may split into separate components. A tree structure is produced representing the splitting up of the blended object. If an object does not split up this technique does not differ from the above single threshold object detection.

3.2. Deblending of the Object

The deblending starts at the leafs of this tree structure, where the object peak information is stored, and uses a fitted Gaussian profile to determine the intensity contribution of each source peak to the low level pixels. The faint pixel intensities are apportioned between competing object components using a distance estimator in the Gaussian fit and then comparing their relative contributions. In cases where no object contributes significant intensity to a given pixel, the pixel is allocated completely to the object contributing most to it. Although a Gaussian profile is used to determine the faint pixel contribution, this profile is not forced upon an object because a source will gather up the faint pixels around it.

As the pattern analyser recognizes the objects cumulative sums of pixel intensity (at all three passbands), and intensity weighted and unweighted pixel positions are saved.

4. Quality Assessment

Tests on the source extraction algorithm were performed using synthetic data of the form as will be produced by the actual data acquisition equipment, and preprocessed at the PDAC. For the description of the density and brightness distribution of stars we used models of Robin (1993) Images with different amounts of stars (200 – 5000) were made with microscanning mimiced by interlacing 9 0.33-pixel-offset subimages.

4.1. Crowding

To assess the crowding loss we applied Monte Carlo methods to create artificial images and tried to retrieve the objects put in. Many different images, with a large range of object densities, were fed into the DENIS source extraction pipeline. The extracted source list was compared, after some massaging, to the input list (Holl and Deul 1993).

4.2. Positional Accuracy

The relative rms errors (within a DENIS frame) will be in the order of $0\overset{''}{.}1$ well below the image pixel dimensions (1'×1'). The positional accuracy of the survey will thus mainly be influenced by the systematic errors of the reference catalog (Guide Star Catalog) (Taff et al. 1990) which on average are about 1".

4.3. Photometric Accuracy

The regime of noise dominated extractions (M > 16.5m) and that of the overexposure effects (M < 9.5m). Generally photometric accuracy of well below 0.1m can be achieved.

Acknowledgments. A. Holl gratefully acknowledges the support of the EEC "Cooperation in Science and Technology with Central and Eastern European Countries" grant No. 2905 and the OTKA grants T4341 and F4343.

References

Epchtein N. et al., 1992, "A Deep Near Infrared Survey of the Southern Sky (DENIS)", ed. N. Epchtein

Robin A.C.: 1993, "Synthesis of galactic stellar populations and expected sources in infrared surveys" in Science with astronomical near-infrared surveys, ed. A. Omont, in press

Holl A., & Deul E.: 1993, "The influence of crowding on DENIS point source detection rates" in Science with astronomical near-infrared surveys, ed. A. Omont, in press

Taff L.G. et al., 1990, "Some comments on the astrometric properties of the Guide Star Catalog", ApJ, 353, L45

Astronomical Data Analysis Software and Systems III
ASP Conference Series, Vol. 61, 1994
D. R. Crabtree, R. J. Hanisch, and J. Barnes, eds.

A Review of the Star*s Family Products

A. Heck
Strasbourg Astronomical Observatory, 11 rue de l'Université, F-67000 Strasbourg, France

Abstract. The current situation of the *Star*s Family* is reviewed, as it is a growing line of products: directories, dictionaries, databases, data sets, etc., on astronomy, space sciences, and related fields and organizations of the world.

1. Generalities

Since more than fifteen years now, we have been compiling directories of organizations involved in astronomy, space sciences and related fields. More recently, we entered the parallel compilation of a dictionary of abbreviations, acronyms, contractions and symbols appearing in the same fields.

Therefore the Star*s Family line of products is structured essentially around two sets of master files: StarGuides, a directory of astronomy, space sciences, and related organizations of the world, and StarBriefs, a dictionary of abbreviations, acronyms, contractions, and symbols in astronomy, space sciences, and related fields.

These lists are made available in various ways with the aim of providing flexible yellow-page services. The products are briefly presented below.

2. The Directory StarGuides

StarGuides (Heck 1993a&b) is the new name of the directory *Astronomy, Space Sciences and Related Organizations of the World (ASpScROW)* (Heck 1991), itself resulting from the merging and scope broadening of the earlier directories IDPAI (International Directory of Professional Astronomical Institutions) and IDAAS (International Directory of Astronomical Associations and Societies), both subtitled 'together with related items of interest' (Heck 1989a&b).

StarGuides gathers together all practical data available on associations, societies, scientific committees, agencies, companies, institutions, universities, etc., more generally organizations, involved in astronomy and space sciences. Many other types of entries have also been included such as academies, bibliographical services, data centres, dealers, distributors, funding organizations, IAU-adhering organizations, journals, manufacturers, meteorological services, national norms and standards institutes, parent associations and societies, publishers, software producers and distributors, and so on.

Besides astronomy and related space sciences, other fields such as aeronautics, aeronomy, astronautics, atmospheric sciences, chemistry, communications, computer sciences, data processing, education, electronics, engineering, energetics, environment, geodesy, geophysics, information handling, management, mathematics, meteorology, optics, physics, remote sensing, and so on, are also covered when appropriate.

Currently more than 5000 entries from about 100 countries have been selected. The information is given in an uncoded way for easy and direct use. For each entry, all practical data available are listed: city, postal and electronic-mail addresses; telephone and telefax numbers; foundation years; numbers of members and/or staff; main activities; titles, frequencies, ISS Numbers and circulations of periodicals produced; names and geographical coordinates of observing sites; names of planetariums; awards, prizes or distinctions granted; and so on.

The entries are listed alphabetically in each country. An exhaustive index gives a breakdown not only by different designations and acronyms, but also by location and major terms in names. Thematic subindices are also provided as well as statistics on the contents (numbers of entries per country, memberships, years of foundation) and a list of telephone and telefax national codes.

3. The Dictionary StarBriefs

Currently StarBriefs (Heck 1993c&d) gathers together about 65,000 abbreviations, acronyms, contractions and symbols encountered in the literature relative to astronomy and related space sciences, as well as to the other fields mentioned earlier.

Sections are devoted to Greek letters, mathematical symbols, special signs and characters, as well as to entries with a numerical beginning.

Abbreviations, acronyms, contractions and symbols in common use and/or of general interest have also been included when appropriate. The traveling scientist has not been forgotten (codes of airlines, locations, currencies, and so on) and humour is not quite absent from this publication either.

4. Associated Databases

Databases associated to StarGuides and StarBriefs are available on-line in various locations. They are regularly updated.

Once in a system, the successive menus are self-explanatory. On-line tutorials and presentations are also available, as well as mailbox facilities.

The databases point to the various entries not only by their different designations and acronyms, but also by locations and major terms in names. Information can also be retrieved according to the categories mentioned above, as well as to thematic subindices corresponding to the various types of data provided for each entry.

Selection criteria are also available, such as the geographical coordinates of a location around which one would like to list the entries (with available coordinates) within a given radius. Any string of characters can also be searched for in most fields.

Besides the databases listed below, other implementations are currently negotiated.

4.1. StarWays

StarWays is associated to StarGuides and has been implemented by the ESIS group at ESRIN, an establishment of the European Space Agency (ESA) located at Frascati, Italy (Heck et al. 1992).

Access to Starways is possible through the ESIS public account reachable via Internet ($telnet 192.106.252.127 with username ESIS, no password required) or SPAN ($set host ESIS or 29617 with username ESIS, no password required). More details on ESIS can be obtained by contacting esis::esis (SPAN) or esis@ifresa51.bitnet.

4.2. StarGates and StarWords

These databases have been set up at the European Southern Observatory (ESO) located in Garching-bei-München, Germany, and correspond respectively to StarGuides and StarBriefs (Albrecht & Heck 1993a,b&c).

They can be accessed through the standard Starcat account reachable via Internet ($telnet stesis.hq.eso.org or 134.171.8.100) or SPAN ($set host stesis). On the top level, select the ESO option and then YellowPages. More details on Starcat can be obtained by contacting archeso@eso.org (Internet) or eso::archeso (SPAN).

4.3. StarWorlds and StarBits

These databases are currently implemented at Strasbourg astronomical Data Centre (CDS) and correspond respectively to StarGuides and StarBriefs (Heck & Ochsenbein, 1993a&b).

5. Other Products and Further Information

The Star*s Family line includes also StarLabels (sets of mailing stickers bearing addresses contained in the permanently updated StarGuides master files) and StarSets (subsets of StarGuides and StarBriefs data on various media).

These other products are made available under some conditions of requirement and usage.

Information on the Star*s Family line can be obtained from the author (see address in the affiliation) or by:

Telephone: +33-88.35.82.22 (direct)
+33-88.35.82.16 (Secretary)
Telefax: +33-88.49.12.55 (direct)
+33-88.25.01.60 (Secretary)
Electronic mail: heck@cdsxb6.u-strasbg.fr (Internet)

Acknowledgments. It is a very pleasant duty to express here our gratitude to the institutions and individuals who have helped in various ways to compile the information presented in the *Star*s Family* files, as well as to Michel Crézé, Director of Strasbourg Astronomical Observatory, for supporting the work.

The implementations as databases of the *Star*s Family* files by the European Space Agency, the European Southern Observatory and Strasbourg Astronomical Data Centre have been strong incentives to continue and always improve these time-consuming compilations.

References

Albrecht, M.A., & Heck, A. 1993a, "StarGates and StarWords — An On-line Yellow Pages Directory for Astronomy", ESO Messenger, 73, 39-40

Albrecht, M.A., & Heck, A. 1993b, "StarGates – An On-line Database of Astronomy, Space Sciences and Related Organizations of the World", A&AS, in press

Albrecht, M.A., & Heck, A. 1993c, "StarWords — An On-line Database of Acronyms, Abbreviations, and Contractions in Astronomy", Space Sciences and Related Fields, A&AS, in press

Heck, A. 1989a, "International Directory of Astronomical Associations and Societies together with related items of interest — IDAAS 1990", CDS Special Publ. 13, vi + 716 pp. (ISSN 0764-9614 — ISBN 2-908064-11-1)

Heck, A. 1989b, "International Directory of Professional Astronomical Institutions together with related items of interest – IDPAI 1990", CDS Special Publ. 14, vi + 658 pp. (ISSN 0764-9614 – ISBN 2-908064-12-X)

Heck, A. 1991, "Astronomy, Space Sciences and Related Organizations of the World – ASpScROW 1991", CDS Special Publ. 16, x + 1182 pp. (ISSN 0764-9614 – ISBN 2-908064-14-6) (two volumes)

Heck, A. 1993a, StarGuides 1993 – A Directory of Astronomy, Space Sciences and Related Organizations of the World", CDS Special Publ. 20, x + 1174 pp. (ISSN 0764-9614 – ISBN 2-908064-18-9) (two volumes)

Heck, A. 1993b, "StarGuides. A Directory of Astronomy, Space Sciences and Related Organizations of the World", Astron. Astrophys. Suppl., in press.

Heck, A. 1993c, "StarBriefs 1993 – A Dictionary of Abbreviations Acronyms, and Symbols in Astronomy, Space Sciences and Related Fields", CDS Special Publ. 21, xl + 868 pp. (ISSN 0764-9614 – ISBN 2-908064-17-0)

Heck, A. 1993d, "StarBriefs. A Dictionary of Abbreviations, Acronyms, and Symbols in Astronomy, Space Sciences and Related Fields", A&AS, in press

Heck, A., Ciarlo, A. & Stokke, H. 1992, "StarWays. An On-line Database of Astronomy, Space Sciences and Related Organizations of the World", A&AS, 96, 565-566

Heck, A. & Ochsenbein, F. 1993a, "StarWorlds – An On-line Database of Astronomy, Space Sciences and Related Organizations of the World", in preparation

Heck, A. & Ochsenbein, F. 1993b, "StarBits – An On-line Database of Acronyms, Abbreviations, and Contractions in Astronomy, Space Sciences and Related Fields", in preparation

Astronomical Data Analysis Software and Systems III
ASP Conference Series, Vol. 61, 1994
D. R. Crabtree, R. J. Hanisch, and J. Barnes, eds.

Homogeneous Access to Data: The ESIS Reference Directory

S. G. Ansari, P. Giommi, A. Micol, P. Natile

ESIS - Information Systems Division of ESA, ESRIN, Italy

Abstract. An important component of ESIS system (Giommi & Ansari, 1993) is the ESIS Reference Directory. This metadatabase contains a variety of information on catalogues and remote data products, such as images, spectra or light curves. The ESIS *Reference Directory* plays a fundamental rôle in the homogeneous access of a variety of astronomical catalogues. It also provides commands which are used by the display and manipulation packages of ESIS to remotely access the data. We discuss in this paper the methods used to set up a reference directory as a central repository and the mechanisms used to access the metadata to resolve ESIS queries.

1. Introduction

The ESIS system (Giommi & Ansari, 1993) is made up of several components. One of these is the *Cats&Logs*, which is the heart of the system. It contains all the available Astronomical catalogues and the *Reference Directory. Cats&Logs* plays a crucial rôle, not only to browse catalogue entries, but along with the *Reference Directory* and the *Query Engine* (Ansari et al. 1993), coordinates the access to local/remote data (e.g., images, spectra, or light curves.)

Cats&Logs is currently setup as a set of ORACLE tables located on computers at ESRIN. In the near future, a more *distributed* approach is possible, where a number of European sites could be involved in providing coordinated access to catalogues and metadata.

The ESIS *Reference Directory* contains, not only metadata: i.e., a description of each catalogue, but also will soon provide the information used within ESIS to access data, depending on the manipulation packages for each type of data (e.g., images, spectra or light curves.) Moreover, the *Reference Directory* interfaces to another Database Management System, which contains all the ESIS bibliographic references (based on FUL/Text) and provides similar information to the bibliographic application.

During the initialization of the *Query Engine* the contents of the *Reference Directory* are read into memory, this is because the *Query Engine* needs to have fast access to essential information to process queries.

2. The Contents of the Reference Directory

Initially, the *Reference Directory* contains the full name of each catalogue, its bibliographic reference, number of entries and the actual ORACLE table name. It also contains the following items used to process each query:

- A homogeneous field (if applicable) containing an object's name, which is to be used by the name search, when SIMBAD is queried. For example: If the HD number is available for all object entries in a catalogue, the HD number extracted from SIMBAD is used to search in that catalogue.
- A default search radius for each catalogue, which is used when a coordinate search (by cone) is made. The radius may vary, depending for example on an instrument's field of view, but a radius of 10 arcmin is normally used for more general catalogues.
- A set of relevant fields are defined per default, that appear in the result table of a query, if the user has not chosen any result fields for a selected catalogue.

A strict naming convention is followed to give access to homogeneous information contained in multiple catalogues:

- Coordinates are stored in J2000.0 equinox and are always designated by RA, DEC for equatorial coordinates and LII, BII for galactic coordinates.
- The primary field containing the name of sources is known as a NAME_LIST.
- If observation dates are available, they always carry the name START_TIME, which is composed if the date and time.
- Exposure times are designated by EXPOSURE.
- Visual magnitudes are designated by VMAG.
- Color indices are designated by BMV (for B-V) and UMB (for U-B) for the Johnson photometric system.
- Flux values are designated by FLUX. If more than one flux value is given in a catalogue (e.g., IRAS_PSC), the flux values are designated by FLUX followed by the frequency (or wavelength range) (e.g., FLUX12 for flux at 12 microns).
- Class usually points to a field in a catalogue containing a spectral, morphological or galaxy type, etc.

Furthermore, catalogues are classified into four different categories:

- Astronomical Catalogues: which comprise general astronomical catalogues like the HD, NGC or the Yale Bright Star Catalogue.
- Mission Logs: are catalogues of observations of missions, such as: HST_LOG, ROSAT_LOG, or EUVE_LOG.

- Result Catalogues: containing derived results from experiments. These include: EXOSAT_CMA, EINSTEIN_IPC, or EINSTEIN_HRI tables.
- Data product Catalogues: that contain information on data datasets of different experiments, such as: EXOSAT_ME, EINSTEIN_IPCIMAGE

Another classification of the catalogues is according to the wavelength region. The following classification has been applied:

- Radio
- Infra Red
- Optical
- Ultra Violet
- X-ray
- Gamma Ray

3. Mechanisms

Because of this strict naming convention, any parametric searches or conditions made on any fields of the selected ESIS catalogues apply universally to ALL catalogues containing these fields. For example: if the user searches for an object in all the mission logs, s/he may give a time condition, which would then be valid for all mission logs containing observation dates. Similarly, catalogue cross correlation is simple, since parametric conditions or field selections are applied to both catalogues regardless of field location or designation.

4. Conclusions

The ESIS *Reference Directory* architecture, in its first implementation has paved the way for a more flexible handling of catalogue searches. Since no preprocessing of catalogue entries is normally necessary (except for the homogenization of coordinates) and any further homogenization (e.g., units of measure of common fields) may be done within the *Reference Directory* and not at the time of loading of the catalogue, it is obvious that the ESIS code can remain general, relying on the information, be it data transforms, or data-access commands, available from the *Reference Directory.*

The contents of the ESIS *Reference Directory* is now directly available online through the NCSA Mosaic ESIS Homepage. In order to obtain information on any catalogue, a description of its contents, its reference, the keywords and wavelength regions defined, use the following direct link to the ESIS databases:

http://mesis.esrin.esa.it:8888/htbin/descat

which will provide you with a list of all the available catalogues within the ESIS system with hyperlinks to pages on each individual catalogue.

References

Ansari, S.G., Giommi, P., Stokke, H., & Preite-Martinez, A. 1993, "Homogeneous Access to Astronomical Data: The ESIS Experience", in Proceedings: Handling and Archiving Data from Ground-based Telescopes, Trieste, 1993

Giommi, P., & Ansari, S.G. 1993, this volume

Astronomical Data Analysis Software and Systems III
ASP Conference Series, Vol. 61, 1994
D. R. Crabtree, R. J. Hanisch, and J. Barnes, eds.

The EUVE Public Archive: Data and User Services

B. A. Stroozas, E. Polomski, B. Antia, J. J. Drake, K. Chen, C. A. Christian, and E. C. Olson

Center for EUV Astrophysics, University of California, 2150 Kittredge St., Berkeley, CA 94720

Abstract. The public science archive for the Extreme Ultraviolet Explorer satellite (EUVE) is now operational. The Archive handles the storage, maintenance and distribution of EUVE data and related documentation, information and software. An additional objective has been to provide new and flexible user interfaces and services to the highly diverse electronic-based community. Some examples which are currently under investigation or development include a spectral data archive server via NASA's Astrophysics Data System, a distributed series of CD-ROMs, the use of various network tools (e.g., *Mosaic*) to allow for convenient data browsing, review and retrieval, and interfaces to allow access to the raw telemetry and data processing software. This paper gives an overview of the existing system capabilities, describes activities currently under development and outlines plans for the future.

1. Introduction

The Extreme Ultraviolet Explorer satellite (EUVE) was launched on 7 June 1992. Its primary objectives are to carry out a six month all-sky (and concurrent deep ecliptic) survey followed by three years of pointed Guest Observer (GO) spectrometer observations in the EUV wavelength region ($\sim 60 - -760\AA$). For a full description of the mission see Bowyer and Malina (1991) and the EUVE special issue of the Journal of the British Interplanetary Society (JBIS 1993). The survey phase of the mission was conducted between 22 July 1992 and 20 January 1993 (with calibration gap-filling during the following six months). At the time of writing, we are ten months into the GO phase.

2. The EUVE Science Archive

The archival and distribution of the huge EUVE data set (expected to exceed 1 TByte over the projected 3.5 year mission lifetime) has required a multi-access approach on the part of the EUVE Science Archive. The task of the Archive group is to make the scientifically interesting data available to the astronomical community as soon as possible after the proprietary data rights expire. (All EUVE science data has associated proprietary data rights. The survey data belongs to the University of California, Berkeley, for a period of one year after the end of gap-filling and will become public in August, 1994. GO data, proprietary for a period of one year, will officially begin to go public in April, 1994.) An additional objective has been to provide new and flexible user interfaces and services to the highly diverse electronic-based community. The sections that follow describe the five major methods of access to the Archive — anonymous FTP, NASA's Astrophysics Data System, the CD-ROM series, an electronic mailserver, and assorted network tools — and the relevant EUVE material available through each.

2.1. Anonymous FTP

The FTP site at the Center for EUV Astrophysics (CEA) lies at the core of the Archive. The site, accessible via anonymous FTP at **cea-ftp.cea.berkeley.edu**, is shared between the EUVE Archive and Guest Observer (EGO) groups. The Archive part of the site, contains the following EUVE-related material:

- Data Products — Contents include 16 1-d spectra of In-Orbit Calibration targets and the survey Bright Source List (Malina, et al. 1994). As data rights expire, additional GO 1-d spectra, as well as catalogs (e.g., Bowyer, et al. 1994) and skymaps from the surveys, will be included.

- Software — Presently available is the Interstellar Medium (ISM) C code which calculates ISM attenuation at EUV wavelengths. In the future we will include additional useful software (e.g., EUV plasma code) contributed from members of the scientific community.

- Documentation — Currently included are calibration and GO target lists, the bibliography of CEA EUVE papers, past issues of the EUVE electronic newsletter, and NASA data rights policy for GO Cycle 1. In the future we will include all EUVE-related abstracts, public relations images and material, and more specific documentation on various aspects of the project (e.g., mission operations, instrument calibrations, hardware, and software).

2.2. NASA's Astrophysics Data System

Another cornerstone of the Archive is NASA's Astrophysics Data System (ADS), a distributed database system employing a client/server architecture (Weiss & Good 1991). The EUVE project is committed to making data and services available via ADS through the operational node at CEA. Although only a limited amount of material is available at this time, a variety of additions are to be included in upcoming ADS releases (the next is scheduled for early 1994):

- Data — At present, the EUVE calibration database and the ROSAT Wide Field Camera Bright Source Catalog (Pounds, et al. 1993) are available. Upcoming additions will include the survey Bright Source List (Malina, et al. 1993) and EUVE catalog (Bowyer, et al. 1994), a small optical test-bed database, and other relevant databases.

- Services — The following services are under development for incorporation into ADS: the ISM server, document and abstract browsing capabilities, and a spectral data archive server which will allow users to easily choose and display EUVE spectra. One long-term project involves incorporating software tools to provide users with remote "on-demand" access to the EUVE raw telemetry and data products.

2.3. CD-ROM Series

The Archive CD-ROM series has been established to provide a better means of distribution for the large EUVE data sets. CD-ROMs are well suited to this task as each can store in excess of 500 MBytes of data; a typical EUVE pointed observation (spanning several days) generates a few hundred MBytes of data. To date, the series includes the following:

- Volume 1, Number 1 (Drake, et al. 1993) — Released at the June, 1993 meeting of the American Astronomical Society (AAS) in Berkeley, CA, it contains the full pointed spectrometer calibration data set for the late-type star AT Mic (observed on 1 July 1992). It also contains the EGO software and reference data used to process this observation, as well as images from the all-sky survey, the ISM software, and other mission-related documents.
- Volume 2, Number 1 (Stroozas, et al. 1994) — This is a set of three CDs to be released at the January, 1994, AAS meeting in Washington, D.C. This set contains the full data sets for eleven pointed spectrometer calibration observations from the In-Orbit Calibration and survey phase of the mission. Also included are the relevant EGO software and reference data, the survey Bright Source List (Malina, et al. 1994) and catalog (Bowyer, et al. 1994) papers, abstracts of other CEA papers, and additional mission-related documents.

Although the long-term distribution plan for the CD-ROM series is still under development, it will undoubtedly include data from the all-sky and deep surveys (e.g., skymaps, catalogs and other data products) as well as additional GO observations.

2.4. Electronic Mail Server

Text data is available via electronic mail server at **archive@cea.berkeley.edu** (Internet). In addition to providing automatic data retrieval via e-mail, it can also be used to remotely run the ISM software. An extensive help file can be retrieved by sending mail to the above address with the word "help" (quotes omitted) as the body of the message. Including the word "index" will retrieve for the user the index of available data.

2.5. Network Tools

There has recently been a significant increase in the amount of network access. The wealth of available on-line information is being utilized with increasing fervor due to excellent new interfaces such as NCSA's *Mosaic*. The EUVE project recognizes the need to expand its accessibility to the astronomical community and the networked world and also to make use of external resources of information. To this end, the Archive group is making a serious effort to expand its accessibility by implementing *gopher*, *WAIS* and, at a higher level, NCSA's *Mosaic*. As these (and other) tools continue to evolve, we plan to take advantage of new enhancements. The *Mosaic* "Home Page" for CEA is available by opening the URL **http://cea-ftp.cea.berkeley.edu/HomePage.html**.

3. Summary

The overall goal for the Archive group is to provide timely and efficient availability, and ease of access, to EUVE material for the entire astronomical community. To accomplish this, a system is currently in place to distribute this material — data, software and documentation — to users through the use of anonymous FTP, NASA's ADS, and the EUVE CD-ROM series. To complement these, a variety of additional easy and user-friendly methods of access are in place or under development including a mailserver and network tools.

The strategy of the EUVE Science Archive group is to provide the astronomical community with (1) the copious quantity of high-quality EUVE data, (2) ease of access to that data using cutting-edge tools and techniques (e.g., CD-ROMs and *Mosaic*), and (3) innovative services to provide analysis tools for maximizing the scientific return.

Acknowledgments. This work has been supported by NASA contracts NAS5-29298 and NAS5-30180. The authors thank the Principal Investigators, Stuart Bowyer and Roger Malina, and the EUVE science team for their advice and support.

References

Bowyer, C.S., et al., 1994, ApJS (submitted)

Bowyer, C.S., & Malina, R.F. 1991, in Extreme Ultraviolet Astronomy, eds. C.S. Bowyer & R.F. Malina (New York: Pergamon Press), 397

Drake, J.J., et al., 1993, EUVE CD-ROM series, eds. C.S. Bowyer & R.F. Malina, 1:1

Journal of the British Interplanetary Society (JBIS), 46, 9, 1993

Malina, R.F., et al., 1994, AJ (in press)

Pounds, K.A., et al., 1993, MNRAS, 260, 77

Stroozas, B.A., et al., 1994, EUVE CD-ROM series, eds. C.S. Bowyer & R.F. Malina, 2:1

Weiss, J.R., & Good, J.C. 1991, in Databases and On-line Data in Astronomy, eds. M.A. Albrecht & D. Egret (Dordrecht: Kluwer), 139

Astronomical Data Analysis Software and Systems III
ASP Conference Series, Vol. 61, 1994
D. R. Crabtree, R. J. Hanisch, and J. Barnes, eds.

Data Analysis and Expected Results of the Tycho Mission

A. J. Wicenec, G. Bässgen, V. Großmann, M. A. J. Snijders, K. Wagner

Astronomisches Institut Tübingen Waldhäuserstr. 64, D-72076 Tübingen

U. Bastian, P. Schwekendiek

Astronomisches Rechen-Institut, Mönchhofstr. 14, D-69120 Heidelberg

D. Egret, J. Halwachs

Centre de Données de Strasbourg, Rue de Université 11, F-67000 Strasbourg

E. Høg, V. V. Makarov

Copenhagen University Observatory, Ostervolgade 3, DK-1350 Kopenhagen K

Abstract. The Tycho Data Analysis Consortium (TDAC) data reduction scheme is presented and the central parts in the reduction chain are described. Figures are shown to verify the photometric stability of the instrument and the rigidity of the astronomical parameters derived from several month of data of the nominal mission.

1. Introduction

In the Tycho project on-board the Hipparcos satellite (Høg et al. 1992) photometric and astrometric data of about 1 000 000 stars to a limit of $B = 12\ mag$ will be derived. The brightest 500 000 stars will obtain magnitudes in two colours B_T and V_T, the fainter 500 000 stars usually only either of the two and one broad-band magnitude T. The accuracy will be about 0.03 arcsec for positions and 0.03 mag for magnitudes, at $B = 10.5$ mag. The standard errors of positions and magnitudes will roughly increase by a factor of two per magnitude, at faint magnitudes. The results are obtained by the Hipparcos satellite's star mapper (which provides simultaneous measurements in two spectral channels), based on predicted star transits (Predicted Group Crossings, PGC's) using its own 'Tycho Input Catalogue' (TIC). The data treatment is carried out by the TDAC, using calibration and satellite attitude information from the two Hipparcos data reduction consortia.

[1]Based on observations made with the ESA Hipparcos satellite

2. Data Analysis

2.1. Reduction Scheme

The TDAC data reduction is divided into two parts, the 'Main Processing' of the whole continuous data stream, i.e., every single countrate measured by the photoncounters of the star mapper, and the 'Reprocessing', where only parts of the data are reduced once again (see Figure 3). During Reprocessing the data of the following objects will be reduced, using enhanced algorithms in order to improve the results for these objects:

- Stars brighter than 9 mag in either of the two bands
- Solar system objects
- Objects which are recognized in the Main Processing, but where not originally in the TIC
- Suspected double stars from the Main Processing
- Individually selected objects

This two-step procedure has been chosen in order to save processing time and to reduce the complexity of the Main Processing and to limit the amount of data in the Reprocessing, while treating the complete raw data collected by the satellite. On the other hand this procedure yields almost online results by using the 'Real Time Attitude Determination' (RTAD, Strada 1989) for the first steps of the Main Processing and the 'On Ground Attitude Reconstruction' (OGAR, Lindegren et al. 1992) for the down-stream tasks of the Main Processing and the complete Reprocessing. Thus the process uses the best attitude available at any stage of the reduction without the need of waiting for auxillary input data (i.e., from outside TDAC), but nevertheless yielding the highest accuracy for the output catalogue. At the expected start of the Reprocessing (beginning 1994) the final attitude will be available and will be used for this reduction step. The tasks of the Main Processing and their outputs are described in Høg et al. (1992), however the amount of data treated by the whole reduction chain and the Reprocessing scheme have not been presented up to now.

2.2. Reduction Status (Nov. 1993)

The Main Processing has been completed for all delivered data in the tasks Prediction and Detection. The task TIC production as a prelaunch task and the task Recognition uses only one year of data, thus these tasks are completed. The Photometric calibration has used almost all available data to produce calibration parameter sets. Prediction Updating, Transit Identification and Astrometry have used approximately 15% of the mission data. The latter tasks are much faster than the first ones and are expected to be ready by mid 1994. The Reprocessing is expected to be ready by the end of 1994. The catalogue production and verification phase is intended to be completed by the end of 1995 and the publication of the catalogues (Hipparcos and Tycho) is expected one year later.

3. Results

3.1. Astrometry

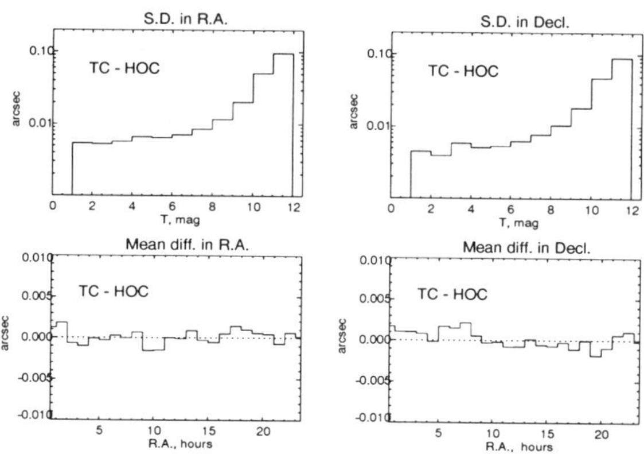

Figure 1. Standard deviation (upper) and mean difference (lower) of the Tycho positions from the Hipparcos positions from five month of mission data, in R.A. and Decl., respectively derived from 31889 stars having at least 30 observations.

3.2. Photometric Calibration

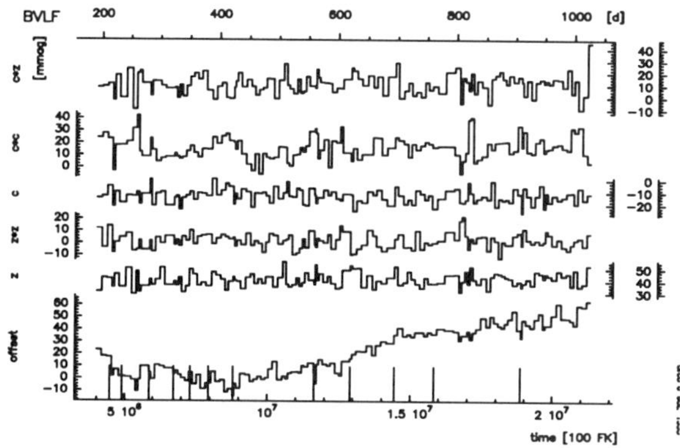

Figure 2. The six photometric calibration parameters calculated for 833 days as a function of time in the combination BVLF (B_T channel, vertical slit system, lower part of grid and following field of view). All parameters in mmag. Lower time axis in filingkey units (1 filingkey = 1/24 s), upper axis days since launch of Hipparcos satellite.

Acknowledgments. Supported by national funding from: Deutsche Agentur für Raumfahrtangelegenheiten (DARA), Centre National d'Etudes Spatiales (CNES) and the Danish Space Board.

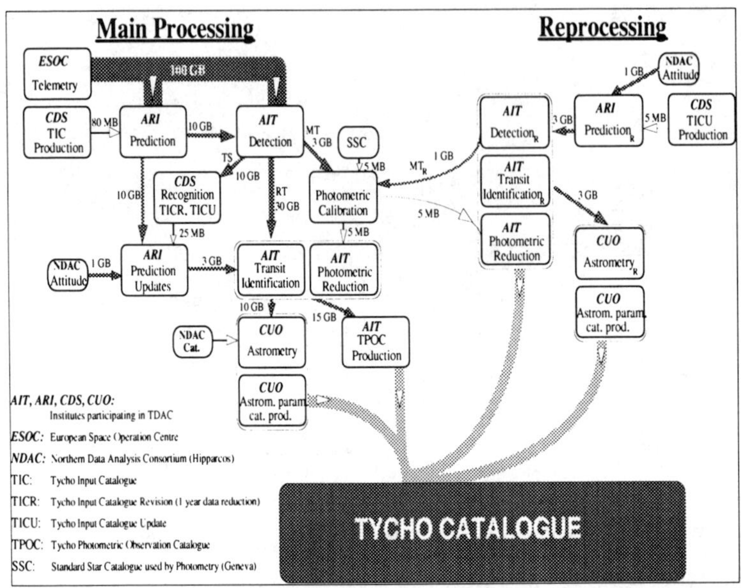

Figure 3. Data flow plan of the complete TDAC reduction chain. The NDAC catalogue used in Astrometry is the preliminary Hipparcos output catalogue from the 1.6 year solution of Hipparcos data. The width of the arrows indicate the amount of data exchange in addition to the written numbers.

References

Høg, E. et al., 1992, A&A, 258, 177

Strada, P. 1989, ESA SP-1111/I, 165

Lindegren, L. et al., 1992, A&A, 258, 18

Astronomical Data Analysis Software and Systems III
ASP Conference Series, Vol. 61, 1994
D. R. Crabtree, R. J. Hanisch, and J. Barnes, eds.

The Hubble Space Telescope Data Archive

Knox S. Long, Stefi A. Baum, Kirk Borne, and Daryl Swade

Space Telescope Science Institute, Baltimore, MD 21218

Abstract. The HST Archive was formally opened for archival research on 1 February 1993. Usage of the archive has increased steadily. By the summer, 3–5 Gbytes of data per month were being retrieved over electronic networks by astronomers outside the Institute. The present archive system, the Data Management Facility, consists of an archive machine with an optical disk jukebox and standalone optical disks drives, two database machines, and two user host machines. The archive currently contains about 1 Tbyte of data. Users are able to query the archive catalog from guest accounts on the host machines. There is a simple registration process for those who want to retrieve data. Once registered, users are able to retrieve data from the public portion of the data archive. In addition to describing the current archive system and our experience in supporting users, we will discuss a new archive system, the Space Telescope Data Archive and Distribution Service, and a new user interface, StarView. These will replace DMF and the current user interface, STARCAT, over the next year.

1. Introduction

Since HST was launched in April 1990, nearly 5000 astronomical targets have been observed. About 1 Tbyte of raw and calibrated data have been produced. These data, an associated catalog, and a group of support personnel comprise the HST archive. Prior to the launch of HST, a prototype optical-disk based archive system — the Data Management Facility — was developed by ST ScI with the support of the Space Telescope European Coordinating Facility and the Canadian Astronomical Data Center. Initially, DMF was used primarily as a storage system. However, as the amount of public data in the archive increased, the retrieval capabilities of DMF were augmented so that researchers could gain ready access to public data in the archive. Today, 200–400 non-ST ScI astronomers access the archive catalog and retrieve 3–5 Gbytes of HST data each month.

2. The Current Archive

The archive system in use at ST ScI today uses a VAX 8600 as the archive engine and a Sun SPARC II as a database server.[1] An optical disk jukebox box and 4 standalone disk drives are attached to the VAX. The data that are archived include the packetized science data stream, uncalibrated and calibrated science data, engineering data, and calibration files, such as flat fields. As data comes from the processing pipeline, the data are copied onto two 2 Gbyte LMSI optical disks in the standalone drives. One of these disks remains at ST ScI and the other is sent to the ST-ECF for use by European astronomers. HST data are reserved for use by the original observers for a period of time, typically one year, after which the data become public. As data become public, a third copy of the data is made for the CADC. Data from HST are currently being archived into DMF at a rate of about 1 Gbyte per day.

An important part of the archive process is the population of several relatively large databases with information extracted from the keyword fields in the header files of the science data. The portions of the catalog obtained directly from the data stream include basic information as well as more detailed engineering information about observations. The observation information includes items such as target, pointing position, observation time, and instrumental configuration. Examples of the engineering data in the catalog include count rate statistics, temperatures, bias levels, and ancillary information about the pointing direction relative to the sun. These data are useful for evaluating the quality of the data and for studying trends in instrumental properties. The relational database also contains proposal abstracts and other information about the observations to help the archive researcher understand the purpose of an observation. Special tables in the database describe updates to calibration files which astronomers may need to recalibrate their data. About 1700 items are catalogued in the database which now totals about 500 Mbytes in size.

Astronomers use the archive via an interface — STARCAT — developed and maintained by the ST-ECF with support from the CADC and ST ScI. Astronomers who want to use the archive log into a Unix or VMS host machine. Once on the host, they can invoke STARCAT, peruse, and then select datasets from the archive on the basis of a wide range of criteria — position, source name, target type, exposure time, PI name, etc. In order to isolate the operational database from the (sometimes heavy) load of external and internal users querying the database, the searches actually take place on a copy of the operational database (which is updated nightly) on a different SPARC II. After an astronomer has identified a group of datasets in which he (or she) is interested, he marks these datasets and submits a request to retrieve the data. The DMF then checks to verify that the datasets are public and retrieves the datasets to a directory on the host machines (or alternatively to 8 mm or 1/2 inch tape if the data volume is large). Once DMF has completed the task of retrieving the data, DMF sends an e-mail message to the astronomer who transfers the data to his home computer via "ftp". The median, mean and 90% retrieval times

---

[1]The ST-ECF and the CADC also operate versions of DMF. Their systems have evolved separately, have somewhat different capabilities, and now are based solely on Sun SPARCstations.

are currently 1, 6, and 13 hours. (This skewed distribution reflects the fact that some retrievals primarily involve the 90 disks in the jukebox, while others require repeated manual intervention to load optical disks into the standalone drives.)

Users who have difficulty in using the archive are encouraged to contact the archive hotseat (archive @stsci.edu). There is a guest account ("guest", password "archive") on both the Unix host (stdatu.stsci.edu) and the VMS host (stdata.stsci.edu). Retrievals cannot be made from the guest account, but one can explore the catalog. Any astronomer can get an account which will enable him to retrieve data by filling out an account request form (by typing "register" from the guest account). Additional information about policies and strategies for using the archive is contained in an "Archive Users Manual" (Baum 1993). At the present time there are 350 registered users outside ST ScI. Users who would like to come to the Institute to use the archive intensively for a short period of time can arrange visits by contacting the archive hotseat.

The HST archive was officially opened for external use in February, 1993. Since then a total of 20 Gbytes of data has been retrieved from the archive using the archive host machines. These retrievals are in addition to the 100 Gbytes of data which have been retrieved inside the Institute — primarily for engineering evaluations and for supporting visitors who come to the Institute to work with their data.

3. A New Archive System

Although DMF has proven to be quite successful as an interim archive system, ST ScI is now preparing for a transition to the permanent archive system — the ST Data Archive and Distribution Service. ST-DADS, which was developed by Loral AeroSys for NASA, was delivered to the Institute in September and a combined ST ScI/Loral team is currently completing the software needed to make it operational in 1994. The functionality that ST-DADS will provide is similar to that of DMF, but ST-DADS performance is expected to be 3 to 10 $\times$ that of DMF. With 6 Gbyte Sony optical disks in 4 optical disk jukeboxes, 2 Tbytes, or $\sim$5 years, of HST data can be maintained on line in ST-DADS. In addition, ST-DADS converts all of the data produced by the HST pipeline to FITS image or table formats as the data are archived.

ST-DADS consists of a cluster of 5 6000-series VAX processors connected through a high speed vector-scalar (APTEC) processor to a disk farm, the optical disk jukeboxes and 4 stand-alone OD drives. Tests of ST-DADS ability to archive data correctly were successfully completed in late September. We are currently planning to begin archiving data to ST-DADS from the HST data pipeline in December, while the software team turns its attention to integrating and completing the software needed to replace DMF.

In conjunction with ST-DADS, ST ScI has been developing a new user interface — StarView — for the HST archive. StarView supports two basic user models: (1) a casual user mode consisting of predefined forms for common queries with results displayed one record at a time on a results form or collectively in a table format, and (2) an expert user mode allowing the user to select and qualify any of the 1700 fields in the database (without needing to understand a structured query language). So that we can support those without good network connections, StarView will support both VT-100 compatible CRTs and X-windows (though not necessarily with the same capabilities). A more detailed discussion of the design and implementation of StarView is presented elsewhere (Pollizzi 1994; Williams 1994; Silverberg 1994). The CRT version of StarView is now ready for release; the X-windowed version will be released in the spring.

The transition to ST-DADS will take place in 1994. Data will continue to be archived to DMF through about March of 1994. Users will continue to be able to access DMF until all major capabilities of DMF have been duplicated in ST-DADS and the data stored in DMF has been copied to ST-DADS. Partly to ease the transition from DMF to ST-DADS, StarView has been adapted recently so that it can be used with the interim archive, DMF. The CRT version will be placed on the DMF host machines for general use by the community in November (as an alternative to STARCAT).

4. Summary

HST now has an active archive research program. The interim archive DMF is functioning and is capable of supporting significant numbers of archive researchers for the near term. The next challenge is to make the transition from an interim archive system to a long term archive system while at the same time trying to respond to the immediate needs of HST users.

Acknowledgments. A large number of engineers and astronomers at ST ScI, the ST-ECF, the CADC, and GSFC have made fundamental contributions to the archive and its current success. The Space Telescope Science Institute and the archive effort are funded by NASA under contract NAS5-26555.

References

Baum, S. (ed.), 1993, "HST Archive Manual (Ver 3.0)", (ST ScI)

Pollizzi, J. 1994, this volume

Silverberg, D. 1994, this volume

Williams, J. 1994, this volume

Astronomical Data Analysis Software and Systems III
ASP Conference Series, Vol. 61, 1994
D. R. Crabtree, R. J. Hanisch, and J. Barnes, eds.

The NRAO VLA Sky Survey

J. J. Condon, W. D. Cotton, E. W. Greisen, and Q. F. Yin

National Radio Astronomy Observatory, 520 Edgemont Road, Charlottesville, VA 22903

R. A. Perley

National Radio Astronomy Observatory, P. O. Box 0, Socorro, NM 87801

J. J. Broderick

Physics Department, Virginia Polytechnic Institute and State University, Blacksburg, VA 24061

Abstract. We are using the compact D and DnC configurations of the VLA to map the sky north of $\delta = -40°$ with $\theta = 45''$ resolution and $S_P \sim 2\,\text{mJy beam}^{-1}$ detection limit at $\nu = 1.4\,\text{GHz}$. New software was developed to help us schedule and monitor $\approx 2 \times 10^5$ snapshot observations, make the snapshot maps automatically, correct and mosaic them into $4° \times 4°$ sky images, and extract a source catalog. All data products will be released to the astronomical community as rapidly as possible.

1. Introduction

The NRAO VLA Sky Survey (NVSS) will map the entire sky north of $\delta = -40°$ at $\nu = 1.4\,\text{GHz}$. A grid of over 2×10^5 partially overlapping snapshot maps will be mosaiced to yield sets of 2326 $4° \times 4°$ sky images in each of the Stokes parameters I, Q, and U with $\theta = 45''$ FHWM resolution and a nearly uniform detection limit $S_P \sim 2\,\text{mJy beam}^{-1} \sim 0.6\,\text{K}$. The final images should contain about 2×10^6 extragalactic sources, including luminous radio galaxies and quasars, most of the galaxies found by *IRAS* at $\lambda = 60\,\mu\text{m}$, ultraluminous starburst galaxies and protogalaxies even at cosmological distances, as well as statistically useful numbers $N \gg \sqrt{N}$ of nearby normal galaxies and low-luminosity active galactic nuclei (AGN). Their rms position uncertainties will range from $< 1''$ for $S > 10\,\text{mJy}$ to $\approx 5''$ at $S = 2\,\text{mJy}$. Since 1.4 GHz falls in the "intermediate frequency gap" between low-frequency variability caused by refractive interstellar scintillations and high-frequency intrinsic source variability, these images will remain accurate pictures of the sky for decades to come.

The NVSS is being made as a service to the astronomical community. We claim no proprietary rights to either the raw data or the finished products because we believe that the full scientific potential of such a large survey will not be realized until all astronomers can use it. The principal data products will the $4° \times 4°$ mosaiced images in FITS format plus ASCII tables of discrete source

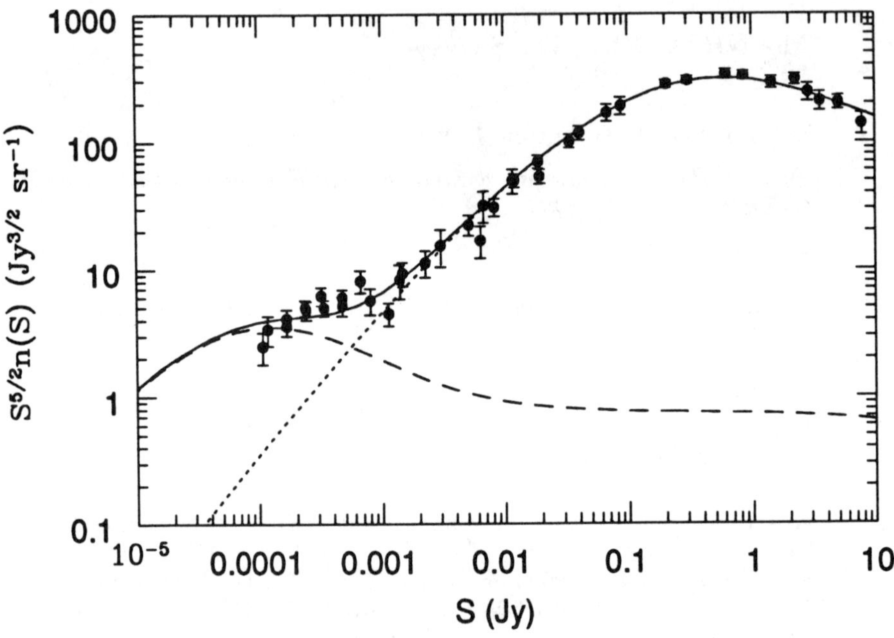

Figure 1. Weighted source count at 1.4 GHz (data points) and models indicating the contributions of evolving "monsters" in true AGN (dotted curve) and "starbursts" (dashed curve).

parameters obtained from the total-intensity maps. They will be released via ftp (ftp 192.33.115.53, login anonymous, password = your name, cd vlass) *as soon as they are made.* To guarantee equal access for all users, we will use only those images that have been placed in this open directory for our own research.

2. Scientific Goals

Both deep ($S < 1\,\mathrm{mJy}$) and large-scale ($\Omega > 1\,\mathrm{sr}$) radio surveys already exist at 1.4 GHz, so we can predict the principal characteristics of a deep large-scale radio survey. Nearly all sources more than a few degrees from the galactic plane are extragalactic. The weighted differential counts of extragalactic sources found at $\nu = 1.4\,\mathrm{GHz}$ are shown in Figure 1.

There are two astrophysically distinct populations of extragalactic radio sources: (1) Over 99% of the sources strong enough ($S \gg 1\,\mathrm{mJy}$) to be detected in existing large-scale surveys are the classical radio galaxies and quasars powered by "monsters" (e.g., supermassive black holes) in AGN. (2) The remaining radio sources are identified with star-forming galaxies containing H II regions ionized by massive ($M \geq 8 M_{\odot}$) short-lived ($\tau \leq 3 \times 10^{7}\,\mathrm{yr}$) stars and relativistic electrons accelerated by their supernova remnants (Condon 1992). These star-forming regions are not usually considered to be "true" AGN, but the most luminous ($L > 2 \times 10^{11} L_{\odot}$ if $H_0 = 75\,\mathrm{km\,s^{-1}\,Mpc^{-1}}$) nuclear "star-

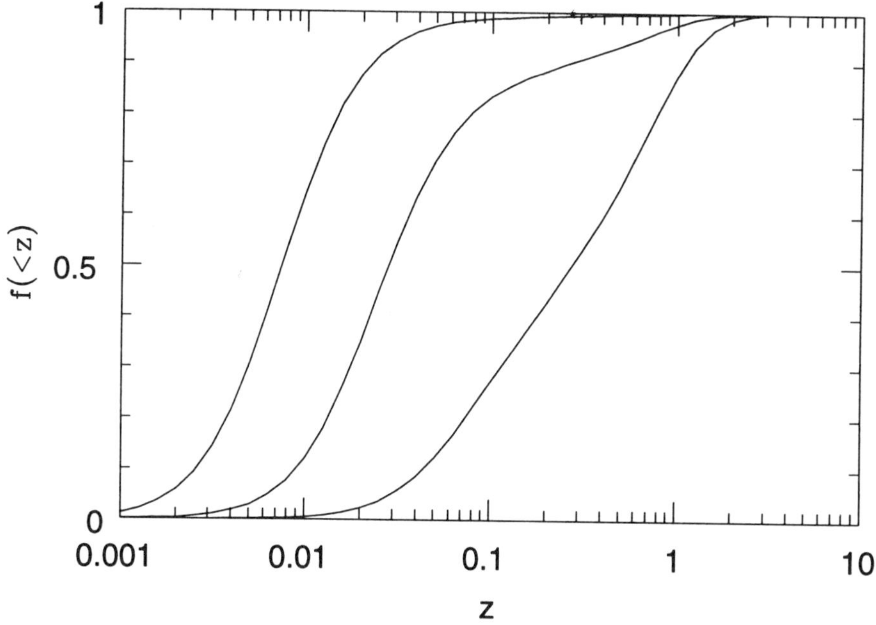

Figure 2. The cumulative fractions $f(< z)$ of radio sources with redshifts $< z$ that are powered by stars change dramatically as the 1.4 GHz flux density falls from 100 mJy (left curve) through 10 mJy (center curve) to 1 mJy (right curve).

bursts" are quite compact ($r \sim 100\,\mathrm{pc}$) and similar in many other observables to AGN powered by monsters. Most radio sources powered by stars are in fairly normal spiral galaxies. Their median face-on disk brightness temperature is only $\langle T \rangle \approx 1\,\mathrm{K}$ at $\nu = 1.4\,\mathrm{GHz}$, so they can be detected only with relatively low-resolution surveys, regardless of their distance. For example, $T = 1\,\mathrm{K}$ corresponds to $S_{\mathrm{P}} \approx 3\,\mathrm{mJy\,beam}^{-1}$ in the $\theta = 45''$ FHWM Gaussian beam of the VLA D configuration but only $S_{\mathrm{P}} \approx 0.04\,\mathrm{mJy\,beam}^{-1}$ in the $\theta = 5''$ beam of the B configuration. The NVSS is being made with the lowest possible resolution to ensure completeness and photometric accuracy.

Radio sources powered by stars obey the remarkably tight far-infrared (FIR) to radio luminosity correlation (Condon, Anderson, & Helou 1991 and references therein). Consequently, flux-limited samples selected in the radio and FIR bands are nearly identical. The median FIR/radio luminosity ratio for a $\lambda = 60\,\mu\mathrm{m}$ sample is $\langle \log(S_{60\,\mu\mathrm{m}}/S_{1.4\,\mathrm{GHz}}) \rangle \approx 2.15$, so the NVSS should detect most of the galaxies above the $S_{60\,\mu\mathrm{m}} = 0.28\,\mathrm{Jy}$ completeness limit of the *IRAS* Faint Source Catalog, Version 2 (Moshir et al. 1989). The cumulative fractions $f(< z)$ of star-forming galaxies with redshifts $< z$ are plotted at three different flux-density levels in Figure 2. The stronger starburst sources are local ($z \ll 1$), but deep radio surveys can detect them even at cosmological distances. Cosmological evolution is negligible above 100 mJy, noticeable at 10 mJy, and dominates the redshift distribution near 1 mJy. The combination of the *IRAS* FSC2 and the NVSS should reveal a number of exotic objects similar to the ultraluminous candidate protogalaxy FSC10214+4724 at $z \approx 2.3$ (Rowan-Robinson et al. 1991) because the NVSS can provide the accurate positions needed to make unambiguous optical identifications of distant, luminous starbursts with optically faint galaxies.

Even the fairly sensitive 4.85 GHz Green Bank survey detected only $\sim$ 30 "starburst" sources sr^{-1} among the $10^4\,\mathrm{sr}^{-1}$ sources stronger than 25 mJy (equivalent to about 60 mJy at 1.4 GHz). At this level, only 1 in 300 extragalactic radio sources is powered by stars—such sources are about as rare as gravitational lenses! There are just enough sources for the simplest statistical tests. For example, Cox et al. (1988) discovered that the FIR/radio correlation is nonlinear using a sample of spiral galaxies brighter than $m \sim 14.5$ detected in the 6C 151 MHz surveys. The NVSS will make a huge difference, increasing the number of sources in the star-forming population by a factor of 300 to $N \approx 10^4\,\mathrm{sr}^{-1}$.

Classical radio sources powered by monsters are typically two orders of magnitude rarer (per unit volume) and more luminous than those in star-forming galaxies. Their median redshift is $\langle z \rangle \sim 0.8$ with surprisingly little dependence on flux density below $S \sim 1$ Jy (Condon 1989), and very few are nearby ($z \ll 1$). Making optical identifications, spectroscopic observations, and other follow-up studies on complete samples of faint radio AGN very difficult even in tiny ($\Omega \ll$ 1 sr) areas (cf. Windhorst, Mathis, & Neuschaefer 1990). The B-configuration FIRST survey (Becker, this volume) was designed to provide the $< 1''$ positions needed to identify weak radio sources with distant AGN.

The scarcity of nearby radio AGN means that there is very little overlap between flux-limited samples of AGN selected in radio and other (e.g., FIR or X-ray) bands. For example, only 1 in 200 AGN stronger than 25 mJy at 4.85 GHz is close enough ($D \sim 100$ Mpc) to appear in the UGC Catalog (Nilson 1973) of galaxies with optical angular diameters $\geq 1'$ (Condon, Frayer, & Broderick 1991). The rare nearby radio galaxies are scientifically valuable because they can be studied in detail—their optical morphologies can be determined, their optical spectra and redshifts are easy to obtain, a given angular resolution yields better linear resolution, etc. The NVSS should detect thousands of nearby AGN for comparison with those detected in other wavebands.

3. Continuous Sky Images from Discrete Snapshots

A large sky image with nearly uniform sensitivity can be made from partially overlapping snapshot maps, each covering one primary beam area. Since the VLA primary beam is circular, the optimum grid of pointing centers is hexagonal. The most uniform sky coverage occurs if the angular separation Δ between pointing centers is much smaller than the FWHM $\theta_{\mathrm{P}} \approx 30'$ of the nearly Gaussian primary beam, but a larger Δ minimizes the observing time lost moving between snapshots and the number of fields that must be mapped. The best compromise minimizes the number of maps without significantly increasing the survey completeness limit.

The sky brightness B at any point in the final sky map is the weighted sum of the sky brightnesses b in the m contributing snapshot maps:

$$B = \sum_{i=1}^{m} W_i b_i \Big/ \sum_{i=1}^{m} W_i \ . \tag{1}$$

The correct statistical weight W_i is proportional to the square of the primary power gain. Thus we *multiply* each snapshot map (*not* corrected for primary beam attenuation) by the primary power pattern $P(\rho)$ before adding it to a large sky map. The square of the primary pattern is added to the large "weight map," and the final sky map is the ratio of these two large maps, as illustrated by Figure 3. Uniform sensitivity requires that the sky be smoothly covered by the sum of these "weight beams" $W(\rho) = P^2(\rho)$, not just by the larger primary beams. For a Gaussian primary beam the weight beam is also a Gaussian but its FHWM is only $\theta_{\rm P}/\sqrt{2}$, so Δ cannot be much larger than $\theta_{\rm P}/\sqrt{2} \approx 21'$.

Ideally each snapshot map would contain all of the information in the primary beam. However, in the skirts of the primary beam the gain is poorly known, frequency dependent, and sensitive to antenna pointing errors. In practice, therefore, the snapshot maps should be truncated at some finite radius $\rho_{\rm m}$ from the pointing center before they can be weighted and summed to produce the final map. A quantitative measure of the information retained is the mapping efficiency $\eta_{\rm m}$ defined by

$$\eta_{\rm m} \equiv \int_0^{\rho_{\rm m}} W(\rho)\rho d\rho \Big/ \int_0^{\infty} W(\rho)\rho d\rho \ . \tag{2}$$

For a Gaussian primary beam, $\eta_{\rm m} = 1 - \exp[-8\ln 2(\rho_{\rm m}/\theta_{\rm P})^2]$. High mapping efficiency requires $\rho_{\rm m} > 0.7\theta_{\rm P}$ and high accuracy requires $\rho_{\rm m} < 0.85\theta_{\rm P}$.

For the NVSS we chose $\Delta = 26'$ and $\rho_{\rm m} = 24' = 0.8\theta_{\rm P}$. "Only" about 2×10^5 snapshot maps are needed to cover the sky north of $\delta = -40°$, and the mapping efficiency is $\eta_{\rm m} \approx 0.98$. If σ_0 is the rms map noise at the snapshot pointing centers, the rms noise over the whole sky is $\sigma \approx 1.09\sigma_0$ and the map completeness is limited by the worst-case noise $\sigma_c \approx 1.15\sigma_0$. Despite their pointillist origin, our final sky maps should be almost as good as those from a perfect survey ($\Delta \ll \theta_{\rm P}$, $\eta_{\rm m} = 1$), in which the final statistical weight for the same total dwell time would be

$$W = \left(\frac{\pi\theta_{\rm P}^2}{8\ln 2}\right) \Big/ \left(\frac{\sqrt{3}\Delta^2}{2}\right) \approx 0.872 \ , \tag{3}$$

yielding $\sigma = \sigma_0/\sqrt{W} \approx 1.07\sigma_0$ everywhere.

The ideal hexagonal pointing grid was distorted so that successive pointings lie along lines of constant right ascension or declination in order to minimize the on-line computing overhead between pointings. Within wide declination bands the declination spacing between successive grid rows grows as $\cos\delta$ decreases; the exact spacing is determined by the constraint that $\sigma_{\rm c}$ remain constant. Since we must schedule about 10^3 snapshots per night of observing, we wrote special FORTRAN programs to create VLA observing files automatically and a Supermongo plotting macro to display the observational status of the $\approx 2 \times 10^5$ survey

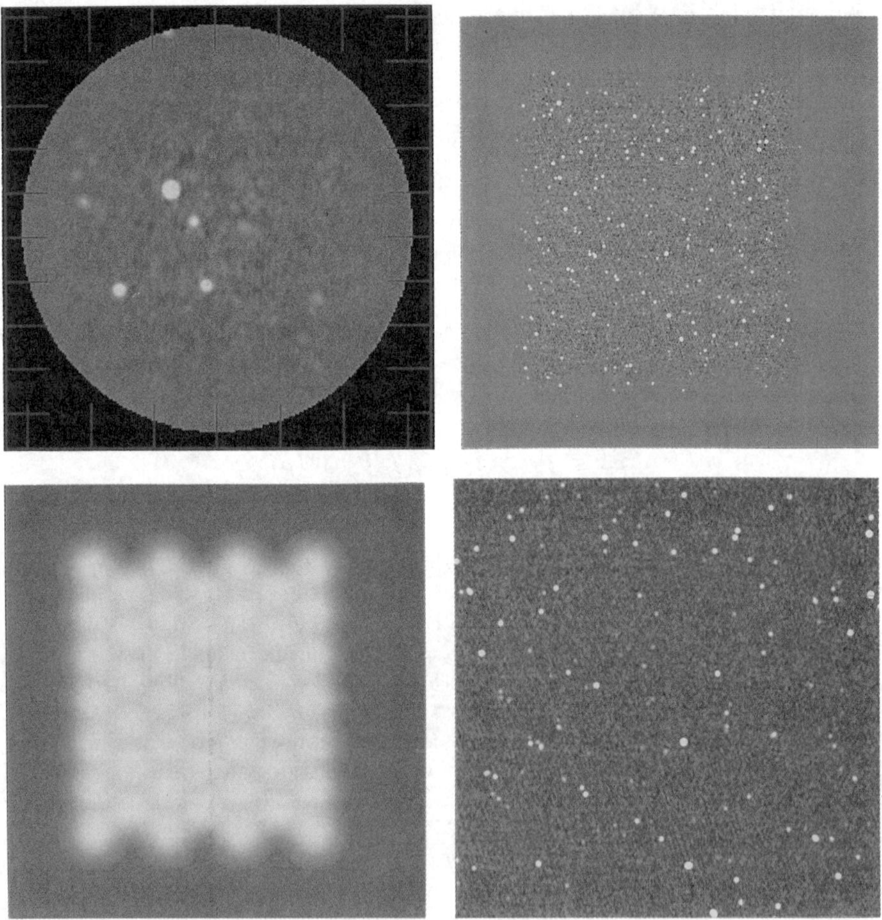

Figure 3. Snapshot maps multiplied by the primary beam (example in top left) are summed (top right), divided by the sum of the squared beams (bottom left) to yield the final sky map (bottom right) with nearly uniform sensitivity.

fields graphically. Not all scheduled observations are successful, primarily due to radio frequency interference. Even though we don't have time to map them, they must be identified for reobservation before the end of each VLA configuration session. We check the calibrated (u, v) data from each observing run with a new AIPS task RFI that prints out those snapshots containing IF/polarization correlator groups in which the rms noise in a certain percentage of correlators exceeds a preset cutoff (e.g., 25% of the correlators have rms noise exceeding the quadratic sum of 125 mJy and 10% of the average flux).

4. Mapping Strategy

The number of snapshot maps is so large that special procedures and AIPS tasks are needed to automate their production, enforce uniform quality standards, and prevent human blunders. Our low resolution, wideband, widefield, snapshot images bend or violate many of the rules used to make aperture-synthesis maps. We make a number of first-order corrections to minimize the resulting errors.

4.1. Automatic Mapping

Most AIPS tasks (e.g., MX, UVSUB, CALIB) were designed for the interactive user who makes maps one at a time. We cannot afford to do this 2×10^5 times; we need an efficient and automatic mapping program. We began its development by writing mapping procedures in the AIPS command language POPS. For example, one set of procedures can map and CLEAN the calibrated (u, v) data, subtract the clean components from the (u, v) data, flag subtracted visibilities with amplitudes much larger than the expected rms noise, add back the clean components to the surviving (u, v) data, remap, self-calibrate and remap if a sufficiently strong source is present, and make some corrections for bandwidth smearing, map distortions, etc. Such POPS procedures are easy to program, but they are computationally inefficient and can call only existing AIPS tasks and verbs. Consequently, we wrote one specialized AIPS task (VLAD) to replace and augment the POPS snapshot mapping procedures. POPS procedures are still used to display and combine the snapshot maps produced by VLAD.

Making the 2×10^5 snapshot maps with VLAD is an "embarrassingly parallel" problem easily distributed among our human workers and their workstations. The computations can be done in batch mode, and 1 Gbyte of disk space is sufficient to store the results from more than 24 hours of computation. A 4 mm DAT tape drive on each workstation is necessary to read the calibrated (u, v) data and store the snapshot maps locally, lest the local area network be overloaded. Just reading and writing these tapes can be quite time consuming.

4.2. Snapshot Deconvolution

The VLA dirty beam for a single snapshot is very dirty indeed. By far the worst sidelobes are three diffraction spikes perpendicular to the VLA arms. They are strongest on grating lobes at angular distances $\approx \pm 0.3, \pm 0.6, \ldots$ deg from the main lobe, corresponding to the $D \approx 38$ m shortest baselines. These lobes cause the notorious hexagon of source "ghosts" often seen in VLA maps of extended sources because their gain solid angle is comparable with that of the main beam. Any source whose projected length along a diffraction line is several synthesized beamwidths may appear brighter in the dirty map at the grating lobe position than at the correct position. CLEAN will therefore put clean components onto the grating lobe positions and subtract flux from the true source position. The usual cure is to clean only the small region surrounding the source.

Unfortunately, this cannot be done in a snapshot survey because the dirty map fluctuations are dominated by sidelobes of sources scattered throughout the primary beam. We have found that even point sources can be weakened by CLEAN, an effect we call "clean bias." Clean bias can be reduced but not completely eliminated. We must map and clean a square 512 pixels $\times$ 12 arcsec/pixel $\sim 2°$ on a side to eliminate the sidelobes of sources even in the first sidelobe of the primary beam, but this clean need not be very deep. We reject weak, isolated clean components as likely due to sidelobes of the dirty beam. We restrict the deeper last clean to the circle of radius $\rho_{\rm m} = 24'$ that is actually used to make the final maps. We minimize the diffraction spikes in the dirty beam by using superuniform, not natural, weighting of the (u, v) data; this downgrades the contribution of short intra-arm baselines. We recommend that designers of future synthesis arrays avoid straight arms!

Two other innovations in the mapping task VLAD improve the dynamic range of NVSS snapshot maps: (1) CLEAN is less accurate if the source being cleaned does not lie precisely at the center of a map pixel. VLAD automatically locates the strongest source $>$ 500 mJy in each field and shifts the map in position by up to 0.5 pixel to put that source at a pixel center. (2) The NVSS observes at both 1.365 and 1.435 GHz simultaneously. The primary beamwidth is inversely proportional to frequency, so sources far from the pointing center appear stronger at the lower frequency by a calculable amount. Also, most sources selected at 1.4 GHz have spectral indices $\alpha \equiv -d \ln S / d \ln \nu \approx 0.7$. VLAD rescales each clean component accordingly and subtracts different amounts from the (u, v) data in each frequency channel.

4.3. Widefield Polarization Correction

We must measure and correct for the instrumental polarization, not only at the pointing center, but over the whole primary beam. The off-axis instrumental polarization is dominated by radiation scattered off the four feed-support legs, so it is stable but rotates with parallactic angle. For a snapshot observation at a single parallactic angle, the instrumental polarization can therefore be removed in the image plane. If `A` is the complex instrumental polarization map rotated to match the snapshot parallactic angle and `I` is the total-intensity image, then the observed Stokes (Q+iU) image `O` can be calibrated to yield the corrected (Q+iU) image `P` by the operation `P = O - I`×`A`. Our preliminary results indicate that the uncorrected instrumental polarization in `P` is small enough that fractional linear polarizations as small as $\sim 1\%$ will be detectable in the final maps.

4.4. Bandwidth Smearing Correction

The NVSS observations are made with a channel bandwidth $\Delta\nu \sim 46$ MHz, so bandwidth smearing (chromatic aberration) blurs the snapshot maps. Bandwidth smearing of a point source can be represented by a radial convolution of the unsmeared restoring beam with a bandwidth-smearing "beam" whose width is proportional to the angular distance ρ from the phase reference position. Variances add under convolution, so the variance of the smeared beam in the radial direction is the sum of $\theta^2/(8\ln 2)$, the variance of the unsmeared Gaussian restoring beam with FWHM θ, and $(\rho\Delta\nu/\nu)^2/12$, the variance of the bandwidth-smearing beam for a rectangular passband. So long as the bandwidth smearing is not severe, the point-source response remains Gaussian, but its radial width is divided by R and its peak height is multiplied by R, where

$$R \approx \left[1 + \frac{2\ln 2}{3}\left(\frac{\rho\Delta\nu}{\nu\theta}\right)^2\right]^{-1/2} . \tag{4}$$

Our method for correcting small amounts of bandwidth smearing is best described in terms of the individual fringe visibilities that are transformed to make the map. The visibility $V(\tau)$ of a unit point source observed on a single baseline with delay $\tau = (ul + vm)/c$ is

$$V(\tau) = \left[\frac{\sin(\pi\tau\Delta\nu)}{\pi\tau\Delta\nu}\right] e^{i2\pi\tau\nu} . \tag{5}$$

Dividing the observed fringe visibility by the quantity $r(\tau) \equiv \operatorname{sinc}(\tau\Delta\nu)$ in square brackets corrects for bandwidth smearing. It also divides the visibility noise by r^{-1}, which is unavoidable. There are many visibilities and many sources contributing to each snapshot map. We correct them all by (1) subtracting the clean components from the (u, v) data used to make a normal smeared map, (2) correcting each visibility for the loss of amplitude [up to a maximum of $(r^{-1} - 1) = 25\%$] estimated by treating each clean component as a point source at the clean component position, (3) adding back the corrected (u, v) data, and (4) remapping. This restricted correction reduces the bandwidth smearing in the final maps to $< 1\%$ ($R > 0.99$) and increases the rms noise by $\leq 2\%$.

4.5. Position Corrections

Mapping a finite portion of the spherical sky with the VLA formally requires a computationally expensive three-dimensional Fourier transform (cf. Perley 1988). The VLA baselines are nearly coplanar, so a two-dimensional transform can map a single snapshot, but the resulting map direction cosines (l', m') differ slightly from the correct values (l, m):

$$l' = l + (\sqrt{1 - l^2 - m^2} - 1)\tan Z \sin\chi \tag{6}$$

$$m' = m - (\sqrt{1 - l^2 - m^2} - 1)\tan Z \cos\chi , \tag{7}$$

where Z is the zenith angle and χ is the parallactic angle of the phase reference position. To first order, $(\sqrt{1 - l^2 - m^2} - 1) \approx -\rho^2/2$, so the map position of a source is pulled toward the zenith by an amount proportional to the square of its angular distance ρ from the phase reference position. This distortion can be as large as several arcsec in the NVSS maps. We correct it by making a compensating geometrical distortion of the finished snapshot map.

The VLA continuum passband centroid frequency ν_c is slightly lower than the nominal central frequency ν_0 used to calculate the (u, v) coordinates. This multiplies the map angular distance ρ from the phase center by a factor $f_1 =$

ν_c/ν_0. Offset maps of a strong point source show that $f_1 = 0.9960 \pm 0.0002$ at 1.4 GHz, so the radial position errors in uncorrected maps are several arcsec near $\rho = \rho_m$.

The effective centroid frequency is also reduced by the frequency dependence of the primary beam pattern P, whose width scales as ν^{-1}, so $P(\rho, \nu) = P(\rho\nu)$. Since the received signal strength is multiplied by P, the effective centroid frequency is multiplied by a second factor f_2

$$f_2 = \int_{\nu_0-\Delta\nu/2}^{\nu_0+\Delta\nu/2} \nu P(\rho\nu) d\nu \Big/ \int_{\nu_0-\Delta\nu/2}^{\nu_0+\Delta\nu/2} \nu_0 P(\rho\nu) d\nu \tag{8}$$

even if the IF passband is a rectangle of width $\Delta\nu$ centered on frequency ν_0. For $\Delta\nu/\nu_0 \ll 1$ and a Gaussian primary beam with FWHM θ_P, this nonlinear map scaling factor is

$$f_2 \approx 1 - \frac{2\ln 2}{3}\left(\frac{\rho\Delta\nu}{\theta_P \nu_0}\right)^2 . \tag{9}$$

Finally, multiplying each snapshot map by the primary beam $P(\rho)$ skews the point-source response (a Gaussian with FWHM $\theta = 45''$) inward since $dP/d\rho \leq 0$. The scaling factor for Gaussian primary and synthesized beams is

$$f_3 \approx 1 - \theta^2/(2\theta_P^2) . \tag{10}$$

The map scaling errors measured from offset observations of strong point sources agree with the calculated errors. We geometrically rescale each snapshot map by $(f_1 f_2 f_3)^{-1}$ to correct for all three effects. The position errors of strong sources on the final maps are well under under $1''$.

Acknowledgments. The National Radio Astronomy Observatory is operated by Associated Universities, Inc. under cooperative agreement with the National Science Foundation.

References

Condon, J. J. 1989, ApJ, 338, 13

Condon, J. J. 1992, ARA&A, 30, 575

Condon, J. J., Anderson, M. A., & Helou, G. 1991, ApJ, 376, 95

Condon, J. J., Frayer, D. T., & Broderick, J. J. 1991, AJ, 101, 362

Cox, M. J., Eales, S. A. E., Alexander, P., & Fitt, A. J. 1988, MNRAS, 235, 1227

Moshir, M. et al., 1989, "Explanatory Supplement to the *IRAS* Faint Source Survey" (Jet Propulsion Laboratory, Pasadena)

Nilson, P. 1973, Uppsala General Catalog of Galaxies (Uppsala Astronomical Observatory, Uppsala)

Perley, R. A. 1988, in Synthesis Imaging in Radio Astronomy, edited by R. A. Perley, F. R. Schwab, & A. H. Bridle (ASP Conference Series), 6, 259

Rowan-Robinson, M. et al., 1991, Nature, 351, 719

Windhorst, R. A., Mathis, D. F., & Neuschaefer, L. W. 1990, in Evolution of the Universe of Galaxies, ed. R. Kron (ASP: San Franscisco), p. 389

Astronomical Data Analysis Software and Systems III
ASP Conference Series, Vol. 61, 1994
D. R. Crabtree, R. J. Hanisch, and J. Barnes, eds.

The VLA's FIRST Survey

Robert H. Becker

Department of Physics, University of California, Davis, CA 95616, and Lawrence Livermore National Laboratory

Richard L. White

Space Telescope Science Institute, 3700 San Martin Drive, Baltimore, MD 21218

David J. Helfand

Astronomy Department, Columbia University, 538 West 120th Street, New York, NY 10027

Abstract. Over the next decade the Very Large Array (VLA) will carry out a systematic survey of the northern sky at 20 cm wavelength in two configurations. We have selected as a moniker the acronym FIRST, which in uncompressed form reads Faint Images of the Radio Sky at Twenty-cm. The high-resolution survey will be done in B configuration and hence will achieve an angular resolution of 5 arcsec. It will cover a 10^4 square degree region centered on the north Galactic pole. Each field will be observed for $\sim$3 minutes with a resulting rms noise level of $\sim$0.2 mJy. The survey will result in a catalog of 10^6 discrete sources as well as 65,000 images each composed of 4×10^6 pixels. To expedite the massive data analysis task this project entails, we have decided to utilize computers.

It is our intention to release to the community compressed, machine-readable copies of all the images as well as an annotated catalog of sources. The challenge will be to accomplish this with a minimum of resources. We hope to automate maximally the data analysis pipeline to achieve this end. This paper describes the history of the project, its current status, and our data analysis methods, followed by a brief resumé of FIRST's expected scientific impact.

1. Introduction

New astronomical surveys almost invariably herald advances in astrophysics. Surveys play a crucial role in identifying new classes of objects, in revealing structure in the Universe, and in expediting studies of individual objects. Each new generation of instrumentation usually warrants a new survey. Hence the ROSAT Sky Survey replaces the HEAO-I Sky Survey, the 2MASS survey (Kleinmann 1992) will replace the AFGRL survey, etc. Since radio telescopes tend to be unique, there are few matched surveys in the northern and southern hemispheres. The only reasonably uniform sky survey in the radio band is the 5 GHz survey carried out at Greenbank and Parkes; it is the best existing radio survey, with a limiting flux density of $\sim$30 mJy and positional accuracy of $\sim$1 arcmin. Both the relatively high flux density limit and the relatively large positional uncertainty limit the usefulness of this 5 GHz database. With this in mind, the National Radio Astronomy Observatory (NRAO) has decided to utilize the VLA to produce a next generation radio sky survey.

2. History of FIRST

There is anecdotal evidence that astronomers have been talking about a VLA Sky Survey since the VLA's inception. In the spring of 1990, having just completed the analysis of a VLA Galactic Plane survey, Rick White, David Helfand, and Bob Becker in a mood of elation were contemplating their next move and, as may have happened many times before to other astronomers, began fantasizing about a VLA Sky Survey. But this time, instead of just going out for a drink, we convinced ourselves that such a project was truly feasible. Three months later, we submitted what was, to the best of our knowledge, the first serious proposal to survey the sky at 20-cm wavelength with the VLA. The proposal called for observing $\sim$125,000 fields for 90 s each in a hybrid BC configuration with the objective of surveying all the sky north of $-30°$ declination. The proposal generated a great deal of discussion among the staff at NRAO, but ultimately was rejected without formal review. But the stage had been set.

Nine months later, NRAO received two all-sky proposals; a revised version of our original proposal, this time in C configuration, and a D-configuration proposal from Jim Condon. In fact, this was a healthy development. In writing a C configuration proposal, we were well aware of the scientific compromises involved. The primary difference among VLA configurations is angular resolution: the FWHM ranges from 54 arcsec in D-configuration to 2 arcsec in A-configuration. A secondary consideration is available bandwidth (high-resolution images must be made in spectral line mode). High resolution images are desirable because they facilitate the identification of optical counterparts. Low resolution images are desirable because they are more sensitive to extended, low surface brightness sources. The submission of both high and low resolution proposals insured a serious discussion of the best way to carry out a VLA survey.

NRAO sent both proposals out for review and there appeared to be a consensus that the time was right for a VLA Sky survey, but a great deal of uncertainty as to which survey to do.

To facilitate an objective, optimal decision, trial observations were scheduled in the spring and summer of 1992 to compare the merits of survey results from C and D-configuration observations. Two scientific panels were convened in October 1992 and March 1993 to review the situation and make recommendations to NRAO. The outcome of these two meetings was a surprise to everyone involved. Rather than sacrifice the advantages derived from either a high or a low resolution survey, it was decided that two surveys would be made, one in D configuration and one in B configuration. The D-configuration survey was chartered to cover 85% of the sky to a 2 mJy flux density limit. The B-configuration survey was chartered to cover the 25% of the sky centered on the North Galactic Pole (NGP; chosen to coincide with the Sloan Digital Sky Survey

[Gunn & Knapp 1993]) to a 1 mJy flux density limit. Observations for both surveys would start in 1993. The rest of this paper will be restricted to discussion of the high-resolution, B-configuration survey.

3. Why a High Resolution Survey?

The VLA is a versatile instrument. It is capable of observing at six wavelengths in four different configurations. It is not sufficient to decide to do a survey, one must also decide which survey to do. The choice of a frequency was an easy one to make. The field of view of the VLA scales linearly with wavelength (inversely with frequency.) A survey at 6 cm would require $\sim$10 times as many observations as a survey at 20 cm covering the same area of sky. The same logic would suggest that the VLA's longest wavelength, 90 cm, would be even better than 20 cm. However, the surface density of bright sources is much higher at 90 cm so that a single snapshot of data is insufficient to achieve the high dynamic range required for reaching the theoretical sensitivity of the VLA. Furthermore, the 90 cm receivers are more than an order of magnitude less sensitive than the 20 cm receivers.

The relatively high resolution of the B configuration ($\sim$5 arcsec) can also be justified despite the extra difficulty that high-resolution imaging entails. There are two primary drawbacks to going to higher resolution. A B-configuration image has $\sim 10^2$ times as many resolution elements per unit area as a D-configuration image, hence 10^2 times as much mass storage is needed to archive images and the computational time to produce each image increases by a factor of $\sim$20. In addition, VLA images will degrade if the bandwidth chosen is too wide (bandwidth smearing). The farther apart the antennas, the smaller is the acceptable bandwidth. In B configuration, this constraint forces the observations to be made in spectral line mode, resulting in the loss of 50% of the available bandpass compared to continuum mode.

However, the scientific advantages are worth the operational disadvantages. In particular, the high resolution results in better positional accuracy for all sources found in the survey and, for a significant fraction of sources, results in morphological classification based on a source's brightness distribution. Much of the power of the radio survey will come from the association of individual radio sources with optical counterparts. But unambiguous association demands a radio error box small enough to exclude random coincidences, where "small enough" depends on both the density of potential optical counterparts and the fraction of optical sources that have radio emission. Approximately 50% of the radio sources in the FIRST survey will have optical counterparts brighter than 23rd magnitude. At this level, sub-arcsecond positions are necessary for reliable identifications.

The morphology of the brightness distribution of a radio source is also important for making a correct optical identification. Many sources are complex; for example, radio triples are common. In such cases, the optical counterpart will be located at the position of a weak central component, far from the brighter radio lobes. Without high resolution, misidentification is likely. The morphology is also important for classifying the nature of the radio emission (e.g., distinguishing between FR I and FR II radio sources.)

4. The Spring, 1993 B Array Observations

During April and May 1993, ~ 120 hours of observations were scheduled. To achieve a flux density limit of 1 mJy (5σ), each field was observed for ~ 3 minutes. The data were taken in spectral-line mode using two 25 MHz bandpasses, each divided into seven 3 MHz channels in two polarizations. The field centers are distributed in the sky in a variable hexagonal grid with an average spacing of 26 arcmin between nearest neighbors. A total of ~ 2200 fields were observed in a narrow declination strip 2.75° wide passing through the NGP and spanning $\sim 9^h$ in RA.

5. Data Analysis

The data analysis procedures required to generate high-resolution, wide-field images from B-configuration VLA data are computer intensive, and at first glance, are beyond the capability of computers available to astronomers. Imaging the full primary beam requires a 2048×2048 image, but ignores the likelihood of bright sources in sidelobes beyond the first zero of the beam pattern.

Proper self-calibration of the data normally requires mapping all the sources contributing to the observed visibility function. To avoid repeatedly creating such large images we have taken advantage of existing radio catalogs, in particular the Greenbank 1400 MHz survey (White & Becker 1992). The first step in self-calibration is to generate a tapered map of the primary beam and satellite maps of all the outlying sources found in the Greenbank catalog out to 10° that will appear brighter than 2 mJy. The resulting maps are searched for sources and a source list is written to a disk file. These sources are then used as input to a self-calibration script which makes small maps around each and every discrete source, thus avoiding the need to make large images. With this methodology we can self-calibrate ~ 50 fields a day using a Sparc 10/41 processor. In general, we do a 3 iteration phase self-calibration on any field containing at least one source brighter than 30 mJy. All the observations from Spring 1993 have already been self-calibrated.

Once the data are self-calibrated, the field is imaged into a 2k by 2k map using the AIPS task WFCLN. We typically clean the images to a flux density level of 0.5 mJy ($\sim 3.5\sigma$). A Sparc 10/41 processor with ~ 25 megabytes of memory can create ~ 25 images/day. The images so produced are distorted because of the 3-D nature of the Celestial Sphere in contrast to the 2-D FFT done by WFCLN. But, since the observations were made as short duration snapshots, the distortion is correctable using the AIPS task OHGEO which transforms the image onto the correct coordinate grid. The overlapping areas of adjacent images will be summed to increase the sensitivity of the survey. All the images from Spring 1993 will be created by January 1994. As of October 1, we have generated 600 images. In Figure 1 we show the distribution of peak flux densities and rms noise levels for these images. The vast majority of the images ($\sim 95\%$) achieve the theoretical noise limit. In Figure 2 we display one of these images to illustrate what the community can expect from the survey.

Although the necessary software does not yet exist, we intend to search the final images to extract discrete sources and construct a survey catalog. The

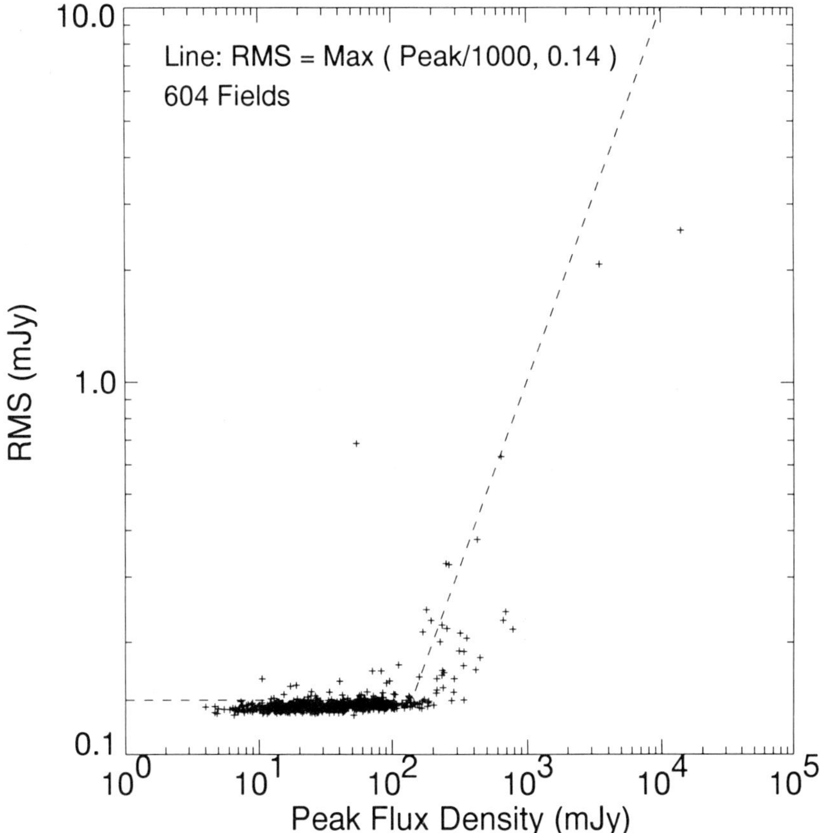

Figure 1. RMS noise in 604 cleaned maps versus peak flux density in the map for fields calibrated and mapped using the automatic procedure described in text. Fields with sources brighter than ~150 mJy are dynamic-range limited, while other fields reach the rms limit set by thermal noise in the 20 cm VLA receivers (~0.14 mJy). Only 1 field out of 600 has failed to reach an acceptable rms level and will require manual intervention; such a low failure rate is certainly acceptable.

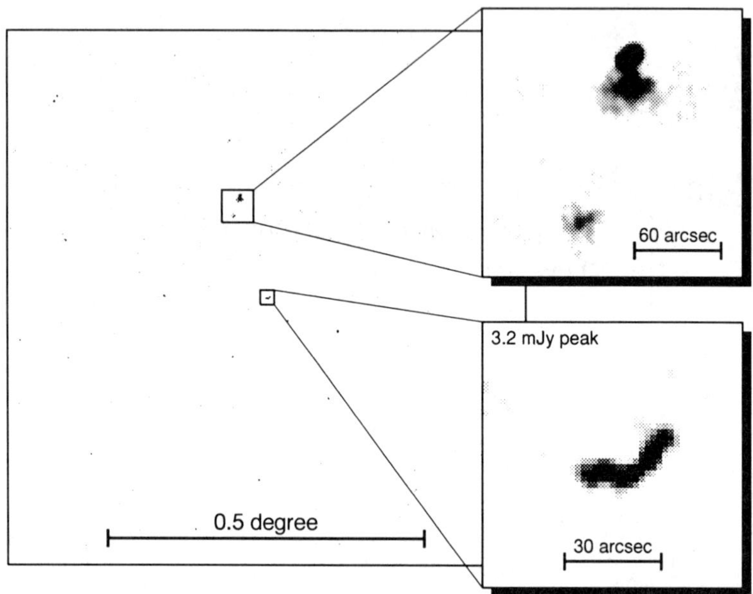

Figure 2. Sample map for the field 10330+30456 generated by our automatic procedures. The large image has 1634 × 1634 1.8 arcsec pixels; the smaller images show details of two interesting sources.

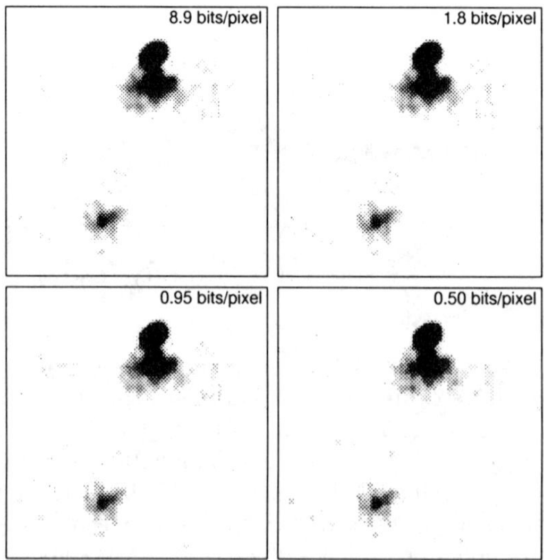

Figure 3. Compressed versions of the image in Fig. 2. The original image has 32 bits/pixel. The 8.9 bits/pixel image differs from the original by at most 0.5 μJy. At 1 bit/pixel the B-configuration images covering 10,000 deg^2 will fit on $\sim$8 CD-ROMs.

catalog will contain source positions accurate to better than an arcsec, flux densities, and morphological information. We also plan to distribute the images in compressed format (White, Postman, & Lattanzi 1992; White 1993) so that the entire survey will reside on several CD-ROMs. A sample compressed image is shown in Figure 3.

6. Scientific Potential of the FIRST Survey

The most important discoveries that have been made using the POSS — quasars, starburst galaxies, the large-scale structure of the Universe — were not anticipated by those who carried out the original survey. And many of these discoveries have come as a result of comparing work done at other wavelengths with this unparalleled optical database. Likewise, we cannot anticipate all of the new source classes and the astrophysical insights that FIRST survey will yield, but we are confident that cross correlation of the maps and catalogs we produce with data from other wavelength regimes will be enormously productive. To maximize the utility of the survey, it has been designed to cover the same 10^4 square degrees of sky as the Sloan Digital Sky Survey (SDSS) which will, by the end of the decade, have produced five-color CCD images of this entire area to $m_v = 24$ and will be in the process of collecting spectra from 10^6 of the objects therein. The subarcsecond positions our survey will produce will allow immediate optical identification of $> 50\%$ of our radio catalog from the SDSS database, further enhancing its utility in infrared and high energy astrophysics. While an exhaustive description of the scientific impact this project will have is beyond the scope of this paper, we outline below a few examples of areas in which the FIRST survey will make fundamental contributions.

6.1. Quasars and Active Galaxies

At 20-cm flux densities above $\sim$ 3 mJy, active galactic nuclei dominate radio source counts. In the thirty years since the discovery of quasars, nearly 10,000 galaxy systems with extraordinarily powerful, compact nuclei have been catalogued, and progress in understanding the mechanisms which power them has been substantial. Nonetheless, we still argue about density vs. luminosity vs. luminosity-dependent density evolution for quasars, and estimates of the typical quasar lifetime are uncertain by roughly two orders of magnitude. The connection between IRAS starburst galaxies and quasars, the importance of mergers, the origin of the powerful γ-ray emission recently discovered by CGRO, the contribution of various classes of objects to the cosmic X-ray background, the importance of relativistic beaming for BL Lacs and their connection to radio galaxies, as well as generalized "unified models" of AGN are all topics of current active research. Work toward answers for each of these outstanding questions will be materially aided by a deep, uniform, large-area radio survey.

Using the Cambridge University Automated Plate Machine (APM) scans of the POSS (McMahon, private communication) and data from a 25 deg^2 pilot for our VLA survey, we have demonstrated that, using optical color selection plus radio identification, we will be able to generate immediately a complete, flux-limited catalog of $\sim 20,000$ radio-loud quasars brighter than $B = 19.5$ for use in evolution studies. Since this sample will be optically bright, it will also be ideal for detailed spectroscopic studies. Cross-correlation with the ROSAT all-sky survey will increase the currently small sample of known BL Lacs by more than an order of magnitude from a little over 100 to several thousand; furthermore, since the 1 mJy limit of the survey is equivalent to the lowest flux density ever found from X-ray selected samples of these objects (Stocke 1990) and is far below the threshold for radio-selected objects, this sample will also be complete and flux-limited. The enormous benefit this new sample will have for the study of all objects in which relativistic beaming is important can be gauged by the progress brought about by the Kuhr et al. (1981) complete radio sample, the flux density limit of which is 1 Jy, 10^3 times our threshold. In all, the FIRST survey will detect over half a million AGN, and most will have optical counterparts in the SDSS. None of the current problems in the field will be immune from attack with this powerful resource.

6.2. Galaxy Evolution

Below about 3 mJy, the radio source $\log N - \log S$ slope changes markedly, indicating the presence of a new population. Work over the past decade (Windhorst et al. 1985; Oort 1987; Thuan & Condon 1987) has demonstrated convincingly that this population consists largely of actively star-forming galaxies and, as such, holds an important key to the subject of galaxy evolution. At 1 mJy, the majority of our sources will be starburst galaxies at redshifts from 0.1 to 0.6 and, perhaps, far beyond (e.g., IRAS 10214+4724 at $z = 2.3$ [Rowan-Robinson *et al.* 1991]). Every galaxy in the IRAS Faint Source Catalog will be detectable and, with the resultant subarcsecond radio positions, immediately identifiable. Many fainter objects, detectable only by future far-IR missions such as ISO and SIRTF, will also be identified, distinguished from AGN by their optical and radio extents. Questions such as the importance of mergers to star formation history, the shape of the IMF, the extent of mass-loss in supernova-driven winds, and the importance of local environment in galaxy evolution will be addressable with this enormous sample. Moreover, the combination of the radio data from our catalog with data from IRAS, CGRO, SDSS, etc., will allow the discovery of many rare objects such as IRAS 10214+4724, which often hold vital clues to the behavior of less extreme systems.

6.3. Large-Scale Structure and Dark Matter

The evolution of large scale structure in the Universe, from the era probed by the COBE observations of the microwave background to the present, is one of the central issues of modern astrophysics and, owing to the important role of dark matter in this evolution, of particle physics as well. The enormous amount of work undertaken over the past decade to map the local large-scale structure has resulted in a relatively clear picture of the distribution of luminous matter at redshifts $z < 0.1$. The challenge now is to extend our knowledge of this structure to much higher redshifts so that its evolution becomes apparent. For high redshifts, our quasar sample will provide a useful start on the largest scales, but it is in the redshift range 0.1–0.5 that real progress will be forthcoming by tracing the matter distribution with our starburst sample. Other avenues of approach include the detection of a large cluster sample through selection of head-tail radio galaxies, and the selection of large-separation gravitational lens candidates via an optical filter on the sample of close doubles (demanding equal relative intensities for the two components in the two wavelength bands).

6.4. Galactic Astronomy

A half dozen categories of radio-emitting stars provide us with information on topics ranging from main sequence and pre-supernova mass loss to surface magnetic activity and its evolution. Only about 200 radio stars are known, and virtually all of these emerged from targeted surveys of optically selected objects. FIRST provides a unique opportunity to produce unbiased, flux-limited, and complete samples of stellar radio sources. The accurate positions of even the faintest radio sources will lead to immediate and unambiguous optical identifications from existing POSS and ESO plate scan databases, as virtually all detectable objects are brighter than these plate limits. (Note that although bright stellar objects are not dense on the sky, radio stars are so rare that one must have subarcsecond radio positions to identify them in a large area survey.) From enriching our understanding of the evolution of dynamo-generated magnetic fields on late type stars (which will emerge from a comparison with the ROSAT database), to identifying nearby, coeval, moving groups of red dwarfs, our survey will open a new era in stellar radio astronomy.

Other Galactic objects may also turn up in the survey. Recently, a 0.5 Jy radio source in the southern hemisphere was identified as the nearest millisecond pulsar yet discovered (Johnston et al. 1993). Comparison of our survey with the Texas 80 cm interferometric survey and the ongoing, low-frequency Westerbork survey will immediately identify good candidates for these fascinating objects from their uniquely steep spectral indices, providing the first wide-area pulsar survey unbiased by the problems of traditional search techniques such as low duty cycles, high dispersion measures, short periods, and binary accelerations.

7. Summary

The FIRST project will produce a radio image of the sky with 20 times the sensitivity and 10 times the angular resolution of the best centimetric surveys now available. If history provides any guide, many new, unforeseen discoveries will emerge both from an analysis of the radio sources themselves and through comparison with catalogs and sky maps in other spectral regimes. Our proposal, accepted by the NRAO, that there be no proprietary period for the survey data, is designed to maximize the scientific return of the project by making the raw data, calibrated UV data, survey images, and source lists avaliable to the community as soon as they are produced. We encourage the widest possible use of the FIRST Survey data products and welcome suggestions from the community concerning their production and distribution.

References

Gunn, J. E., & Knapp, G. R. 1993, in Surveys in Astronomy, ed. B. T. Soifer, ASP Conference Proceedings, in press

Johnston, S., et al., 1993, Nature, 361, 613

Kleinmann, S. G., 1992, in Robotic Telescopes in the 1990's, ed. A. V. Filippenko (San Francisco: Astron. Soc. Pacific), p. 203

Kuhr, H., et al., 1981, A&AS, 45, 365

Oort, M. J. A., 1987, Ph.D. thesis, University of Leiden

Rowan-Robinson, M., et al., 1991, Nature, 351, 719

Stocke, J. T., et al., 1990, ApJ, 348, 141

Thuan, T. X., & Condon, J. J. 1987, ApJ, 322, L9

White, R. L, & Becker, R. H. 1992, ApJS, 79, 331

White, R. L., Postman, M., & Lattanzi, M. G. 1992, "Digitized Optical Sky Surveys", eds. H. T. MacGillivray & E. B. Thomson (Amsterdam: Kluwer), p. 167

White, R. L. 1993, in Space and Earth Science Data Compression Workshop, NASA Conference Publication 3183, ed. James C. Tilton, p. 117

Windhorst, R. A., et al., 1985, ApJ, 289, 494

Astronomical Data Analysis Software and Systems III
ASP Conference Series, Vol. 61, 1994
D. R. Crabtree, R. J. Hanisch, and J. Barnes, eds.

The Westerbork Northern Sky Survey (WENSS:) A Radio Survey Using the Mosaicing Technique

M. A. R. Bremer[1,2]

Sterrewacht Leiden, Leiden University, Niels Bohrweg 2, 2333 CA Leiden, The Netherlands

Abstract. This report explains how the WENSS survey is carried out. The mosaicing concept is explained which is essential for doing such a large survey.

1. Introduction

The Westerbork Northern Sky Survey is a survey presently carried out with the Westerbork Synthesis Radio Telescope (WSRT). The principle investigators are G.K. Miley and A.G. de Bruyn and the project manager is E. Raimond. The reduction of the data is done in Dwingeloo by Y. Tang and the catalogue production by M.A.R. Bremer (Leiden). Follow-up science is done in Leiden by M.N. Bremer and R.B. Rengelink.

The whole sky above declination $+30^{\circ}$ is mapped at 92 cm (10,000 square degrees). At 49 cm we hope to map about one third of that area (3,000 – 4,000 square degrees); mapping the whole region as we have done at 92 cm is just not possible. The amount of observing time would be 4 times larger as with 92 cm since the field of view is twice as small. At the present time the reduction and catalogue production can keep pace with the datastream from the telescope (observing, initial calibration and archiving of the raw data). The limiting flux density at both wavelengths should be between 10 – 15 mJy (5σ level).

The result will be a catalogue with about 300,000 sources at 92 cm and 60,000 sources at 49 cm. Because the catalogue contains two frequencies, we have spectral information available for a large fraction of the sources. With the database we can select and study these sources more than one order of magnitude deeper in flux than was previously posssible.

The positional accuracy will be better than in previous surveys (five arcseconds for the faintest sources and about one arcsecond for the brighter ones).

---

[1]Leiden Observatory

[2]NFRA/ASTRON (Netherlands Foundation for Research in Astronomy)

2. The Mosaicing Concept

Since 1990 the WSRT can work in a new observing mode, using the mosaicing technique. Recent on-line and off-line software developments by R. Braun, H. van Someren-Gréve and W. Brouw have enabled very efficient, in observing and dataprocessing, mosaic observations. Only because of these developments it was possible to make this survey.

Imaging an extended region requires fully sampling the area at pointing distances less or equal than HWHM intervals, with the primary beam. At 92 cm the HPBW is 2.67°. This ensures complete sampling of the visibilities in the (spatial frequency) u-v plane, with D/2 intervals in both coordinates. The regular east-west array configuration of the WSRT, gives a radial sampling interval of nominally 72 m (this is declination and position angle dependant). By using six different array configurations (moving the telescopes A, B, C and D) a radial sampling of $D/2 = 12$ m can be obtained. Together with the assumption that the data are subsequently used in parallel to constrain the sky brightness the sensitivity is guaranteed to be uniformal within a few percent.

Through careful optimization of the telescope slew and brake parameters, it has been possible to limit the move time between each pointing centre to 10 seconds. Within this time interval moves can be made between about 10 and 120 arcminutes.

To reach a uniform sensitivity after correcting for the primary beam attenuation the grid interval between pointing centres has to be about one half of the primary beam half-power width. At 92 cm this means a stepsize of 1.33° and at 49 cm 0.7° (the WSRT has dishes of 25 m diameter). Table 1 and 2 give an overview of the different grid sizes at both frequencies. At 92 cm 75 mosaics are needed to cover 10,000 square degrees.

Table 1. Mosaic characteristics at 92 cm.

Declination strip (°)	RA step (°)	DEC step (°)
+37	1.651	1.333
+50	2.045	1.333
+66	3.186	1.333

With an integration time of 20 or 30 seconds the net observing efficiency becomes resp. 67 and 75 %. Each declination strip has a different number of pointing centres.

As an example, a few mosaic patterns of the WENSS survey are discussed below. A dutycycle of 40 minutes is used to cover one "round" in a mosaic pattern. This means that the pattern is scanned 18 times in a full 12 hour synthesis. This yields about 700 different UV points as well as 500 redundant UV points. At these relatively low frequencies (325 and 608 MHz) sidelobe confusion in the images becomes a very serious problem, because the large field of view intercepts a large number of sources. One way of dealing with this problem is to make the same observations at different array configurations (the 4 movable telescopes A, B, C and D are stepped at regular intervals). At 92 cm each mosaic is observed 6 times with 12 meter baseline increment. For the 49 cm

Table 2. Mosaic characteristics at 49 cm.

Declination (°)	RA step (°)	DEC step (°)
+32	0.783	0.667
+36	0.818	0.667
+40	0.865	0.667
+44	0.918	0.667
+48	0.989	0.667
+52	1.071	0.667
+56	1.176	0.667
+60	1.314	0.667
+64	1.500	0.667
+68	1.748	0.667
+72	2.118	0.667
+76	2.687	0.667
+80	3.673	0.667
+84	6.000	0.667
+88	16.364	0.667

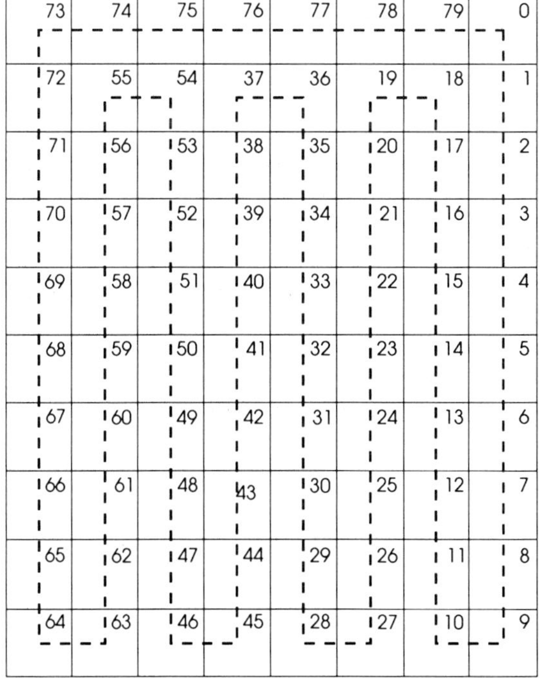

Figure 1. Mosaic pattern at 92 cm (declination=37°).

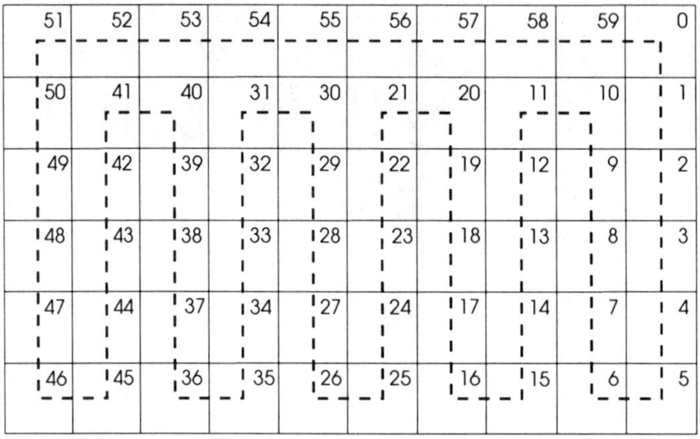

51	52	53	54	55	56	57	58	59	0
50	41	40	31	30	21	20	11	10	1
49	42	39	32	29	22	19	12	9	2
48	43	38	33	28	23	18	13	8	3
47	44	37	34	27	24	17	14	7	4
46	45	36	35	26	25	16	15	6	5

Figure 2. Mosaic pattern at 49 cm (declination=66°).

observations only 3 different configurations with 24 meter baseline increment are used. The starting hour angles for a given pointing centre are phased such that a very regular UV coverage results. The total number of UV points becomes about 4000 at 325 MHz and about 2000 independent UV points are available at 608 MHz. Immediately following the observations the data are sorted per pointing centre to permit efficient further off-line processing.

Acknowledgments. Thanks to A.G. de Bruyn for reading this manuscript and giving useful comments.

The Westerbork Synthesis Radio Telescope is operated by the Netherlands Foundation for Research in Astronomy (NFRA) with financial support from the Netherlands organization for scientific research (NWO).

References

Braun. R, de Bruyn, A.G., 1990, ASTRON/NFRA newsletter Dec. 1990, 1

Astronomical Data Analysis Software and Systems III
ASP Conference Series, Vol. 61, 1994
D. R. Crabtree, R. J. Hanisch, and J. Barnes, eds.

Facilities for Retrieval of Radio-Source Data

Heinz Andernach

Observatoire de Lyon, F-69561 Saint-Genis-Laval Cedex, France

Daniel E. Harris and Carolyn Stern Grant

Center for Astrophysics, 60 Garden St., Cambridge, MA 02138, USA

Alan E. Wright

ATNF Parkes, PO Box 276, Parkes, NSW 2870, Australia

Abstract. There are now 100 radio-source catalogs available in electronic form, with altogether over half a million entries. We describe how these data are presently documented, archived, distributed and searchable over the network. We stress the need to use these data to complement existing multi-waveband integrated databases of astronomical objects.

1. Introduction

Bibliographical studies indicate that well over 800,000 radio-source measurements have been published so far in hundreds of papers (Andernach 1992a). Although data collection and processing in radio astronomy has relied on electronic equipment over most of its history, it was very difficult to find radio-source data in electronic form until recently. The dramatic growth in number and size of radio-source catalogs (Andernach 1992b) prompted one of us in 1989 to collect all source lists available electronically, in order to make these publicly accessible. Until now some 100 such tables with a total of ~550,000 entries were secured. Discussions have begun within the IAU Working Group on *Radioastronomical Databases* (Andernach 1994) as to how the data should be distributed and eventually integrated into other multi-waveband databases.

2. Status and Completeness of Electronic Data

The collected source tables (in ASCII format) vary in size from only ~40 to ~84,000 entries. The data files and an index with a minimal description of them are available on request from the first author. This index was provided to the data centers at NASA-GSFC and CDS Strasbourg from early on, but lack of electronic documentation (usually not provided by the authors) entailed that so far only a dozen larger catalogs are distributed by these centers.

To assess the completeness of our collection we prepared a list of the 50 largest published radio-source catalogs containing a minimum of ~1000 sources.

These comprise a total of 435,000 entries. The 35 of these 50 catalogs which are available electronically contain ~90% of the data, and 9 of these 35 catalogs (~40% by data volume) are currently available from the astronomical data centers. The 15 other large catalogs (summing >40,000 entries) were not available in electronic form, according to their authors (e.g., 8C, MDS1, MDS2, CUL1-3, etc.). We also prepared lists of all systematic surveys obtained with the Westerbork (W), Cambridge (5C) and Penticton (P) interferometers. While most of the 5C- and P-survey source lists were collected by us, we were unable to obtain electronic copies of the vast majority of the W-surveys. Recovery of the medium-size lists would require page-scanning and subsequent optical character recognition (OCR) while the smaller lists could be ingressed manually, e.g., within the updating efforts of databases like NED or SIMBAD.

We made tests with OCR for some large printed tables. A.E.W. page-scanned the original PKS 2700 MHz survey parts 11–14 and recovered the data in a week's work. However, poor printing quality of many other papers calls for tedious "teaching" of the OCR software and often prevents data recovery within reasonable effort. Though some help was offered from Beijing Observatory to key-punch some of those data sets which are more important and/or difficult to scan, the collaboration of data centers with their experience (e.g., in electronic preparation of old star catalogs) would be highly desirable. Happily the electronic preservation of future data looks better since the editors of some major journals agreed to have larger tabular data sets published on either CD-ROM (for ApJ, ApJS, AJ) or accessible via FTP from the CDS archive in Strasbourg (for A&A, A&AS). However, this option is not yet mandatory for authors.

The fraction of published source tables available in electronic form decreases rapidly with decreasing size of the tables. This limits the parameter space covered by electronic tables as the more numerous smaller lists would add complementary frequencies, observing epochs and angular resolutions to those contained in the fewer big lists. Thus the smaller and complementary lists in our collection of tables could provide useful information on many sources, including those to be detected in the large-scale sky surveys with the Westerbork and VLA interferometers (see Bremer; Becker et al.; Condon et al., these proceedings).

3. Search Facilities for Radio-Source Catalogs

In mid-1992 the *Einstein On-Line Service* (EOLS) at CfA (Karakashian et al. 1992) began the incorporation of the (then) 50 electronically available source lists. By mid-1993 almost all of these tables (totaling ~428,300 rows) were available in EOLS for on-line search and some 9000 lines of documentation were written mostly by one of us (C.S.G.). With EOLS the data sets can be searched individually in circular or rectangular areas on the sky, either using a canned query mode (`SQQ`) or more sophisticated SQL-like queries (`DBQ` mode) in which any of the data columns may be used as a search criterion. A third `MQQ` option allows searches through user-specified groups (or even *all*) of the EOLS catalogs. Data format and documentation is compatible with that of NASA's Astrophysical Data System (ADS, cf. Eichhorn, these proceedings) and all EOLS catalogs are made available in ADS. At present, the radio-source catalogs from EOLS make up some 20% of the ~200 catalogs managed by ADS.

We anticipate to add at least the major radio catalogs into EOLS. However, during the year of incorporation of the first 50 source lists, another 44 lists with >100,000 entries were collected and await ingression into EOLS. Since only catalogs with proper ADS-style documentation will be put on line, and EOLS no longer has the funding to compose the documentation, authors *must* submit such documentation to ensure incorporation of their data.

EOLS currently offers far more radio-source data than any other database, but it does not provide the cross-identifications and the bibliography of systems like e.g., NED or SIMBAD. Also, EOLS cannot precess source positions on the fly and catalogs lacking B1950 positions are currently excluded from the `MQQ` option. Batch-type queries around many positions supplied from a file are not supported.

Databases like NED and SIMBAD have also folded in several of the larger radio-source catalogs. In SIMBAD e.g., one can use the `info` command to check whether a certain catalog (by catalog acronym or author) has been folded in. These systems are strictly object-oriented and thus establish explicitly and carry forward all identifications with all other database entries. Clearly, the folding of all collected radio data into such systems presents an ambitious project, and would provide a most powerful tool for radio astronomy.

About 150 astronomical object catalogs, including some of the larger radio-source lists, are searchable remotely with DIRA (*Distributed Information Retrieval from Astronomical files*) of the ASTRONET group in Italy (Nanni & Tinarelli 1994). DIRA allows to project the distribution of objects from various catalogs on the same screen and thus offers a way of cross-identification between them. Incorporation of some of our radio source data has just started.

The software package SURSEARCH, prepared in 1991 by C. Purton and P. Durrell of DRAO, allows to query a database of 380 radio-survey fields for overlap with a user-specified sky position. Updating this database (currently not foreseen) would require the input of only about a dozen basic parameters for every new published survey. It could thus be maintained with minimum effort (e.g., at a data center) and would still provide an efficient and yet unavailable tool to find references to radio measurements in a given sky region.

One of us (A.E.W.) has prepared a first draft of the "COMpendium of RADiosources" which will be known as COMRAD. The aim is to give the basic information on radio sources in a *single* format. As a first step, twelve of the largest source surveys were included (4C, 6C1-4, B2, B3, MRC, UTRAO, WB92, PKSCAT, RATAN, MG1-4, 87GB, PMN). For a total of ~303,000 entries this version provides the catalog name, source position (J2000) and error, flux density and error, observing frequency and original source name as published. The new master-base would permit access on any low-level IBM/PC compatible using the `dBase` format, and be distributed on diskettes with suitable software. It would allow searches for large numbers of positions from a file. Alternatively, one of us (H.A.) is preparing a routine to search COMRAD remotely, allowing simple interrogations and retrieval of results via FTP.

Discussions are underway to include our collection of radio-source tables on a CD-ROM, e.g., on a successor to the two CD-ROM's already distributed by NRAO. The number of tables (~100) with their widely different format would call for a reasonably comfortable source-search and retrieval software to be provided with them. This was done e.g., for NASA-ADC's first selection of the most popular astronomical catalogs on CD-ROM (Brotzman & Gessner 1991). However, the necessary preparation of all tables for their use with a FITS table browser is a major undertaking (documentation, data consistency checks, etc.) for which the participation of experienced personnel of data centers would be essential.

4. Conclusions

We described only some of the many different approaches and compilations available. What is clearly needed is a strategy to coordinate these efforts and define priorities for a type of service that has become common for non-radio wavebands. To achieve this, the radioastronomical community (e.g., as represented in the IAU-WG of Comm. 40) should provide the seeds and define its requirements to the data centers and database managers.

During a splinter ("BOF") session of the present conference the issue was discussed among some 16 interested delegates, representing several large radio astronomy institutions and multi-waveband databases. The idea was supported that published data, of either tabular, image or spectral type, should be archived in a publicly accessible data center on a *mandatory*, not voluntary basis. Despite the impressive developments in database-software, user interfaces and information retrieval tools presented on ADASS'93, it was felt that there is still a lack of user-tunable utilities for cross-identification between catalogs at different (radio and non-radio) wavelengths. Neither is there a master index of machine-readable astronomical catalogs which would locate an item in the different collections offered by the data centers and various database systems.

Acknowledgments. The presentation of this work (by H.A.) was made possible with grants from NATO (kindly made available by R.E.M. Griffin) and NRAO. All participants of the BOF session are thanked for the lively discussion.

References

Andernach, H. 1992a, in Highlights of Astronomy, 9, 731, Dordrecht: Kluwer

Andernach, H. 1992b, in Astronomy from Large Data Bases – II, A. Heck & F. Murtagh, ESO Conf. & Workshop Proc., 43, 185, Garching: ESO

Andernach, H. 1994, in Handling and Archiving Data from Ground-based Telescopes, M. Albrecht & F. Pasian, ESO Conf. & Workshop Proceedings, Garching: ESO, in press

Brotzman, L., & Gessner, S.E. 1991, Selected Astronomical Catalogs, Vol. 1, CD-ROM issued by NASA-ADC, Greenbelt, USA

Karakashian, T., Harris, D.E., Garcia, M., Fabbiano, G., Flanagan, J.M., & Stern, C.P. 1992, in Astronomical Data Analysis Software and Systems I, D. Worrall, C. Biemesderfer, & J. Barnes, A.S.P. Conf. Ser., 25, 44, San Francisco

Nanni, M., & Tinarelli, F. 1994, Proc. ASTRONET Conference, Trieste, April 1993, Mem. Soc. Astr. It., in press

Astronomical Data Analysis Software and Systems III
ASP Conference Series, Vol. 61, 1994
D. R. Crabtree, R. J. Hanisch, and J. Barnes, eds.

Guide Star Catalog Data Retrieval Software III

O. Yu. Malkov and O. M. Smirnov
Institute of Astronomy of the Russian Academy of Sciences
48 Pyatnitskaya St., Moscow 109017 Russia

Abstract. The Guide Star Catalogue (GSC) is widely used by the astronomical community for all sorts of applications. Unfortunately, the actual GSC data is not very easily accessible, due to its internal format. To facilitate its retrieval, we have created GUIDARES, a user-friendly program that lets you look directly at the data in the GSC, either as a graphical sky map, a plot, a histogram, or a simple text table.

GUIDARES can read a sampling of GSC data from a given sky region, store this sampling in a text file, and display a graphical map of the sampled region either in projected celestial coordinates (perfect for finder charts) or in the Aitoff projection. It supports rectangular and circular regions defined by coordinates in the equatorial, ecliptic (any equinox), galactic or supergalactic systems. New features include magnitude cutoffs, different output formats (with or without multiple entry filtering), and the ability to make quick plots and histograms of various attributes of the data being retrieved.

1. Introduction

The Guide Star Catalog (GSC) was initially created at ST ScI to support the operational requirements of the Hubble Space Telescope (Lasker et al. 1990; Russell et al. 1990; Jenkner et al. 1990), specifically, an all-sky set of reference coordinates for pointing and tracking. A new version of the catalog, GSC 1.1, includes bright stars from the HIPPARCOS INCA database. The GSC is widely used by the astronomical community for many different applications, such as:

- statistical studies of sky regions. The extremely large size and scope of the data set (17 million objects) opens up research possibilities that were previously impossible.
- searches for candidate optical counterparts of astronomical objects found at other wavelengths.
- mapping of object vicinities (in particular, finder charts for variable stars).

The catalog's distribution format (a set of two CD-ROMs) requires minimum hardware and is well suited for all sorts of conditions, especially observations.

Unfortunately, the actual data in the catalog is not easily accessible. It is in the form of FITS tables, and the coordinates are given in one standard system (J2000.0). The included software (Priou 1990) doesn't help much—it can only search the GSC for objects from a rectangular region specified by coordinates in J2000.0, and list the objects on the screen, without any possibility to store them in a file for later use. Thus, even generation of a simple finder chart is no trivial undertaking.

When we recently started some statistical research of our own using the GSC (see Malkov & Smirnov elsewhere this conference), we became aware of more serious problems with the catalog. Near the edges of the original survey plates, the images become blurred, the photometric flux increases, and a lot of stars are misclassified as non-stars (Mink 1993, private communication). These problems have to be dealt with if statistical surveys are to cover areas larger than the central region of a GSC plate (about 30 square degrees).

To help PC users solve the problem of data retrieval, we have created the Guide Star Catalog Data Retrieval Software, or GUIDARES (Smirnov & Malkov 1992). This is a user-friendly program which lets one easily produce text samplings of the catalog and sky maps in Aitoff or celestial projections. This paper describes GUIDARES version 2.

GUIDARES 2 requires a PC with a CD-ROM drive and CGA, EGA or VGA graphics (for plotting sky maps only). It will work without a CD-ROM if the necessary GSC files are copied onto the hard disk, preserving the GSC's directory structure.

The main function of GUIDARES is to produce an ASCII table of object entries from a specified region, and, optionally, a graphical sky map of the region. It can handle rectangular and circular regions in four different coordinate systems.

2. User Interface

To the user, GUIDARES 2 looks like an intuitive and self-explanatory menu that is used to enter the following information:

- The logical drive where the GSC files are located.
- The coordinate system in which the region is specified: equatorial, ecliptic, galactic or supergalactic. For equatorial and ecliptic coordinates, the equinox may also be set (the default is 2000).
- The shape of the region: rectangular or circular.
- Magnitude limits for the objects to be retrieved.
- The actual coordinates of the region (minimal and maximal or central and radius).
- Which types of objects to retrieve. The user can selectively disable and enable retrieval of stellar, non-stellar, or mixed objects (mixed objects are those objects that have several GSC entries with conflicting classification, i.e., they are classified as stars in some entries and as non-stars in others).

- How to treat multiple entry objects. GUIDARES 2 can replace them with one entry in the output file containing weighted averages for position and magnitude, or it can just dump the multiple entries as multiple ones in the output file (previous versions did not have the latter option).

The software performs a lot of error control along the way. Illegal values of coordinates are impossible to enter, the command to start retrieval is unavailable until all the necessary data has been specified, inverted coordinates are implicitly swapped if necessary, etc.

3. Data Retrieval

During retrieval, the evolution of the process is shown on the screen. In addition, a sky map in the celestial or Aitoff projection is simultaneously plotted. Internally, GUIDARES must perform the following steps:

- Convert the region coordinates to the standard GSC system.
- Determine which GSC files overlap the user's region. This is done algorithmically. In contrast, Priou's software uses a lookup table, and for this reason it is bigger in size than GUIDARES 2, while providing much less functionality.
- Scan the GSC files for objects that fall within the user's region, convert the objects' coordinates back into the user's system.

If errors occur, or if the CD-ROM must be changed, GUIDARES notifies the user via a message box with standard "Abort, Retry, Ignore" or "Continue, Cancel" options.

4. Output and Sky Maps

For every region, GUIDARES 2 creates an output *sampling* file. This is an ASCII table with the following data for each object:

- GUIDARES ID of object, obtained by appending the number of the object in its GSC file to the GSC file number.
- position and position error; magnitude and magnitude error.
- classification.

A GUIDARES SkyMap can be produced both during retrieval and afterwards, by reading the sampling file. A coordinate grid with labels is displayed, and objects are plotted over it. Stellar objects are represented by disks, non-stellar ones by circles, and mixed objects by semi-filled circles; the radius of the disks and circles is proportional to the magnitude. In the Aitoff projection, all objects are plotted as dots (this projection is intended for relatively large sky areas).

GUIDARES 2 automatically selects the highest resolution available and takes into account the screen aspect ratio, so that square regions look truly square at any resolution.

5. Future Plans

More GUIDARES functionality will be implemented through the concept of *filters*. Data from the GSC passes through a filter before being placed in the output file. GUIDARES 2 has two filters, a fully transparent one which just places the entries into the file as they are, and one which computes weighted averages for multiple-entry objects. The filters are switched via an option in the menu (see above).

Some applications of the GSC may require more specialized filters. For example, our statistical survey (Malkov & Smirnov, this conference) will require us to deal with large areas, where the already mentioned edge effects can cause an artificial deficit in the GSC stellar distributions. We propose to overcome this problem by bulding an "intelligent" (i.e., using methods of artificial intelligence) filter that takes advantage of the fact that the plates overlap at the edges. The blurred objects are likely to have multiple catalog entries originating from different plates, and the filter can hopefully analyze this abundance of information to come up with correct classifications and magnitudes for the objects.

6. Distribution

To avoid the hassle of separate "Read Me" files, all the documention has been built into the executable module. By selecting an option from the GUIDARES 2 menu, the user can produce a documentation file whenever he needs. Therefore, the whole package is distributed as a single file.

GUIDARES is a shareware product. An unregistered copy of the latest version can be picked up by anonymous ftp from `iraf.noao.edu` (`140.252.1.1`), directory `contrib`, file `g2.exe`.

If you have any questions, comments, or even nasty criticisms, we can be contacted by e-mail either at `oms@airas.msk.su` (Oleg Smirnov), `omalkov@airas.msk.su` (Oleg Malkov), or `guidares@airas.msk.su`.

Acknowledgments. We would like to thank NOAO and SAO for the financial support that made this presentation possible.

References

Jenkner, H., Lasker, B.M., Sturch, C.R., McLean, B.J., Shara, M.M., & Russell, J.L. 1990, AJ, 99, 2081

Lasker, B.M., Sturch, C.R., McLean, B.J., Russell, J.L., Jenkner, H., & Shara, M.M. 1990, AJ, 99, 2019

Priou, D. 1990, "Guide Star Catalog Listing Program", Institute Géographique National

Russell, J.L., Lasker, B.M., McLean, B.J., Sturch, C.R., & Jenkner, H. 1990, AJ, 99, 2059

Smirnov, O.M. & Malkov, O.Yu. 1993, PASPC, 52

Astronomical Data Analysis Software and Systems III
ASP Conference Series, Vol. 61, 1994
D. R. Crabtree, R. J. Hanisch, and J. Barnes, eds.

Testing the Galaxy Model with the Guide Star Catalog

O.Yu. Malkov, O.M. Smirnov

Institute of Astronomy of the Russian Academy of Sciences
48 Pyatnitskaya St., Moscow 109017 Russia

Abstract. Bahcall and Soneira(1980) constructed a model for the disk and spheroid components of the Galaxy. We have developed software for the detailed statistical analysis of the GSC data versus the B-S model, and used it to process selected areas distributed over the whole sky.

1. Introduction

A detailed model for the Galaxy and a theoretical distribution of visible stars derived from it is very important for interpreting observational data, detecting unexpected phenomena, determining areas for future research, etc. The primary goal of our project is the construction of an improved Galaxy model, specifically: testing of the Bahcall-Soneira Galaxy model using the over 10^7 stars included in the HST Guide Star Catalog (GSC); a more accurate determination of galactic structure parameters from the GSC data; extension of the model into low galactic latitudes using an improved obscuration model. A secondary goal is a detailed investigation of the photometric and statistical properties and irregularities of the GSC (limiting magnitude, degree of completeness, etc.), as well as development of methods and software to deal with them. This can be very valuable to other applications of the GSC.

2. The Model and the GSC

The Bahcall-Soneira model for the disk and spheroidal components of the Galaxy gives results that are consistent with well-established V-band star counts for galactic latitudes above 20° and magnitudes up to 20^m. Comparisons have been made for data sets of up to 10^5 stars taken from various small sky areas (Bahcall & Soneira 1980, referred to as the Model paper from now on). The existing model uses relatively effective estimates for the galactic structure parameters such as disk scale length, spheroid star density and ellipticity; they can possibly be improved if larger data sets are used. There are indications that there exists a third Galaxy component, the massive halo, but it hasn't been conclusively established with available star counts. Some researchers suggest that yet other Galaxy components also exist.

The GSC, containing data on over 10^7 stellar objects up to 15^m—16^m, can provide the basis for some very extended comparisons, covering large areas of the sky. One has been performed for a 40 square degree plate (about 4000 stars) near the north galactic pole, and has shown good agreement with the model for magnitudes up to $14^{m}.5$, above which the GSC becomes incomplete (Jenkner et al. 1990).

The two features of the GSC vital to our project are its extremely large size and the fact that it is drawn from the whole sky. With correct statistical treatment, they should allow us to reach important conclusions about the overall distribution of visible stars and the validity of the investigated models, not possible with conventional observations.

3. Preliminary Studies and Initial Comparisons

For preliminary comparisons, we selected a set of circular areas with $R = 3°$ (about 28 square degrees), uniformly distributed across the whole sky. To avoid dealing with blurred plate edges and plate overlap at this stage, each area was shifted to the nearest plate center, and the effective radius decreased to leave out any irregularities. We prepared GSC samplings for the areas using our GUIDARES package (Malkov & Smirnov, this volume), and modified the B-S model to produce results in the GSC photometric bands of the corresponding plates.

The GSC samplings for our areas contain on the order of 5000 stars. To analyze the data, we developed a suite of IDL programs, and performed standard statistical tests (chi-square for binned star counts, Kolmogorov-Smirnov for cumulative distributions).

3.1. Determining the Limiting Magnitude

Past a certain magnitude (which is somewhat less than the highest magnitude in the catalog), the GSC becomes incomplete. It is very important to accurately determine this value. We studied plots of the residuals produced by subtracting cumulative model distributions from the GSC data. There are several types of residual curves, but each shows a distinctive upward bend (second derivative crosses through zero) near the faint end (see Figure 1). Assuming that the GSC becomes incomplete rapidly, the limiting magnitude should be just before this bend. See Table 1 for a summary of the results.

4. Interpretation of Initial Results

4.1. Limiting Magnitudes

The limiting magnitudes for non-equatorial areas do not show any discernible trends related to latitude or longtitude. For most areas, they are within 14^{m}—15^{m}. The equator presents a radically different picture. The limiting magnitude there grows from the galactic center (where it is about 12^{m}) towards the anticenter at a rate of roughly 1^{m} per 90° longtitude. This is quite natural and not contrary to expectations. But the area covering the anticenter shows a surprisingly low limit: $12^{m}.5$ instead of the expected 15^{m}. We can suggest at least two explanations for this phenomenon: a non-standard GSC plate, or a peculiar obscuration effect. Both possibilities will be investigated further.

	Plate	B	l	b	R	m		Plate	B	l	b	R	m
01	01PM	1	170	87	180	14.3	22	00DI	1	179	1	148	12.6
02	03QS	1	4	62	164	14.0	23	0445	0	225	2	81	14.4
03	037Z	1	46	58	154	14.1	24	0617	6	270	−3	67	13.2
04	01AL	1	93	57	180	14.3	25	0687	6	320	−2	101	12.6
05	00LK	1	137	63	151	14.1	26	01CE	0	0	−29	91	13.9
06	000B	1	177	59	121	14.4	27	03U3	0	45	−31	84	14.3
07	01AJ	1	227	61	132	14.3	28	008Q	1	88	−29	151	14.4
08	00WV	0	269	58	118	14.3	29	037O	1	138	−30	145	14.8
09	012I	0	319	62	102	14.6	30	01X0	1	180	−28	154	15.1
10	044C	0	1	31	85	14.8	31	01CG	0	222	−29	80	15.0
11	02HC	1	47	29	146	14.4	32	0398	0	269	−32	84	14.9
12	036B	1	89	32	138	14.3	33	00Z6	0	313	−32	49	13.8
13	00LQ	1	137	30	95	14.2	34	0125	0	0	−59	110	13.9
14	00XT	1	179	32	131	14.8	35	020O	0	41	−61	101	14.7
15	02V6	1	223	30	151	14.1	36	04N1	0	84	−62	133	15.0
16	00GU	0	269	29	158	15.0	37	00WT	0	139	−62	67	14.3
17	02IZ	0	318	31	105	14.2	38	012H	0	178	−62	94	14.6
18	067N	6	359	0	80	12.0	39	0126	0	226	−60	93	14.1
19	02V0	1	47	0	144	13.1	40	03BY	0	269	−60	93	15.0
20	02RW	1	89	1	145	13.0	41	02PU	0	308	−62	92	15.3
21	01MW	1	134	0	121	13.9	42	025A	0	317	−87	97	14.6

(*Plate* is the ID of the GSC plate centered on the area, B is the GSC photometric band, l and b are the galactic coordinates of the center of the area, R is its radius in arc minutes, m is the established limiting magnitude)

Table 1. Areas used for initial comparisons

4.2. Model Discrepancies

Practically all areas showed inadequacies in the model, which was either below or above the actual data. Our statistical tests produced near-zero (< 0.01) probabilities, which means that both the star counts (as measured by the chi-square test, see selected plots in Figure 1) and the relative distributions of stars among magnitudes (modified Kolmogorov-Smirnov test) are substantially different from the model. We have noted a few trends:

A deficit of bright stars 11^{m}—12^{m} in the GSC relative to the model exists at high latitudes (see, for example, Figure 1, region 42: the south galactic pole). Such localization can't be explained by obscuration effects. It is more likely that the model overestimates the contribution of the disk component. At this stage, we are reluctant to consider the spheroid component as a source of any discrepancies, as even at high latitudes its contribution can only be detected at relatively faint magnitudes (according to the Model paper, spheroid star counts become comparable to the disk ones only after 17^{m}, whereas at 14^{m} their level is at about 25% of the latter).

A significant deficit of faint stars in the GSC is present at the central galactic (prime) meridian ($l = 0°$). Possible explanations have to do with inadequte estimates of the stellar distributions in the disk component along the radius, and will be investigated using more areas along this meridian. There is one curious phenomenon: the same picture can be seen at $l = 180°$, $b = -30°$, i.e., at low latitudes of the anticentral meridian. A closer look did not show this area to be any different from its neighbors. However, we have established that a symmetrical comparison area above the galactic equator (showing a picture that's consistent with the rest of the sky), overlaps with a spot of very low obscuration (Arenou et al. 1992). With the already mentioned strange limiting magnitude at the anticenter in mind, we suggest that a deficit of faint stars is characteristic for the anticentral meridian as well.

A surplus of faint stars in the GSC can be seen at most of the other areas. Note that the Bahcall-Soneira model is not considered to be correct at latitudes

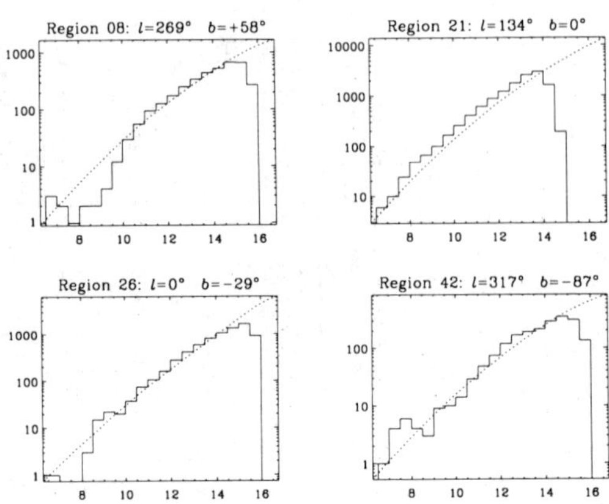

Figure 1. GSC star counts vs. model counts for four areas.

below 20°, due to its simplified obscuration model. We have not yet established any systematic differences between the northern and southern hemispheres.

5. Future Research

For further research, we plan to increase the size and number of our comparison areas, with the ultimate goal being to cover the whole sky with larger areas containing on the order of 10^5—10^6 stars. Edge effects in the GSC (increased photometric flux and blurred images, which lead to many misclassified objects, Mink 1993, private communication) have to be dealt with in order to extend the comparison. However, since GSC plates overlap at the edges, the misclassified objects have a surplus of information originating from several overlapping plates. We are currently investigating if some methods from the field of artificial intelligence can take advantage of this extra data to derive correct classifications.

Acknowledgments. We would like to thank NOAO and SAO for financial support which made this presentation possible. We are also very grateful to Drs. A.E. Piskunov and D.J. Mink for stimulating discussions, as well as Prof. J.N. Bahcall for his model software and constant interest in the work. The project is partially supported by the Russian Foundation for Fundamental Research (grant RFFR 93-02-2942).

References

Arenou, F., Grenon, & M., Gómez, A. 1992, å, 258, 104

Bahcall, J.N., & Soneira, R.M. 1980, ApJS, 44, 73

Jenkner, H., Lasker, B.M., Sturch, C.R., McLean, B.J., Shara, M.M., & Russell, J.L. 1990, AJ, 99, 2081

Russell, J.L., Lasker, B.M., McLean, B.J., Sturch, C.R., & Jenkner, H. 1990, AJ, 99, 2059

Astronomical Data Analysis Software and Systems III
ASP Conference Series, Vol. 61, 1994
D. R. Crabtree, R. J. Hanisch, and J. Barnes, eds.

Picturing the Guide Star Catalog

D. Mink

Harvard-Smithsonian Center for Astrophysics, Cambridge, MA 02138

Abstract. The information contained in the two CDROMs of the Hubble Space Telescope Guide Star Catalog has been condensed into a series of all-sky maps. The magnitudes of Guide Stars of the northern and southern hemispheres, which were taken on different types of plates have been calibrated to produce a statistically accurate V-filter flux map of the Milky Way galaxy. Individual object fluxes cannot be calibrated accurately, but by using a large number (16,000 for the first pass, 180,000 for the second) of sources near the equator which were measured on both Palomar and SERC plates, an average calibration can be fit. An intermediate step in the calibration produced a catalog of 180,524 objects in the GSC for which measurements at two colors exist. A catalog binning program, SKYPIC, developed from SKYMAP was used to produce two-dimensional FITS images from the Guide Star Catalog. Maps of source density, as well as flux maps of sources flagged as non-stellar, show excesses at the edges of plates.

1. Introduction

The Hubble Space Telescope Guide Star Catalog (described by Lasker et al. 1990, Sturch et al. 1990, and Jenkner et al. 1990, and Space Telescope Science Institute 1992) is a valuable resource for astronomers. In addition to helping to point the Hubble Space Telescope, it can be used to aid earth-based observations. A program, SKYMAP, has been developed to search the catalog and make finder charts for astronomical observation (Mink 1992). Software was also developed to create images from flux points for the Spacelab 2 Infrared Telescope (Kent et al. 1992). It was a logical step to combine these programs and make all-sky flux images of the Guide Star Catalog. The resulting SKYPIC program adds the improved map projection and catalog access features of SKYMAP to the flux-binning capabilities of the Spacelab 2 IRT software to produce flux images from any point source catalog.

2. Image Creation

To make an image from a typical star catalog, the first step is to turn the magnitudes in the catalog into energy flux, which is proportional to 10 to the -2.5V power. For the Guide Star Catalog, the "V" magnitude from the Northern

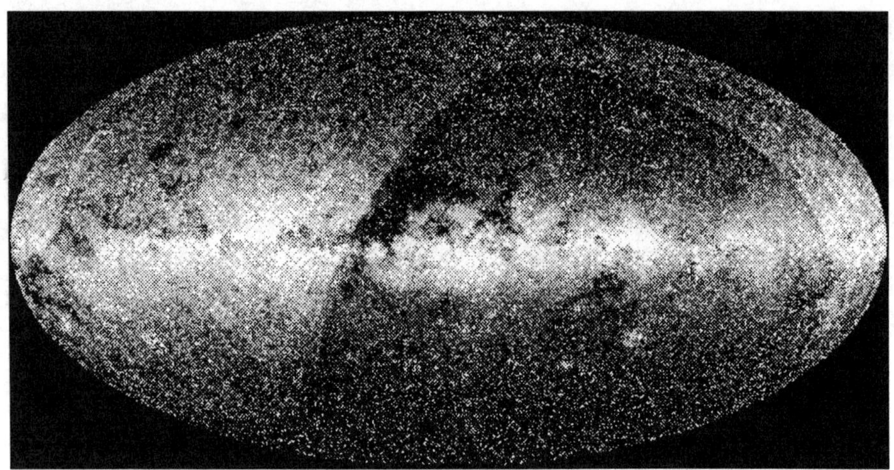

Figure 1. Aitoff projection of HST Guide Star Catalog flux in 30-minute bins with no magnitude recalibration.

Hemisphere Quick V Palomar plates was used as the standard, and a scale factor was chosen such that the smallest non-zero bin value was one.

Each sky position was then projected into an image array. For all-sky maps, the Aitoff projection was used, using the same algorithm as the IRAS All-Sky survey plates (IRAS Explanatory Supplement 1988). For the first pass, a 30-arcminute pixel size was used, giving an image with 720 pixels in longitude and 360 pixels in latitude. Positions and magnitudes for selected sources were read from the Guide Star Catalog CDROMs. For objects with multiple entries, error-weighted mean positions and magnitudes were used. The image levels are scaled so that the faintest star gives one count. For the first attempt, all magnitudes were assumed to be at the same bandpass. Figure 1 shows the result, scaled using SAOIMAGE's histogram method to bring out low flux levels without diminishing bright pixels. The brightness demarcation at the equator shows that magnitudes from the SERC-J plates in the Southern Hemisphere do not mean the same thing as those on the PAL-V Northern Hemisphere plates.

3. Calibration

There are 180,524 Guide Star sources which appear on both SERC-J and PAL-V. Since an image of the whole sky is being created, only statistical colors matter, not colors of individual sources. A second-order polynomial, chosen to speed the millions of evalutations needed, was fit to the plate magnitudes of the two-color sources as shown in Figure 2. An upper magnitude limit of Mag(PAL) + Mag(SERC) less than 29.5 was chosen to keep the magnitude limit difference between the two plate types from biasing the fit. To eliminate misidentifications, sources whose positions differed by more than one arcsecond were also ignored. The fit was used to convert SERC-J magnitudes to PAL-V magnitudes.

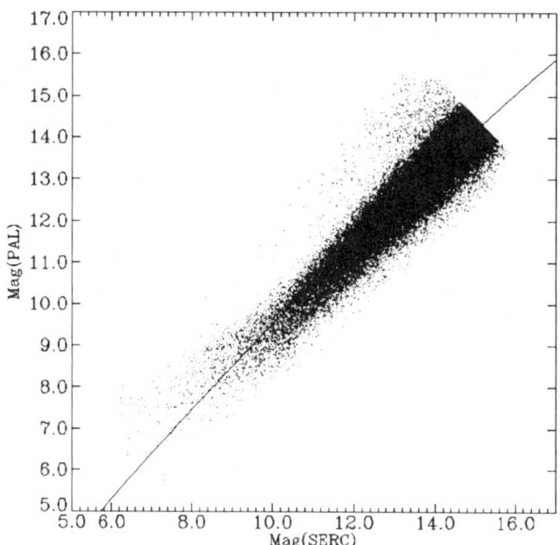

Figure 2. Polynomial fit of UK SERC Schmidt J magnitudes to Palomar Quick V magnitudes for 97,927 Guide Star Catalog sources appearing on both types of plates.

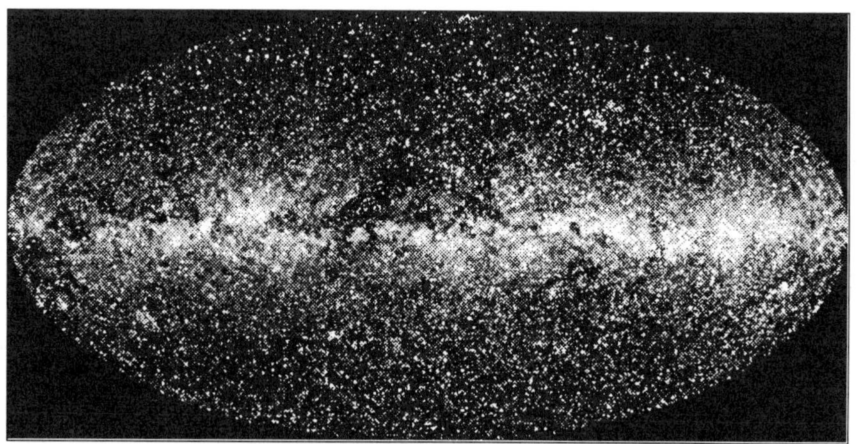

Figure 3. Aitoff projection of HST Guide Star Catalog flux in 15-minute bins with magnitude recalibration.

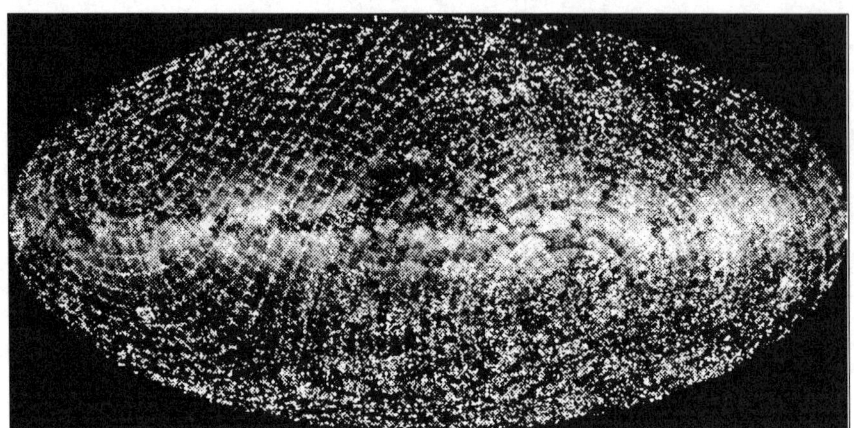

Figure 4. Aitoff projection of HST Guide Star Catalog flux in 15-minute bins with magnitude recalibration.

4. Images and Interpretation

Figure 3 shows the calibrated image with 15-arcminute bins bringing out more structural detail. Note that there is some faint structure following plate edges. This excess flux may be due to unidentified duplication of sources due to errors in astrometry near plate edges.

Figure 4 shows only sources flagged as non-stellar in the Guide Star Catalog. The structure is more pronounced, indicating that there is more flux from nonstellar sources in plate overlap regions, especially in the northern hemisphere (on the upper left). Fuzzy images at plate edges may cause an excess number of nonstellar sources. There may be multiple entries for the same sources due to bad plate solutions near the edges of Schmidt plates.

References

Guide Star Catalog Version 1.1, 1992, CD-ROM, Baltimore: Space Telescope Science Institute.

Neugebauer G., Habing H., Clegg P., & Chester T. 1988, "IRAS Explanatory Supplement", eds. Washington, D.C.: Government Printing Office.

Jenkner, H., Lasker, B., Sturch, C., McLean, B., Shara, M., and Russell, J. 1990, AJ99, 2081.

Kent, S.M., Mink, D., Fazio, G., Koch, D., Melnick, G., Tardiff, A., Maxson, C. ApJS78, 403.

Lasker, b., Sturch, C., McLean, B., Russell, J., Jenkner, H., and Shara, M. 1990, AJ, 99, 2019.

Mink, D.J. 1993, in Astronomical Data Analysis Software and Systems II, ASP Conf. Ser., Vol. 52, eds. R.J. Hanisch, J.V. Brissenden, and J. Barnes, 499.

Russell, J., Lasker, B., McLean, B., Sturch, C., and Jenkner, H. 1990, AJ, 99, 2059.

Astronomical Data Analysis Software and Systems III
ASP Conference Series, Vol. 61, 1994
D. R. Crabtree, R. J. Hanisch, and J. Barnes, eds.

Processing and Analysis of the Palomar – ST ScI Digital Sky Survey Using a Novel Software Technology

S. Djorgovski and N. Weir

Palomar Observatory, 105-24 Caltech, Pasadena, CA 91125, USA

U. Fayyad

Jet Propulsion Laboratory, 525-3660 Caltech, Pasadena, CA91125,USA

Abstract. We describe the design and implementation of a software system for producing, managing, and analyzing catalogs from the digital scans of the Second Palomar Observatory Sky Survey (POSS-II). The system (SKICAT) integrates new and existing packages for performing the full spectrum of tasks from raw pixel processing, to object classification, to the matching of multiple, overlapping Schmidt plates and CCD calibration sequences. It employs modern machine learning techniques, such as decision trees, to perform automatic star-galaxy classifications with a $> 90\%$ accuracy down to $\sim 1^m$ above the plate limit. The system also provides a variety of tools for interactively querying and analyzing the resulting object catalogs.

1. Introduction

Digitization of the Second Palomar Observatory Sky Survey (POSS-II; see Reid et al. 1991 for a description) is now in progress at ST ScI (Djorgovski et al. 1992, Lasker et al. 1992, Reid & Djorgovski 1993). So far, only a subset of the green (IIIa-J emulsion) and red (IIIa-F) plates have been scanned and processed, with the near-infrared ones (IV-N) expected to follow shortly. Both the photographic survey and the plate scanning are estimated to be $> 90\%$ complete circa 1997. The resulting data set, the Palomar-ST ScI Digital Sky Survey (DPOSS), will consist of ~ 3 TB of pixel data: ~ 1 GB/plate, with 1 arcsec pixels, 2 bytes/pixel, 20340^2 pixels/plate, for ~ 900 survey fields in 3 colors. ST ScI will provide an astrometric solution for each plate accurate to within approximately one arcsec r.m.s. We are also conducting an intensive program of CCD calibrations using the Palomar 60-inch telescope, using Gunn-Thuan gri bands. These CCD images serve both for magnitude zero-point calibrations, and as training and test data for star-galaxy object classifiers. These scans will be the highest quality set of images covering the entire northern sky produced to date, and their potential scientific value is enormous, if only the relevant information can be extracted quickly and efficiently.

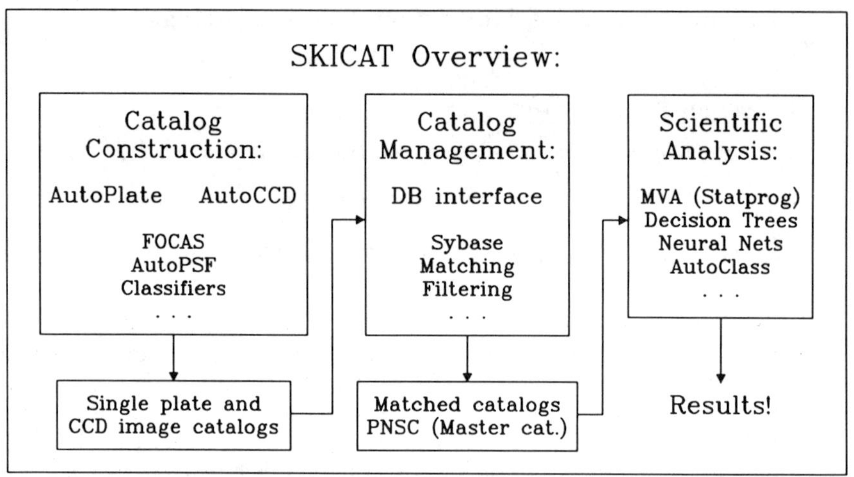

Figure 1. A schematic overview of the SKICAT system.

Caltech Astronomy and the JPL Artificial Intelligence Group have been engaged in an effort to integrate state-of-the-art computing methods for the scientific utilization of DPOSS, and in particular, to produce a Palomar Northern Sky Catalog (PNSC). The traditional means of extracting useful information from imaging surveys is through the construction of object catalogs. Thanks to developments in the fields of pattern recognition and machine learning, it is possible to reliably construct such a catalog objectively and automatically with a high degree of accuracy.

The Sky Image Cataloging and Analysis Tool (SKICAT) is a suite of programs designed to facilitate the maintenance and analysis of astronomical surveys comprised of multiple, overlapping images. More generally, it provides a powerful, integrated environment for the manipulation and scientific investigation of catalogs from virtually any source. SKICAT incorporates machine learning software technology for classifying sources objectively and uniformly, and facilitates handling the enormous (by present-day astronomical standards) data sets resulting from DPOSS (Fayyad et al. 1992, 1993*abc*; Weir et al. 1992, 1993*ab*, 1994; Weir 1994). It is a collection of new and borrowed, commercial and public domain, software products which have been integrated for a common purpose, with consistent command line and X-windows interfaces.

The SKICAT utilities fall into three main categories: catalog construction, catalog management, and catalog analysis. The relationship of these processes is illustrated schematically in Figure 1.

The first step, catalog construction, results in individual plate and CCD image catalogs. These, in turn, may be registered within the SKICAT database management system and matched, object by object, with other catalogs to create a 'matched catalog'. A matched catalog, or individual catalogs, may then be queried to create object catalogs, custom-designed for specific analysis purposes.

2. Catalog Construction

SKICAT provides two methods at present for constructing catalogs from digital sky images: one for DPOSS plate images, the other for CCD calibration sequences. While only these data sources are currently supported, SKICAT was purposefully designed to facilitate the construction and use of other types of catalogs in the future, e.g., catalogs of radio, infrared, or X-ray sources, etc.

2.1. Processing of Plate Scans

The heart of SKICAT is a collection of programs for the quasi-automatic processing of DPOSS plates from raw pixel to classified catalog form. Starting with a 1-GB digitized plate scan from ST ScI, SKICAT provides the tools for transferring the pixel data to SKICAT format, measuring the plate sky level and image boundaries, and determining a photographic density-to-intensity relation. The user then initiates a script, AutoPlate, which automates the process of cataloging the plate as a set of overlapping 2048^2 pixel 'footprints'.

The three most critical elements of plate processing are detection, photometry, and classification. By using the Faint Object Classification and Analysis System (FOCAS; Jarvis & Tyson 1979, Valdes 1982) for image detection and measurement, SKICAT is able to reach close to the faintest reliable limits of the plate scans, i.e., down to the typical equivalent limiting B magnitudes of $\sim 22^m$. In addition, by measuring quasi-asymptotic rather than isophotal magnitudes, using local sky estimates from annuli surrounding each object, and adapting the measurement thresholds within and across each plate to adjust for differences in sky level, noise, and pixel-to-pixel correlation, we are able to obtain very consistent photometry within and across plate boundaries. SKICAT thereby takes advantage of some of the latest advances in CCD image processing for the purpose of reducing digitized Schmidt plate data.

For classification, SKICAT benefits from the application of recent developments in machine learning. In particular, it utilizes the GID3* and O-Btree decision tree induction software, together with the Ruler system for combining multiple trees into a robust collection of classification rules (Fayyad et al. 1993*abc*) . These algorithms work by using measurements of a training set of classified objects and inferring an efficient set of rules for accurately classifying each example. The rules are simply conjunctions of multiple "if ... then ..." clauses, which condition upon any of eight different object parameters to determine an object's classification. The real advancement in using this type of classifier relative to those used in most large-scale surveys to date is twofold: first, we are able to condition upon a larger and more diverse set of attributes; second, we allow the computer to decide what are the optimal number and form of the rules. This naturally selects the more important attributes for classification.

We have also experimented with Neural Nets, and found their performance to be no better that that of decision trees, with the additional disadvantages of slow training and difficulty in interpreting their results (but see Odewahn et al.

1992, for a related work). Decision trees are constructed very quickly, and there is no problem of convergence, unlike with neural nets.

We create separate sets of rules for objects from the J (green) and F (red) plates. We use the CCD calibration data, which generally have superior image quality, to construct the training sets used to train the plate object classifiers. Classifications derived from the CCD data, more reliable than "by eye" estimates from the plates themselves, are matched to plate measurements to form the training sets. For measurements we use a set of robust, renormalized object parameters that we found to be distributed in a stable fashion within and across plates. By training the algorithms to classify based on these attributes, we are able to effectively remove the effect of PSF variation across a given plate, or even between different plates. Average accuracy of star-galaxy classifications as a function of magnitude is determined from tests using independent CCD-classified plate data. In both g and r bands (corresponding to the J and F plates), the accuracy drops below $\sim 90\%$ at about the same equivalent magnitude level, $B \sim 21.2^m \pm 0.2^m$. This is $\sim 1^m$ above the plate detection limits, and $\sim 1^m$ better than what was achieved in the past with similar data. This increase in depth effectively doubles the number of galaxies available for scientific analysis, relative to the previous automated Schmidt surveys.

Plate X,Y to RA,Dec assignment, like object classification, is automatically performed in the final stages of catalog construction. As both of these steps use existing catalog measurements, not raw pixel data, they may be repeated at later times using a different set of classification rules or astrometric solution coefficients. SKICAT easily facilitates the continuous improvement of its catalogs as better calibration, or even entirely new algorithms, become available.

2.2. CCD Calibrations

CCD catalogs are constructed using most of the same tools as are applied to plate data. A script called AutoCCD, analogous to AutoPlate, is used to quasi-automatically process an image from pixel into catalog form. The primary differences are in the forms of pre- and post-processing that are applied. In particular, a whole host of standard CCD calibration procedures (e.g., de-biasing, flat-fielding, photometric calibration, etc.), far different from those for plates, must be followed before running AutoCCD. In addition, we found FOCAS's built-in classifier to provide very accurate results on the CCDs down to the plate detection limit, which is our magnitude limit of interest. We were, therefore, able to let FOCAS automatically classify each object, with just a quick follow-up check by eye, producing excellent quality data without the need for much human interaction or more sophisticated classification algorithms.

As mentioned previously, CCD data are used for two purposes in constructing the DPOSS. First, they provide "true" object classifications, at very faint levels, for our classifier training sets. Because the CCD images are of higher resolution and signal to noise ratio (SNR) than digitized plates, we are able to assign accurate classifications to objects whose morphology is not reliably distinguishable, even by an expert, when looking at the plate image alone. Through the machine learning process, the aim is to train the computer to consistently classify these faint objects, thereby enabling it not just to mimic a human's performance, but actually *improve* upon it.

The second, probably most important, purpose for the CCD measurements is to provide photometric calibration for the plate catalogs. We use Gunn g, r, and i measurements to calibrate the IIIa-J, IIIa-F, and IV-N plate data, respectively. These CCD bandpasses provide a reasonable match to the photographic emulsion plus filter passbands.

3. Catalog Management

Once individual catalogs are constructed, the next steps are to combine, query, or more generally, manage them. Before describing the catalog management tools within SKICAT, brief sections on the system environment and terminology are in order.

3.1. Environment

The SKICAT system is implemented in the Unix operating system. The software is largely written in C and C-shell scripts, although portions of it are written in Fortran. As mentioned before, SKICAT is built around and incorporates a number of pre-existing software packages: FOCAS routines for image detection and measurement; the GID3*/O-Btree/Ruler induction software for object classification; and the Sybase commercial relational database management system for maintaining and accessing the data. While SKICAT was developed using these packages, none are irreplaceable. Each package serves its purpose, and because of the modularity of the system, could be substituted for another which performs the same function. In addition, SKICAT provides quick and easy access to most system utilities through a common X Windows graphical user interface, while users familiar with Unix can access the same utilities directly from the Unix command line.

3.2. Some Terminology and Definitions

A brief description of various terms we use may be helpful here:

A *feature* is the set of measurements (magnitude, surface brightness, position angle, etc.) of a unique object contained in a catalog. For example, a star may be a feature within a catalog, as might be a galaxy or a satellite trail detected on a plate.

A *table* is a collection of data organized by row and column, where each row has a value (or space for a value) for every column in the table. For example, a list of galaxies may be organized in the form of a table, with one row per galaxy (feature) and one column per galaxy measurement. SKICAT tables are stored and manipulated using Sybase. Therefore, all references to tables refer specifically to the Sybase data structures of the same name.

A *catalog* consists of a features table and a header table. These are data sets produced by Autoplate and AutoCCD. A features table contains one row for each feature appearing in the catalog. The header table contains information relevant to the entire catalog (image source, date of creation, etc.) and is generally used for reference purposes.

An *object* is a unique image artifact or physical sky object (i.e., star, galaxy, etc.) to which there may correspond multiple features within distinct catalogs.

For example, the *object* M87, which lies in the overlap of two plates, would appear as a *feature* within both plates' catalogs.

A *matched features table* is a table containing features from multiple, matched catalogs. Features at the same RA and Dec position (within astrometric uncertainty) are considered to be different measurements or features of the same object. They are assigned a common object ID during the matching process.

A *matched catalog* is a data set which consists of a matched features table and a table listing those catalogs comprising it. New catalogs are added to it by matching each new feature with existing matched features (objects). The user controls which subset of measurements to include in the matched features table and also specifies parameters affecting the matching algorithm. In a reverse operation, selected columns within catalog features tables may be updated from their corresponding entries in the matched features table.

Objects tables are produced by filtering and outputting selected columns of object entries from any individual catalog or the matched catalog. They might be generated for catalog calibration, specialized scientific analysis, or as distributed data products (such as the PNSC). These tables may also be queried and manipulated using the SKICAT table manipulation tools.

3.3. Catalog Construction and Management Methodology

For each plate or CCD image, the catalog construction scripts generate a Sybase header and features table, together comprising what we term a SKICAT catalog. The header table consists of columns of parameters used to guide the catalog construction process, the name of the image from which the catalog was derived, the location of the image on offline storage, comments, and other information necessary to identify the data source and reconstruct the catalog from scratch, if necessary. The features table contains one row for each detected feature in the image. The columns represent the measured attributes of each feature. Approximately 50 parameters per object are measured and saved in the individual plate and CCD catalogs.

After the construction process, catalogs within SKICAT must be registered in the SKICAT system tables, where a complete description and history of every catalog loaded to date is maintained. Catalog revisions, that might result from deriving new and improved plate astrometric solutions or photometric corrections, are also logged. The system is thereby designed to manage a data base constantly growing and improving with time. The SKICAT system tables also keep track of which catalogs are currently loaded on-line. Those which are on-line exist on the specially allocated Sybase disk drive(s) and are not visible from the host operating system. SKICAT provides tools for quickly and easily saving/loading catalogs off-line/on-line. Only registered catalogs may be moved to/from off-line storage or matched with other catalogs.

Multiple, overlapping catalogs can be matched into a special SKICAT data structure called the matched catalog. This process is illustrated in Figure 2. The matched catalog consists of a matched features table and a table of those catalogs comprising it. The matched features table contains independent entries for every measurement of every object detected in the constituent catalogs. Because of size and speed considerations, not every attribute may feasibly be saved within the matched catalog, but a sufficiently small subset of parameters is generally

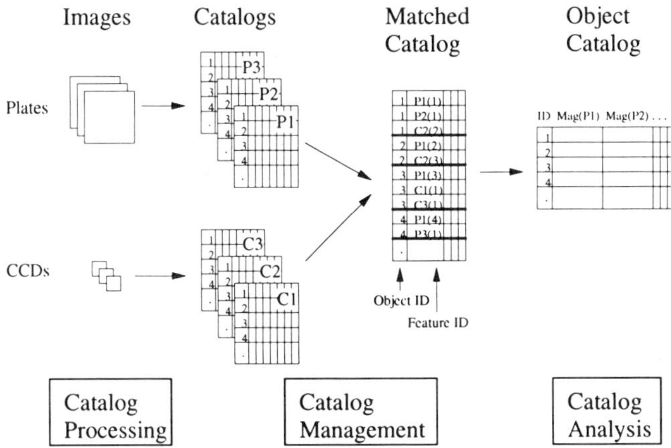

Figure 2. SKICAT data flow and catalog organization.

more than adequate for most uses of the data. (Of course the saved catalogs themselves provide a complete archive of the full list of parameters if they are ever needed.) In its present version, SKICAT allows for only one on-line matched catalog, though it may be saved and loaded to/from off-line storage and a new one created at any time. There is no fundamental practical or philosophical reason for this limitation, and it may be removed in a future version.

The matched catalog may be queried using a sophisticated filtering and output tool to generate a so-called object table (see Figure 2), which contains just a single entry per matched object. With this tool, the user may, for example, generate a distributable data product, such as a galaxy list, from the current set of matched plate catalogs. The tool may also be used to perform consistency checks within catalog overlap regions, or to perform specialized scientific analysis over large survey regions. For example, a user may request a listing of all stars within a well-defined section of sky covered by multiple J and F plates, specifying exactly which object attributes to report (e.g., magnitude, RA, Dec, etc.) and from which source (specific J plates, average of all F plates, etc.).

Catalogs may be easily altered using a procedure that allows arbitrary operations on table columns. In this way, catalogs may be recalibrated or otherwise adjusted in light of new or improved data. Such updates might include applying a field-effect correction to a plate's list of magnitude or performing new classifications using an improved rules set. A catalog may also be modified by updating selected columns from corresponding columns in the matched catalog.

This procedure would be appropriate if, for example, the entries in a matched catalog were calibrated, and the calibrated measurements needed to be passed back to the original catalogs for archival purposes. An updated catalog could subsequently be re-registered as a new version of the existing catalog. Both the original and new header information would now be saved in the system, maintaining a complete history of catalog revisions. Via this mechanism, SKICAT is designed to maintain a "living," growing database, instead of a data archive fixed for all time. Using SKICAT, we hope to build a new kind of an astronomical catalog, one which is continuously being improved and extended, and to provide a ready set of software tools for its scientific exploitation.

A typical SKICAT reduction procedure is to construct a catalog, register it, save it, match it into the matched catalog, then delete it from on-line storage. The matched catalog is then queried to establish photometric, classification, and astrometric consistency in light of the new data. If discrepancies are found, one or more measurements within the matched catalog are corrected, and the corresponding individual catalog(s) updated. The updated catalog(s) are then registered as new versions before being saved to tape. Once a matched catalog of suitable size and quality is constructed, a list of object positions and average magnitudes across multiple plates may be produced, which could be used for scientific investigations.

4. Catalog Analysis

The third layer of SKICAT, which is still under development, will consist of a powerful toolbox of modern data analysis algorithms to be applied for survey data space exploration and the scientific analysis of the catalogs. It will facilitate more sophisticated scientific investigations of these expanding survey data sets, including a multivariate statistical analysis package, and a wide variety of Bayesian inference tools, objective classifiers, and other advanced data management and analysis packages and algorithms.

The analysis tools included in the current version of SKICAT are the GID3*/O-Btree decision tree induction software and Ruler program for classification learning, as well as the extremely useful collection of stream processing routines included in the standard FOCAS distribution. The very same classification learning software which was used to create the classifiers in SKICAT's plate cataloging script are available for use on any SKICAT data set, or even data from external sources. SKICAT provides an environment for implementing these tools to train and produce classifiers for scientific uses of the DPOSS, or any other catalogs, that we had never anticipated.

We also intend to explore the potential of *machine-assisted discovery*, where modern, AI-based software tools automatically explore large parameter spaces of data and draw a scientist's attention to unusual or rare types of objects, or non-obvious clusters of objects in parameter space. We have begun applying the AUTOCLASS (Cheeseman et al. 1988) unsupervised classification software to DPOSS, with plans to implement this and other Bayesian inference and cluster analysis tools within SKICAT in the future. The results of our initial experiments are very encouraging: given a set of object parameters, but no pixel information whatsoever, AUTOCLASS decided that there are four distinct types of objects in the data. Upon subsequent examination, these turned out to be stars, stars with a fuzz (FOCAS sf object type), and two kinds of galaxies, differing in the apparent concentration (perhaps early and late Hubble types?). Note that these classifications, which obviously make physical sense, are produced from the data themselves, with no human input and no training data examples. We see this as a step towards a fully automatic, objective exploration of very large astronomical data sets. There may be even a potential here for a *discovery of new types of astronomical objects*, which may have escaped notice, e.g., due to their rarity, but would occur in sufficient numbers in a data set consisting of billions of sources.

SKICAT represents our attempt to introduce modern machine-learning based software tools into astronomy in a meaningful way. The resulting Palomar Northern Sky Catalog (PNSC), when completed, is expected to contain $\sim 5 \times 10^7$ galaxies, and $> 2 \times 10^9$ stars, in 3 colors (photographic JFN bands, calibrated to CCD gri system), down to the limiting magnitude equivalent of $B \sim 22^m$, with the star-galaxy classification accurate to $\sim 90 - 95\%$ down to the equivalent of $B \sim 21^m$. The catalog will be continuously upgraded as more calibration data become available. It will be made available to the community via computer networks and/or suitable media, probably in installments, as soon as scientific validation and quality checks are completed. Analysis software (parts of SKICAT) will also be freely available. The first, partial releases may be available within a year or two from now.

A vast variety of scientific projects will be possible with this data base, including studies of large-scale structure, Galactic structure, automatic identifications of sources from other wavelengths (radio through X-ray), generation of objectively defined catalogs of clusters and groups of galaxies, searches for quasars, variable or extreme-color objects, low surface brightness galaxies, etc., to name just a few. Some of this work is now starting at Caltech.

Our work on DPOSS provided an initial motivation for the development of SKICAT, but this is just a beginning: these tools are quite general, and applicable to a broad range of digital sky surveys, and other information contexts. Digital imaging data sets of many Gigabytes (or even Terabytes) in size are becoming more common, and may become standard in a very near future. Their exploration and full scientific utilization call for a new generation of data processing and analysis tools. We see our work as a step in this direction.

Acknowledgments. This work was supported at Caltech in part by the NASA AISRP contract NAS5-31348, Caltech President's fund, and the NSF PYI award AST-9157412, and at the JPL under a contract with the NASA. The POSS-II is partially funded by grants to Caltech from the Eastman Kodak Co., the National Geographic Society, the Samuel Oschin Foundation, the NSF grants AST 84-08225 and AST 87-19465, and the NASA grants NGL 05002140 and NAGW 1710. We acknowledge the efforts of the POSS-II team at Palomar, and the scanning team at ST ScI.

References

Cheeseman, P., et al., 1988, in Proc. Fifth Machine Learning Workshop, San Mateo: Morgan Kaufmann, 54

Djorgovski, S., Lasker, B., Weir, N., Postman, M., Reid, I.N., & Laidler, V. 1992, BAAS, 24, 750

Fayyad, U., Doyle, R., Weir, N., & Djorgovski, S. 1992, in Proceedings of the ML-92 Workshop on Machine Discovery (MD-92), ed. J. Zytkow, San Mateo: Morgan Kaufmann, 117

Fayyad, U., Weir, N., Roden, J., Djorgovski, S., & Doyle, R. 1993*a*, in ed. K. Krishen, Sixth Annual Workshop on Space Operations, Applications, and Research (SOAR-92), NASA CP-3187, 340

Fayyad, U.M., Weir, N. & Djorgovski, S., 1993*b*, in Proc. Tenth International Conference on Machine Learning, San Mateo, CA: Morgan Kaufmann, 112

Fayyad, U., Weir, N., & Djorgovski, S., 1993*c*, in Proc. Second International Conference on Information and Knowledge Management (CIKM-93), Washington: ISCA/ACM, in press

Jarvis, J., & Tyson, J.A. 1979, Proc. SPIE, 172, 422

Lasker, B., Djorgovski, S., Postman, M., Laidler, V., Weir, N., Reid, I.N., & Sturch, C. 1992, BAAS, 24, 741

Odewahn, S.C., Stockwell, E.B., Pennington, R.L., Humphreys, R.M., & Zumach, W.A. 1992, AJ, 103, 318

Reid, I.N. et al., 1991, PASP, 103, 661

Reid, I.N., & Djorgovski, S. 1993, in Sky Surveys: Protostars to Protogalaxies, ed. B.T. Soifer, ASPCS, 43, 125

Valdes, F. 1982, Proc. SPIE, 331, 465

Weir, N., Djorgovski, S., & Fayyad, U. 1992, BAAS, 24, 1139

Weir, N., Djorgovski, S., Fayyad, U., Roden, J., & Rouquette, N. 1993*a*, in Astronomy from Large Data Bases II, ed. A. Heck & F. Murtagh, ESO CWP-43, 513

Weir, N., Fayyad, U., Djorgovski, S., Roden, J. & Rouquette, N. 1993*b*, in Astronomical Data Analysis Software and Systems II, A.S.P. Conf. Ser., Vol. 52, eds. R. Hanisch, R. Brissenden & J. Barnes, 39

Weir, N., Djorgovski, S., Fayyad, U., Smith, J.D., & Roden, J. 1994, in Astronomy From Wide-Field Imaging, IAU Symp. #161, ed. H. MacGillivray, Dordrecht: Kluwer, in press

Weir, N. 1994, Ph.D. thesis, California Institute of Technology

Astronomical Data Analysis Software and Systems III
ASP Conference Series, Vol. 61, 1994
D. R. Crabtree, R. J. Hanisch, and J. Barnes, eds.

Sloan Digital Sky Survey

Stephen M. Kent, Chris Stoughton, Heidi Newberg, Jonathan Loveday, Don Petravick, Vijay Gurbani, Eileen Berman, Gary Sergey

Experimental Astrophysics Group, Fermi National Accelerator Laboratory, MS 127, PO Box 500 Batavia, IL 60510 USA

Robert Lupton

Princeton University, Peyton Hall, Ivy Lane, Princeton, NJ 08544 USA

Abstract. The Sloan Digital Sky Survey will produce a detailed digital photometric map of half the northern sky to about 23 magnitude using a special purpose wide field 2.5 meter telescope. From this map we will select $\sim 10^6$ galaxies and 10^5 quasars, and obtain high resolution spectra using the same telescope. The imaging catalog will contain 10^8 galaxies, a similar number of stars, and 10^6 quasar candidates.

1. Introduction

The partners in the SDSS are the University of Chicago, Princeton University, the Institute for Advanced Study, Fermilab, Johns Hopkins University, and the Japanese Promotion Group. The specific goals of the survey are as follows:

1. In π steradians of the North Galactic Pole: (a) obtain a photometric survey in 4 or 5 filters to $R = 23$ (5σ); (b) obtain redshifts for all galaxies to $B = 19$; (c) obtain redshifts for all quasars to $B = 20$.

2. In a strip $2° \times 50°$ of the South Galactic Pole: (a) obtain a deep photometric survey to $R = 25$; (b) obtain redshifts for all galaxies to $B = 20$; (c) obtain redshifts for all quasars to $B = 21$.

3. If feasible, obtain a best effort imaging survey of the Galactic Plane.

The survey will use a new $f/5$ 2.5 meter telescope of altitude-azimuth design that is under construction at Apache Point Observatory (APO) in New Mexico. The telescope is a modified Ritchey-Cretien design that uses two corrector lenses near the focal plane to achieve a 3° field of view with no distortion.

Imaging will be done using a camera that consists of a mosaic of 52 CCDs. Thirty of these CCDs are Tektronix 2048×2048 arrays that are used for the primary imaging observations, arranged in an array of 6 columns with 5 CCDs per column (Figure 1). Each CCD in a column has a different filter with the following wavelengths: u: 3506; g: 4734; r: 6270; i: 7691; z: 9247.

The imaging survey will be conducted in drift scan mode. The telescope will be actively tracked so that a given piece of the sky trails along the 5 CCDs of a column in succession. The transit time of a single CCD will be 55 seconds and the time to cross the array will be about 7 minutes. The columns of CCDs are spaced by slightly less that one CCD width; thus, 2 successive interlaced scans of the telescope will produce a completely filled image of a strip of the sky 2.5 degrees wide.

A total of 22 small CCDs, leading and trailing the main imaging CCDs, provide the astrometric calibration. These CCDs will tie bright ($V < 9$) stars with known astrometric positions to fainter ($V = 14$) secondary stars. The desired accuracy is $0.2''$ rms in each coordinate.

A separate 0.61 meter monitor telescope (MT) will be used for the photometric calibration. This monitor telescope will have a single CCD camera and a filter wheel box. The functions of this telescope are fourfold. First, it will set up a set of standard stars for the photometric calibration (the SDSS filters are not on any standard system). Second, during imaging observations, it will repeatedly observe the standard stars to monitor the atmospheric transparency. Third, it will observe a large number of patches in common with the 2.5 meter telescope, to calibrate the main imaging survey. Fourth, it will observe spectrophotometric standards to calibrate spectra.

Galaxy, quasar, and star targets will be selected from the imaging data for follow-up spectroscopy. The spectroscopy will be done with two multifiber spectrographs, each with a blue and red channel. The two spectrographs combined can measure 600+ objects simultaneously in the 3° field. The fibers will be positioned using drilled plates. The spectroscopic resolution is 3 Å, allowing velocity dispersions to be measured for the brighter galaxies. The exposure times will be on the order of 1 hour.

The galaxy survey is intended to be as complete as possible. Galaxies will be skipped only if they are so close so as to cause interference between fibers. Since the distribution of galaxies on the sky is highly variable, the plate centers will not be placed on a uniform grid but rather will be adjusted to increase overlap in regions of high target density.

The results from the survey will be distributed in the most convenient form available. The products include: tables of all objects found in the survey ($\sim 200 \times 10^6$) with parameters; postage stamps of all objects; tables of redshifts for all objects with spectra; reduced 1 dimensional spectra; and the 2 dimensional images from which spectra were extracted.

Writing and running the code to build this archive requires coordinated effort of approximately 10 computer professionals at Fermilab and 20 or so scientists at the six institutions. In the following sections, we discuss the organization of these software products, as well as the tools we have developed and are using to work together.

2. Online Systems

The on-line systems are the hardware and software at APO that collect and record data from the instruments. Figure 2 shows the major pieces of the online systems.

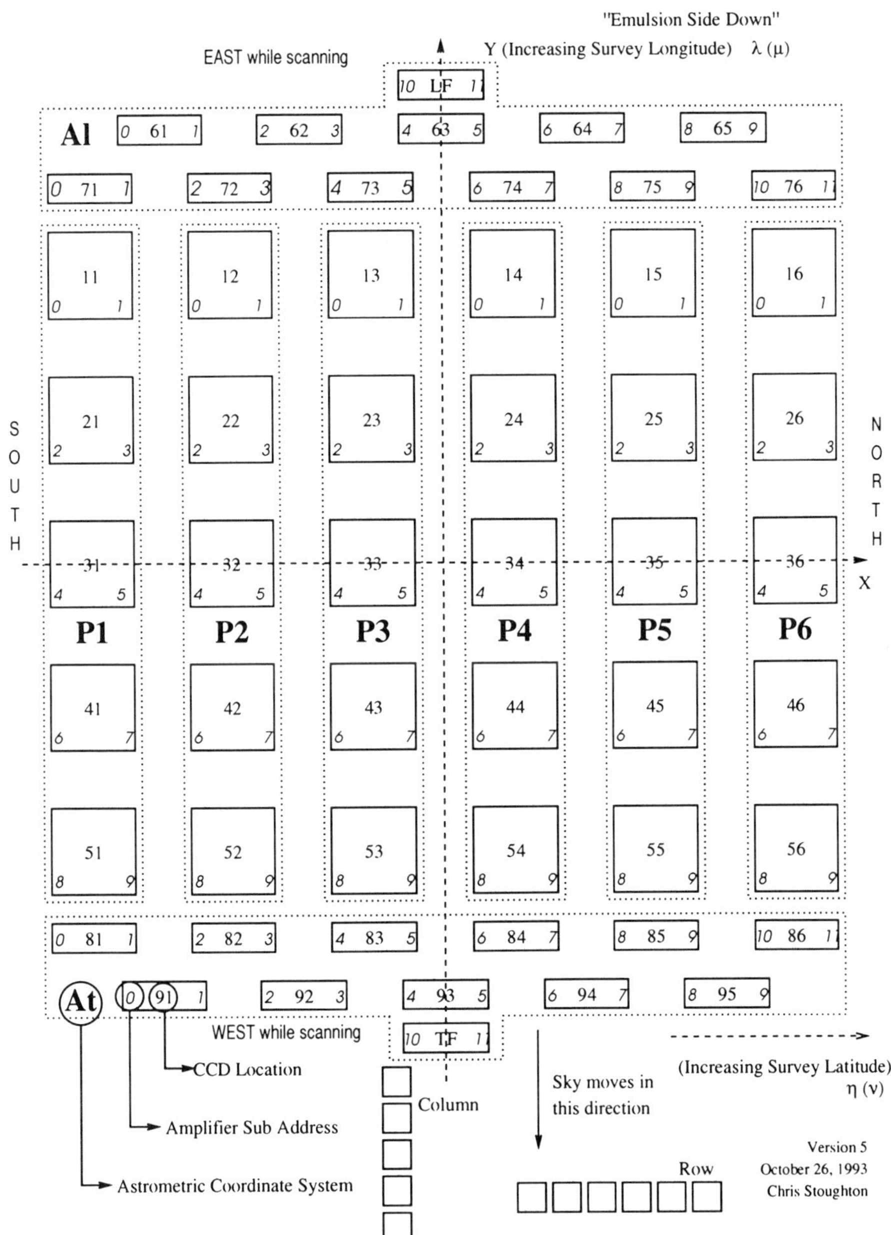

Figure 1. Focal plane layout for the CCD imaging camera.

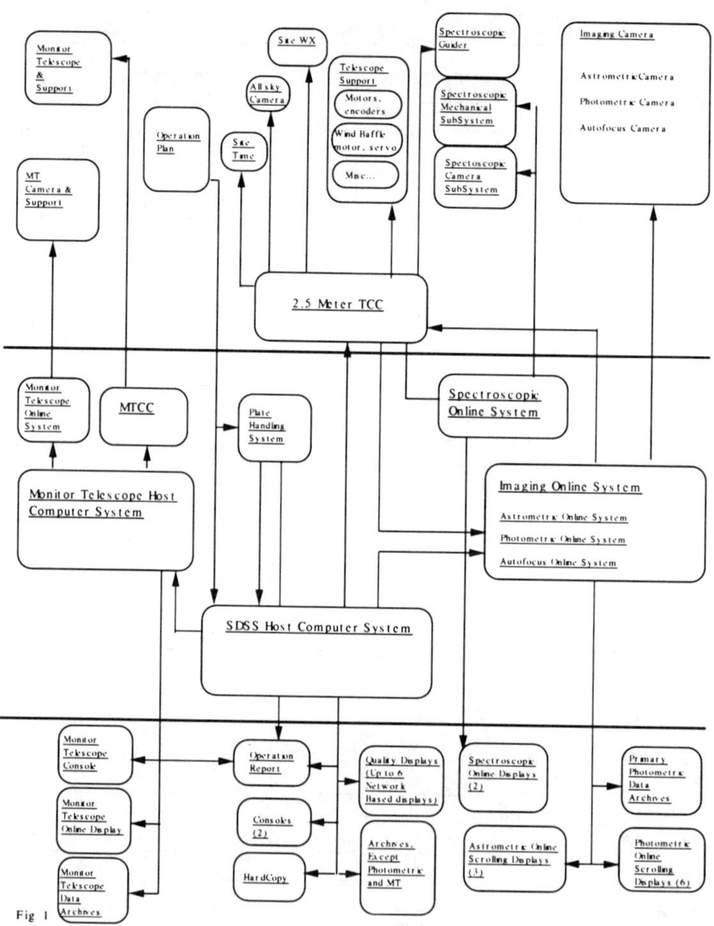

Figure 2. Online systems at Apache Point Observatory.

2.1. Imaging

The imaging camera produces the largest amount of data at the highest rate of all the instruments. MacKinnon (1993) describes the development of this system in some detail. We concentrate here on the organization of the data stream, on-line monitoring, and the data products written to tape.

The data acquisition and on-line processing are performed in MVME167 processor boards embedded in the data acquisition system. In addition to controlling the data flow between the CCD electronics, disk buffer, and tape drives, these processors will do a basic analysis of each frame. First, a histogram of all pixel values for each column of *pixels* is continuously computed. Periodically, the statistics of each histogram (the 25%, 50%, and 75% quartiles) are reported. This allows us to track the sky level. Second, bright objects are detected and copied out of the frames, allowing us to monitor the point spread function, object counts, and telescope tracking. Data are also sent to scrolling display monitors.

Before going to tape, the data are written to a hard disk buffer. We delay writing data from the leading chip in the column to tape until the corresponding position in the sky has scanned past all of the CCDs in the column. The five frames of different color are then recorded sequentially on tape. This disk buffer also allows us to selective copy frames to a Unix environment for more detailed analysis and viewing. Although we do not support the full bandwidth in this mode, the flexibility enables us to track down problems as they arise.

At sidereal rate the CCDs are clocked at 37.5 rows per second. Including the overscan region, the data rate is 168.6 kbyte/sec for each CCD. The array of photometric chips (6 columns of 5 CCDs) produces data at just over 5 Mbyte/sec. To record at this rate we write data in parallel streams. The data from each column of 5 CCDs are written to two Exabyte tape drives to make primary and backup copies of the data simultaneously. If a single drive fails, we continue the scan and make the backup copy later. A total of 12 drives will record two copies of the photometric imaging data. The data products written by the imaging system are:

photometric frames: the pixel values from all of the photometric CCDs.

postage stamps: the pixel values just around bright objects, extracted and measured on-line. This list is not guaranteed to be complete.

quartile arrays: the summary of the pixel intensity in each column, extracted from the histograms each frame.

imaging report: a summary of the imaging observing for the run — starting and stopping time, and summaries of the data quality (PSF profile, sky values) to give an overview of the observing conditions for the run.

instrument report contains data about special maintenance, such as replacing a CCD in the camera, that need to be tracked.

2.2. Monitor Telescope

The camera electronics are a high speed version of those for the main imaging camera. This system is distinguished by its automated observing mode.

We will write the frames to disk during the night, and copy them (twice) to tape at the end of the night. The data products written by the monitor telescope system are:

monitor telescope frames are the frames for all the exposures.

monitor telescope report is the summary of the monitor telescope observations during the night, including which fields were completed, and also a summary of the atmospheric extinction measured throughout the night.

2.3. Spectroscopic

The spectroscopic system controls the spectrograph, and acquires and monitors the spectra and associated calibration frames. We will extract and calibrate at least a subset of the spectra during the night, using IRAF.

The guiding system will be tightly coupled to the telescope control computer. It will read out small CCDs illuminated by coherent bundles of fibers placed on guide stars. The overall performance of this system, along with the measured flux and image shape, will also be recorded with the spectroscopic data. The data products written by the spectroscopic system are:

spectroscopic frames are the two-dimensional spectra (two or three exposures), along with any flats and calibration arcs that are routinely taken.

plugging report gives the correspondence from fiber number to hole number, to match spectra with objects.

spectroscopic report is a summary of the spectroscopic observations for the night, including the exposure time for each plate, as well as summary information from the guiding system.

3. Data Processing Pipelines

Data recorded on the mountain will all be shipped to Fermilab for subsequent processing. Figure 3 gives the overall organization of the off-line data processing. We organize the processing into four pipelines: photometric, astrometric, monitor telescope, and spectroscopic. Our goal is to have the data processed "automatically" through the pipelines. The results are checked before committing them to the database. This check is then used to guide future observing plans to keep the survey uniform.

3.1. Astrometric Pipeline

The astrometric pipeline reads the postage stamp files from the astrometric and photometric CCDs. At the beginning of the survey we will construct a "great circle star catalog" of known astrometric standards (from the Hipparcos catalog, for example) converted to our survey coordinate system. Standards from this catalog are matched with stars measured in the astrometric CCDs to determine a primary calibration. Fainter stars on the astrometric CCDs transfer this calibration to the photometric CCDs. Because the final positions of fainter stars will not be available until after the photometric pipeline has run, provisions are made to recalibrate the output of the photometric pipeline at any time.

3.2. Monitor Telescope Pipeline

The MT frames pipeline processes the two dimensional frames and produces lists of detected stars. The major steps are: flat field and bias correction, find the sky, locate the stellar images, measure instrumental aperture magnitudes, align frames in different colors, and (for the primary standards) identify the field. The

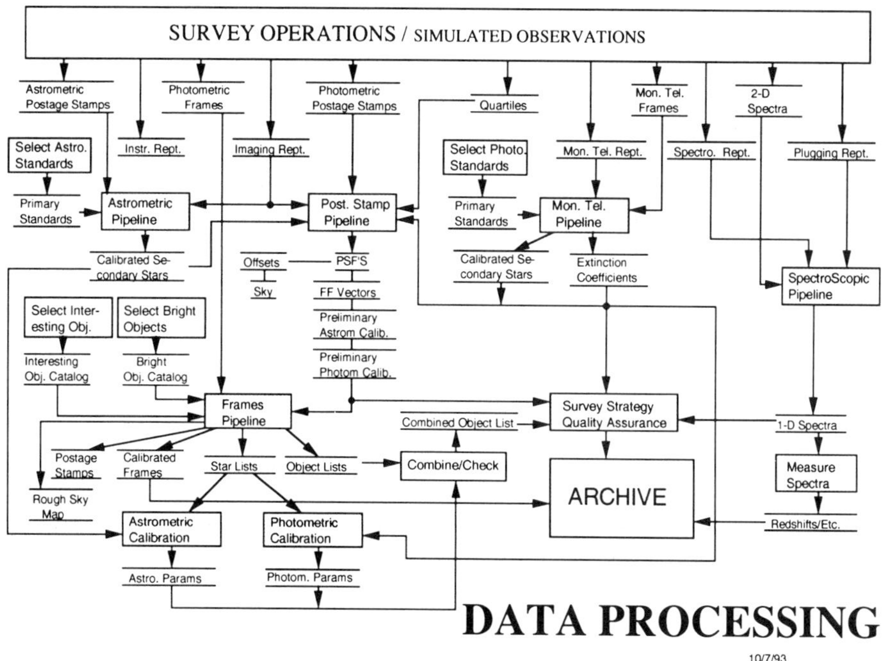

Figure 3. Overall design of the off-line data processing.

output is a list of stars with instrumental magnitudes and colors. Then, the MT calibration pipeline runs on the primary standards to compute the extinction and instrumental coefficients.

3.3. Photometric Pipeline

The basic job of the photometric pipeline is to take the imaging data from the photometric CCDs and produce a list of objects. The "postage stamp pipeline" reads the postage stamps of bright stars generated by the imaging on-line system and determines: sky value, flat field, PSF, preliminary astrometric and photometric parameters. The "frames pipeline" reads all of the photometric frames to find and measure objects in each frame.

The processing steps are: correct frames; reject bad frames; find, measure, merge, bright objects; identify known objects; find objects; merge colors; and measure objects.

The output of the photometric pipeline is a list of the objects, their measured parameters, and the postage stamps (now a complete set of all detected objects) for each frame. The lists from all frames in a scan are combined. Final astrometric and photometric calibration parameters are calculated in a separate step once the astrometric and monitor telescope pipeline results are found satisfactory.

3.4. Spectroscopic Pipelines

The spectroscopic pipeline takes the data and calibration exposures and, along with the spectrophotometric standard measurements from the monitor telescope, extracts and measures the spectra.

We extract with following steps, using IRAF routines with minor modification to handle our data format: bias and throughput correction; aperture tracing; spectra extraction; dispersion correction; sky spectra extraction; sky subtraction; and median averaging.

These calibrated 1-d spectra are measured by: combining red and blue spectra; masking night sky lines; finding emission and absorption lines; identification with cross correlation.

4. Software Tools

To coordinate the efforts of over two dozen people, computer professionals and scientists, at 6 institutions we define standards for the software development environment, and then support products for building the framework. A guiding principle is to make all the applications look as similar as possible.

4.1. Standards

The standards for the hardware and software environment help us build robust, portable, documented code that is readily available to all members of the collaboration.

Platforms: We attempt to make our software reasonably portable. All code (with the exception of some on-line software) must run on both IRIX and SunOS operating systems. These are the operating systems in use at our member institutions.

Compilers: Virtually all new code is written in either ANSI C or C++. Some legacy code is written in Fortran. Code common to many applications is required to compile in 5 different compilers. This not only ensures that the code is portable, but also catches many subtle programming errors.

Software Distribution: Software distribution is via three tools: UPS (Unix Product Support) and UPR (Unix Product Remote) from Fermilab, and RCVS (Remote Concurrent Versioning System) from Stanford. UPS is a facility to maintain and support multiple versions and flavors (IRIX or SunOS) of our software products. UPR is a menu-driven facility that allows remote users at other institutions to fetch copies of the current versions of software products. RCVS is a source code management system that allows several people to "check out" a module for work simultaneously. Although this sounds a bit frightening at first, our experience (and the experience of others) is that when people continue to check in their changes regularly, any conflicts are detected by the system and are easy to resolve. The big advantage is that we avoid a gridlock situation where a small number of people are working on a module, locking out everyone else.

Documentation: We use TeXand LaTeXconverted to Postscript for printed documents, and the hyper text markup language (HTML) with a browser (Xmosaic) for reading on-line documents. A simple perl script converts HTML to

LaTeX, so we need to maintian only one set of documents. This documentation is distributed or updated whenever the product is distributed.

4.2. Framework

While it might have been natural to use IRAF as the framework for new code development, we chose not to do so for a couple of reasons. First, there was little IRAF experience within the collaboration, particularly at Fermilab. Second, IRAF is not well suited to the photometric pipeline where efficient control of system resources is needed. We created our own framework, drawing upon public domain software in many places. Virtually all applications, from the data acquisition systems to the database interface routines, are developed within this framework.

TCL (Berkeley) TCL (Tool Command Language) is the backbone. TCL is a programmable command line interpreter, akin to cl in IRAF. The major advantages are that it is highly portable (it even runs on the MVME 167 microprocessors in the on-line systems) and that it explicitly provides a clean interface for adding new commands implemented as blocks of C code. Online systems and data reduction pipelines are written as TCL verbs and scripts.

Tk (Berkeley) Tk (Took Kit) is a graphical user interface built with TCL commands. Simple screens to run on-line systems and, for example, to query the database, are built without compiling.

pgplot (Cal Tech) is a plotting package which we integrated as a set of TCL verbs and c modules.

FSAOImage (SAO) is an extension of the venerable SAOImage display program, implemented as a set of TCL verbs.

Libfits (Johns Hopkins) is basic FITS input and output.

SHIVA: Written at Fermilab, it binds these pieces together and supports a variety of data structures, such as linked lists and image regions.

IRAF: We will use IRAF in one place, the spectroscopic pipeline, for extraction of 1-d spectra from 2-d frames.

5. Databases and Archives

The data archives are critical both for the operation of the survey and for the analysis of the final catalogs. All of the persistent data will be kept in an object oriented data base (OODB), with the possible exception of the primary photometric data and the corrected photometric frames.

We are currently using a commercial OODB (Versant) for the archive. We chose this company after a small evaluation of other alternatives, but are working to keep Versant-specific code isolated to facilitate upgrades and to facilitate moving to another vendor if necessary in the future.

There are many features of Versant, and of OODB technology in general, that are attractive. First, the structure of the database allows a natural interface to a high level language, in this case C++. The objects and methods of the database are implemented as C++ classes. Versant currently supports data bases on multiple machines, which is important as we access from different institutions and APO. Schema evolution is also supported to some extent, allowing us to extend the definition of an object class without having to completely rebuild the database.

As an example of how Versant performs, we stored the ACRS catalog of 250,000 stars in Versant. Creating the database is done with simple Unix commands. A user program was written to read the ACRS catalog from a FITS file and write it to the new database. During compilation, schema are generated which are put in the database to define which kinds of objects it will accept. It took the program 15 minutes on a modest Unix workstation to load the full database. Another user program executed a Versant command to build an index on right ascension in the database. A third user program allows a user to query the database for objects in a specific RA and DEC range and return the results in a file. The resulting access time for the astrometric standards from one scan of the imaging camera is acceptably short. We are now in the process of loading the entire Guide Star Catalogs into a database to test Versant's performance on very large datasets.

6. Conclusion

These systems work because we have defined all of the software systems to be modular. Each of the steps in the pipelines, as well as the parts of the data acquisition systems, are developed independently once the interfaces are defined. We can install better versions of algorithms as they become available, test the robustness of each module before integration, and share modules between systems, because of this design.

The Sloan Digital Sky Survey has a rather difficult goal: to construct a large, calibrated, consisted set of catalogs. Only by working together with such a set of software tools will it be possible for us to succeed.

Acknowledgments. Work at Fermilab supported by the U. S. Department of Energy under contract No. DE-AC02-76CH03000. The SDSS is partially supported by the Alfred P. Sloan Foundation.

References

Gunn, J. E., & Knapp, G. R. 1992, PASP, 43, 267

Kron, R. G., 1992, ESO Conference and Workshop Proceedings, 42, 635

Bryan MacKinnon, B. et al. 1993, Conference Record of the Eighth Conference on Real Time Computing Applications in Nuclear, Particle, and Plasma Physics (IEEE), 329

Astronomical Data Analysis Software and Systems III
ASP Conference Series, Vol. 61, 1994
D. R. Crabtree, R. J. Hanisch, and J. Barnes, eds.

Aladin: Towards an Interactive Atlas of the Digitized Sky

F. Bonnarel, Ph. Paillou, F. Ochsenbein, M. Crézé

Centre de Données astronomiques de Strasbourg - CDS, Observatoire Astronomique de Strasbourg, France

D. Egret

IPAC, Mail 100-22, Caltech, Pasadena CA 91125, U.S.A.

Abstract. The purpose of the ALADIN project is to develop an interactive atlas of the digitized sky allowing the user to visualize on his/her workstation digitized images of any part of the sky, to superimpose entries from astronomical catalogs or user data files, and to interactively access the related data and information from the SIMBAD database for all known objects in the field.

The software architecture of ALADIN is based on the client–server philosophy. Each set of stored data (astronomical catalogs, SIMBAD database, and image pixels) will be accessed through a dedicated server.

We expect this new tool to be specifically useful for a multi–spectral approach (searching for counterparts of sources detected at various wavelengths), and for a number of applications related to the quality control and the cross–identification of observational data.

1. Introduction

The development of software and hardware technology now allows to develop new kinds of interactive access to all sort of astronomical data. In the same time, astronomers of 1993 looking for optical counterparts of γ-ray to radio sources, need to go beyond already existing catalogs or databases. From these two main considerations came the idea to propose a complete "on-line" digitized *sky atlas*.

This atlas should be a public interactive tool, available for all laboratories through networks. It should allow any astronomer to point a region of the sky, to display the corresponding digitized image (with accurate positions and fluxes), and to overlay cataloged data from the SIMBAD data base (Egret et al. 1991) and CDS catalogs.

More details about this project initiated by the CDS in 1992 can be found in Paillou et al. (1993, 1994). We will focus here on the most recent developments on the project, and on the use of the client/server approach in the overall architecture of the system.

2. The Aladin Project

ALADIN will provide images of the complete sky. These images will allow, when necessary, to perform processing or re-calibrations specifically adapted to the environment of a given object.

Several digitized surveys will be made available at CDS in a near future. The first one is the survey completed at ST ScI for the needs of HST program (Lasker 1994): the integration of compressed SERC-J (factor 10) into the ALADIN project is expected to take place at the end of 1993, and the integration of compressed POSS-IE (factor 10) at the end of 1994, thus providing a complete, digitized sky for the end of 1994, occupying about 100 GB.

From 1994 on, the CDS will also integrate selected fields provided by the French CAI (MAMA, Guibert 1992). In the future, it is thought to integrate the POSS-II survey digitized by ST ScI (from 1995), and CCD images generated by the DSS Sloan and LITE projects (from 1998).

The completed interactive sky atlas will require a storage capacity of the order of one Terabyte, for a complete sky coverage (in one color with a resolution of 1 arcsec and coding on 16 bits; this can be less when data compression is used).

CDS was funded by INSU to start the project with a juke-box of 12 inch WORM optical disks, allowing the storage of 520 GigaBytes (ATG-Cygnet Juke-Box, with the DOROFILE Unix File system emulation).

3. Design and Software Development

After a preliminary study (1992/1993) and the installation of a local prototype (1993), the project has now started its development phase.

The architecture is based on the client-server philosophy which, optimizes access to distributed computing resources and offers modularity and flexibility, thus making much easier further hardware and software evolutions of the system.

ALADIN is composed of 5 main pieces of software : a graphical user interface, the CDS multimedia server, the Image database server, the Catalogue server and the SIMBAD server.

Communications between the user interface and the CDS server, and between servers and databases, are designed according to the client/server model.

The access to the system from the outside world is managed by the "CDS multimedia server" which is connected to each dedicated server through the CDS local network.

The SIMBAD server: The SIMBAD software was not originally (1988) written following this client/server approach: a server mode has been recently added to the "user interface class" (Bonnarel et al. 1994). In order to make easier the development of new user interfaces to the database, a set of client routines has been developed on top of the communication layer of SIMBAD. These routines allow to retrieve objects on the basis of identifiers or coordinates, and to extract the corresponding data (defined by simple "astrotypes") and the full bibliographical references.

The Catalogue server: The CDS maintains since 1972 an archive of astronomical catalogs in machine-readable form, currently yielding a collection of over 650 catalogs. The requests issued by ALADIN for extraction of data from the astronomical catalogs are essentially based on the following two criteria: choice of the catalogs to be scanned, according to a predefined purpose (e.g., catalogs for photometric calibration, with IR data, etc.), and region of the sky to be examined.

The Image Database is currently under development. Its main function is to retrieve sets of images according to astronomical criteria: survey, color, region of the sky, etc. The selection of the images is made using Image Qualifiers stored in a relational database. In order to give a great modularity to each step of the request, an object oriented design has been retained for the software. Image Set, Image Qualifier, Image Archive and Image are the basic classes of the software.

The CDS multimedia server is the common gateway to all the databases and archives of the CDS. An object oriented design has also been retained for this software.

The User Interface is an X Window/Motif distributed client, providing Image Display functionalities and various menus for querying the CDS databases and archives and displaying the data and images.

Astronomical packages, such as astrometric and photometric calibration tools, source extraction, classification, will also be provided.

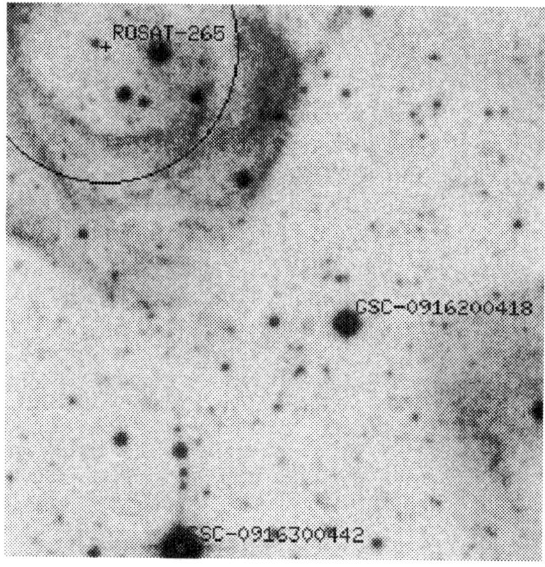

Figure 1. Close-up view of a sample image used for cross-identification of ROSAT sources in the Large Magellanic Cloud.

4. Status of the Project

A prototype of ALADIN has been developed at the CDS, using some well defined test regions of the sky: the Large Magellanic Cloud and the Galactic anti-center. The corresponding images are stored on the ATG optical disk system (9 Giga bytes/disk) of the CDS.

For the LMC, the CAI-MAMA group digitized two ESO plates in two colors, B and R. Each image is made up of 29000×29000 pixels coded on 16 bits.

It is currently possible to visualize the images of these two ESO plates of the LMC, and to overlay GSC stars and Simbad objects, using an interactive graphical user interface: GIFAD (Paillou and Ochsenbein 1992). The user can also overlay objects coming from his/her own data files.

This is currently used, at the Strasbourg Observatory, in a project of cross-identifications of ROSAT X-ray sources in the Large Magellanic Cloud.

Distribution of a functional version of ALADIN is foreseen for the end of 1995. Several milestones are defined: Southern hemisphere integrated in ALADIN for the end of 1993, Northern hemisphere integrated for the end of 1994, distribution of version 1.0 of the user interface for the end of 1995.

References

Bonnarel, F., Divetain, E., Ochsenbein, F., Paillou, Ph., & Wenger, M. 1994, in Astronomy from Wide-Field Imaging, Postdam, Germany, H.T. MacGillivray Ed., Kluwer Acad. Publ., in press

Egret, D., Wenger, M., & Dubois, P. 1991, in "Databases & On-line Data in Astronomy", Albrecht & Egret (Eds.), Kluwer Acad. Publ., 79

Guibert, J. 1992, in Digitised Optical Sky Surveys, H.T. MacGillivray & E.B. Thomson (Eds), Kluwer Acad. Publ., 103

Lasker, B. 1994, in Astronomy from Wide-Field Imaging, Postdam, Germany, H.T. MacGillivray Ed., Kluwer Acad. Publ., in press

Paillou, Ph., & Ochsenbein, F. 1992, in Astronomy from Large Databases, II, A. Heck & F. Murtagh (Eds), ESO Proceedings 43, 417

Paillou, Ph., Bonnarel, F., Ochsenbein, F., & Crézé, M. 1993, ALADIN Deep Sky Mapping Facility, Project Report, CDS, Strasbourg, May 1993

Paillou, Ph., Bonnarel, F., Ochsenbein, F., & Crézé, M. 1994, in Astronomy from Wide-Field Imaging, Postdam, Germany, H.T. MacGillivray Ed., Kluwer Acad. Publ., in press

Astronomical Data Analysis Software and Systems III
ASP Conference Series, Vol. 61, 1994
D. R. Crabtree, R. J. Hanisch, and J. Barnes, eds.

The APS On-Line Database of POSS I

G. Aldering, R. M. Humphreys, S. Odewahn, P. Thurmes

University of Minnesota, 116 Church St, SE, Minneapolis, MN 55455

Abstract. We present details of our database constructed from POSS I plates digitized by the Automated Plate Scanner (APS) at the University of Minnesota. Both POSS I E and O plates covering those fields with $|\, b \,| > 20°$ (644 fields) have now been digitized. We have written a custom database engine, *STARBASE*, which is used to construct databases from reduced scan data. It can then be used to query the resulting databases.

The astronomical community will have access to *STARBASE*, and hence to all fully reduced POSS I scan data. This access is being provided via the Astrophysical Data System (ADS) interface. Direct Internet use of *STARBASE* and its companion X-window plotting package, *STARPLOT*, will also be provided in the near future. The release of plate databases will be incremental and is expected to take 1 - 2 yrs to complete. By the time of this conference we will be offering a fully calibrated test database covering 9 fields at the NGP through ADS. The test NGP database alone contains positions, magnitudes, colors, star/galaxy classification, and numerous other parameters for $\approx 2 \times 10^5$ stars and $\approx 10^5$ galaxies.

1. Introduction

For the past several years, the Automated Plate Scanner (APS) project has been working towards the goal of digitizing the O and E plates of the first Palomar Sky Survey (POSS I) and creating object and pixel databases from these scans. Digitization of all POSS I plates with $|\, b \,| > 20°$ has been completed. The final object database will consist of $\approx$ one billion stars and several million galaxies, having astrometry, photometry, and star/galaxy classifications as good as can be attained from POSS I copy plates. This APS database of POSS I will be available over the Internet through NASA's Astrophysics Data System (ADS) in the very near future. We expect this database to be a rich source of original data for Galactic and extra-galactic studies, as well as a valuable adjunct to modern spacecraft and ground-based observations.

2. The Digitization Process

We have digitized POSS I using the Automated Plate Scanner, a flying laser-spot scanning machine constructed for use in the proper motion survey Luyten conducted with the Oschin Schmidt Telescope. Because of its use for proper motion work, it is able to scan two 14-inch square plates simultaneously. The scan speed we have used for POSS I allows a pair of plates to be scanned in about 6 hours. The (spatially filtered) pixels with transmission less than 65% of sky are saved at an in-scan resolution of 12 μm and cross-scan resolution of 5 μm. The 65% transmission transit locations are also saved, at an in-scan resolution of 0.375 μm. The transit data are used to determine object centers, diameters, ellipticities, and position-angles. The pixel data are used for performing star/galaxy classification. Consideration is being given to making greater use of the pixel data for object parameterization now that sufficient computing power is available. Further details of the scanning procedure can be found in Pennington et al. 1993.

3. APS POSS I Processing Pipeline

The procedures required to create object databases from the raw scan data are shown in Figure 1. Flattening corrects for variations in the throughput of the optics and for the changing beam profile along the 12 mm height of a scan. Standardization of the gain transforms the raw transmissions so that a given data-number corresponds to the same physical transmission for our entire POSS I survey. This transformation utilizes scans of a reference field taken before and after each plate-pair is scanned.

The reassembly of the transit and pixel data into objects uses standard algorithms for the segmentation of thresholded data. Once reassembly is done, numerous parameters (x, y, diameter, ellipticity, position-angle, etc.) are calculated using the transit and pixel data. Next, the 30 stripes making up a full plate are aligned, and redundant detections in the 1 mm of overlap flagged. Once this is accomplished, the O and E plate-pairs are aligned within 12 mm x 12mm processing boxes in order to allow objects to be accurately matched between plates. These tasks are performed as the scan data are being collected, which means that a simple object catalog is available shortly after a scan is completed. Extensive software modifications have been made to this portion of the processing pipeline since the initiation of POSS I scanning, therefore the raw scan data will be reprocessed for the purposes of database construction.

Astrometric calibration is carried out next. A plate model consisting of 6 linear terms plus a cubic radial term is used to determine an astrometric solution from the plate positions of ACRS astrometric standards. The astrometric standards are bright, and therefore badly saturated on the POSS I plates. Off-axis, they become asymmetric due to optical aberrations and ghost reflections. To overcome these difficulties, we measure centers using the intersection of the diffraction spikes. A typical solution retains 90% of the available astrometric standards ($\sim$ 190 per plate), and has an RMS of $\sim 0''.56$ in X and $\sim 0''.42$ in Y. Schmidt plates are notorious for having significant systematic astrometric errors near their edges. Such systematic errors are present in our data at the plate edges. They are highly reproducible and we plan to correct for them, possibly in a manner like that being used for the 2nd GSC.

Our photometric calibration for stars on POSS I is based on a magnitude-diameter relation. This relation is determined from CCD sequences we have obtained for over 266 fields. The photometry obtained in this manner is better than 0.20 mag (O) and 0.25 mag (E) for stars brighter than magnitude 20. The

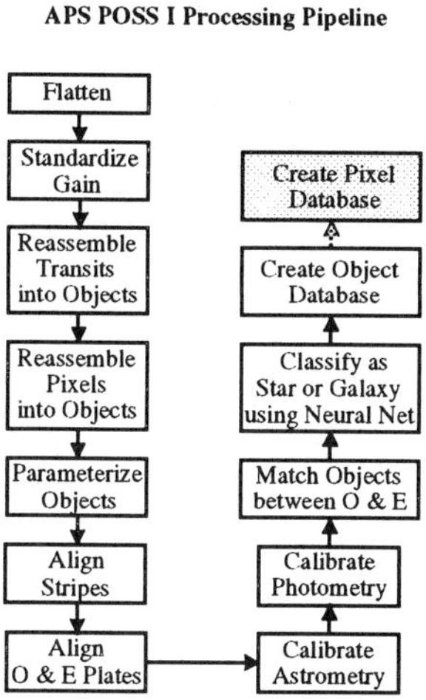

APS POSS I Processing Pipeline

```
Flatten
  ↓
Standardize Gain                      Create Pixel Database
  ↓                                     ⇡
Reassemble Transits into Objects      Create Object Database
  ↓                                     ↑
Reassemble Pixels into Objects        Classify as Star or Galaxy using Neural Net
  ↓                                     ↑
Parameterize Objects                  Match Objects between O & E
  ↓                                     ↑
Align Stripes                         Calibrate Photometry
  ↓                                     ↑
Align O & E Plates  ───────────→      Calibrate Astrometry
```

Figure 1. Flow chart of the APS POSS I Processing Pipeline.

remaining fields are calibrated using a mean magnitude-diameter relation which is scaled to match photometry from the Guide Star Photometric Catalog. A magnitude-diameter relation is also used to assign magnitudes to galaxies, however the physical justification for this approach is poor. We hope to be able to determine the density-intensity transformation for the POSS I plates and thereby determine magnitudes directly from our pixel data. Initial tests using published galaxy surface photometry show that this approach is quite promising. Moreover, we have developed a physical model for the stellar magnitude-diameter relation which may allow us to determine the density-intensity transformation from the magnitude-diameter relation.

An unavoidable consequence of pushing the POSS I plates to their limits is the detection of a large population of spurious objects (emulsion defects, scratches, etc.) near the plate limit. We attempt to maximize the reliability of our database entries by accepting only those images which are present on both the O and E plates. A match is made with the nearest source having $r < 4''$. This radius provides the maximum ratio of correct/spurious matches over the size range for which contamination presents a problem. If the star/galaxy classifier can be taught to recognize most of the spurious sources, our reliability will be increased even further. This would also allow us to consider the inclusion of a larger fraction of those sources with extreme colors.

The last step in the pipeline prior to the actual creation of an object database is the star/galaxy classification. This step is second only to the photometric calibration in enhancing the power of the database for quantitative scientific study. We use a neural network classifier trained on a large set of sources having reliable classifications. Sources brighter than magnitude 16 can be correctly classified with a success rate of 95% or greater. The success rate is maintained in the high-80's to low-90's down to magnitude 20. Details of the implementation and testing of the neural network can be found in Odewahn et al. 1992 and Odewahn et al, 1993.

Once these processing steps are complete, *STARBASE* is invoked to create O and and E plate databases. The database records are fixed-length binary records, and hash tables are constructed for all searchable parameters. This maximizes the speed with which *STARBASE* is able to service subsequent queries to the database.

The processing step following the creation of the object databases is shaded in Figure 1, indicating that it has not yet been implemented. The pixel database will consist of all pixels with transmission less than 65% than sky (typically $\mu_B = 24.5$). Implementation of this processing step awaits the acquisition of the necessary $\sim$ 1 Tb of on-line or near-line storage.

4. On-line Access

Our database engine, *STARBASE,* is used to access the object databases resulting from the processing pipeline. In order to make these databases available through ADS, we wrote a Unix C-shell script which captures the SQL query generated by the ADS SQLserver and redirects it to a file which is then read by *STARBASE.* The *STARBASE* query results are written to a temporary file which is then flushed to the standard output and captured by the ADS SQLserver for transmission back to the user. The modifications to *STARBASE* required were the addition of command-line arguments specifying the input and output files, and expansion of the *STARBASE* parser vocabulary so that a limited set of SQL commands could be recognized. Since the original query language used in *STARBASE* was very similar to SQL, this was not particularly difficult.

It will be some 1 – 2 yrs before our full POSS I database is available on-line due to the length of time required to send a plate-pair through the entire processing pipeline. We hope to be able to add 1 – 2 plate-pair databases per day. These will be made available through ADS as they are processed, so a vast number of plate databases will already be present by the time ADS has incorporated our catalog. Databases from a set of 9 plates in the direction of the NGP are now being used to test our ADS interface.

Other avenues for providing on-line access to our POSS I database are being explored. These might include allowing users to logon to our machines and run *STARBASE,* and our X Windows plotting program, *STARPLOT,* directly. Another promising avenue would be a query interface to X-mosaic.

References

Odewahn, S. C, Humphreys, R. M., Aldering, G., & Thurmes, P. M. 1993, PASP, in press

Odewahn, S. C, Stockwell, E. B, Pennington, R. L., Humphreys, R. M., & Zumach, W. A. 1992, AJ, 103, 318

Pennington, R. L., Humphreys, R. M., Odewahn, S. C, Zumach, W., & Thurmes, P. M. 1993, PASP, 105, 521

Astronomical Data Analysis Software and Systems III
ASP Conference Series, Vol. 61, 1994
D. R. Crabtree, R. J. Hanisch, and J. Barnes, eds.

Calibrating the USNO PMM

Arne A. Henden, Jeffrey R. Pier, David G. Monet and Blaise Canzian
U.S. Naval Observatory, Flagstaff Station, P. O. Box 1149, Flagstaff, AZ 86002-1149

Abstract.
The accuracy of the USNO Precision Measuring Microdensitometer(PPM) in matching the calibration data is discussed.

1. Introduction

The USNO Precision Measuring Microdensitometer (PMM) hardware and software have been under development for a number of years. The PMM is now in the process of measuring plates for calibration purposes.

The machine is an Anorad granite x-y stage mounted on air bearings and has a 30x40inch useful area. Two Videk MegaPLUS Cameras with 1320x1035 pixel CCD cameras are mounted above the stage, and two uniform-illumination light sources are mounted beneath the stage. A 4-plate plateholder is mounted on the stage in such a manner that two plates can be measured simultaneously. Each camera is attached to its own Silicon Graphics 4D/440S workstation for real-time data processing. More information regarding the hardware and software is given elsewhere (Monet 1993).

The PMM will be used in the near future to measure POSS-I (Minkowski and Abell 1963) and POSS-II (Reid et. al. 1991) Palomar Sky Survey plates as well as the USNO quick-J plates to determine positions, proper motions, and photometry of most stars and galaxies north of the Celestial Equator that are visible on the plates. Approximately 5000 plates will be measured with the majority of the processing carried out in real time. The goals are ± 0.1 arcsec error for astrometry and ± 0.1 mag error for photometry. If time permits, the pixel data will be stored on Exabyte tapes.

The software will be kept at a fixed revision level once the measuring begins so that all of the processing is performed in a uniform manner. Therefore, two test fields are being studied in detail to investigate both the behavior of the hardware and software of the PMM and the field errors of the Schmidt plate material. These fields are: a high galactic latitude (HGL) region centered at α=14:55 δ=0 (1950); and a low galactic latitude (LGL) region centered at α=22:24 δ=45 (1950). The HGL field has much overlap between the POSS-I and POSS-II surveys, is sparsely populated so that crowding effects are not present, and contains a number of galaxies. The LGL field is densely populated with severe crowding and blending of images, and the POSS-II plate falls almost exactly at the boundary of four POSS-I plates. The relationship between the various plate boundaries and CCD fields is shown in Figure 1.

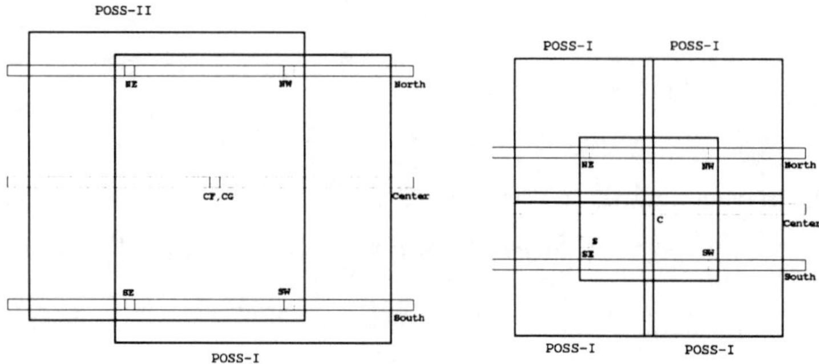

Figure 1. Diagram of plate boundaries for test fields.

Three types of calibration data are being taken. Deep CCD sequences in several colors will be used to check the software photometry algorithm and to investigate color transformations onto a standard system. Even deeper CCD frames taken in good seeing will be used as a check on the star/galaxy classification. Strip scans with the USNO 0.2m transit telescope (Stone 1993) will be used for checking the PMM astrometry algorithms, investigating the geometrical distortion of the Schmidt plates, and to examine the vignetting function of the Palomar Oschin Schmidt telescope. These calibration products are being used for the two test fields. The final PMM catalog will use the best astrometric and photometric calibrators available at the time of its production.

2. Deep CCD Sequences

The POSS-I plates were 103aO (blue) and 103aE (red), approximating the bandpasses of Johnson B and R filters. However, the 103aO bandpass is slightly blueward of B, and the 103aE response is narrower than R and centered at Hα.

The POSS-II plates are IIIaJ (blue), IIIaF (red) and IVN (near-IR). The approximate bandpasses are B,R,I. The IIIaJ response is slightly redward of the Johnson B; the IIIaF is similar to the 103aE response; and the IVN response is narrower than the I filter.

We decided to use a hybrid filter system for the CCD photometry. We selected filters to match the plate bandpasses, therefore making transformations to/from the natural plate wavelength system easier and more reliable. The filters chosen are: B, V, Hα line continuum (40nm wide, centered on H-alpha) and WF/PC F814W (a near-IR filter with similar characteristics to IVN). For the purposes of this project, these filters are labeled BVRI respectively.

For each test region, six CCD fields were chosen. Four fields define the corners of the overlap segment of the POSS-I and POSS-II plates. One field is located near the center of the overlap region. An additional field is taken near a GSPC-I (Lasker et. al. 1988) or GSPC-II (Postman, et. al. 1992) field to compare our photometry with that of the Guide Star Photometric Catalogs.

The POSS-I plates have magnitude limits of approximately B=21 and R=20. The POSS-II plate limits are about J=22.5, R=20.8 and I=19.5. Therefore, we selected exposures such that the limiting magnitude is approximately the same as the POSS-II, and that photometric errors are in the range of $\pm 0.1^m$ at V=19.

3. 0.2m Transit Telescope Strip Scans

The USNO 0.2m transit telescope is being used to take CCD strip scans of the sky, with each strip being 20 arcmin high. For our test fields, strips were taken at the north, center, and southern extremes of the overlap region and slightly longer in Right Ascension than the total width of the plate material. A wide V filter (WF/PC F606W) is used, giving a limiting magnitude of about V=17. The astrometry is calibrated with ~50 members of the extragalactic reference frame reachable with this system, and has an accuracy of ±0.1arcsec at V=16. Photometry is performed using a digital aperture and compares well with the deep CCD V-filter photometry to about V=16. Several thousand stars are located in each strip, giving a dense local astrometric data set for the test fields.

These strips yield two important calibration data: the astrometry is used both to test the PMM algorithms and hardware and to remove the geometrical distortion (unbending) of the photographic plate; the photometry can be used to remove the vignetting of the telescope.

4. High Angular Resolution CCD Frames

Accurate knowledge of which survey plate objects are truly galaxies is essential for building a good star/galaxy separator. We have obtained deep 2048×2048 V-band images of the center fields with the 1.55 m telescope during good seeing ($1''$FWHM). The images cover an area 11 arcmin square. These images will be used to test the star-galaxy classifier. Because the images are much deeper than the POSS survey limit, inspection of the images by eye can accurately classify most galaxies that appear on the survey plates. 73 galaxies with V-band magnitudes (in a $10''$ aperture) ranging from 17.1 to 21.6 were identified in one field. A very distant, faint cluster (members > 22 mag) of several dozen galaxies with two subclusters is evident in the field as well. The identifications were based on obvious extended morphology for the brighter galaxies. Identification of faint galaxies sometimes required verification that the FWHM of the radial surface brightness profile was significantly larger than that of a typical star. Data from our CCD imagery will provide the information necessary to allow us to identify galaxies accurately to the plate limit.

5. Preliminary Results

Please note that all results reported here are preliminary and only reflect the status of a rapidly evolving system as of the date of this paper.

Shown in Figure 2a is a comparison between the CG field magnitudes obtained with the 1m deep CCD frames and those determined from the 0.2m strip scans. The 0.2m magnitudes have not been adjusted for color effects and show

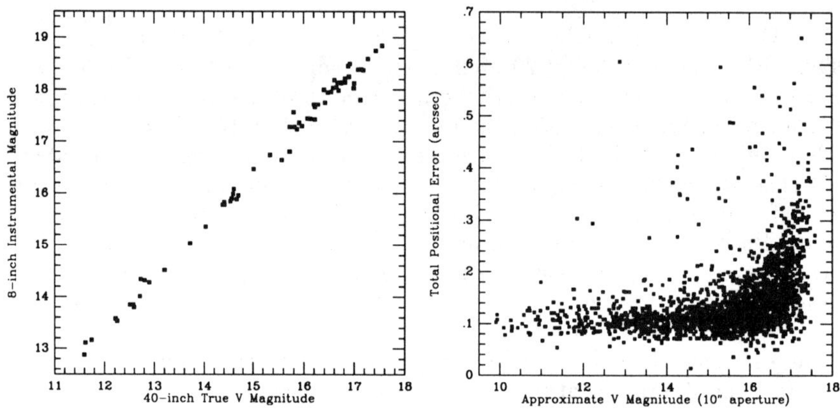

Figure 2. 0.2m photometric and astrometric results.

the typical scatter in using the raw instrumental magnitudes. Figure 2b indicates the internal astrometric error for one of the 0.2m strip scans.

The calibration project is well underway with reasonable results considering the preliminary nature of the current comparisons.

References

Lasker, B. M., Sturch, C. R., Lopex, C., Mallama, A. D., McLaughlin, S. D., Russell, J. L., Misniewski, W. Z., Gillespie, B. A. Jenkner, H., Siciliano, E. D., Kenny, D., Baumert, J. H., Goldberg, A. M., Henry, G. W., Kemper, E. & Siegel, M. J. 1988, ApJS, 68, 1

Minkowski, R. L., & Abell, G. O. 1963, in Basic Astronomical Data, ed. K. Aa. Strand (Chicago: Univ. Chicago Press), 481

Monet, D. G. 1993, ASP Conference Series, 43, 139

Postman, M., Siciliano, L, Shara, M., Rehner, D., Brosch, N., Sturch, C., Bucciarelli, B. & Lopez, C. 1992, Conference on Digitized Optical Sky Surveys, ed. H. T. MacGillivray and E. B. Thomson (Dordrecht: Kluwer Acad. Publ.), 61

Reid, I. N., Brewer, C., Brucato, R. J., McKinley, W. R., Maury, A., Mendenhall, D., Mould, J. R., Mueller, J., Neugebauer, G., Phinney, J. Sargent, W. L. W., Schombert, J., & Thicksten, R. 1991, PASP, 103, 661

Stone, R. S. 1993, in Development in Astrometry and their Impact on Astrophysics and Geodynamics, IAU Symposium 156, ed. I. I. Mueller & B. Kolaczek (Dordrecht: Kluwer Academic Publ.), 65

Astronomical Data Analysis Software and Systems III
ASP Conference Series, Vol. 61, 1994
D. R. Crabtree, R. J. Hanisch, and J. Barnes, eds.

The UIT Bright Objects Catalog

Eric P. Smith

Laboratory for Astronomy & Solar Physics - NASA/Goddard Space Flight Center, Code 681, Greenbelt, MD 20771

A. J. Pica

Department of Physics, Salisbury State University, Salisbury, MD 21801

R. C. Bohlin

Space Telescope Science Institute, 3700 San Martin Drive Baltimore, MD 21218

M. K. Fanelli[1]

Laboratory for Astronomy & Solar Physics - NASA/Goddard Space Flight Center, Code 681, Greenbelt, MD 20771

R. W. O'Connell

Astronomy Department, University of Virginia, Charlottesville, VA 22903

M. S. Roberts

National Radio Astronomy Observatory, Edgemont Road, Charlottesville, VA 22903

A. M. Smith

Laboratory for Astronomy & Solar Physics - NASA/Goddard Space Flight Center, Code 681, Greenbelt, MD 20771

T. P. Stecher

Laboratory for Astronomy & Solar Physics - NASA/Goddard Space Flight Center, Code 681, Greenbelt, MD 20771

Abstract. We have created a catalog of UV bright objects observed with the Ultraviolet Imaging Telescope (UIT). The catalog contains nearly 2200 entries and consists of a list of ra, dec, near-UV magnitude, near-UV$-V$ color, and cross-identification (if extant) for each object, most of which are local stars. These data were extracted from UIT images using the FOCAS software. Once object positions and magnitudes were

[1]National Research Council Postdoctoral Fellow

obtained they were cross correlated with many astronomical catalogs to produce our final product.

1. Introduction

The Ultraviolet Imaging Telescope (UIT) is a 38cm, $f/9$ Ritchey-Chrétien design telescope optimized for imaging and spectroscopy in the spectral range 1300Å $< \lambda <$3000Å. It was flown aboard the Space Shuttle as part of the ASTRO payload in December 1990. Some of its primary goals were to:

- study young stars and regions of star formation,
- search for hot stars in globular clusters and other evolved stellar populations,
- study the interstellar medium via narrow-band imaging.

Details of the UIT instrumentation and initial batch data reduction may be found in Stecher et al. (1992). The UIT images a wide field of view nearly 40 arcminutes in diameter with a resolution of ~2.5 arcsec FWHM. Therefore, each field contains not only the primary target but many other sources. While the UIT did not execute a sky survey these additional data in each frame can be used to:

- search for and study optically faint, UV bright stars in the Galaxy,
- search for optically faint, UV bright galaxies,
- study stellar populations within the Galaxy,
- provide a list of UV bright targets for subsequent HST observation.

2. Catalog Production

During its mission the UIT obtained 361 near-UV ($\lambda \sim$ 2500Å) and 460 far-UV ($\lambda \sim$ 1500Å) images of 66 targets. Since the limiting magnitude for the far-UV images was ~2 magnitudes brighter than that for the near-UV images we chose to include only the near-UV data in the catalog. The limiting magnitude for the longest near-UV exposures was in the range $19 < m_{2500\text{Å}} < 21$ depending upon the exposure time. Not all UIT images were suitable for inclusion in the catalog. The final catalog contains entries from 55 pointings (rejecting M31, LMC, Solar system and short exposures). The digitized near-UV images were analyzed with FOCAS (Valdes 1982) employing a 4σ detection threshold (chosen to minimize the detection of noise near the image edges) and a minimum area per object of 45 pixels ($\simeq$ UIT point-spread-function area). This high detection threshold implies that the brightest sources are included in the catalog while some faint, but possibly real objects are excluded. The resultant catalogs were compared positionally with catalogs of stars and galaxies (e.g., HST Guide Star, SAO, RC2, Principal Catalog of Galaxies). In addition, we individually inspected unmatched objects to asses their astronomical "reality".

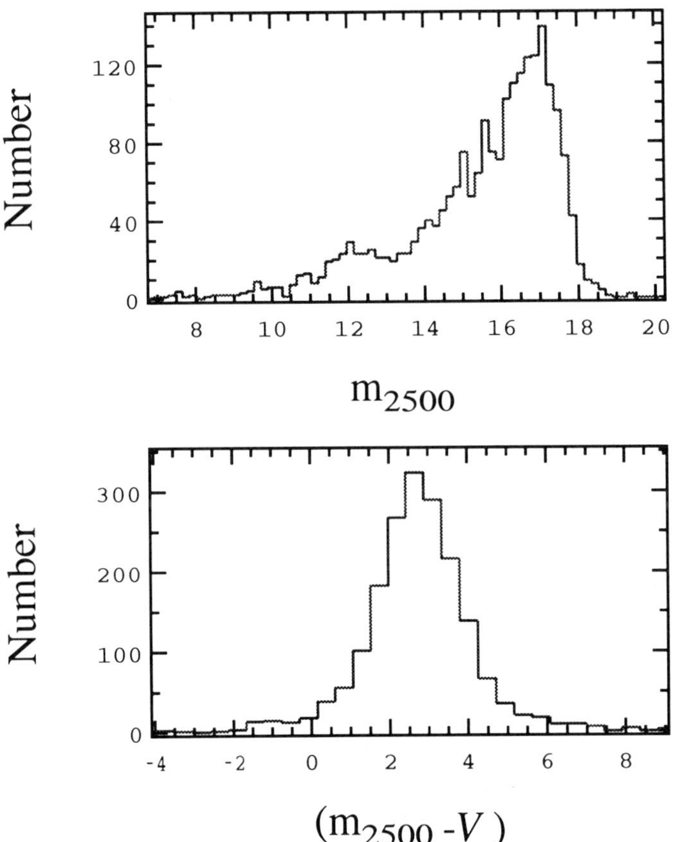

Figure 1. The ($m_{2500\mathring{A}}$) aperture magnitude distribution for objects in the UIT bright objects catalog [upper panel] and the ($m_{2500\mathring{A}} - V$) color distribution for objects found both in UIT frames and various astronomical catalogs.

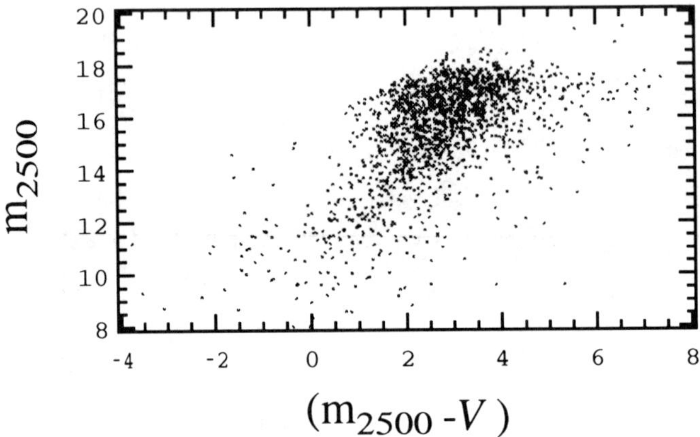

Figure 2. The color magnitude diagram for the catalog. The sharp boundary in the upper left results primarily from the limiting magnitude of the HST Guide Star catalog.

3. Catalog Features

The catalog contains 2184 sources of which 322 are unmatched in published catalogs. The mean near-UV magnitude for these unmatched objects is $< m_{2500} >=$ 15.5. The color distribution of the matched objects is show in Figure 1. The mean color of this distribution is $< m_{2500} - V >= 2.7 \pm 1.3$ (roughly the color of a mid G V star). The color magnitude diagram for the matched objects in show in Figure 2. The broad cluster of points represents, for the most part, the local main sequence.

References

Stecher, T. P. et al., 1992, ApJ, 395, L1

Valdes, F. 1982, "FOCAS User's Manual", Kitt Peak National Observatory, Computer Support Office, Tucson, AZ

Astronomical Data Analysis Software and Systems III
ASP Conference Series, Vol. 61, 1994
D. R. Crabtree, R. J. Hanisch, and J. Barnes, eds.

The Design of an Intelligent FITS File Database

A. H. Rots
Universities Space Research Association, XTE Guest Observer Facility, Code 668, Goddard Space Flight Center, Greenbelt, MD 20771

Abstract. The sophistication of XTE's on-board data systems is such that the handling of the telemetry poses new challenges in the areas of data identification and data management. Our solution has been the development of a multi-mission data description language and a hierarchical FITS table database.

1. Introduction

The X-ray Timing Explorer (XTE) is a High Energy Astrophysics mission intended for launch in the second half of 1995. It carries two pointed instruments (PCA and HEXTE) that together cover the range 2-250 keV at μs time resolution and moderate spectral resolution, and one instrument (ASM) that will monitor the X-ray sky continuously over the 2-10 keV range.

XTE's on-board science data systems provide considerable processing power and unprecedented flexibility in telemetry data modes. Events are processed on-board in several simultaneous data modes, chosen from a large repertoire. New data modes may be added during the mission. Consequently, keeping track of the collected data in the database and providing a mechanism to select data that satisfy selection criteria expressed in physical terms is a challenging problem.

The XTE Guest Observer Facility, in conformity with the practices at the Office of Guest Investigator Programs, will provide the data in FITS format. The design of these FITS files includes two new features that address the cataloging and data selection issues.

First, a hierarchy of FITS tables will be used to navigate the database. A master index will allow software to browse through the catalog with the granularity of individual observations, and find references to instrument indices (one index per instrument or subsystem per observation), as well as source information. An instrument index table will contain references to data files generated by data system components for various time intervals during the observation. The emphasis for the data tables is on those containing raw data, but there will be additional ones holding, for instance, data products and calibration information. Given access to the master index and a set of selection criteria, extractor software will be able to determine the location of the requested data.

Second, a Data Description Language (DDL) has been developed to label each data item unambiguously and to facilitate data selection browsing. Through the use of tokens along a number of predefined axes the DDL describes the structure of the data (dimensionality of arrays), the physical placement of the data along the axes (e.g., time increments), and the selection criteria (e.g., single layer events, which detector, calibration events, veto layer events) applied when the data were collected. The eventual goal is to develop a data extractor or browser that is capable of parsing these data descriptors and matching data requests with available data.

Both schemes can easily be generalized for multi-mission use.

2. Design of the FITS Data Tables

The central part of designing FITS data tables is developing criteria for hierarchical structuring of the incoming telemetry data. This involves defining the scope of the three hierarchical entities that constitute a table: the table itself, the rows, and the columns.

Tables: The items contained in a single table will represent all data from one Application (CCSDS jargon for a physical telemetry data source). The parallel science data streams referred to above will each come down from separate Applications, as will the "Housekeeping" (monitoring) data from each of the detectors. The time span covered by a table is determined by events that force the start of a new table: the start of a new observation; a change in the telemetry format of the Application; or a reset (reboot) of the Application.

Rows: Each row will contain the contents of one telemetry packet or logical group of packets and will thus consist of all the data that an Application sends down for a given time stamp when it flushes its buffer. The associated time stamp is made part of the row.

Columns: Data from different data sources in the Application (such as detectors or layers) will be separated into different columns. The data from a single data source, in a single table cell, may consist of arrays with the following axes: energy, wire position, time, frequency, phase, or lag.

3. Data Description Language

As indicated in Section 1., the SOC needs a mechanism that can keep track of the contents of each stream and that will accommodate future, unknown data modes. For this purpose we have developed the Data Description Language (DDL) which has two main functions: data identification and data selection.

The DDL has to provide a tag for each data object or table cell that identifies the contents unambiguously and provides information in four areas:

- The selection criteria that were applied in collecting the data.
- The structure of the data.
- The bit packing of the data in the telemetry stream.
- The meaning of the data.

The DDL also has to provide a mechanism to query the database:

- As a means of specifying data selection criteria.

- As a means of specifying desired properties of the data structure to be returned.

Data descriptors are built from a number of different tokens, separated by logical operators. Each token can be thought of as representing a coordinate axis in data space. The argument of each token specifies its value. The values are, in most cases, integer numbers but individual bits can be addressed by "name", if appropriate. The usual bit-wise logical operators may be applied. Table 1 lists the possible tokens. It should be noted that, in addition, there are L (lag), F (frequency), and P (phase) tokens that are all related to the T (time) token; these four are the only ones that allow floating point arguments.

The X token is defined in order to allow future expansion. To understand it, one should realize that XFF obtains its data by passing a data descriptor to an Accessor object that acts as a communicator for a data object. The X token allows direct access to any Accessor methods, if the need would ever arise.

Table 1. DDL Tokens.

Token	Explanation	Example
O	Observatory (spacecraft, mission)	`O[XTE]`
I	Instrument	`I[HEXTE]`
G	Detector group (cluster; HEXTE only)	`G[0]`
D	Detector	`D[3]`
E	Element (layer, bit in an event word)	`E[X1L & CAL]`
H	Housekeeping item	`H[hvSh3]`
S	Status information in science data stream	`S[liveTime]`
C	Pulse height channel	`C[0:255]`
T	Time [start;increment;number of items]	`T[0;1;16]`
A	Anode position (ASM only)	`A[0 8091;32]`
Z	Defined by instrument team (emergency use)	`Z[myItem]`
X	Direct Accessor method access	`X["getValue"]`

4. Design of the FITS Database

The FITS standard provides for the exchange of tables, as well as a table hierarchy (through the EXTLEVEL keyword). We can take advantage of this by using auxiliary tables to keep track of the data tables. In other words, we can build a database out of a hierarchy of FITS tables.

Our basic design contains a three level hierarchy: Master Index, Subsystem (or Instrument) Index, and Data Table. Table 2 outlines this structure.

Looking at the system from the top down, each row in the Master Index contains, for a single observation, references to all the Subsystem Indices that contain references to Data Tables belonging, or pertaining, to that observation. There will be Subsystem Index tables for each of the three science instruments, for the spacecraft attitude control system, for the clock corrections, for the orbit ephemerides, for the system of calibration files, etc.; in addition, there will be

Table 2. Elements of the FITS Database.

Table	Type	Rows	Columns
Master Index	ASCII	Observations	Subsystem Indices
Subsystem Index	ASCII	Observation segments	Data Tables
Data Table	Binary	Time stamps	Data items

columns containing observational parameters, such as Observation ID, start and stop times, and source information.

The Instrument Index tables contain rows that correspond to segments of the observation during which all telemetry data for that instrument was deposited in the same set of Data Tables. Each row contains references to the Data Tables generated by all Applications associated with the Subsystem, as well as data mode and configuration information, and references to relevant calibration files.

The Data Tables have been described in Section 2.

It will be clear that extracting a sub-database (e.g., all observations for a given source) not only is a simple operation, but also yields a new database that has an architecture identical to that of the original one. It involves lifting the relevant rows out of the Master Index, depositing them into a new Master Index table, and copying all the Subsystem Index and Data Tables directly and indirectly referenced in those Master Index rows. Consequently, the system can also function in the user's home environment.

5. Data Retrieval

Given a set of selection constraints, such as selected source(s) and time range(s), and a data descriptor, it becomes fairly simple to navigate through the system and find the data items one is looking for. The Master Index acts as an observation catalog with references to Subsystem Indices, while the latter contain configuration information and references to the Data Tables. Beyond that, it becomes a matter of matching data descriptors.

A graphical user interface will be provided to construct data descriptors based on menu and button selections made by the user. These descriptors will be matched against the data descriptors that are attached as tags to the data. There are four levels of implementation for the data descriptor matching:

- Literal match.
- Equivalent match: two data descriptors may have the same meaning (e.g., through the use of wildcard characters).
- Inclusive match: return a collection of data items that, together, contain the information requested in the data descriptor (and, possibly, more).
- Intelligent match and transformation: the retrieval system will collect the necessary data and transform it to conform with the data descriptor. This implementation requires an object-oriented environment.

Astronomical Data Analysis Software and Systems III
ASP Conference Series, Vol. 61, 1994
D. R. Crabtree, R. J. Hanisch, and J. Barnes, eds.

CADC Optical Disk Tools

Séverin Gaudet and Norman Hill
Canadian Astronomy Data Centre, Dominion Astrophysical Observatory, Victoria, B.C., V8X 4M6

Abstract. This paper describes tools developed at the CADC for accessing the WORM optical disks in the DMF format. These range from low-level block access through to Unix—like reading and writing tools through to the retrieval system used to process user requests for off-line data.

1. Introduction

The Canadian Astronomy Data Centre (CADC) was established in 1986 to distribute HST data to the Canadian astronomical community. The initial software system used was the VMS-based Data Management Facility (DMF) from the Space Telescope Science Institute (ST ScI) which included the archiving of data onto 12" Write-Once Read-Many (WORM) optical disks in its own format. In 1991, the CADC began two projects which required the development of our own optical disk tools — the migration to Unix and the archiving of other data onto optical disk, in particular that of the Canada-France-Hawaii Telescope (CFHT).

2. Optical Disks

The optical disks used at the CADC are 12" WORM disks. The disks are vendor specific, i.e., a SONY disk can only be read by a SONY drive. The disks are double-sided but are only accessed one side at a time (side A and side B). The current optical disk setup at the CADC is described in Tables 1 and 2.

Table 1. Optical Disks.

Manufacturer	Model	Block size (bytes)	Blocks per side	Disk capacity (Gbytes)	Medium	Cost per Mbyte ($)
Laser Magnetic Storage	LaserDrive Media	1024	1024000	2.048	glass	0.20
SONY	WDM-6DL0	1024	3276000	6.552	polycarbonate	0.06

The current generation of 12" WORM optical disk drives have new features which make these devices even more attractive. Firstly, disk capacities are now in the 10 – 15 Gbytes range. Read and write speeds have generally gone up by

Table 2. Optical Disks Drives.

Manufacturer	Model	Year[a]	Read speed (Kbytes/second)[b]	Disk Model
Laser Magnetic Storage	LD1200	1986	203.6	LaserDrive Media
SONY	WDD-600	1990	304.4	WDM-6DL0
SONY	WDD-931	1992	537.5	WDM-6DL0

[a]Year in which the drives were generally available.

[b]Based on I/O tests done at the CADC using the Perceptics driver on a SPARCstation 1+.

half. And lastly, drives are now configured with 2 laser heads, enabling them to access both sides of the disk simultaneously. The result from the archive perspective is that a retrieval manipulation will have access to four times the amount of data in a single disk mount.

3. Formats

The DMF system developed at the ST ScI is responsible for the access to and maintenance of the Hubble Space Telescope catalog and archive. Part of the DMF system was an optical disk archive currently based on the LMS LD1200 drive and on an optical disk format now referred to as the DMF format (McGlynn et al. 1988). The main reason why the ST ScI chose to implement its own disk format is to be independent of vendor formats. Most vendors have their proprietary formats with a software layer to emulate VMS or Unix file systems. The DMF format is quite simple. It partitions the disk into three areas:

- A disk header area which is typically the first 10 blocks on a disk. The header contains the disk name, location and size of directory. It allows for renaming a disk and for a software write-lock.
- The disk directory area. This area varies in size depending on the disk capacity but contains several tens of thousands of blocks reserved for directory entries. There is a header and a trailer block to this area.
- The data area is the remainder of the disk. Once a file's contents are written in this area, the file's directory entry will contain the location of the blocks containing the file's data.

4. Optical Disk Tools

When the CADC began the CFHT archiving project on Unix, we realized that we had to develop our own optical disk tools. Although the DMF system was initially designed to run on both VMS and Unix platforms, only the VMS version was supported. In addition, the optical disk subsystem was tied very much to the whole DMF environment and could not be easily isolated. Also we had decided to use the newer (with higher capacity and speed) SONY optical disk drives for the CFHT archive and no support for these existed within the DMF system.

4.1. General Purpose Tools

The general purpose tools developed and currently in use at the CADC are summarized in Table 3. These tools are implemented on both VAX/VMS and on Sun SPARCstation platforms. The VAX versions use the DMF device driver supported on the LMSI drives. The Sun versions use a device driver from Perceptics, Inc. supporting all our models of optical disk drives.

Table 3. Summary of general purpose optical disk tools.

Name	Unix analogy	Description
`odcopy`	cp	copies a file from optical disk to optical disk
`oddf`	df	disk space summary
`odfsck`	fsck	disk consistency check
`odls`	ls	list of files on disk
`odname`	format	labeling and formatting of disk
`odread`	cp	read a file from optical disk
`odwrite`	cp	write a file to optical disk

4.2. Special Purpose Tools

We have also written a set of specialized tools to automate certain tasks associated with optical disks but require communication with the archive database to manage the archive. These tools are summarized in Table 4 and described below.

Table 4. Summary of special-purpose optical disk tools.

Name	Description
`odingest`	Adds a disk to the archive directory
`odpreview`	Generates preview data
`odretrieve`	retrieves files for a user request
`cadcod`	copies public HST data onto CADC disks

odingest When an optical disk arrives either from the ST ScI or from the CFHT, the disk must be ingested into the archive. What `odingest` does is read the disk's directory, making entries into the whole archive directory describing the location of each file and then verifies for the data source that the expected files are located on that disk (i.e., all the files of a dataset are indeed located on the disk). This provides a correlation between what was expected and what files were actually received.

odpreview `odpreview` is the tool used for the automatic generation of preview graphics and images from archive data (Hill et al. 1993). Part of the task is to determine which datasets are *preview-able*. Once determined, the tool displays a list of disks with candidate datasets and the prompts the operator to mount one of them. It then reads the necessary files from the optical disk, processes them into the preview files and loads the preview files into the database.

odretrieve CADC users can request datasets through the STARCAT interface. The requests are stored in database tables where the CADC operator can monitor the pending requests and execute `odretrieve` to process a request. The operator is prompted to mount specified optical disks from which the requested files are read into a magnetic disk staging directory.

cadcod `cadcod` is a hybrid program. It runs at the ST ScI on VMS. It copies public HST data from ST ScI optical disks onto CADC optical disks. As in the tools above, `cadcod` communicates with a database to determine what needs to be copied and prompts the ST ScI operator to mount specified disks containing the files to be copied.

5. Future Developments

STDADS The Space Telescope Data Archive and Distribution Service is scheduled to replace the DMF within the next year. The STDADS system uses the SONY optical disk in a new format (Loral 1992). The CADC will add the read capability to its tools but have as yet no plans to add write capability in the STDADS format.

odtape A facility to go directly to or from magnetic tape is required to create an off-site backup of the optical disk archive. `odtape` would copy one side of an optical disk to a single 8mm or DAT tape. The magnetic tape format has not yet been determined. Like `odcopy`, this tool would bypass the need for large magnetic staging disk space when moving optical disk files to tape and vice versa.

Optical Disk Server Currently a process using an optical disk drive must be executed on the platform on which the drive is physically connected. An optical disk server would remove that limitation.

References

Hill, N., Crabtree, D., Gaudet, S., Durand, D., Irwin, A., & Pirenne, B. 1993 this volume

Loral Aerosys 1992, "STDADS Distribution Interface Control Document"

McGlynn, T., & L. Hunt 1988, "DMF File Handler Guide — Design and Software Specifications", ST-ECF O-02 Document, Vol. IX

Astronomical Data Analysis Software and Systems III
ASP Conference Series, Vol. 61, 1994
D. R. Crabtree, R. J. Hanisch, and J. Barnes, eds.

Dbsync: A Computer Program for Maintaining Duplicate Database Tables

N. Hill and S. Gaudet

Dominion Astrophysical Observatory/Canadian Astronomy Data Centre, 5071 W. Saanich Road, Victoria B.C. V8X 4M6

Abstract. This paper is a description of the `dbsync` computer program which was developed at the Canadian Astronomy Data Centre (CADC) for maintaining duplicate copies of database tables on different SYBASE^{tm} relational database servers.

1. Introduction

The Canadian Astronomy Data Centre (CADC) and the Space Telescope-European Coordinating Facility (ST-ECF) are the two centres responsible for the distribution of Hubble Space Telescope (HST) data to the scientific community outside of the United States. One of the requirements for the external data centres is the maintenance of an accurate, up to date subset of the Space Telescope Science Institute (ST ScI) HST archive database. To simplify this task, the CADC undertook to develop a method for automatically maintaining a copy of the HST archive database at each of the external data centres.

The computer program `dbsync` was developed by the CADC to maintain both the content and the structure of the tables in the HST database. `Dbsync` is run at the external archive sites, accessing the source database over the Internet. `Dbsync` updates the content of selected tables and rebuilds any tables whose structure has changed. The external archive sites retain control over the transfer allowing complete flexibility in the selection of data from the source database.

2. The HST Archive DataBase

The HST data archive database is stored using the SYBASE^{tm} relational database management system. The archive consists of nearly 200 tables in 2 databases, with a total size at time of writing of 550 MB. Data are being added to the tables continuously, and the structure of the tables and the subset of the data required by the remote archive sites are subject to change.

The flow of data between archives sites is shown in Figure 1. The main flow of data is from the ST ScI operations server to the ST ScI user server and the servers at ST-ECF and CADC. In addition to the primary flow of data, ST-ECF and CADC both produce original data that is required at both sites, and some original data is produced on the ST ScI user server.

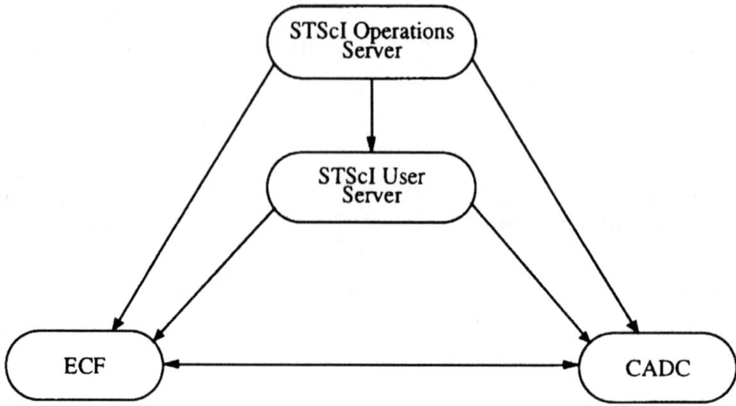

Figure 1. The HST archive data flow.

Initial attempts at maintaining the duplicate database's made it clear that an automated system for maintaining the structure of the tables was necessary, and prompted the development of `dbsync`.

3. `Dbsync` Design

`Dbsync` was designed to automatically maintain a subset of the ST ScI archive database with a minimum of operator intervention at the destination site, and no operator intervention at the source site. `Dbsync` has the following design features:

- It is a single program that can run on either the source or destination system, using the Internet to connect to the remote system.
- Control over the data transfer resides in the destination database.
- Incremental updates transfer only records not already in the destination table.
- New tables added to the source database are noted for the operator.
- Changes to the structure of tables in the source database are automatically incorporated into the destination database.

`Dbsync` is based on the ability of a SYBASE client to connect to a server across the Internet. `Dbsync` makes a connection to the source and destination servers simultaneously which allows the program to run at the destination site and to maintain the control information in the destination database. The simultaneous connections also allow direct comparison of the data in the source tables, and the data to be transferred from the source database to the destination database through program memory without the necessity of using scratch disk space.

`Dbsync` is able to detect differences between the structure of tables by comparing the contents of the SYBASE system tables which contain the information about the structure of the tables. If structural differences are found, `dbsync` uses information about the table structure from the source database, and information about the indexes and protections from the destination database to generate SQL code to create a new table. After the table structure has been updated the table is populated from the source database.

Control over the data transfer is provided by a "table of tables" which resides the in destination database, and which contains control information on every table in the source database. Any table in the source database which does not have an entry in the table of tables is flagged with a warning. The table of tables contains information which allows tables in the source database to be ignored, specifies a processing order for tables, allows tables to be renamed in the destination database, indicates if incremental updates are permitted, and allows a subset of the fields in a table to be copied.

The initial implementation of `dbsync` was found to be useful as a general database maintenance tool and several changes have been made to make it more usable in this role. The table of tables was made optional, and an optional list of tables to process can be supplied on the command line.

4. Efficiency of `Dbsync`

Since `dbsync` uses the same mechanisms to transfer data to and from the SYBASE server as the SYBASE bcp program, but does not have the overhead of reading and writing the data files, `dbsync` can copy files on a single server in less time than is required by bcp to copy a table to a file and back into a table. `Dbsync` does not match the efficiency of the SQL "select into" command for copying tables on a single server, but the ease of use of `dbsync` often outweighs the performance penalty. Table 1 contains some table copy times using various methods. Two database tables were used to simulate a range of conditions.

Table 1. `Dbsync` performance (times are in seconds). Database table A has 2.8 Megabytes of data in 88000 rows, and database table B has 2.2 Megabytes of data in 175 rows.

Table	`dbsync`	bcp [a]	bcp[b]	select into[b]
table A	247	690	298	206
table B	41	48	48	40

[a]With indexes in place

[b]With indexes dropped and rebuild separately. Includes index build time

5. Current use of `Dbsync`

`Dbsync` is now used nightly to update the data tables of the HST data archives at the ST-ECF, the CADC and the ST ScI. It has been performing this task for approximately one year.

In addition to its operational uses, `dbsync` has proven to be a useful tool for general database maintenance, and is useful whenever it is necessary to move data from one database to another. `Dbsync` has also been used to copy entire databases from an old version of the SYBASE server to a new version, considerably reducing the difficulty of upgrading to new releases of the database server.

6. Limitations

`Dbsync` currently has several limitations. These limitations are:

- `Dbsync` currently cannot detect modified rows.
- `Dbsync` requires each table to have a row number field before it can perform incremental updates.
- User defined data types are mapped into the underlying data types in the destination table.

7. Future Enhancements

Several enhancements have been suggested for `dbsync` which aren't required for the maintenance of the HST archive but will add to its usefulness as a database administration tool. Following is a list of changes that may be incorporated into `dbsync` in the future:

- Allow the optional copying of table protections, ownership and indexes.
- Copy database objects other than tables.
- Copy user defined data types instead of mapping them into their base types.

8. Conclusions

`Dbsync` is now in use at all three HST archive sites, and due to its ease of use it has encouraged data transfer between sites. `Dbsync` was developed to perform a specific task, but has since evolved into a general purpose tool for copying and updating tables between databases.

SYBASE corporation has announced a new product called the Database Replicator. This may or may not make `dbsync` obsolete. That will depend on the configurability of the Database Replicator, its cost and its performance. On paper the Database Replicator does not have the limitations outlined above.

Part 4. Data Analysis
Section A. Image Analysis

Astronomical Data Analysis Software and Systems III
ASP Conference Series, Vol. 61, 1994
D. R. Crabtree, R. J. Hanisch, and J. Barnes, eds.

Astronomical Image Processing on the PC with PCIPS

O. M. Smirnov

Institute of Astronomy of the Russian Academy of Sciences
48 Pyatnitskaya St., Moscow 109017 Russia

N. E. Piskunov

Astronomy Department, University of Western Ontario, Elgin Field Observatory, London, Ontario NGA 3K7 Canada

Abstract. Our PCIPS image processing system runs on both 486 and Pentium-based PCs and satisfies a broad range of data analysis needs, while providing maximum expandability.

1. Introduction

PCIPS (Smirnov & Piskunov 1993a) is an image processing package for the PC environment that we have been developing for the past three years. The main goal of the project was to overcome the limitations of the real-mode architecture of Intel processors, and develop a flexible and capable image processing platform on the PC (due to past embargoes, and recent economic difficulties remain the primary hardware base in Russia). With PCIPS, modern 486- and Pentium-based PCs (and even 386-based ones) can provide performance that is the equivalent of workstations costing much more. It's main capabilities are:

- Support of one- and two dimensional images (up to 16384 pixels, or up to 2048×2048) in eight data types: 8-, 16- and 32-bit signed and unsigned integers, 4- and 8-byte floating point, support for world coordinate scales.
- Easy-to-use and intuitive mouse- and keyboard-driven graphical user interface (GUI), including an on-line help facility; built-in interactive visualization tools (display and color table manipulation, etc.)
- Image Database with an extended directory structure for storage of images and miscellaneous scalar data.
- Support for various external formats (FITS, ASCII, binary, Photometrics, GIF, etc.), hard copy output in PostScript and PCL.
- Several astronomical application packages currently in use
- Unlimited functionality via external application modules, which are seamlessly integrated into the system's graphical interface.
- Application Program Interface (API) facilitates easy implementation of new application modules

2. Current Application Packages

2.1. The Basic Package

PCIPS includes a general-purpose application package for elementary image processing. This package provides all the necessary tools for importing images into the Database, performing simple arithmetic, geometric and statistical operations, producing a comprehensive visual and hard copy of the results:

Elementary mathematics: simple binary and unary operations (arithmetic, exponential, logarithmic) over images and constants. Includes arithmetics in the world coordinate scale as well as pixel-by-pixel ones.

Statistics, filters and fits: an assortment of applications that compute statistics (mean, mode, median, etc.) within an image window or over several rows or columns of a 2D image; boxcar, Gaussian, and median filters; edge enhancement filters, least-squares polynomial and Gaussian fits.

Geometric transforms: histogram equalization of 2D images, resample image to new grid (i.e., zoom or shrink), rebin or resize image, cut fragment, rotate, transpose or mirror an image, extract row or column of 2D image.

Export and import: applications for reading and writing various data formats. Includes support for input and output in FITS, ASCII, plain binary, Photometrics, GIF, and hard copy output in PCL or PostScript. The application program interface allows easy extension to other formats.

2.2. Stellar Photometry with PCDAOPHOT II

Via P.B. Stetson, one of us (Smirnov) was provided with a copy of the sources for his DAOPHOT II package (Stetson 1987), and permission to port it to PCIPS. This was quite a challenging task, as the original DAOPHOT II is a large stand-alone Fortran program unsuitable for the PC e.g., it declares several megabytes of static arrays, which can not be done on Intel processors running in real mode, due to their lack of a virtual memory system. The code had to be converted to C or C++ (a Fortran API for PCIPS is still in development), and integrated with PCIPS's own virtual memory mechanism. At the same time, we wanted to avoid modifying the code at all costs. The porting was performed in two stages:

- First, we developed a set of C++ classes that encapsulated PCIPS's virtual memory system, by providing "virtual arrays" that could be used to address images. These classes turned out to be so convenient and useful that they are now an optional part of the API.
- Each non-obsolete DAOPHOT II command was converted to C++ (using a source-code conversion system[1], and integrated into PCIPS

A key feature of PCIPS's API is automatic support of a graphical user interface and visualization tools. This made it possible to improve the interface

[1] f2c, under development by AT&T and Bellcore. Available from Netlib.

of some DAOPHOT commands a great deal. Most of them now automatically produce plots of stars, and allow the user to deselect or select additional stars.

To complete the package, we added some applications, mostly for preprocessing of images: flat field correction, cosmic ray hit detection, estimation of objects' FWHM, etc. We are also working on a set of post-processing applications to handle the star lists produced by PCDAOPHOT II.

2.3. Échelle Spectra Reduction

The CCD/Échelle package was developed by Piskunov under PCIPS. The package facilitates high-quality reduction of echelle spectra, and can handle very complex echelle spectrograms with several geometrically incorrect spectral orders registered in one CCD frame. Each step of the reduction is a a separate application:

FIND locates spectral orders on a frame and establishes their boundaries with maximum precision. It also estimates the bias level.

CRMASK identifies cosmic ray hits using an adjustable criterion. A two-dimensional histogram of the criterion distribution function is displayed, on which an appropriate cut-off level is set by the user. All pixels above the cut-off level are marked as CR hits on a special CR mask image generated by **CRMASK**.

EXTR is the workhorse of the package. It subtracts the bias level, removes cosmic ray hits according to the CR mask, does flat-field correction and performs geometric corrections, before extracting individual spectral orders. The output is a set of one-dimensional spectra, which can be passed to the Spectral package for further processing.

2.4. Spectral Analysis Package

The Spectral analysis package continues where CCD/Échelle left off:

DISP is used to build dispersion curves for spectra. The user identifies the comparison lines and enters their laboratory wavelengths. **DISP** determines the center of each line using one of four methods: Gaussian fit, parabolic fit, maximum pixel or gravity center. After enough comparison lines have been identified, **DISP** does a least-squares fit of the dispersion curve with a polynom of a specified order and shows the deviation and predicted positions of non-identified lines from the user's list.

CONT determines the continuum level of a spectrum. It has automatic and interactive options for selection of continuum points. The selected points are fitted with a polynomial continuum level, which is then divided by the original spectrum to produce a normalized one.

RLAM uses the dispersion curve computed by **DISP** to convert the normalized spectrum into residual intensity–wavelength scale.

VELCOR performs correction of spectra for radial velocity.

CRHIT detects and removes cosmic ray hits from one-dimensional spectra.

MultiProfile performs simultaneous multiple spectral line profile fits. It automatically locates most lines, and lets the user add more before fitting a Gaussian. To help separate blended lines, **MultiProfile** has a set of optional parameter clamps, and a "rigorous" fitting mode, in which a fit is attempted with all lines having one common full-width at half-maximum (FWHM), or with the lines being grouped among two possible FWHMs.

2.5. Fourier Analysis Package

The Fourier package is the most recent addition to the PCIPS family of applications, and as such, it is still in adolescence (but growing rapidly). Currently, it contains a set of basic tools for Fourier analysis: direct and inverse FFTs in one or two dimensions, power spectrum estimation, and interactive filtering in the Fourier domain. In conjunction with the standard application package, the Fourier package can be used for all kinds of sophisticated tasks.

3. Future Development

The next version of PCIPS is targeted at Windows NT. Among others, the system will run on Intel (486 and Pentium) PCs, IBM's RISC workstations, and the DEC Alpha. It will include a next-generation visual language (Smirnov & Piskunov 1993b), with a LEGO-like concept of building large applications out of smaller ones. A similar concept is currently implemented as the Cantata language under the *Khoros* system (Rasure & Williams 1991, Rots 1993).

4. Feedback

PCIPS 2.0 is a commercially available product. We'll be happy to answer any questions and inquiries about the system. Please contact us by e-mail at:
`oms@airas.msk.su` (Oleg Smirnov) or
`pcips@airas.msk.su` (PCIPS feedback)

Acknowledgments. This presentation was made possible by a travel grant from SAO. Also many thanks to our main testers, T. Ryabchikova and A. Ipatov.

References

Rasure, J., & Williams, C. 1991, J. of Visual Languages and Computing,2, 1

Rots, A.H. 1993, in Astronomcical Data Analysis Software and Systems II, A.S.P. Conf. Ser., Vol. 52, eds. R.J. Hanisch, R.J.V. Brissenden & J. Barnes, 194

Smirnov, O.M., & Piskunov, N.E. 1993a, in Astronomcical Data Analysis Software and Systems II, A.S.P. Conf. Ser., Vol. 52, eds. R.J. Hanisch, R.J.V. Brissenden & J. Barnes,259

Smirnov, O.M., & Piskunov, N.E. 1993b, PASPC, 52, 208

Stetson, P.B. 1987, PASP, 99, 191

Astronomical Data Analysis Software and Systems III
ASP Conference Series, Vol. 61, 1994
D. R. Crabtree, R. J. Hanisch, and J. Barnes, eds.

Radially Symmetric Fourier Transforms

M. Birkinshaw

Harvard-Smithsonian Center for Astrophysics, 60 Garden Street, Cambridge, MA 02138-1596

Abstract. Fourier transforms of radially-symmetric functions can be performed efficiently using the Hankel transform of order zero. Convolutions of radially-symmetric functions can also be performed simply. Illustrations of the method are presented, and the Gibbs' phenomenon associated with the *ROSAT* PSPC PRF is discussed.

1. Introduction

A model is often convolved with an instrument response at some stage in the analysis of astronomical data. Frequently both functions are radially symmetric, for example when fitting the X-ray surface brightness of a cluster of galaxies. At times there is a direct need for the Fourier transform of a radially-symmetric function, for example in calculating the response of an optical interferometer to a limb-darkened star. When convolutions or Fourier transforms of radially-symmetric functions are to be calculated, the one-dimensional Hankel transform of order zero (the radial Fourier transform, RFT) is a useful alternative to the two-dimensional Fourier transform. This paper demonstrates the use of the RFT and describes some of its properties.

2. Mathematical development

2.1. Fourier Transforms

The Fourier transform of the two-dimensional function $f(\mathbf{r})$ is

$$\tilde{f}(\mathbf{k}) = \frac{1}{2\pi} \int d^2r\, f(\mathbf{r}) \exp\left(-i\mathbf{k}.\mathbf{r}\right) \tag{1}$$

which, if f is radially symmetric, becomes

$$\tilde{f}(\mathbf{k}) = \frac{1}{2\pi} \int_0^\infty dr\, r\, f(r) \int_0^{2\pi} d\phi \exp\left(-ikr\cos\phi\right) \tag{2}$$

where ϕ is the angle between $\mathbf{k}$ and $\mathbf{r}$. The integral representation of the J_0 Bessel function is

$$J_0(z) = \frac{1}{\pi} \int_0^\pi d\theta \cos(z\cos\theta) \tag{3}$$

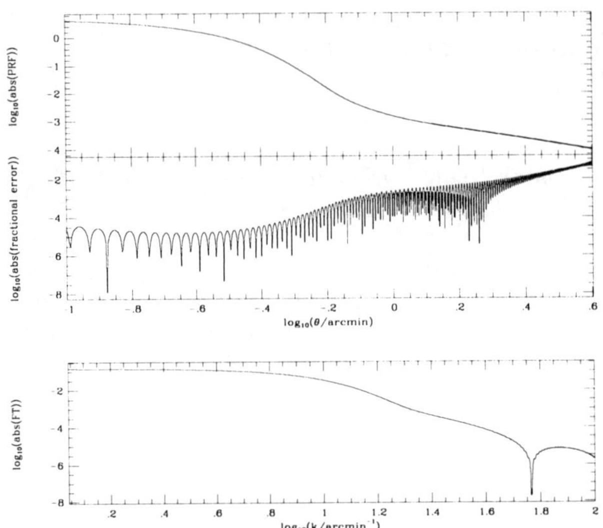

Figure 1. The RFT of the regularized point response function (PRF) of the *ROSAT* Position Sensitive Proportional Counter (PSPC). The top panel shows the 1-keV PRF and the back-RFT of its RFT (indistinguishable at this scale). The absolute value of the fractional difference is shown in the second panel: the error nowhere exceeds 10^{-4} of the peak. The bottom panel shows the RFT of the PSPC PRF. Note that these log – log plots suppress the sign of the difference and the RFT.

so that equation (2) can be rewritten in RFT form

$$\tilde{f}(k) = \int_0^\infty dr\, r\, J_0(kr)\, f(r) \tag{4}$$

where $\tilde{f}(k)$ is radially symmetric in Fourier space. The inverse relation,

$$f(r) = \int_0^\infty dk\, k\, J_0(kr)\, \tilde{f}(k) \tag{5}$$

is easily proved, and demonstrates that the forward and back RFTs are identical operations on functions $f(r)$ and $\tilde{f}(k)$. The relationships (4) and (5) are well known (e.g., Bracewell 1965) and are examples of the Hankel transform of order zero. Sample RFTs calculated using (4) are shown in Figure 1 and Figure 2.

2.2. Convolutions

The convolution $c(\mathbf{r})$ of two functions $a(\mathbf{r})$ and $b(\mathbf{r})$ is

$$c(\mathbf{r}) \equiv a * b = \int d^2u\, a(\mathbf{u})\, b(\mathbf{r} - \mathbf{u}) \tag{6}$$

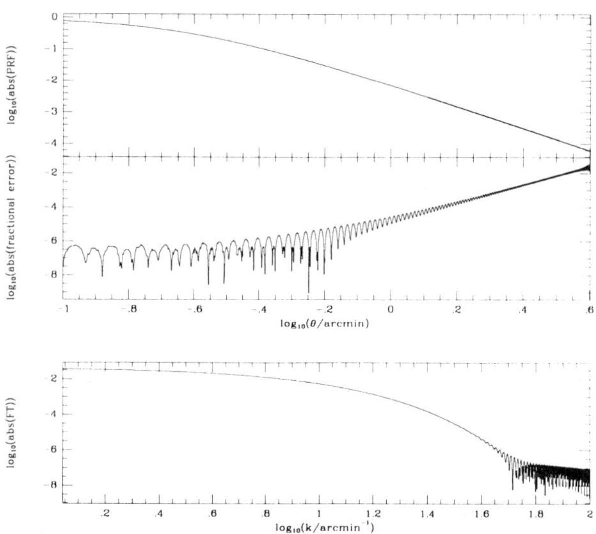

Figure 2. The RFT of an isothermal β model X-ray surface brightness function with $\beta = 0.5$ and core radius $\theta_{cx} = 15$ arcsec. The panels are as in Figure 1.

and it is well known that the Fourier transform of c is related to the Fourier transforms of a and b by

$$\tilde{c}(\mathbf{k}) = 2\pi\, \tilde{a}(\mathbf{k})\, \tilde{b}(\mathbf{k}) \ . \tag{7}$$

If both a and b are radially-symmetric functions, with Fourier transforms radially symmetric in Fourier space, then $\tilde{c}$ is radially symmetric in Fourier space, and c is radially symmetric in real space with

$$c(r) = 2\pi \int_0^\infty dk\, k\, J_0(kr)\, \tilde{a}(k)\, \tilde{b}(k) \ . \tag{8}$$

Thus to evaluate the convolution of two radially-symmetric functions, we need to evaluate two integrals like (4) and one like (8). This scheme requires only two evaluations of the J_0 Bessel function per k-space value per r-space value, because the calculations of $\tilde{a}$ and $\tilde{b}$ can be performed together. Accurate algorithms for $J_0(z)$ are readily available. An example of a convolution performed using RFTs is given in Figure 3.

3. Gibbs' phenomenon

As in other Fourier transform techniques it is important to be aware that jumps in the functions, or their derivatives, cause the appearance of the Gibbs' phenomenon. This is apparent in the fractional error panels of Figs 1 and 2, where truncation of the functions at large θ has caused low-amplitude oscillations. In Figure 1, which shows the RFT of the point response function (PRF) of the

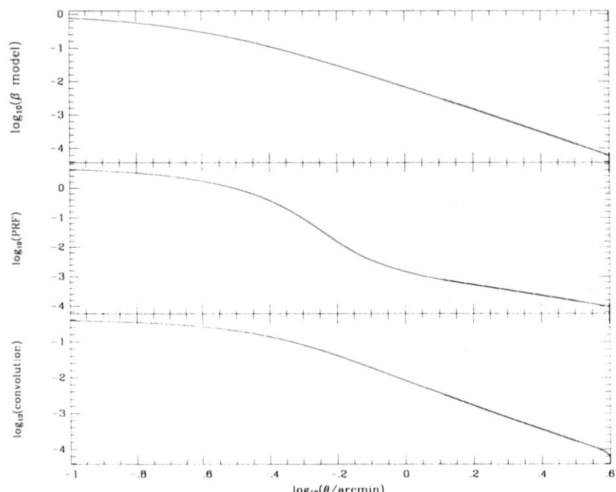

Figure 3. The appearance of the PSPC PRF, an isothermal β model with $\beta = 0.5$ and $\theta_{\mathrm{cx}} = 15$ arcsec, and the RFT-derived convolution of the functions. The turn down in the convolution at $\log_{10}(\theta/\mathrm{arcmin}) \approx 0.6$ is caused by the Gibbs' phenomenon.

ROSAT Position Sensitive Proportional Counter (PSPC), further oscillations are generated at small θ. The amplitude of these oscillations was intially large, but was reduced to the level seen in Figure 1 by making a small (< 1 per cent) alteration of the standard form of the PRF (Hasinger et al. 1992) to replace the cusp at $\theta = 0$ (from the "scattering term") with a higher-order singularity at $\theta = 0.07$ arcmin.

In Figure 3, which shows the convolution of the functions of Figs 1 and 2 the Gibbs' phenomenon has little effect. Thus the RFT is useful for performing accurate convolutions of arbitrary-length and irregularly-sampled arrays quickly: the 8027-element convolution in Figure 3 used a 1395-point Fourier transform array and took under 10 min on a SPARCstation 2. This technique has been used to fit *ROSAT* images of elliptical galaxies (Worrall & Birkinshaw 1994).

This work was supported by NASA grant NAG5-2312 and contract NAS8-39073.

References

Bracewell, R. 1965, "The Fourier Transform and its Applications": McGraw-Hill, New York

Hasinger, G., Turner, T.J., George, I.M., & Boese, G. 1992, NASA/GSFC Office of Guest Investigator Programs, Calibration Memo CAL/ROS/92-001

Worrall, D.M., & Birkinshaw, M. 1994, this volume

Astronomical Data Analysis Software and Systems III
ASP Conference Series, Vol. 61, 1994
D. R. Crabtree, R. J. Hanisch, and J. Barnes, eds.

Partition Based Point Pattern Analysis Methods for Investigation of Spatial Structure of Various Stellar Populations

L. Pásztor

MTA TAKI, Budapest Herman Ottó út 15., H-1022, Hungary

Abstract. A primary piece of information on pointlike astronomical objects is their location. Essential operations, like identification of physical stellar systems and their members, are based on this information, which often makes the application of statistical methodology necessary. Application of various raster and quadtree data representation based methods, furthermore the Voronoi model for the investigation of point patterns are presented. They can be used for studying the spatial distribution, homogeneity/heterogeneity and isotropy/anisotropy properties of various stellar populations. The presented statistics provide information on spatial structure and on significance of the identified effects.

1. Introduction

Generally, analysis of spatial data involves the usage of either the connectivity or the similarity of spatial objects. From this point of view analysis of point data, where only the location of points is given without any attributes, is a rather peculiar task, which often requires special tools.

A data set consisting of irregularly distributed points within a region is referred as (spatial) point pattern. The objective of analysis of a point pattern may involve test of complete spatial randomness (CSR), estimation of intensity, stochastic model fitting etc., to provide an explanation of the underlying (astronomical) processes. For detailed discussion of point patterns see e.g., Ripley (1981), or Cressie (1991). In this paper simple, univariate, 2D point patterns are discussed. Simple, that is single points occur and univariate, that is the points do not posses any attributes. In the case of simple, univariate point patterns, where is neither connectivity between objects, nor their similarity is interpretable due to the lack of attributes, the analysis of spatial structure often requires sophisticated methods. Generally, definition of neighborhood structure is focussed on. Various distance as well as quadrat methods differ in definition of this "generalized connectivity". A frequent approach is the tiling of the point pattern, often with variable spatial resolution, according to the estimated intensity of the underlying point process.

Various statistics of spatially partitioned point patterns are presented. They provide useful information on spatial structure. They can be used for testing complete spatial randomness, measuring spatial autocorrelation and they provide measure of significance as well. Using reference point samples, basic classification may be carried out, which can be used in stochastic modeling of underlying processes. Additionally, intensity estimations and characteristic dimensions can be derived.

2. Raster Data Representation

In the raster model of space the sample area is broken into regular, discrete units called cells. Raster data representation of point patterns means that the sample area is divided into equal subregions, generally squares, and their population (number of points belonging to the quadrants) is given.

2.1. Hierarchical Counting

Completely random spatial point patterns are treated as realizations of a homogeneous Poisson point process. One of their features is that the number of points in any two disjoint regions is independent. Deviation from this characteristic is attributed to deviation from CSR. The Poisson behavior can be tested with the relation of variance and parameter of the process, since for a Poisson process the two values are identical. Consider a grid laid on a point pattern and the distribution of number of points falling into each cell. Denoting area of the sample by A, number of its elements by N and number of raster units by M, intensity of the process can be estimated by N/A, while the area of a single cell is obviously A/M. Thus for the estimation of the parameter of Poisson process N/M is derived. The variance is estimated by the empirical variance. As the resolution of the grid is refined, the correlation of estimated variance and parameter can be studied. For a CSR point pattern linear relationship is expected between variance and $1/M$, while the effects of various spatial structures on this relationship ought to be determined. From numerous Monte Carlo simulations the following results issued.

(i) For regularly aggregated samples the $var - M^{-1}$ curve is composed of two linear parts indicating the two levels of the hierarchical cluster structure. The breakpoint of the curve provides a critical resolution and consequently a critical density value which reflects the density in the aggregates. Beyond this limit the cluster structure of the pattern is destroyed by the overlying grid and so the cluster process behaves as an ordinary homogeneous Poisson process.

(ii) Adding CSR pattern to the former regularly aggregated samples, as noisy background, the structure of the $var - M^{-1}$ curve is conserved, merely the slope of the non-Poisson part decreases, which indicates the inclination toward a random behaviour as opposed to the original aggregated feature.

(iii) Composing clustered patterns with different density parameters, the $var - M^{-1}$ curve breaks into further parts with breakpoints corresponding to the original within cluster densities.

(iv) Finally, different $var - M^{-1}$ slopes indicate the inclination from random pattern toward either the aggregated or the regular behaviour.

Feature (i) makes the identification of cluster processes and their scales, feature (ii) the separation of noisy background and cluster structure, feature (iii) the identification of hierarchical cluster structure and the characteristic scales, feature (iv) the classification of samples, into the basic categories, possible.

2.2. Samples

Hereafter five point patterns are discussed. Three of them represent the three basic point pattern categories, namely aggregated, random and regular. The fourth sample (as well as the former ones) is the result of a Monte Carlo simulation, and possesses the same large scale spatial structure as the fifth, real sample, but without its small scale fluctuations. This latter is a thoroughly studied sample of IRAS point sources in the galactic region $110^o < l < 130^o$ and $5^o < b < 25^o$ (Pásztor et al. 1993).

2.3. Moran Statistics

One of the simplest measures of spatial autocorrelation, join count statistics, was introduced by Moran (1948) for testing nominal scale data. In this context spatially structured data sets are treated as mosaics of areas with different colors. Elements with common boundary are said to be linked by a join. The basis of join count statistics is: the distributions of joins between areas of different and same colors respectively under the null hypothesis, H_0, of no spatial autocorrelation in the sample, are asymptotically normal with moments determined by the geometry of the data set. The first two moments can be used to test whether the number of various joins departs significantly from random expectations.

Moran statistics can be applied to the spatial analysis of point patterns, too, using partition based representations of the point-like data sets. If we consider raster representation of a point pattern, the populated and empty cells can be coded with black and white respectively. Then the resulted binary mosaic can be tested by the aid of Moran statistics. A critical resolution can be derived for each pattern, namely the least value for which all the elements of the pattern belong to different squares. Then the result of join count statistics for the colored raster units is attributed to the spatial structure of the point pattern. The method provides fairly good discrimination among the different point patterns but its efficiency and applicability can be improved applying

3. Quadtree Data Representation

PR quadtree representation of point patterns also associate data points with quadrants (Samet 1990). In a PR quadtree leaf node is either empty or contains one point. As for color coding, empty and engaged leafs are said to be white and black respectively. A modified version of PR quadtrees (MPR quadtree) is introduced by Pásztor (1993). A given partition of a point pattern also provides a mosaic of tiles colored with B or W respectively, just according to the coding process. Spatial autocorrelation of this tiles may be determined by their Moran statistics. Spatial structure of MPR quadtree represented point patterns is conserved in their quadtree partitions, the standard normal deviate of BB join count statistic on different quadtree levels provides an efficient tool for the discrimination among various point patterns (Pásztor 1993). As for the application of join count statistics, the advantages of quadtree over raster representation can be summarized in three items.

(i) Applicability on different levels (resolution), since there is no critical resolution determined by the pattern.

(ii) Less ambiguous and so more efficient discrimination at a given level (resolution).

(iii) Comparability problems do not emerge, as Moran values, used for discrimination, are derived from equivalent partitions.

4. Tessellation, the Voronoi Model

The Voronoi region of an object is the region of space closest to the given object than to any other object of the sample. The set of Voronoi regions for a set of spatial objects, called Voronoi diagram (also known as Dirichlet tessellation or Thiessen polygons), provides a partition of a point pattern according to its spatial structure, too. Features of this kind of decomposition also can be used for analysis of the underlying point process. For a CSR sample the expected values of various characteristics of polygons can be given: $E(\text{number of sides}) = 6$; $E(\text{length of the perimeter}) = 4 * \beta^{-1/2}$; $E(\text{area}) = \beta^{-1}$ (Upton and Fingleton 1985). Unfortunately, there are no analytical results for the second moments.

(i) Usage of the result for the number of sides cannot be proposed for spatial analysis, since it has (deterministic) geometrical reasons.

(ii) Merely the difference between expected values and empirical averages for the length of perimeter and area distribution is neither can be proposed for test of CSR due to the lack of available significance measurements.

(iii) The distribution themselves, however, are useful for spatial analysis. By homogeneity tests classification of point patterns into the basic categories can be carried out. Random and regular patterns have unimodal, normal distributions with different variances. The multimodality indicates hierarchical (cluster) structure, the number of modes is determined by the number of scales in the sample.

(iv) Probably the most useful application of Voronoi model is its aid in compilation of surface density maps from point data. As the area of tiles is the estimation of the inverse value of the local intensity, assigning $1/t$ to the point locations as local densities, the map can be easily compiled with the aid of any kind of (either deterministic or stochastic) interpolation method.

Acknowledgments. The author is grateful to SAO and the Hungarian State Research Found (Grant No. OTKA-F 4239) for the travel grants.

References

Cressie, N. 1991, "Statistics for Spatial Data", Wiley

Moran P.A.P., 1948, J. Roy. Stat. Soc. ser. B 10, 248

Pásztor L., Tóth L.V., & Balázs L.G., 1993, A&A268, 108

Pásztor, L., 1993, "Revealing Spatial Structure of Point Patterns by Quadtree Statistics", submitted to COSIT '93, Elba, Italy

Ripley B.D., 1981, "Spatial Statistics'. Wiley

Samet H., 1990, "The Design and Analysis of Spatial Data Structures'. Addison-Wesley

Upton G., & Fingleton B., 1985, "Spatial Data Analysis by Examples", Wiley

Astronomical Data Analysis Software and Systems III
ASP Conference Series, Vol. 61, 1994
D. R. Crabtree, R. J. Hanisch, and J. Barnes, eds.

Cosmic Ray Hit Detection with Homogenous Structures

O. M. Smirnov
Institute of Astronomy of the Russian Academy of Sciences
48 Pyatnitskaya St., Moscow 109017 Russia

Abstract. Cosmic ray (CR) hits can affect a significant number of pixels both on long-exposure ground-based CCD observations and on Hubble Space Telescope frames. Methods of identifying the damaged pixels are an important part of the data preprocessing for practically any application. This paper presents an implementation of a CR hit detection algorithm based on a *homogenous structure* (also called *cellular automata*), a concept originating in artificial intelligence and dicrete mathematics. Each pixel of the image is represented by a small automaton, which interacts with its neighbors and assumes a distinct state if it "decides" that a CR hit is present. On test data, the algorithm has shown a high detection rate (≈ 0.7) and a low false alarm rate (≤ 0.1) in relatively little time per frame. A homogenous structure is extremely trainable, which can be very important for processing large batches of data obtained under similar conditions. Training and optimizing issues are discussed, as well as possible other applications of this concept to image processing.

1. Introduction

Cosmic ray (CR) hits can affect a significant number of pixels both on long-exposure ground-based CCD observations and on the Space Telescope frames. Thus, methods of identifying the damaged pixels are an important part of the data preprocessing for practically any application. Various algorithms exist for detecting CR hits, see Murtagh & Adorf 1991 for an excellent review.

In this paper I investigate an approach to CR hit detection based on the concept of a *homogenous structure* (Kudryavtzev et al. 1990). On test data, the algorithm (henceforth referred to as HSCR) has shown a good detection rate $R_D \approx 0.7$ and a low false alarm rate $R_F \leq 0.2$. I will also discuss an implementation of HSCR, as well as questions of optimization and trainability of the algorithm.

2. Homogenous Structures

Homogenous structures (then called cellular automata) were first introduced in the works of Von Neuman. He considered them in a mostly philosophical light, in an attempt to construct a self-replicating automaton. In later years, they were investigated as a mathematical concept, and a whole theory was developed, dealing with such questions as complexity, completeness and computations in homogenous structures (Kudryavtzev et al. 1990).

Mathematically speaking, a homogenous structure is an n-dimensional grid of identical finite automata. A finite automaton can be considered to be a computer (in fact, computers are finite automata in the most rigorous sense), with inputs, outputs, and internal states. The automaton runs in discrete time, with its internal states and outputs at moment $t+1$ determined via a *response function* by its states and inputs at moment t. If we place identical automata at each node of an n-dimensional grid, and link their inputs and outputs in a homogenous pattern, what we get is a homogenous structure. When we let the automata run, they interact with each other, and can demonstrate very interesting, complex, surprising, and useful behaviour. Some examples:

- A very simple automaton: four binary inputs and outputs (north, south, east, west), one binary state. The automata are connected into a 2D grid. If we plot the 1-states as black pixels and the 0-states as white pixels, we'll see a very interesting picture: any initial pattern is replicated over time indefinitely: 4 times, 16 times, 64 times, etc.

- David Conway's game of *Life* (recently implemented in screen savers on Suns) is also a homogenous structure. With a very simple automaton, it imitates pseudo-living organisms that live, fight, mutate, replicate and die. In fact, only slightly different structures were used to successfully model some real-life biological processes.

Both examples serve to demonstrate one vital feature of homogenous structures: even a very simple base automaton can lead to very complex behaviour of the structure in whole. Thus, algorithms based on the concept are very easy to implement.

3. Cosmic Ray Hit Detection

For cosmic ray hit detection, I used a more general form of homogenous structure, called a homogenous structure with inputs and outputs. Its main difference is that every automata has an additional set of auxiliary unconnected inputs and outputs that "stick out" of the structure (see figure). Such a structure can perform computations: we feed data to the auxiliary inputs, and read results at the auxiliary outputs.

The main ideas here is that a CR hit is a purely local event. So we place certain automata at each image pixel, thus producing a homogenous structure, and let each pixel-automaton interact with its neighbors and determine whether it contains a CR hit.

For input data, HSCR uses a transformed *pseudo-gradient* image g_{ij},

$$g_{ij} = \sqrt{d'_{ij} + \lambda d''_{ij}}$$

where d_{ij} is the original image, d'_{ij} and d''_{ij} are its first and second gradient (taken as the maximal difference from neighboring pixels), and λ is an adjustable parameter. Then each pixel g_{ij} is fed into the corresponding auxiliary input of a 2D homogenous structure.

The structure. Of course, most of the algorithm's functionality is concentrated in the structure's base automaton. The automaton here is relatively simple. It has four primary inputs/outputs, linked to its northern, southern, eastern and western neighbors. Its output on each side is a weighted average of the input from the opposite side and the auxiliary input (which is equal to a pixel of the pseudo-gradient image). Thus, information about the local distribution of g_{ij} is propagated in each direction, with the weights determining the extent of this propagation.

To detect CR hits, each pixel compares its pseudo-gradient with the mean pseudo-gradients received from its neighbors via the inputs (as follows from above, each input is a weighted average of the pseudo-gradients at several consecutive pixels in its direction). If it is higher than the mean by some threshold (the threshold is determined dynamically from the overall pseudo-gradient statistics), the automaton considers itself to be a CR hit, sends this information to its auxiliary output, and from then on substitutes its pseudo-gradient value for the minimum value received from the neighbors. Thus the damaged (high) value is excluded from future estimates of the local distribution propagated across the structure. All of the above can be expressed in formal mathematical terms. If you're interested (and if you know Russian), refer to Smirnov 1993.

The output of the structure is a binary image of the same format as the input, where detected CR hits are flagged by set bits. The image changes at each cycle corresponding to one moment of time, and converges to a good result in 5–10 cycles.

Why it works. In simplified terms, HSCR compares the pseudo-gradient of each pixel to some weighted average of the pseudo-gradients in its vicinity. Since CR hits usually have a somewhat higher pseudo-gradient, they are effectively detected. This weighted average is higher where the image has sharp features, so parts of them are rarely misflagged. Note that the weights are a very flexible parameter of the algoritm. For example, they can be set to propagate the values in one direction more actively than in another, which is useful for images with structure of a known orientation (i.e., echelle spectra). And because a pixel's pseudo-gradient is automatically excluded from the local statistics once the pixel is flagged as a hit, even extended (tangential) and closely-spaced hits do not affect the statistics enough to slip through undetected.

The simple automaton makes HSCR very flexible and easily modifiable. Also, it is quite fast, with little computation done per each pixel. Another advantage of the algorithm is that it is naturally concurrent, since each pixel is processed independently.

4. Implementation

The beauty of a homogenous structure is the fact that you only have to program one pixel of it (assuming you have a module that processes the pixels). A prototype HSCR has been implemented in C++ under the PCIPS image processing system (Smirnov & Piskunov, this volume). As of now, the program is intended to help further investigate the method. When running, it shows the distribution of the values at each cycle, which demonstrates the propagation of statistics, and it can also plot histograms of this distribution, where the contribution of CR hits becomes plainly visible.

5. Optimization and Training

Optimization. Observational data usually comes in batches obtained under similar conditions. It is thus desirable to be able to optimize an algorithm for a particular batch. HSCR allows this via the manipulation of several parameters:

- The pseudo-gradient can be calculated using a few other formulas with one parameter, λ. This can optimize the algorithm with respect to the general character and shape of CR hits.
- The weights used in propagation of statistics can be adjusted to make the propagation anisotropic. This can optimize the algorithm with respect to the structure of the image.
- The detection threshold can be adjusted, also optimizing for the character of the cosmic rays.

Training. A trainable algorithm must be able to optimize itself automatically, given a set of training data. Thus, CR hits can be detected on a few frames manually and with the help of conventional algortihms, then these frames are fed into the trainer module, which examines them and estimates optimal values for the parameters that would produce such results. The rest of the batch is then handled fully automatically. I have not yet experimented with a trainer module for HSCR, but it should not be too complex. The output of HSCR shows a smooth dependence on variations in the parameters, so some sort of downhill optimization should be possible to implement.

Acknowledgments. This presentation was made possible by a travel grant from SAO. Also many thanks to Prof. Kudryavtzev for stimulating discussions.

References

Kudryavtzev, V.B., Podkolzin, A.S., & Bolotov, A.A. 1990, "Basic Theory of Homogenous Structures", Moscow: Nauka Publishers

Murtagh, F., & Adorf, H.-M. 1991, in *3rd ESO/ST-ECF Data Analysis Workshop*, Munich: ESO, 51

Smirnov, O.M. 1993, in *International Seminar on Discrete Mathematics, Moscow University, 1992–93*, eds. V.B. Kudryavtzev, Yu.M. Gorchukov, Moscow University Press, in press

Astronomical Data Analysis Software and Systems III
ASP Conference Series, Vol. 61, 1994
D. R. Crabtree, R. J. Hanisch, and J. Barnes, eds.

Searching Procedures for Groups, Clusters and Superclusters of Galaxies

M.Kalinkov and I.Kuneva
Institute of Astronomy, Bulgarian Academy of Sciences
72 Tsarigradsko Chausse blvd, 1784 Sofia, Bulgaria

Abstract. A unified procedure for searching of aggregations of galaxies — from pairs to superclusters is applied for several catalogs of galaxies and of clusters of galaxies.

1. Introduction

There are many catalogs of aggregations of galaxies — pairs, groups, clusters and superclusters. Almost so many are the searching procedures.

The first objective 3D algorithms for groups of galaxies are proposed by Materne (1978) and Paturel (1979). A similar method was used by Tully (1987). A method (friends of friends) which is widely used was developed by Huchra & Geller (1982) — cf. Morgan & Hartwick (1988), Maia et al. (1989). Vennik (1984) and Gourgoulhon et al. (1992) and Fouqué et al. (1992) applied algorithms very close to Tully's, while Garcia (1993) selected groups of galaxies with methods based on Huchra & Geller and Tully procedures.

2. Searching Procedure

We have developed an algorithm for searching superclusters of galaxies in the 3D case (e.g., Kalinkov & Kuneva 1985, 1986). That is a percolation method with two parameters — a density contrast exceeding the mean density for a sphere with a given radius.

That procedure was applied for groups of galaxies for CfA survey (Kuneva & Kalinkov 1990). An advantage is that nearly spherical systems were found. A fault is that 3D cartesian coordinates are directly evaluated from redshift. There is a possibility of tuning the searching algorithm with the luminosity function of galaxies, as in the HG procedure.

A very important advantage of the algorithm is the determination of the term *isolation* with the help of quantiles (say 90%) of the distance distribution to the nearest neighbour out of the system.

3. Some Results

We have searched for groups in samples from CfA, SSRS, zCAT and PGC catalogs. We have compared our lists of groups, found for different searching radii

r_0 and density contrast ρ_c, with other catalogs — HG, GH, Maia et al., Fouqé et al., Garcia ... It was established that the first version of the algorithm with appropriate $r_0 = 5$ Mpc and $\rho_c = 20$ leads to finding groups which are central parts of groups in other catalogs. Using the new refined procedure we find groups more extended along the redshift coordinate , nevertheless they are not so elongated as according to other catalogs. It is very interesting that the kinematical and dynamical properties of groups for various searching algorithms are not so different. The procedure allows to find not only groups, but pairs of galaxies too.

The main aim of this study is to examine critically all lists of group candidates in order to complete the project Reference Catalog of groups of galaxies. The RCGG will contain information for members of the groups, kinematical and dynamical properties, isolation, observations, references.

The procedure is applicable for clusters of galaxies, then the groups found are indeed superclusters of galaxies according to the current terminology. A comparison of superclusters from different catalogs (Bahcall & Soneira 1984, Batuski & Burns 1985, Postman et al. 1992, Cappi & Maurogordato 1992, Zucca et al. 1993) shows again than the properties of superclusters are very similar independent by the searching procedure.

Acknowledgments. The authors would like thank Mr. I. Valtchanov for valuable help. This work was supported by the National Science Research Funds of the Bulgarian Ministry of Education and Science (contract Ph 260/1992).

References

Bahcall, N.A., Soneira, R.M. 1984, ApJ, 277, 27

Batuski, D.J., Burns, J.O. 1985, AJ, 90, 1413

Cappi, A., Maurogordato, S. 1992, A&A, 259, 423

Fouqé, P., Gourgoulhon, E., Chamaraux, P., Paturel, G. 1992, A&A, 93, 211

Garcia, A.M. 1993, A&A, 100, 47

Gourgoulhon, E., Chamaraux, P., Fouqé, P. 1992, A&A, 255, 69

Huchra, J.P., Geller, M.J. 1982, ApJ, 257, 423

Kalinkov, M., Kuneva, I. 1985, AN, 306, 284

Kalinkov, M., Kuneva, I. 1986, MNRAS, 218, 40^P

Kalinkov, M., Kuneva, I. 1990, in Paired and Interacting Galaxies, eds. J. W. Sulentic, W. C. Keel, C. M. Telesco (Washington: NASA), p. 143

Maia, M.A.G. & da Costa, L.N. 1989, ApJS, 69, 809

Materne, J. 1978, A&A, 71, 106

Morgan, C.G. & Hartwick, F.D.A. 1988, ApJ, 328, 38

Paturel, G. 1979, A&A, 71, 106

Postman, M., Huchra, J.P. & Geller, M.J. 1992, ApJ, 384, 404

Tully, R.B. 1987, ApJ, 321, 280

Vennik, J. 1984, Tartu Astr. Obs. Teated, No 73

Zucca, E., Zamorani, G., Scaramella & R., Vettolani, G. 1993, ApJ, 407, 470

Astronomical Data Analysis Software and Systems III
ASP Conference Series, Vol. 61, 1994
D. R. Crabtree, R. J. Hanisch, and J. Barnes, eds.

Multiple Regression Redshift Calibration for ACO Clusters of Galaxies

M.Kalinkov, I.Kuneva, I.Valtchanov

Institute of Astronomy, Bulgarian Academy of Sciences
72 Tsarigradsko Chausse blvd, 1784 Sofia, Bulgaria
e-mail: markal@bgearn, valtch@bgearn

Abstract. A new procedure for redshift estimate of ACO clusters of galaxies has been developed.

1. Introduction

There are about 12 000 clusters of galaxies known, but the redshifts are measured for no more than 2000. The number of rich clusters of galaxies in the whole sky catalog of Abell et al. (1989), henceforth ACO, is 4073 and only 1155 have a measured redshift. For the other ACO clusters, not rich enough or too distant to be included in the main catalog, 1174 clusters, redshifts are known for 192. Clusters with measured redshifts are used for calibration of the relations for redshift estimates.

There are two philosophies for the estimation procedure. The first one consists of calibration of a magnitude-redshift relation — e.g., m_{10}-log z, where m_{10} is the tenth brightest galaxy. The second one is more preferable from the theoretical point of view — the apparent magnitude of a standard candle is a linear function of log R_L, where R_L is the luminosity distance, and the calibration have to be based on a magnitude-log R_L relation. That approach was used by Scaramella et al. (1991) and Zucca et al. (1993).

Nevertheless, we follow the first philosophy since the internal error is smaller.

A magnitude-redshift relation was used by Abell (1958) and according to Fullerton & Hoover (1972) it is

$$\log z = -8.5332 + 0.72123\, m_{10} - 0.015834\, m_{10}^2.$$

That one might be used when there is need to restore Abell's distance scale.

Other relations are derived by Rowan-Robinson (1972), Corwin (1974) — in a very thorough study, Kalinkov & Kuneva (1975), Mills & Hoskins (1977). Leir & van den Bergh (1977) have used three different relations of log z on the K-corrected m_1 and BM, on the K-corrected m_{10} and RG, and RG and log r_c. The correction for K-dimming is according to Sandage (1973) for the Abell magnitude-redshift relation. BM is the Bautz-Morgan class usually coded as 1 = I, 2 = I-II, 3 = II, 4 = II-III, 5 = III, and r_c is the apparent radius of cluster (in mm on the PA). A final redshift estimate is a weighted mean value from the three relations.

Postman et al. (1985) obtained a new magnitude-redshift relation and determined the Scott effect. They insist that the Malmquist bias is small compared to the Scott effect.

A very detailed calibration with linear multiple regressors are carried out by Kalinkov & Kuneva (1985). However, the st.dev. of $\hat{z}$ is not small — about 15% for $z \approx 0.15$. But in log z the st.dev. of Corwin (1974) relation is s = 0.10 (26% in z). Compare s = 0.09 (23% in z) of Mills & Hoskins (1977) and s = 0.14 (38%) of Postman et al. (1985).

All above results are for the Abell (1958) catalog. For ACO other relations have to be used — see Abell et al. (1989), Couchman et al. (1989), McGill & Couchman (1989), Plionis & Valdarnini (1991), Scaramella et al. (1991), Plionis et al. (1992), Zucca et al. (1993). Redshift estimate for ACO clusters is worse than for A-clusters.

2. Reference Catalog of ACO Clusters of Galaxies

We have been working on a project for a new type of catalog — Reference Catalog (RC) of Abell-Corwin-Olowin clusters of galaxies (Kalinkov & Kuneva 1990). The RC contains information for ACO clusters of galaxies gathered from catalogs, lists, papers, which is reduced in a homogeneous system. The cosmologically dependent properties are computed for a model with $q_0 = 1/2$ and $H_0 = 100$ km s^{-1} Mpc^{-1}. The cluster properties include coordinates, optical, radio and X-ray observations, redshifts, types, substructure, supercluster membership ...There are also cross-references between all known catalogs or lists of clusters. Many papers containing observations or theoretical results are cited. A key word system is developed to give an idea for the essence of each reference. Morphological description of the clusters and/or of interesting objects towards the clusters are given too.

The primary aim of the RC is to give a picture for the existing observations for each cluster as well as to choose clusters for future investigation. The secondary aim is to generate lists of reliable clusters that might be used for studying the large scale structure in the Universe.

A considerable part of the RC compilation is finished, especially for the 4073 rich ACO clusters. But there are many problems concerning radio sources and their cross-identification towards the clusters. Now we are gathering information for the poor ACO clusters. We hope to incorporate many data in the ACO database project (Kalinkov et al. 1993).

We use the RC for the new redshift calibration. It is impossible to find calibration relation for the northern and southern clusters together. That is why the calibration procedure is different for them.

3. Northern Clusters

There are 2712 clusters in the A-catalog, 966 of which have a measured redshift. Preliminary considerations show that five clusters have to be excluded — namely A34, 465, 484, 536 and 2661. Scott effect corrections Δm_{Sc} are given in Table 1, following the prescription of Postman et al. (1985). First ranked magnitude is from Leir & van den Bergh (1977).

Table 1.

RG	$\Delta m_{10,Sc}$	n	$\Delta m_{1,Sc}$	n
0	-0.191	296	-0.121	107
1	0.000	403	0.000	386
2	0.253	188	0.133	178
3	0.470	65	0.169	61
4+5	0.620	9	0.187	8

Our efforts to improve the magnitude-redshift relation with Δm_{Sc} failed. Moreover we have found some indications for a Malmquist bias.

We recommend the following regressions for redshift estimate.

A preliminary estimate of $\hat{z}$ is computing from

$$\log z = (-4.7473 \pm 0.0735) + (0.22625 \pm 0.00438)\, m_{10}, \; r = 0.858, \; s = 0.122.$$

Then compute one Abell's radius r_A from

$$\hat{r}_A = 0.014\,334 \left((1+\hat{z})^{3/2}\right) \left((1+\hat{z})^{1/2} - 1\right)^{-1} \text{ deg},$$

Mean apparent density $\hat{D}_A$ from

$$\hat{D}_A = N_A / \left(\pi \hat{r}_A^2\right) \text{ g deg}^{-2},$$

where N_A is the Abell count of galaxies, and mean apparent density $\hat{D}_H$ from

$$\hat{D}_H = N_H / \left(\pi \hat{r}_A^2\right) \text{ g deg}^{-2}.$$

Here N_H is the corrected Abell's count. The meaning of N_H is to correct N_A for the error due to unreliable Abell's distance scale.

Then depending on the original RG of the A-catalog the five regressions must be applied to reach the final estimate:

$$\log \check{z} = a_0 + a_1 \hat{r}_A + a_2 m_{10} + a_3 \log \hat{D}_A + \log \hat{D}_H.$$

The coefficients are given in Table 2 together with st.dev. from the multiple regression and the number of observations. Note that all regressions with ˆ are corrected for the preliminary redshift estimate $\hat{z}$ (on m_{10} only). Thus $s_{\log \check{z}} =$ 0.017 or $s_{\check{z}} = 0.0073$.

4. Southern Clusters

There are 1361 rich southern clusters in ACO and for 186 of them redshift is known (we exclude several clusters with evident foreground measurements).

Table 2.

RG	a_0	a_1	a_2	a_3	a_4	s	n
0	-3.4911	-0.2227	0.15642	1.7167	-1.6997	0.019	296
	±996	±156	±781	±243	±159		
1	-3.6315	-0.2016	0.16782	1.8338	-1.8402	0.019	403
	±738	±125	±599	±184	±137		
2	-3.3418	-0.2907	0.15147	1.9797	-1.9794	0.014	188
	±838	±107	±728	±286	±195		
3	-3.1923	-0.3266	0.1406	2.1642	-2.1480	0.008	65
	±2315	±979	±131	±343	±274		
4+5	—	-2.1484	—	2.2027	-2.2870	0.005	9
		±496		±787	±797		

Table 3.

RG	$\Delta m_{1,Sc}$	$\Delta m_{3,Sc}$	$\Delta m_{10,Sc}$	n
0	0.158	0.103	-0.046	84
1	0.000	0.000	0.000	61
2	0.143	0.179	0.350	32
≥ 3	1.104	0.944	0.881	9

Again we follow Postman et al. (1985) prescription for the Scott effect and the results are given in Table 3.

Magnitudes m_1, m_3 and m_{10} in the RC are corrected with $K = 4.14z - 0.44z^2$ (Ellis 1983; Shanks et al. 1984) and with galactic absorption law $A = -0.149 + 0.207 \csc|b|$ (Fisher & Tully 1981).

As might be seen in Table 3 the Scott effect is revealed for m_{10}, and contrary with respect to m_1 and m_3 — see RG0.

It is the main fault of ACO — the background corrections made assuming an "universal" luminosity function for field galaxies, maybe, that causes the reducing of the Scott effect.

With $\hat{z}$, which here is the ACO estimate — their formulae (11), again the most appropriate is separation according to RG. We search regressions in the form

$$\log \check{z} = a_0 + a_1 \log \hat{z} + a_2 m_1 + a_3 m_3 + a_4 \hat{M}_1 + a_5 \hat{M}_3.$$

Regression coefficients are given in Table 4.

We have $s_{\log \check{z}} = 0.111$ and there is no prospect to forthcoming improvements of the st.dev.

5. Supplementary Southern Clusters

There are 1174 supplementary southern clusters in the ACO catalog, which are not rich enough or too distant to be included in the main catalog. It is very risky

Table 4.

RG	a_0	a_1	a_2	a_3	a_4	a_5	s	n
0	-26.00	-2.390	0.621		-0.576		0.124	84
	±4.51	±553	±104		±106			
1	-6.490		0.1677		-0.1260		0.114	61
	±664		±135		±236			
2	-30.03	-3.022		0.6896		-0.6797	0.080	32
	±3.72	±474		±805		±910		
3+4	-3.394			0.1505			0.072	9
	±150			±140				

to propose a magnitude-redshift relation for this clusters, because many of them have "negative" counts of galaxies (due to the main fault of ACO mentioned above).

Nevertheless, we recommend the next procedure (after exclusion of clusters S182, 230, 773, 861 and 863)

$$\hat{z} = \text{dex}\,(0.5\,(\log\, z_1 + \log\, z_2))\,,$$

where

$$\log\, z_1 = (-3.603 \pm 0.136) + (0.15157 \pm 0.00923)\left(\frac{m_1 + m_3}{2}\right)$$

with $s = 0.162, n = 190$ and

$$\log\, z_2 = (-3.986 \pm 0.181) + (0.1618 \pm 0.0112) m_{10},\ \ s = 0.175,\ \ n = 190.$$

6. Conclusion

Multiple regression formulae for the A-catalog allow to estimate the redshifts very sufficiently. For the southern clusters and especially for the supplementary clusters there are not so many redshift measurements and the calibration is uncertain. However, the absence of reliable counts of galaxies (as Abell's) for the southern clusters make the progress in sufficient calibration very slow.

This work was supported by the National Scientific Research Funds of the Bulgarian Ministry of Education and Science (contracts Ph 107/1991 and Ph 260/1992).

References

Abell, G.O. 1958, ApJS, 3, 211

Abell, G.O., Corwin, H.G., Olowin, R.P. 1989 ApJS, 70, 1

Corwin, H.G. 1974, AJ, 79, 1356

Couchman, H.M.P., McGill, C., Olowin, R.P. 1989, MNRAS, 239, 513

Ellis, R.S. 1983, in The Origin and the Evolution og Galaxies, eds. B.T.J. Jones & E.J. Jones (Dordrecht: Reidel), 253

Fisher, J.R., Tully, R.B. 1981, ApJS, 47, 139

Fullerton, W., Hoover, P. 1972, ApJ, 172, 9

Kalinkov, M., Kuneva, I. 1975, AZhL, 1, Nr. 2, 7

Kalinkov, M., Kuneva, I. 1985, AN, 306, 283

Kalinkov, M., Kuneva, I. 1990, in Superclusters and Clusters of Galaxies and Environmental Effects, eds. G. Giurcin, F. Mardirossian, M. Mezzetti (Trieste)

Kalinkov, M., Kuneva, I., Paturel, G. 1993, in Handling and Archiving Data from Ground-Based Telescopes, eds. M. Albrecht & F. Pasian (ESO)

Leir, A.A., van den Bergh, S. 1977, ApJS, 34, 381

McGill, C., Couchman, H.M.P. 1989, MNRAS, 236, 51

Mills, B.Y., Hoskins, D.G. 1977, Austr. J. Phys., 30, 509

Plionis, M., Valdarnini, R. 1991, MNRAS, 249, 46

Plionis, M., Valdarnini, R., Coles, P. 1992, MNRAS, 258, 114

Postman, M., Huchra, J.P., Geller, M.J., Henry, J.P. 1985, AJ, 90, 1400

Rowan-Robinson, M. 1972, AJ, 77, 543

Sandage, A.R. 1973, ApJ, 183, 711

Scaramella, R., Zamorani, G., Vettolani, G., Chicarini, G. 1991, AJ, 101, 342

Shanks, T., Stevenson, P.R.F., Fong, R., McGillivray, H.T. 1984, MNRAS, 206, 767

Zucca, E., Zamorani, G., Scaramella, R., Vettolani, G. 1993, ApJ, 407, 470

Astronomical Data Analysis Software and Systems III
ASP Conference Series, Vol. 61, 1994
D. R. Crabtree, R. J. Hanisch, and J. Barnes, eds.

An Alternative Algorithm for CMBR Full Sky Harmonic Analysis

C. A. Wuensche[1] and P. Lubin

Department of Physics, University of California, Santa Barbara CA, 93106 USA

T. Villela

Divisão de Astrofísica, INPE São José dos Campos, SP, 12201 Brazil

Abstract. We discuss an alternative algorithm to deal with many-point data sets and show that, for the kind of problem we want to solve, it is more robust, although a little slower, than the currently used algorithm and can always present a numerical solution, based on the maximum likelihood test. We describe the algorithm and the kind of data it is suitable for. The basic idea is to avoid the inversion of large matrices, usually the core of the algorithms currently chosen, by using the Singular Value Decomposition method. Results of Monte Carlo simulations of full sky maps and a comparison of both methods are discussed.

1. Introduction

The recent advances in space astronomy and astrophysics have created a demand for more powerful computers and more "robust" data analysis softwares. As an example, astronomical observations which map large portions of the sky, and eventually the whole celestial sphere, will become natural, following the development of the so-called satellite astronomy, complementing ground-based observations. These missions generate very large data sets that demand large amounts of memory to be handled, because the finer the information per sky patch, the larger the storage matrices required by the data reduction/analysis software.

Satellites have been used in various ranges of the electromagnetic spectrum, covering from microwaves to γ-rays, with great success. For instance, one of the most exciting questions in cosmology is concerned with the angular distribution of the Cosmic Microwave Background Radiation (hereafter CMBR). By studying the CMBR angular distribution and its properties, one can learn about the density fluctuations that may have generated the large scale structures seen today. Since the discovery of the CMBR in 1965 (Penzias and Wilson, 1965), a large number of experiments have been contributing to the studies of CMBR spectrum and angular distribution. Particularly, the COBE satellite FIRAS,

[1]On leave of absence from Instituto Nacional de Pesquisas Espaciais — INPE São José dos Campos, SP, 12201 Brazil

DIRBE and DMR's experiments have mapped the entire celestial sphere, studying the spectrum and the angular distribution of CMBR (Mather et al. 1990, Smoot et al. 1992). The software developed to analyze the data from the DMR's has to deal with very large data sets, when binning the raw data and also doing a least square fit to a spherical harmonics series (Torres et al. 1989, Jansky and Gulkis 1990). We present here an alternative algorithm using the Singular Value Decomposition method, hereafter referred to as SVD, to deal with these many-point data sets, show that it is more robust than the currently used algorithm (known as the Normal Equations method, hereafter NE), and can always present a numerical solution, based on the standard maximum likelihood test.

2. The Algorithm

One way to do CMBR anisotropy experiments is to collect data that are difference temperatures of two sky regions, as the DMR's do. In order to convert the measurements of temperature differences in a map of the sky it is necessary for some manipulation of the data. The general idea is to find a temperature distribution T_i which minimizes the χ^2 of the map compared to the actual difference data. One can use, for instance, a linear least squares algorithm used to fit the data set $(pixel_i, T_i)$ to a spherical harmonic series. The data set we deal with is a set of sky temperatures in 6144 pixels distributed over the 6 faces of a cube — the quadrilateralized sphere (Chan and O'Neil 1986), and these pixels are easily converted to the true spherical angle pair (θ,ϕ) (or right ascension and declination; galactic latitude and longitude, for example). The figure of merit defined to analyse the goodness-of-fit is the usual χ^2 test:

$$\chi^2 = \sum_{i=1}^{N} \frac{1}{\sigma_i^2} \left[T_i - \sum_{k=1}^{M} a_k Y_k \right]^2 , \tag{1}$$

where T_i is the sky temperature at a pixels (obtained as a combination of various DMR measurements around a given position in the sky), $\sum_{k=1}^{M} a_k Y_k$ is the general spherical harmonics expansion and the pixel k is associated to θ and ϕ, the true spherical angles. What one usually sees in the literature is the use of NE when trying to fit a temperature distribution into its eigenmodes (Lubin and Villela 1986, Smoot et al. 1991). However, the reason that led us to search for an alternative algorithm was the instability in the matrix inversion behavior in our least square fits. Specially after removing part of the celestial sphere, the basis functions become non-orthogonal and there is a mode mixing among different multipole components. Nevertheless, SVD handles this problem quite elegantly, while keeping the χ^2 low. We also discovered that, for low-order fits, both algorithms run at about the same speed. Actually, for $l \leq 5$, SVD is faster than NE (for l=2, $t_{SVD} = 0.089min$ and $t_N E = 0.178min$). Both algorithms can run fits up to order l=18 (limited only by the machine's memory), which, naively, represents the division of the celestial sphere into 10° patches. Following Press et al. (1992) and Strang (1982), we define the weighted basis functions and the weighted data points as

$$A_{ij} = \frac{Y_j(x_i)}{\sigma_i}; b_i = \frac{y_i}{\sigma_i}. \tag{2}$$

(with A generally $M \times N, M \geq N$). The basis functions chosen are the spherical harmonics and, among various possible ways to generate the P_{lm}, we chose one described in Abramowitz and Stegun (1972), and coded in Press et al. (1992). The chosen recurrence relation is:

$$(l-m)P_l^m = x\,(2l-1)\,P_{l-1}^m - (l+m-1)\,P_{l-2}^m, \tag{3}$$

for $l \geq m$, $m \geq 0$ and $-1 \leq x \leq 1$ ($x \equiv cos\theta$, as usual). Please see Press et al. (1992) for a detailed discussion of the stability test for recurrence relations.

The difference between NE and SVD comes mainly from the singularity handling branch in the algorithm. Both algorithms solve a set of simultaneous linear equations that map the data vector b into the solution space, through the A matrix performing the linear transformation from one space to the other. However, if A is singular, the vector x will be partially mapped onto b, and the "misbehaved" part of it will be mapped in a *null* subspace. SVD constructs orthonormal basis for both the *null* space and the mapping space, and the set of solutions we obtain tells us if b lies in the range of A or not. If it does, then the problem will have a solution coming from the A mapping, plus a linear combination of vectors in the *null* space (more than one solution). If we want to single out a specific one, the best approach is to throw away the linear combination of vectors in the *null* space. Following this approach we minimize the residuals of the fitting $r = |A \cdot x - b|$, and in throwing away the solutions included in the *null* space, we discard either round-off errors or non-suitable sets of basis functions.

3. Monte Carlo Simulations

We have done a number of Monte Carlo simulations to verify the stability of the algorithm. We simulate the sky as it is seen by the DMR experiment, including higher order multipole terms, run the program and check the χ^2 of each fit. The residuals from the multipole recovery are very small ($\simeq 1.5\%$ in the worst case), and the $\chi^2/\mathrm{DOF} \simeq 1$, for a sky fitted to standard amplitude spherical harmonics (as defined in Jackson 1975), using different noise levels. A noise level of 1 means the noise has roughly the same amplitude as the dipole and quadrupole. We also observed the behavior of both algorithms when fitting a simulated DMR map to quadrupole and higher order multipoles. The behavior of both algorithms for different l's was tested both with uniform ($\sigma = 1$) and non-uniform, simulated sky coverage. No galaxy cuts were performed on these tests, and we found that the χ^2/DOF for both lies between 0.9 and 1.1. We also studied the stability with the galactic cut, and found that the SVD fit is reasonably more stable to the removal of portions of the celestial sphere, as opposed to NE. For a 30° cut out off the equator plane, the NE χ^2/DOF is 3.6, more than 3 times the SVD value (1.16). On the other hand, considered as a major disadvantage, the time spent in high order fittings ($l \geq 10$) can be as much as 5 times longer for SVD than for NE. A more detailed study of the method, including the covariance matrix and mode mixing analysis will be published elsewhere (Wuensche et al. 1994).

4. Conclusions

One of the interesting features of SVD is the automatic computation of an orthonormal set of basis functions. Trying to do a straight Gram-Schmidt orthonormalization procedure is a slow and error-prone process, as pointed out by Wright (1993). Using the SVD method, it is possible to decompose the input matrix A into three "orthogonal" matrixes $U(NxM, N \geq M$, returns the desired orthogonal functions), $W(MxM$, diagonal, contains the singular values), and V^T $(MxM$, used to compute the covariance matrix and the fitting coefficients), and the new orthonormalized functions are generated automatically during the execution of the SVD process.

SVD was proven to be efficient and robust, and is a good alternative to the generally used NE. To compensating the time disadvantage, it offers the stability of the algorithm, and the automatic computation of a new set of orthonormalized basis functions (analogous to a Gram-Schmidt method) in a single pass. Further studies are necessary to better understand the mode mixing problem, and will represent a major step towards doing harmonic analysis on sky maps.

References

Abramowitz & Stegun. 1972, "Handbook of Mathematical Funtions"

Chan, F.K. & O'Neill, E.M. 1976, EPRF Technical Report 2-75 (CSC)

Gulkis, S. & Janssen, M. 1981 NASA COBE report, 4004

Jackson, J. D. 1975, "Classical Electrodynamics", 2nd. Edition, John Wiley & Sons

Janssen, M. & Gulkis, S. 1992, in Infrared and Submillimetre Sky after COBE, M. Signore & C. Dupraz, Dordrecht: Boston, 1992

Lubin, P. & Villela, T. 1986, in Galaxy Distances and Deviations from Universal Expansion, B. Tully, New York: Plenum, 169

Mather et al., 1990, ApJ, 354, L37

Penzias A. & Wilson, R. 1965 ApJ, 771

Press et al., 1992, "Numerical Recipes - The Art of Scientific Computing", 2nd. Edition, Cambridge University Press

Smoot et al., 1991 ApJ, 371, L1

Smoot et al., 1992 ApJ, 396, L1

Strang, G. 1980, "Linear Algebra and its Applications", Academic Press

Torres, S. et al., 1989, in Data Analysis in Astronomy III, V. Di Gesu et al., New York: Plenum, 319

Wright, E. 1993 preprint

Wuensche et al., 1994, Experimental Astronomy, submitted

Astronomical Data Analysis Software and Systems III
ASP Conference Series, Vol. 61, 1994
D. R. Crabtree, R. J. Hanisch, and J. Barnes, eds.

A New Mosaic Task for WF/PC Images

J.C.H̃su

Space Telescope Science Institute, 3700 San Martin Drive, Baltimore, MD 21218

Abstract. A new task WMOSAIC, now installed in STSDAS version 1.3, produces a mosaic (1600×1600 pixels) of the four groups that comprise the full wide-field/planetary camera (WF/PC) field of view. This task corrects for geometric distortions in each chip as well as rotations, offsets, and scale differences among the CCD detectors.

WMOSAIC does a careful delineation of the boundaries of each group by using the Kelsall spots data. This procedure minimizes the unsightly "cross" caused by the pyramid shadows between groups.

1. Introduction

The geometric distortion caused by the WF/PC optics has been carefully studied by Roberto Gilmozzi and Shawn Ewald at ST ScI. They measured stars' positions in a cluster from multiple exposures at slightly different pointings of the same field. From these measurements they derived a set of bi-dimensional polynomial coefficients for each detector. Each WF detector uses a 6th order polynomial and each PC detector uses a 3rd order polynomial. From the same data, Gilmozzi and Ewald also derived the coefficients of offsets, rotations, and magnification differences among the detectors. All corrections applied in the WMOSAIC task use these coefficients.

The pyramid shadow boundaries of each chip are determined from the Kelsall spots exposures. After correcting for geometric distortions, the Kelsall spots of each axis lie almost on a straight line. These fitted boundaries are then used to match the images together.

After the completion of this task, there may be a cross-shaped dark area at the center. This is due to imperfect matches between chip boundaries and over-masking of pyramid shadows in the original images. A separate task called SEAM can be used to "airbrush" this for a more visually pleasant appearance.

It takes about 6 minutes to run WMOSAIC on a lightly loaded Sun SPARC 2 machine. The task SEAM takes much shorter time to run.

It should be mentioned that a task also named WMOSAIC existed in STSDAS releases before version 1.3. This task, now called QWMOSAIC, still resides in the WFPC package. Task QWMOSAIC simply does four image copies and rotations of 270, 0, 90, and 180 degrees for each of the four groups to make a 1600×1600 output image. It is useful for a quick look of the panoramic view of all four exposures.

2. Geometric Corrections and Boundary Matching

The equations used to calculate the geometric corrections are:

$$dx = \sum_{j=0}^{n}\sum_{i=0}^{n} Ax_{ij} P_i(x) P_j(y) \quad (1)$$

$$dy = \sum_{j=0}^{n}\sum_{i=0}^{n} Ay_{ij} P_i(x) P_j(y) \quad (2)$$

where

$$x = (x_0 - 400)/400 \quad (3)$$

$$y = (y_0 - 400)/400 \quad (4)$$

(x_0, y_0) is the pixel coordinate of the original data. The new position becomes $(x_0 + dx, y_0 + dy)$. P_i is the ith order Legendre polynomial, and Ax_{ij} and Ay_{ij} are the distortion coefficiencies. The wide field camera uses 6th order polynomials, so each WF chip has 49 Ax and 49 Ay coefficiencies. Likewise, each PC chip has 16 Ax and 16 Ay coefficients since PC uses 3rd order polynomials

The geometric correction amounts to about 5 pixels at the corners of each detector where the distortion has the largest effect. In general, the correction is less than 0.1 arc second for the central 80 percents of the area in each chip, i.e., less than one pixel for the WF and 2 pixels for the PC.

After the geometric correction, corrections due to offset, rotation and magnification differences among the detectors are performed. In each camera, we use the second group (WF2 or PC6) as the reference detector. The standard FITS keywords CRPIX, CRVAL, and CD matrix of this reference group are used in the output image. After this second step of corrections, most original pixels centers will no longer be at pixel centers of the new coordinate system. A linear interpolation of the fluxes is then performed to calculate the flux for each pixel in the new grid. Photometric uncertainties can be introduced during such interpolation.

The boundaries between chips are derived from Kelsall spots exposures. It is found that the chip positions of these spots are quite stable for our purposes. Movement at the level of one pixel or less over long periods of time has been observed for these spot positions. We fit a second order polynomial for the Kelsall spots on each axis of each chip. When this fitted boundary cuts through a pixel, only a fraction of the flux, corresponding to the fractional area outside the pyramid shadow is attributed to the final flux of that pixel.

3. Further "Airbrushing" of the Boundaries

After using WMOSAIC to mosaic the four WF/PC groups into one image, brightness discontinuities may still show up near the boundaries between the four frames. A separate task called SEAM is trying to 'airbrush' these discontinuities to render the whole image a better overall appearance. On the other hand, this also implies that fluxes near the seams after running this task may have doubtful scientific validity and should not be used for precise photometric data reduction.

As an example to explain the algorithm of this task, imagine there are two frames, one on top of the other and the 'seam' between them runs horizontally. The smoothing is carried out for each column, i.e., each column contains a 'stitch' which is a segment of pixels vertical to the direction of the seam. The boundary between these two frames crosses all stitches at the middle points.

We start from the bottom of each stitch: the bottom N pixels of the stitch are used to determine the 'background' flux level. The median flux of these pixels is regarded as the 'background' and we then examine each pixel upward till one having a flux level of either greater or less than this background flux by a user-specified fraction. This pixel will be marked as the beginning of the flux discontinuity. Repeat the same from the other end and we get the ending point of the discontinuity. Now take the median flux of the N pixels just below the beginning point (call it F1) and the median flux of the N pixels just above the ending point (call it F2), replace all fluxes of the 'discontinuity pixels' by linearly interpolating between F1 and F2.

4. Conclusion

The new WMOSAIC task should generate an astrometrically accurate mosaic image for WF/PC. However this accuracy is *relative* to the Guide Stars used in that observation. The absolute position is tied to the accuracy of the Guide Star Catalog which is, in general, only accurate to about one arc second. Photometric measurements from the final mosaic image should be put under careful scrutiny. Fluxes of individual pixels are subject to errors due to resampling. Fluxes near the boundaries are less accurate than those near the central region of each detector.

Acknowledgments. I am grateful to John MacKenty for useful inputs and discussions.

Astronomical Data Analysis Software and Systems III
ASP Conference Series, Vol. 61, 1994
D. R. Crabtree, R. J. Hanisch, and J. Barnes, eds.

Deriving the Flat Field Response for the Faint Object Camera from a Nonuniform Source

Perry Greenfield

ST ScI 3700 San Martin Dr. Baltimore MD 21218

Abstract. Described is a method to derive the flat field response of an imaging instrument from observations of a nonuniform extended target that has an unknown surface brightness distribution. The method is based on obtaining overlapping exposures of the target using a carefully planned set of offsets so that the underlying surface brightness distribution of the target may be derived. This method has successfully been used to derive UV flat fields for the Faint Object Camera on the Hubble Space Telescope using observations of the inner region of the Orion Nebula.

1. Introduction

The Faint Object Camera has no on-board calibration source from which it can obtain UV flat fields. Since there are no spatially uniform astronomical sources of UV radiation suitable for flat fields, we must find some means of determining the flat field response from nonuniform sources. Early exposures of the inner region of the Orion Nebula suggested that although it was not completely uniform, it was reasonably smooth enough that its brightness distribution could be determined by a series of spatially offset, but overlapping, exposures. The basic principle involved is the comparison of the measured flux from same location on the detector between the offset exposures thus providing a way of deriving the relative brightness of different places in the Nebula. This strategy has been successfully used to provide flat fields in the UV for the FOC; this paper describes the process used to derive the flat fields.

2. Observational Strategy and Analysis

The basic strategy for the observations and analysis consists of the following steps. 1) Choose a set of spatial offsets having x and y components that are integral multiples of a basic offset for their respective axis. The extended source should not have any significant structure on a scale size comparable or smaller than the basic offsets. The set of offsets should be chosen to sample the range of offsets that span an image in a way that minimizes the number of comparisons needed. 2) Section the camera's field of view into a regular grid of contiguous rectangular apertures whose sizes are the same as the basic offsets with the axes of the grid aligned with the direction of the offsets (see Figure 1). Note that the axes of the offsets do not necessarily match that of the image. This grid of apertures has the property that an aperture in an image will always align exactly with apertures in all the other images. 3) A new "aperture" image is constructed that has pixels which consist of the total flux in each aperture in the aperture grid for each offset image. 4) Since the aperture grid in the offset images maps into a regular grid on the sky, the extended source can be represented by an "aperture" image with pixels the same size as those used for the offset aperture images. Of course, this image of the external source must be larger than the individual images to allow for all the offsets. This aperture image for the external source represents the true brightness distribution of the source. This model presumes that the external source is effectively constant over an aperture (actually, the model works if either the source or the flat field is effectively constant over the aperture). 5) Likewise, we can represent the flat field response on the same grid used for the offset aperture images. 6) The exact form of the model can also depend on other parameters. In this case, it was assumed that there was a spatially constant background for each image which varied between images. Mathematically we have

$$I^l{}_{mn} = f_{mn}e_{ij} + b_l \tag{1}$$

where I, f, and b refer to the expected image, flat field, source, and background values, m, n refer to the "x, y" offset in an aperture image, i, j refer to the aperture offset on the sky, and l refers to the lth image. The m, n and i, j are related as follows

$$i = m + \Delta i_l \tag{2}$$

$$j = n + \Delta j_l \tag{3}$$

where $\Delta i_l, \Delta j_l$ are the offsets for each image. 7) A weighted nonlinear least squares fit is carried out for the parameters f_{mn}, e_{ij}, and b_l. Normally it is necessary to fix one of the f_{mn} or e_{ij} parameters and one of the b_l. 8)The resulting e_{ij} can be interpolated to produce source images with the same scale as the original images to remove the source's spatial variation in each of the offset images. The resulting corrected images can be stacked to produce the flat field.

The details of this method as applied to the FOC were as follows: 1) As mentioned the inner region of the Orion Nebula was used. The $f/96$ relay used the F140W filter (1400Å wideband) while the $f/48$ relay used the F195W, F342W filter combination. The $f/48$ filter bandpass was dominated primarily by the [0II] 3727 Å line. The set of offsets used for $f/48$ is shown in Figure 1. There were 10 $f/96$ exposures and 11 $f/48$ exposures of 1000 s each using the "full" format mode (512 zoomed pixels by 1024 pixels) with an approximate fields of view of $22'' \times 22''$and $44'' \times 44''$respectively. The basic offset (for both x and y) was $1\farcs41$ and $1\farcs67$ for the respective relays resulting in total counts per aperture on the order of 100,000 and 34,000 respectively. 2) Since the FOC images are geometrically distorted, it is necessary to geometrically correct them so that the apertures are properly aligned. Since the offset axes did not align with the image axes, a rotation was included in the geometric correction to align the resulting geometrically corrected image with the offset grid. 3) FOC images suffer measurable nonlinearity in count rate for flat fields even at the relatively low count rates of 0.05 counts pixel^{-1} s^{-1} (approximately 10% loss of counts

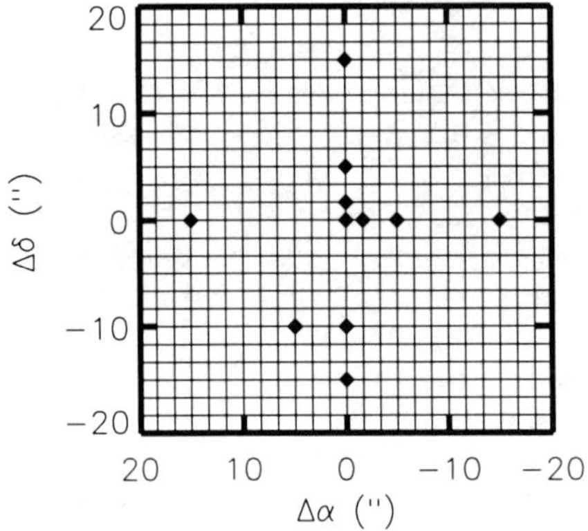

Figure 1. The diamonds show the relative positions of the offsets used for the *f*/48 flat fields. The overlayed grid corresponds to the aperture grid discussed in the text for which the nebula's brightness distribution was fit.

at those count rates). The raw images were used to determine the appropriate correction to apply to the apertures using the known intensity transfer function for FOC flat field response. 4) The fitting problem involves a substantial number of parameters (approximately 1000 for *f*/96 and 3000 for *f*/48). The fitting method used—Levenberg-Marquart—requires the repeated inversion of a 3000×3000 element matrix for a solution for the *f*/48 relay. Since the matrix to be inverted is sparse, a sparse matrix package (called appropriately enough, *SPARSE*) developed by Kundert and Sangiovanni-Vincentelli (1988) was used to perform the inversions. The package was combined by means of a C program with a set of IDL procedures to set up and solve the system of equations.

3. Results

Figure 2 shows the results for the fit to the Orion Nebula brightness distribution for *f*/48. The peak-to-peak variations in the brightness distribution for *f*/48 and *f*/96 were on the order of 30% and 10% respectively. Although the structure is clearly visible in the *f*/48 fit, it is consistent with the brightness distribution being smooth relative to the aperture size since there are no sudden pixel to pixel changes. Figure 2 also shows the stacked set of corrected exposures for *f*/48 where a bicubic spline has been used to interpolate the Orion Nebula surface brightness. This represents the flat field response for the observed wavelength. The rms residual for *f*/48 was about 1.5% and for *f*/96 it was about 0.75%. These values are approximately twice those predicted from Monte Carlo simulations. For both relays there are spatially correlated residuals for the first image which are apparently a result of small variations in flat field response occurring shortly after high voltage turn-on. All the fit background values were consistent with

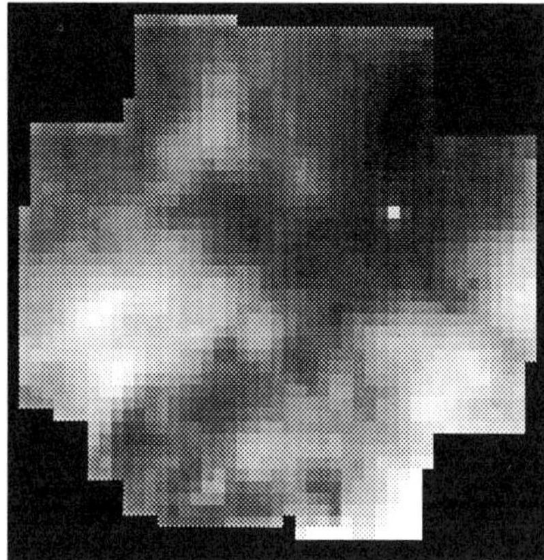
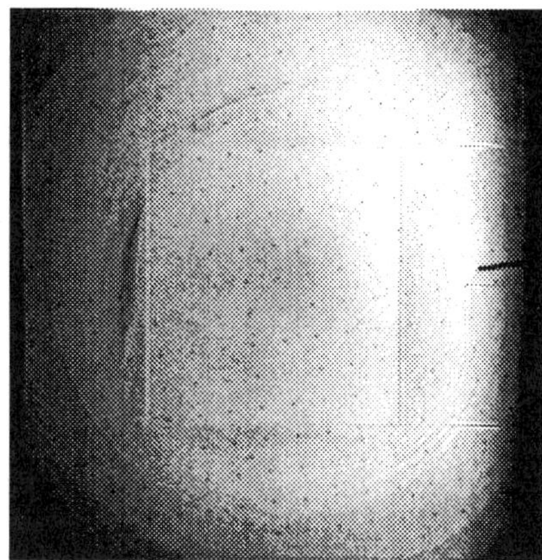

Figure 2. These figures show the fit for the Orion Nebula surface brightness (left) and the the derived flat field response (right) for the *f*/48 relay. The pixels displayed for the surface brightness fit correspond to the aperture grid shown in Figure 1, i.e., they have a size of 1.″667. The bright isolated pixel corresponds to a faint star. An interpolated version of the surface brightness fit was used to correct the individual exposures which were subsequently averaged to produce the geometrically corrected flat field on the right. The area around the star was excluded from the average in those images in which it appeared.

previously observed levels of background (both detector and sky) as well as variability in the background.

Acknowledgments. All FOC images were taken using the NASA/ESA Hubble Space Telescope, obtained at the Space Telescope Science Institute, which is operated by AURA, Inc., under NASA contract NAS 5-26555. The author acknowledges support from ESA through contract 6500/85/NL/SK.

References

Kundert, K. & Sangiovanni-Vincentelli, A. 1988, Sparse Users Guide: A Sparse Linear Equation Solver, Version 1.3a

Astronomical Data Analysis Software and Systems III
ASP Conference Series, Vol. 61, 1994
D. R. Crabtree, R. J. Hanisch, and J. Barnes, eds.

Psfmeasure/Starfocus: IRAF PSF Measuring Tasks

Francisco Valdes

IRAF Group, NOAO[1]*, PO Box 26732, Tucson, AZ 85726*

Abstract. This paper describes two new IRAF tasks `psfmeasure` and `starfocus` for measuring parameters of the point-spread function in astronomical images and estimating the best focus from images with multiple focus settings.

The IRAF[2] tasks `psfmeasure` and `starfocus` measure the point-spread function (PSF) width of stars or other unresolved objects in digital images. The measured width is based on the circular radius which encloses a specified fraction of the background subtracted flux. The algorithmic details of this are described in another paper of these proceedings. The two tasks use the same source code and differ only in whether multiple focus values are analyzed. For data from a single focus setting extraneous input parameters are eliminated. `Psfmeasure` can measure a single object or analyze variations in PSF widths and shapes from multiple images and multiple positions. `Starfocus` analyzes one or more objects at multiple focus settings. When there are multiple objects spatial variations can be examined and compromise best focus values estimated. This paper concentrates on the more complex case of multiple focus values but much of the discussion also applies to the PSF measuring task.

The tasks have three stages; selecting objects and measuring the PSF width and other parameters, an interactive graphical analysis, and a final output of the results to the terminal and to a logfile. The input begins with a list of images. The list may consist of explicit image names, wildcard templates, and @files. The images may consist of a single exposure at some focus setting or multiple exposures in which the telescope or detector are shifted between the exposures at a sequence of focus settings. The multiple exposures can be shifted on the image along lines or columns and there may be a double step at either end of the sequence (a standard focus pattern used at many observatories).

Focus values are associated with each image. The focus values may be specified in several ways; a list of values, an @file, or an image keyword.

Identifying the object or objects to be measured may be accomplished in several ways. One may tell the tasks to automatically look for an object near the center of the image. This does not require any user input. Alternatively one may select objects with the image cursor. This is usually taken from an

---

[1]National Optical Astronomy Observatories, operated by the Association of Universities for Research in Astronomy, Inc. (AURA) under cooperative agreement with the National Science Foundation

[2]Image Reduction and Analysis Facility, distributed by the National Optical Astronomy Observatories

image display (and the program can display the image automatically if desired) though a file containing the positions can be substituted. The latter is useful if an automatic finding task is used. The selection of objects with an image cursor or file has two variants; objects are selected in the first image of a sequence and other images are assumed to have objects in the same positions or the selection can continue independently for all images. In the former case only one image needs to be displayed and interactively marked.

The tasks accumulate PSF data for each object selected. These are all analyzed together when all the objects have been selected. However, one may also graphically examine the PSF information for each object as it is marked. The PSF information measured is described in a separate paper. The graphical examination and analysis features of these tasks in described in the next section.

When the task finishes it prints the results to the user's terminal and also to an optional log file. The tabulated results may be previewed during the execution of the task with a colon command. The results begin with a banner and the overall estimate of the best focus and PSF size. If there are multiple stars measured at multiple focus values the best focus estimate for each star is printed. A star is identified by its position (the starting position for multiple exposure images). The average size, relative magnitude, and best focus estimate are then given. If there are multiple focus values the average of the PSF size over all objects at each focus are listed next. Finally, the individual measurements are given. The columns give the image name, the column and line position, the relative magnitude, the focus value, the PSF size as either the enclosed flux radius or the FWHM, the ellipticity, and the position angle.

The task estimates a value for the best focus and PSF size at that focus for each star. This is done by finding the minimum size at each focus value (in case there are multiple measurements of the same star at the same focus), sorting them by focus value, finding the focus value with the minimum size, and parabolically interpolating using the nearest focus values on each side. When the minimum size occurs at either extreme of the focus range the best focus is at that extreme focus; in other words there is no extrapolation outside the range of focus values.

The overall best focus and size when there are multiple stars are estimated by averaging the best focus values for each star weighted by the average flux of the star as described above. Thus, when there are multiple stars, the brighter stars are given greater weight in the overall best average focus and size. This best average focus and size are what are given in the banner for the graphs and in the printed output.

The log output also includes an average PSF size for all measurements at a single focus value. This average is also weighted by the average flux of each star at that focus.

1. Interactive Graphics

The graphics part of `starfocus` and `psfmeasure` consists of a number of different plots selected by cursor keys or window buttons and menus. The available plots depend on the number of stars and the number of focus values. The various plots are a spatial plot at a single focus, a spatial plot of best focus values, enclosed flux for stars at one focus and one star at all focuses, size and ellipticity vs. focus for all data, size and ellipticity vs. relative magnitude at one focus, radial profiles for stars at one focus and one star at all focuses, size and ellipticity vs. radius from field center at one focus, and a *zoom* plot of a single measurement.

If there is only one object at a single focus the only available plot is the zoom plot. This plot may be selected with the cursor for any object from any other plot. It has three graphs; a graph of the normalized enclosed flux verses scaled radius, a graph of the intensity profile verses scaled radius, and the equivalent Gaussian full width at half maximum verses enclosed flux fraction. The latter two graphs are derived from the normalized enclosed flux profile as described in the algorithms paper. An example of this plot is shown in Figure 1.

There are three types of symbol plots showing the measured PSF size (either enclosed flux radius or FWHM) and ellipticity. These plot the measurements verses focus (see Figure 2), relative magnitude, and radius from the field center. The focus plot includes all measurements and shows dashed lines at the estimated best focus and size.

Grids of enclosed flux vs. radius, intensity profile vs. radius, and Gaussian FWHM vs. enclosed flux fraction give a visual overview of PSF changes with focus or star. Any of the graphs with enclosed flux or intensity profiles vs. radius may have the profiles of the object with the smallest size overplotted.

The final plots give a spatial representation. These require more than one object. One gives a spatial plot at a single focus. The space bar can be used to advance to another focus. This plot has a central graph of column and line coordinates with symbols indicating the position of an object. The objects are marked with a circle (when plotted at unit aspect ratio) whose size is proportional to the measured PSF size. In addition an optional asterisk symbol with size proportional to the relative brightness of the object may be plotted. On color displays the circles may have two colors, one if the object size is above the average best size and the other if the size is below the best size. The purpose of this is to look for a spatial pattern in the smallest PSF sizes.

Adjacent to the central graph are graphs with column or line as one coordinate and radius or ellipticity as the other. The symbols are the same as described previously. These plots can show spatial gradients in the PSF size and shape across the image.

The second type of spatial plot shows the best focus estimates for each object. This requires multiple objects and multiple focus values. As discussed previously, given more than one focus a best focus value and size at the best focus is computed by parabolic interpolation. This plot type shows the object positions in the same way as the other plot except that the radius is the estimated best radius. Instead of adjacent ellipticity plots there are plots of best focus verses columns and lines. Also the two colors in the symbol plots are selected depending on whether the object's best focus estimate is above or below the overall best focus estimate. This allows seeing spatial trends in the best focus. An example of this type of plot is shown in Figure 2.

In addition to the keys and buttons which select plots there are other keys which do various things such as print help, delete or undelete a star or stars, print information about one measurement, adjust the normalization of the enclosed flux profile, and change parameter values. There is, of course, also a key to exit the program.

The deletion of bad measurements, usually due to being too faint or too near another object, is an important feature. When an object is deleted it is not included in any averages or output. Objects may be deleted individually, by image, or over all focus values and they may be undeleted as well.

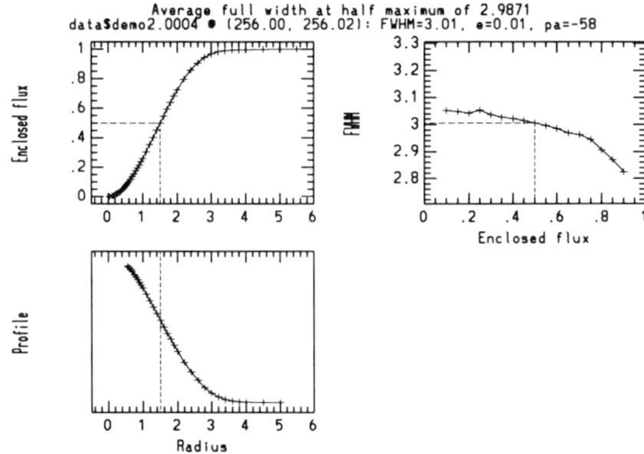

Figure 1. Plots of the enclosed flux vs. radius, pixel flux vs. radius, and equivalent Gaussian FWHM vs. enclosed flux for a single PSF measurement. The dashed lines mark the radius enclosing 50% of the flux.

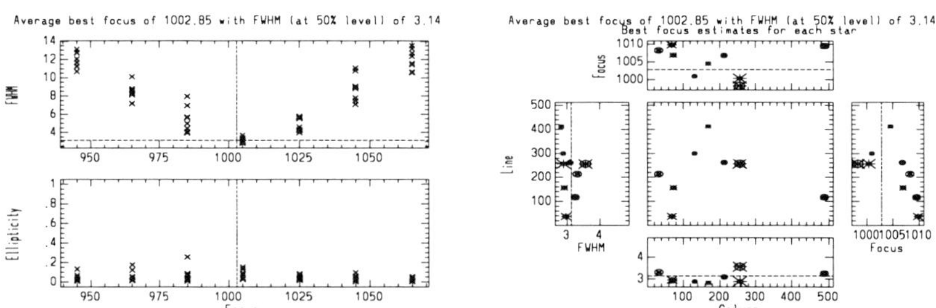

Figure 2. Two plots from `starfocus`. The left plot shows the FWHM and ellipticity vs. focus value for all measurements. The right plot shows the position, relative best focus size, magnitude, best FWHM, and best focus as a function of object position in the image. The size of the circles represent the FWHM and of the asterisks the relative magnitude.

Astronomical Data Analysis Software and Systems III
ASP Conference Series, Vol. 61, 1994
D. R. Crabtree, R. J. Hanisch, and J. Barnes, eds.

Psfmeasure/Starfocus: PSF Measuring Algorithms

Francisco Valdes

IRAF Group, NOAO[1], PO Box 26732, Tucson, AZ 85726

Abstract. This paper describes algorithms used in the IRAF tasks `psfmeasure` and `starfocus` for accurately measuring parameters of the point-spread function in astronomical images.

1. Introduction

This paper describes the major algorithms used to measure the point-spread function, PSF, in astronomical images using the IRAF[2] tasks `psfmeasure` and `starfocus`. The functionality of these tasks is described in another paper in these proceedings. The algorithms are designed to provide accurate measures of various properties of the PSF, such as width, extending into the difficult regime of marginally sampled images where the PSF width is of order of a single pixel.

The PSF is characterized by the azimuthally symmetric enclosed flux profile of objects such as stellar images. To measure the enclosed flux profile requires algorithms for determining the center of an object, determining the background, accurately sampling the profile and image pixel values, defining the normalization, and measuring radial flux profiles and widths. There are additional algorithms for measuring relative magnitudes, ellipticities, position angles, and combining multiply measurements at varying positions and focuses which are given in the technical documentation for the IRAF tasks.

2. Center Determination

The center of a PSF object is determined starting with an initial estimate. In the tasks this is either given by the user as an image cursor coordinate (either interactively or from a cursor coordinate file) or assuming the object is near the center of the image. A subraster containing the object is extracted. The half size of this subraster is set by a fitting radius, plus a background region buffer distance, plus a background region annulus width. The lines and columns of the subraster are summed to form marginal distributions. The mean value of each distribution is computed and then the centroid of the data above this mean

[1]National Optical Astronomy Observatories, operated by the Association of Universities for Research in Astronomy, Inc. (AURA) under cooperative agreement with the National Science Foundation

[2]Image Reduction and Analysis Facility, distributed by the National Optical Astronomy Observatories

defines a new estimate of the object center. If the center moves to a different pixel the process is repeated using a raster about the new center. The iterations continue for a maximum of three times.

3. Background Determination

A constant background is determined from an annulus centered on the object with inner edge given by a specified fitting radius plus a buffer distance. The outer edge is then specified by an annulus width. All pixels with pixel center radii within the annulus limits are sorted by value into a one dimensional array using the *qsort* algorithm. The point of minimum slope in this array, using a numerical derivative length of 50% of the pixels, defines the background level. For integer valued images this algorithm is modified by first adding a uniform random dither value between -0.5 and 0.5 in order to avoid multiple regions of constant values; i.e., slope of zero. This algorithm is a bin-free method of measuring the mode.

4. Enclosed Flux Profile and Intensity Profile

The background subtracted, azimuthally symmetric enclosed flux profile is determined at a set of nonuniformly spaced radial intervals, dr. Similarly, the image pixels are subsampled with different subsample sizes, dx and dy, depending on the radius of the pixel center. The sampling intervals and sizes are given by

$$\left.\begin{matrix} dr \\ dx \\ dy \end{matrix}\right\} = \begin{cases} 0.05 & r < 1 \\ 0.10 & r < 2 \\ 0.20 & r < 4 \\ 0.50 & r < 9 \\ 1.00 & r > 9 \end{cases}$$

This nonuniform sampling is done to give fine profile resolution near the PSF center, particularly for narrow PSFs, but avoid increased memory and computation time requirements for very large PSFs.

Bi-cubic spline image interpolation is used for evaluating the pixel value at each subpixel. This value is background subtracted and added to all points in the enclosed flux profile which contain the pixel. Even with pixel subsampling there are points where a subpixel straddles a particular radius in the enclosed flux profile. When this happens the fraction of the subpixel with radius interior to the enclosed flux radius is used to define the fraction of the subpixel value included in the enclosed flux.

Because of errors in the background determination due to noise and contaminating objects it is sometimes the case that the enclosed flux is not completely monotonic with radius. The enclosed flux normalization is the maximum of the enclosed flux profile even if it occurs at a radius less than the maximum radius. It is possible to change the normalization and subtract or add a background correction interactively.

A cubic spline function is fit to the normalized enclosed flux profile at the measured sample radii. This interpolation function is then used for such things as determining the radius enclosing a specified fraction of the flux.

The intensity radial profile, $P(R)$, is related to the enclosed flux radial profile, $F(R)$, as given below. The derivative is estimated using simple numerical differentiation.

$$\begin{aligned} F(R) &= \int_0^R P(r) r dr \\ P(R) &= (dF/dR)/R \end{aligned}$$

5. PSF Radii and FWHM

One of the key measures of the PSF is a radius or width. The width is determined at a specified fraction, f, of the enclosed flux; for example the width enclosing 50% of the flux. A radius, $R(f)$, is measured straightforwardly from the continuous interpolation function of the enclosed flux profile. Another common measure used in the literature for characterizing the PSF width is the full width at half maximum, $FWHM$, of the intensity radial profile. This is also often determined by assuming a Gaussian profile. Making this assumption the radius at the specified fraction can be converted to an equivalent Gaussian $FWHM$ using the equation

$$FWHM(f) = 2R(f)\sqrt{\ln(2)/\ln(1/(1-f))} \tag{1}$$

By varying f one can also make plots of $FWHM$ verses f. This gives an interesting way to see departures of the PSF from Gaussian. A Gaussian PSF will have a constant $FWHM$. Increasing or decreasing $FWHM$ will indicate wings, sharp cores, out of focus, or other shapes.

6. Gaussian Subtraction Algorithm

Even with fine pixel subsampling, using an image interpolation function, and partial pixels at the boundaries, the enclosed flux profile has significant errors when the PSF is narrow. This is because the image interpolation function may only have a few pixels to fit and determine a shape that is very rapidly varying. The solution to this is to use a physically realistic analytic function to model the rapid variation.

The analytic model is an azimuthally symmetric Gaussian. The Gaussian width is estimated using equation (1) with a fraction of 80%. The normalization is set such that the integral of the model over the central pixel is equal to central pixel value. The model is then subtracted from the image subraster and the cubic spline interpolation function is fit to the residuals which are now much more slowly varying. The enclosed flux profile is then re-evaluated with subpixel values given by the sum of the model at that point plus the interpolation function value from the residuals.

This Gaussian subtraction is only done if the first $FWHM$ estimate is less than 4 pixels. If the $FWHM$ is less than 2 pixels even the Gaussian model

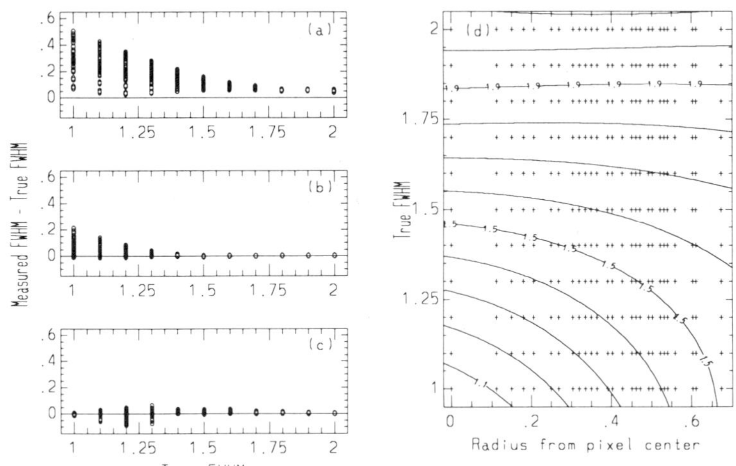

Figure 1. Results of simulations with noisefree Gaussian profiles at varying *FWHM* and center. (a) The *FWHM* error without model subtraction. (b) The *FWHM* error with model subtraction but no empirical *FWHM* correction. (c) The *FWHM* error with both empirical *FWHM* correction and model subtraction. (d) The correction surface showing the measured uncorrected *FWHM* as a function of true *FWHM* and pixel center. The symbols mark the simulated profiles.

subtraction is not adequate since the initial *FWHM* is significantly biased to broader values. What is done in this case is that an empirical correction is applied to this measured *FWHM*. The correction is determined from simulations with known *FWHM* by comparing the true *FWHM* with the measured *FWHM*. The correction is actually a function of the measured *FWHM* and the center of the object relative to a pixel center. This is because a narrow PSF centered on a pixel will appear sharper than those centered near the boundary with other pixels. This algorithm allows accurately recovering PSF widths down to a *FWHM* of 1 pixel.

In Figure 1 a series of Gaussian profiles with varying *FWHM* and center are sampled as a noisefree image (each point at the same true *FWHM* has a different subpixel center). Figure 1a shows the error in the *FWHM* as a function of the true *FWHM* when the Gaussian subtraction algorithm is not used. In Figure 1b the algorithm is applied without correction to the initial *FWHM* estimate. Figure 1c shows the result with the correction to the *FWHM* applied. The correction surface is shown in Figure 1d where the marks are the simulated objects and the contour lines are at constant initial measured *FWHM*. This is basically a surface fitted to the points of Figure 1a. The surface shows that the greatest correction is needed at the smallest true *FWHM* and for centers furthest from a pixel center.

Astronomical Data Analysis Software and Systems III
ASP Conference Series, Vol. 61, 1994
D. R. Crabtree, R. J. Hanisch, and J. Barnes, eds.

Iterative/Recursive Image Deconvolution. Method and Application to HST Images

L. K. Fullton, B. W. Carney, and J. M. Coggins

University of North Carolina at Chapel Hill, Department of Physics and Astronomy, CB #3255, Phillips Hall, Chapel Hill, NC 27599-3255

K. A. Janes

Boston University, Department of Astronomy, 725 Commonwealth Ave., Boston, MA 02215

P. Seitzer

University of Michigan, Department of Astronomy, 830 Dennison Bldg., Ann Arbor, MI 48109-1090

Abstract. Image restoration results are presented using a new iterative/recursive method for removing a linear, spatially-invariant blur from an image. Our new algorithm is linear and flux-conserving. It converges quickly and is relatively insensitive to image noise or error in the point spread function (PSF). We have used it to restore Hubble Space Telescope (HST) Planetary Camera images of the globular cluster NGC 6352. The resulting color-magnitude diagram illustrates the photometric accuracy which can be obtained from images deconvolved using our technique with the PSFs currently available from the Space Telescope Science Institute. If better PSFs become available, we believe the color-magnitude diagram could improve significantly. For comparison, we have analyzed the unrestored images with PSF-fitting photometry.

1. Algorithm

Our algorithm is a modification of a classic image restoration technique known as basic iterative deconvolution (BID; van Cittert 1931, Jansson 1970). The first step in this technique is to make a guess as to what true object when blurred with the PSF produces the observed image. If this guess is correct, then blurring it with the PSF will produce the observed image. If the guess is wrong, it can be corrected based on the difference between the observed image and the blurred guess. The correction might be simply the difference between the observed image and the blurred guess. The correction is applied, producing a new estimate of the true object, and the method is iterated as many times as desired.

Our modification of the BID algorithm hinges on the realization that the image difference used to correct the object estimate is the difference between *blurred* images and must be *deblurred* before being used to correct the estimate.

Therefore we recursively invoke this enhanced procedure using the difference image as the observed image along with the same PSF.

The observed difference image is our initial guess of the deblurred difference image. We blur it and compute the difference as before. We now need to deblur *this* difference, so we recur again. This continues through as many levels as we desire. Finally, at the deepest recursion level, we invoke the unmodified BID algorithm for some small number of iterations and pass the result back up the recursion stack. At each level, the deblurred result becomes the correction applied to the estimate of the true object at the next higher level.

The net effect of this algorithm is to convolve the image with a filter that rapidly converges to the inverse PSF (Coggins, Fullton and Carney 1993, hereafter CFC). The method produces no artifacts, conserves flux and is noise-resistant. Each difference image down through the recursion levels is a higher order difference (i.e., higher order derivative). The larger spatial extent of the higher order derivative kernels regularizes the procedure by smoothing out small scale noise (Blom 1992). The BID algorithm provides a robust estimate of the image correction at the deepest recursion levels since repeated blurring with the PSF produces derivative images that are quite smooth.

In simulations, images deconvolved with our algorithm produced linear photometric results. The intensity of point sources in the restored images was conserved to within 1-2% over a luminosity range of 6 dex.

2. Observations

We deconvolved HST Planetary Camera F555W and F785LP filter images of the globular cluster NGC 6352. The image scale is approximately 0.04 arcsec/pixel. We preprocessed the images applying the A/D correction, bias and preflash subtraction, dark current and flat field corrections. Three 300-second F555W exposures and four 1000-second F785LP exposures from CCD #7 were averaged using the IRAF imcombine task with cosmic ray rejection to produce separate F555W and F785LP master frames. These master frames were deconvolved with the iterative/recursive algorithm invoked with 3 recursion levels and 2 iterations at each level. Variation of the PSF across the image was accounted for by dividing the 740x724 image into nine 300x300 overlapping sections and restoring each section separately with a PSF generated by the Tiny Tim software (Krist 1992) for the section center. The signal-to-noise ratio of the fainter stars in the deconvolved image was much improved over those in the original. There were no artifacts in the image, and the residuals produced by the slight mismatch between the actual PSF and the PSF used in the restoration were smaller compared to the central stellar fluxes than the diffuse halos in the original data. These residuals have only a minimal effect on photometry of the image.

3. Photometry

We performed photometry on both the original and deconvolved images of NGC 6352. The Stellar Photometry Software (SPS) developed by Janes and Heasley (1993) was used to perform PSF-fitting photometry on the original (blurry) data. A single spatially-constant PSF was generated from a large sample of bright

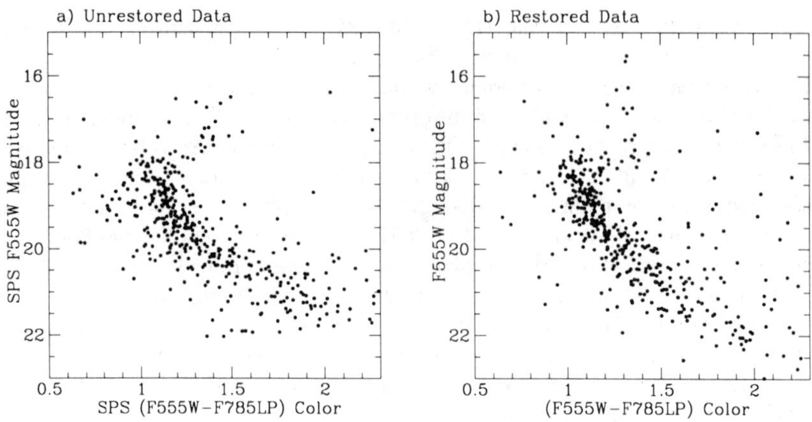

Figure 1. Color-magnitude diagram of NGC 6352. a) SPS Photometry of the unrestored image. b) Aperture photometry of the restored image.

stars scattered uniformly about the frame. This PSF was used to measure the magnitudes of all stars found using the SPS automatic star-finding algorithm. After this first pass, a few stars were added to the list from visual inspection, and the fits were repeated. The resulting color-magnitude diagram (CMD) is shown in Figure 1a. The photometric zero-points have been adjusted to correspond to the F555W and F785LP magnitudes determined from the restored image.

Aperture photometry was performed on all stars identified in the deconvolved images with aperture radii of 3 and 5 pixels for the F555W and F785LP images respectively. Aperture corrections were determined from the brightest uncrowded stars in each subsection and applied to the magnitudes of all the measured stars. The CMD for all stars whose nearest neighbor was greater than 5 pixels away is plotted Figure 1b. Comparison with the CMD from the PSF-fits to the unrestored images indicates that the main-sequence is better defined in the deconvolved data, especially at the faintest limit. Comparison of these results with those from the simulation (CFC) indicate that the precision of the NGC 6352 data is limited mainly by the mismatch between the actual PSF and that used in the deconvolution. We believe we could obtain even better photometric results from a PSF which matched the observed PSF more closely.

To illustrate the relative quality of the two different photometric reductions, we have plotted histograms of the color difference between a star's position in the CMD and the mean main-sequence defined by a parabolic fit to the data below the level of the main-sequence turn-off. These histograms are shown in Figure 2. Note that our deconvolution procedure allowed for the spatial variation of the PSF while the PSF-fitting software did not. Two histograms are plotted, which represent the data for main-sequence stars of decreasing brightness in 1 magnitude bins. The solid lines indicate the scatter about the mean sequence determined from the aperture photometry on the restored images. Since the largest aperture used was 5 pixels in radius, only data from stars whose nearest neighbor was more than 5 pixels distant are included in the figures to minimize

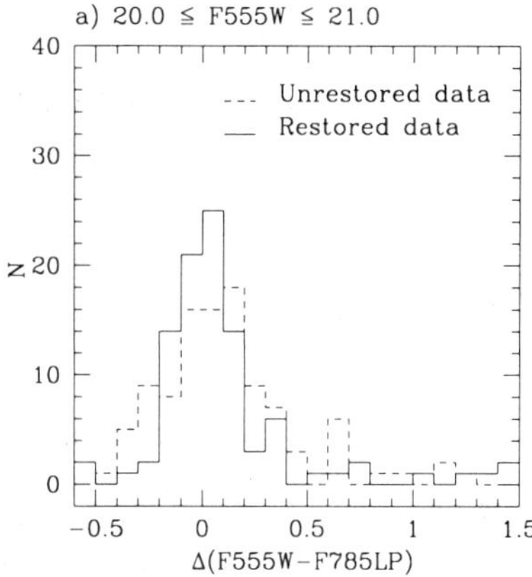

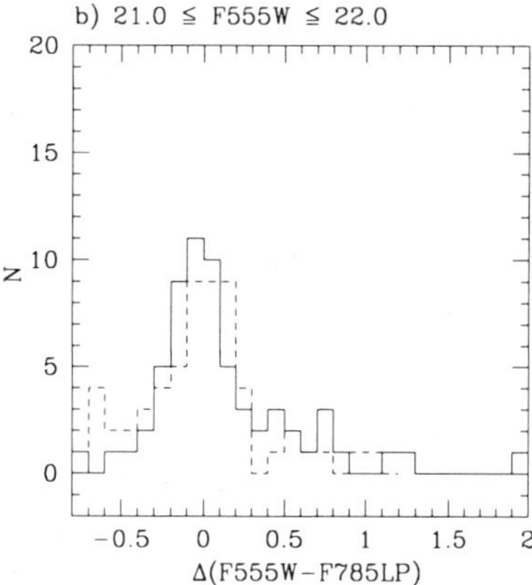

Figure 2. Histograms of the color difference between a star's position in the CMD and the mean main-sequence. a) Stars with F555W filter magnitudes between 20 and 21. b) Stars with F555W magnitudes between 21 and 22.

scatter in the diagram due to the use of aperture photometry in crowded fields. This scatter is obviously not attributable to the effects of the deconvolution algorithm but rather depends on the density of stars in the original image. The dotted lines represent the scatter about the mean sequence determined from the PSF-fits on the unrestored data.

167 stars were measured fainter than 20th magnitude with the profile-fitting software. Because of the higher signal-to-noise ratio in the deconvolved images, more stars were measurable in both filters of the deconvolved data than in the original images, especially at the faint end. Thus, even after omitting stars with close neighbors from the histograms (about 10 percent of the total sample), there are a total of 162 stars represented in the histograms of the deconvolved data.

Figure 2 indicates that the scatter about the mean faint star sequence is much smaller for the aperture photometry of the deconvolved images than for the PSF-fits on the original images. Since this effect is small at the brightest magnitudes, they are not plotted. The figures indicate the superiority of the use of our deconvolution technique for constructing CMDs of HST data.

References

Blom, J. 1992, Ph. D. Dissertation, University of Utrecht

Coggins, J. M., Fullton, L. K., and Carney, B. W. 1993, CVGIP: Graphical models and image processing, preprint (CFC)

van Cittert, P. H. 1931, ZAp, 69, 298

Janes, K. A., & Heasley, J. N. 1993, PASP, 105, 527

Jansson, P. A., 1970, J. Opt. Soc. Am., 60, 184

Krist, J. 1992, The Tiny Tim User's Manual, Space Telescope Science Institute, Baltimore

Astronomical Data Analysis Software and Systems III
ASP Conference Series, Vol. 61, 1994
D. R. Crabtree, R. J. Hanisch, and J. Barnes, eds.

Image Restoration Using the Damped Richardson-Lucy Method

Richard L. White

Space Telescope Science Institute, 3700 San Martin Drive, Baltimore, MD 21218

Abstract. This paper describes an iterative image restoration technique that closely resembles the RL iteration but that avoids the amplification of noise that occurs in the RL iteration. I call this the damped RL iteration because the modification appears as a damping factor that slows changes in regions of the model image that fit the data well while allowing the model to continue to improve in regions where it fits the data less well.

1. Introduction

The Richardson-Lucy (RL) algorithm (Richardson 1972; Lucy 1974) is the technique most widely used for restoring HST images. The RL iteration converges to the maximum likelihood solution for Poisson statistics in the data (Shepp & Vardi 1982), which is appropriate for optical data with noise from counting statistics. The RL iteration can be derived from the imaging equation and the equation for Poisson statistics:

$$I(i) = \sum_j P(i|j)\, O(j) \quad , \tag{1}$$

where O is the unblurred object, $P(i|j)$ is the PSF (the fraction of light coming from true location j that gets scattered into observed pixel i), and I is the noiseless blurry image. The joint likelihood $\mathcal{L}$ of getting the observed counts $D(i)$ in each pixel given the expected counts $I(i)$ is

$$\ln \mathcal{L} = \sum_i D(i) \ln I(i) - I(i) - \ln D(i)! \quad . \tag{2}$$

The maximum likelihood solution occurs where all partial derivatives of $\mathcal{L}$ with respect to $O(j)$ are zero:

$$\frac{\partial \ln \mathcal{L}}{\partial O(j)} = 0 = \sum_i \left[\frac{D(i)}{I(i)} - 1 \right] P(i|j) \quad . \tag{3}$$

The RL iteration is

$$O_{new}(j) = O(j) \sum_i P(i|j) \frac{D(i)}{I(i)} \Big/ \sum_i P(i|j) \quad . \tag{4}$$

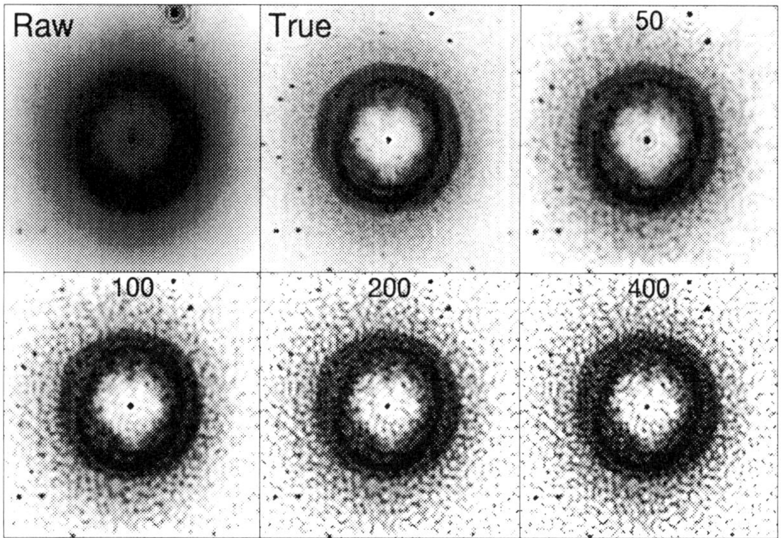

Figure 1. RL restoration of simulated 85×85 pixel HST PC observation of a planetary nebula. Restored and true images are 256×256 (method of White 1990 was used to restore image with finer pixels than data). As the number of iterations (shown at top of each image) increases, the images of bright stars improve, but noise is amplified unacceptably in the nebulosity.

It is clear from a comparison of eq. (3) and (4) that if the RL iteration converges (as has been proven by Shepp & Vardi 1982), meaning that the correction factor approaches unity as the iterations proceed, then it must indeed converge to the maximum likelihood solution for Poisson statistics in the data.

Despite its advantages, the RL method has some serious shortcomings. In particular, noise amplification can be a problem. This is a generic problem for all maximum likelihood techniques, which attempt to fit the data as closely as possible. If one performs many RL iterations on an image containing an extended object such as a galaxy, the extended emission usually develops a "speckled" appearance (Figure 1). The speckles are not representative of any real structure in the image, but are instead the result of fitting the noise in the data too closely. In order to reproduce a small noise bump in the data it is necessary for the unblurred image to have a very large noise spike; pixels near the bright spike must then be very black (near zero brightness) in order to conserve flux.

The usual practical approach to limiting noise amplification is simply to stop the iteration when the restored image appears to become too noisy. However, the question of where to stop is a difficult one. The approach suggested by Lucy (1974) was to stop when the reduced χ^2 between the data and the blurred model is about 1 per degree of freedom. Unfortunately, one does not really know how many degrees of freedom have been used to fit the data. If one stops after a very few iterations then the model is still very smooth and the resulting χ^2

should be comparable to the number of pixels. If one performs many iterations, however, then the model image develops a great deal of structure and so the effective number of degrees of freedom used is large; in that case, the fit to the data ought to be considerably better. There is no criterion for the RL method that tells how close the fit ought to be.

Another problem is that the answer to the question of how many iterations to perform often is different for different parts of the image. It may require hundreds of iterations to get a good fit to the high signal-to-noise image of a bright star, while a smooth, extended object may be fitted well after only a few iterations. Thus, one would like to be able to slow or stop the iteration automatically in regions where a smooth model fits the data adequately, while continuing to iterate in regions where there are sharp features.

2. The Damped RL Iteration

An effective approach to such an adaptive stopping criterion is to modify the likelihood function of eq. (2) so that it becomes flatter in the vicinity of a good fit. The approach I have taken is to use a likelihood function that is identical to eq. (2) when the difference between the blurred model I and the data D is large compared with the noise, but that is essentially constant when the difference is smaller than the noise.

The *damped RL iteration* starts from the likelihood function

$$\ln \mathcal{L} = \sum_i f(U_i) \quad , \tag{5}$$

where

$$U_i = -\frac{2}{T^2}\left[D(i)\ln\frac{I(i)}{D(i)} - I(i) + D(i)\right] \tag{6}$$

and

$$\begin{aligned} f(x) &= \tfrac{N-1}{N+1}\left(1 - x^{N+1}\right) + x^N \quad , & x < 1 \\ &= x \quad , & x \geq 1 \quad . \end{aligned} \tag{7}$$

$f(x)$ is the "damping function", chosen to be a simple function that is linearly proportional to x for $x > 1$, is approximately constant for $x \sim 0$, and has continuous first and second derivatives at $x = 1$. The constant N determines how suddenly the function f becomes flat for $x < 1$. The results reported in this paper use $N = 10$, but the value of N has little effect on the results as long as it is larger than a few.

U_i is a slightly modified version of the $\ln \mathcal{L}$ function of eq. (2). The constants and multiplicative factors are chosen so that the expected value of U in the presence of Poisson noise is unity if the threshold $T = 1$. If the threshold $T = 1$ then the damping turns on at 1σ, if $T = 2$ at 2σ, etc.

With this new likelihood function, we can follow the steps outlined above to derive the damped RL iteration:

$$O_{new}(j) = O(j)\frac{\sum_i P(i|j)\left[1 + \tilde{U}_i^{N-1}[N - (N-1)\tilde{U}_i]\frac{D(i) - I(i)}{I(i)}\right]}{\sum_i P(i|j)} \quad , \tag{8}$$

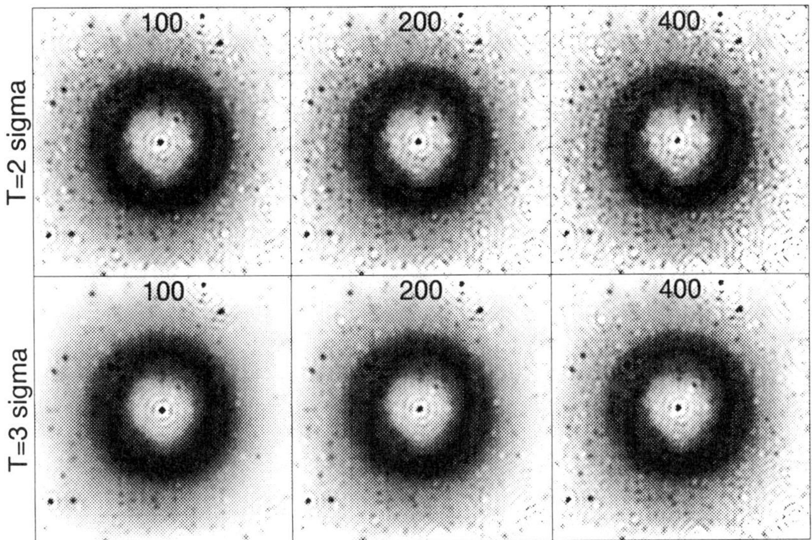

Figure 2. Restoration of data from Figure 1 using the damped iteration with noise thresholds of 2 and 3σ. Bright stars are still well-sharpened, but noise amplification is much better controlled than for standard RL iteration.

where $\tilde{U}_i = \min(U_i, 1)$. Note that in regions where the data and model do *not* agree, $\tilde{U} = 1$ and so this iteration is exactly the same as the standard RL iteration. In regions where the data and model do agree, however, the second term gets multiplied by a factor which is less than 1, and the ratio of the numerator and denominator approaches unity. This has exactly the desired character: it damps changes in the model in regions where the differences are small compared with the noise. Figure 2 shows the results of applying the damped iteration to the planetary nebula data.

It is unlikely that the ad hoc approach described here represents the best solution to the noise amplification problem, but it does have the advantages that it is easy to implement and robust, and it appears to produce better results than the RL method in many cases. The method produces restored images that have good photometric linearity and little bias.

References

Lucy, L. B. 1974, AJ, 79, 745

Richardson, B. H. 1972, J.Opt.Soc.Am., 62, 55

Shepp, L. A., & Vardi, Y. 1982, IEEE Trans. Medical Imaging, MI-1, 113

White, R. L. 1990, in The Restoration of HST Images and Spectra, eds. R. L. White & R. J. Allen (Baltimore: ST ScI), p. 139

Astronomical Data Analysis Software and Systems III
ASP Conference Series, Vol. 61, 1994
D. R. Crabtree, R. J. Hanisch, and J. Barnes, eds.

Implementation of the Richardson-Lucy Algorithm in STSDAS

E. B. Stobie, R. J. Hanisch, R. L. White

Space Telescope Science Institute, 3700 San Martin Drive, Baltimore, Md 21218

Abstract. The Richardson-Lucy algorithm (Richardson 1972 and Lucy 1974), a popular technique for restoring HST images, was implemented as the **lucy** task in **STSDAS** in late 1990. With widespread use of the task feedback from users helped to identify several of its shortcomings. Many of these were addressed with the release of V1.1 in April, 1993 and V1.2 in November, 1993. Two significant enhancements include the masking of bad pixels and the implementation of the the accelerated iteration algorithm (Hook & Lucy 1992). A brief history of the task's evolution is presented including plans for future development.

1. Introduction

The Richardson-Lucy algorithm (Richardson 1972 and Lucy 1974) with modifications by Snyder (1990) has been a popular technique for restoring HST images. Many of its characteristics make it well suited to HST data.

- The Richardson-Lucy iteration converges to the maximum likelihood solution for Poisson statistics in the data (Shepp & Vardi 1982), which is appropriate for optical data with noise from counting statistics.
- With the inclusion of the noise (read-out) and adu (conversion factor between electrons and DN) the noise characteristics of the image can be modeled properly.
- This method forces the restored image to be non-negative and flux is conserved both locally and globally at each iteration.
- The restored images are robust against small errors in the point spread function (PSF).
- The algorithm is easy to implement with reasonable memory requirements and is efficient in execution. A single iteration of a 512x512 image can be performed in a few seconds on a Sun SparcStation 1.

The Richardson-Lucy algorithm was first implemented in **STSDAS** as the **lucy** task in the **restore** package in late 1990. It is the purpose of this paper to describe the evolution of this task and the design considerations used to maximize execution speed while minimizing memory requirements. Future enhancements are also discussed.

2. lucy 1.0 - October 1990

The initial SPP implementation of the **lucy** task for both 1- and 2-dimensional data was modeled after a prototype IDL procedure developed by R.L. White. There were five required input parameters: input, psf, and output image names, read-out noise, and adu. Three optional parameters included a model image for resuming iterations from a previous result and parameters for controlling the number of iterations to be performed or the χ^2 to be achieved.

In the interest of execution speed all data arrays are stored in memory. By juggling the use of some arrays the memory requirements are limited to two complex and five real arrays. The total array space required for the restoration of a 512x512 image is 9 MB. An additional factor in achieving efficient task performance is choosing image dimensions that are factors of 2 or that factor well.

As the popularity of the **lucy** task increased users quickly became aware of its limitations. When large negative pixels were present in the input image the entire image was elevated so that these pixels did not fall below the value of $-(\text{noise}^2/\text{adu})$ to insure a non-negative image. This caused large positive offsets in the sky brightness. Also, artifacts often appeared at the opposing side of an image when a bright source was located near the image boundary. This wrap-around effect is caused by the use of FFT's in the algorithm which assume the data is periodic. Users found it annoying that the only means of monitoring the results of the restoration was by stopping and restarting the task.

3. lucy 1.1 - April 1993

The next release of the **lucy** task in April, 1993, addressed these concerns and included additional capability as well. These improvements were:

- **Pixel masking**. The user can optionally specify a mask or data quality file to exclude bad pixels. Although masks are stored internally with bad pixels set to 0, data quality files that express bad pixels with non-zero values may be used when the value of **goodpixval** is set to 0. Additionally, the task checks for any negative pixels below the user controlled threshold (default = 5σ). These pixels are automatically masked.

- **Extending the output image size**. Artifacts introduced by wrap-around effects when strong sources are near the image edge can be significantly reduced by extending the size of the output image. Typically images are extended by at least half the size of the PSF. These pixels are automatically masked in the restoration. Also, the user can achieve better restorations when a strong source is located just outside an image boundary but visible by the edge of its PSF as shown in Figure 1.

- **Saving intermediate results**. The user can retain the results of every nth iteration specified by the **update** or **nsave** parameters, by overwriting the output image or creating new images with every update.

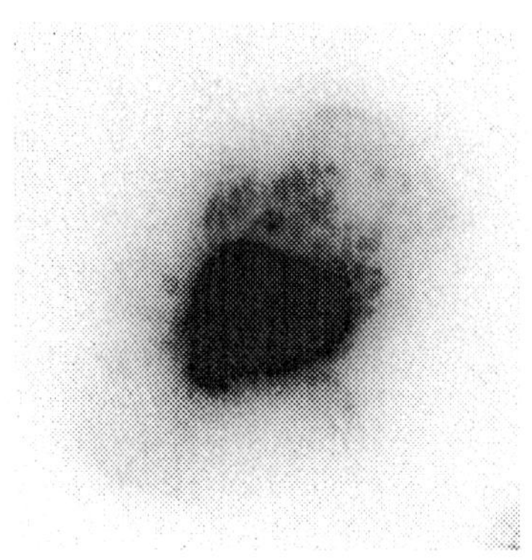

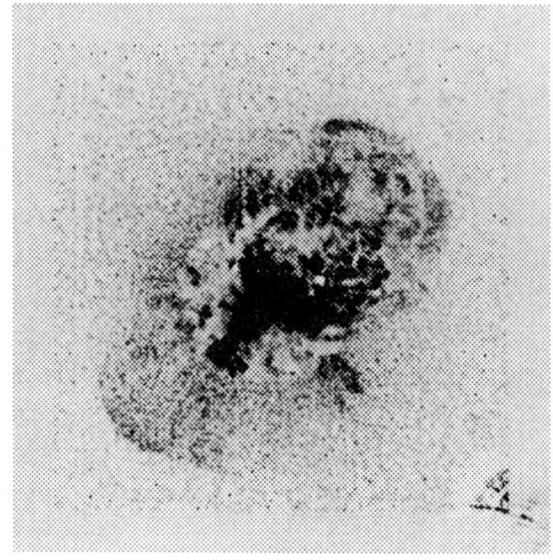

Figure 1. Image 1 is courtesy D.Ebbets, R.White et al. Image 2 (restored) shows a star clearly visible in the lower right corner outside the bounds of the input data.

- **Background**. The user can optionally specify the background image or constant that was subtracted from the input image to be included in the restoration for proper computation of the image statistics.

Two new arrays, a short integer array for the mask and a real array for the background image, are required for the implementation of these features. The solution is a tradeoff between memory requirements and execution speed. By reading the input image into memory twice per iteration (during the deconvolution step and the computation of χ^2) instead of storing it continuously in memory the memory requirements of the task actually decreased without the specification of the optional background image. The increase in execution time incurred by reading in the input image for each iteration is negligible.

4. lucy 1.2 - November 1993

The greatest concern for the **lucy** task at this point is its lengthy execution time, especially as some images require 1000 or more iterations. Version 1.2 addresses this performance issue with the implementation of the accelerated iteration algorithm (Hook & Lucy, 1992). Processing time is typically reduced by factors of 3 by selecting **accelerate=yes**. Figure 2 illustrates the relative performance of the accelerated and the standard algorithms.

An additional feature is the inclusion of an optional weight array. Flat-field variations can affect the noise characteristics of an image (Snyder 1990). If these are significant, the flat field should be included in the restoration as the **weight** parameter. Since the weight image is multiplicative, reference flat fields should be inverted before being used in the restoration. Also, in cases where short and long exposure images are combined to correct for saturated pixels or bad columns, the relative pixel weight can be adjusted to be proportional to the exposure time for each pixel.

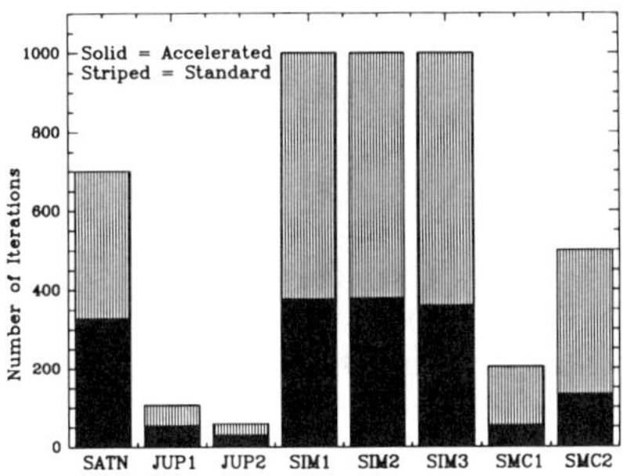

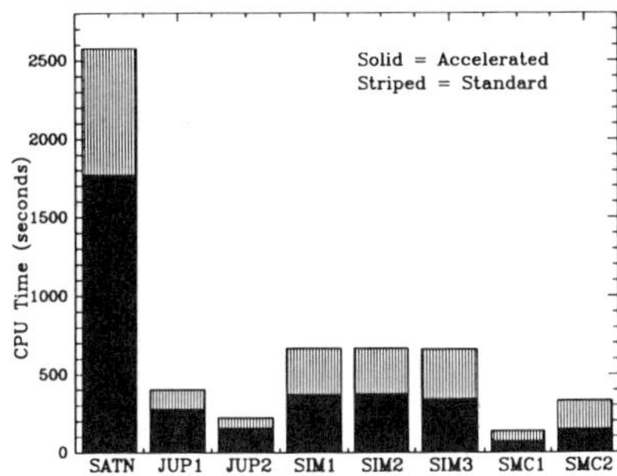

Figure 2. Graph 1 displays the number of iterations required by the Standard and Accelerated Algorithms to reach the same χ^2 and Graph 2 shows the relative CPU time for both.

The acceleration algorithm requires the addition of four real arrays to the lucy task which severly limits its use on workstations with less than 32 MB of memory. By overlaying some of the intermediate results on the general purpose real arrays and overlaying two real arrays on the general purpose complex array, essentially no additional memory was required. The mask array is deallocated after being multiplied by the weight array. The total memory requirment for a 512x512 image without the optional background image is now 9 MB (exactly the requirement of version 1.0). With the background image 10 MB of array space is needed.

5. Future Plans

The **turbo** acceleration algorithm developed by Rick White, which typically gives a factor of 3 in improved speed over the accelerated algorithm, is currently being implemented in the **lucy** task. Both the turbo and accelerated algorithms will be released with **STSDAS V1.3.1** in late 1993. The next project will be the implementation of a damping algorithm to significantly reduce the noise amplification. See paper, Image Restoration Using the Damped Richardson-Lucy Iteration by R.L. White in this publication.

References

Hook, R.N., Lucy, L.B. 1992, in Science with the Hubble Space Telescope, eds. P.Benvenuti & E. Schreier (Garching: ESO), 245

Lucy, L.B. 1974, AJ, 79, 745

Richardson, W.H. 1972, J. Opt. Soc. Am. 62, 55

Shepp, L.A., Vardi, Y. 1982, IEEE Trans. Medical Imaging M1-1, 113

Snyder, D. 1990, in The Restoration of HST Images and Spectra, eds. R.L.White & R.J.Allen, Space Telescope Science Institute, 56

Astronomical Data Analysis Software and Systems III
ASP Conference Series, Vol. 61, 1994
D. R. Crabtree, R. J. Hanisch, and J. Barnes, eds.

Experiments on Resolution Enhancement in HST Image Restoration

Nailong Wu[1]

Space Telescope Science Institute, 3700 San Martin Drive, Baltimore, MD 21218

Abstract. Experiments have been carried out to assess quantitatively the effectiveness of various methods for resolution enhancement in HST image restoration. In this paper the design of experiments is described first, then the main results are presented, and finally conclusions are drawn.

1. Introduction

Resolution enhancement is an important aspect of image restoration. Quantitative studies are highly desirable, and have been carried out theoretically as well as experimentally (Lucy 1992a, Lucy 1992b, Terebizh 1993). However, those studies were restricted to the cases of Poisson noise only, point spread functions (PSFs) having analytical forms, and mostly one-dimensional images.

Quantitative studies on resolution enhancement in HST image restoration can only be done experimentally due to the lack of explicit solutions (i.e., algorithms are iterative) in the two-dimensional case in general, particular forms of the PSFs of HST cameras, and readout and Poisson noises in WF/PC images.

2. Methods

The methods used for HST image restoration in the experiments are: the Richardson-Lucy method, Maximum Entropy Method (MEM), and in some cases Wiener and inverse filters.

The programs are: lucy, mem (referred to as mem0 hereafter), wiener (including inverse filter) in the IRAF/STSDAS package, and the commercial package MEM/MemSys5 (referred to as M5).

Descriptions of these programs can be found in *Restoration – Newsletter of ST ScI's Image Restoration Project*, Summer 1993 / Number 1.

3. Implementation

3.1. Generation of Synthetic Images

Point spread function is generated by Tiny Tim v.2.4 for the WFC, without filter, $\lambda = 555$ nm, subsampling factor = 8, 480x480 in size and rounded up to

[1]Also with Beijing Astronomical Observatory, CAS, Beijing, 100080, China

512x512 (0.0125 asec/pixel). Its full width at half maximum (fwhm) is about 4.6 pixels.

To generate blurred images, the PSF is used for *one peak image*. For *two peak image*, an image having two point sources of equal amplitude and with desired separation in a row is convolved with the PSF. For *extended source image*, an image having a desired number (width of the source) of contiguous pixels of equal value in a row is convolved with the PSF.

All the blurred images generated above are scaled so that the peaks are of the desired value (in DNs) maximal 4000 (saturation) or minimal 15.625 (nominal 15). Then readout and Poisson noises are added to get the final desired images in the experiments. The readout noise *rms* is 13 electrons or $13/7.5 = 1.8$ DNs. The gain or adu for calculating the Poisson noise is 7.5 electrons/DN. In the case of high signal-to-noise ratio (SNR), only one noise seed is used. For low SNR, on the other hand, three (unless specified otherwise) noise seeds are used for the statistical purpose.

3.2. Resolution Criteria

For *one peak sharpening*, fwhm is used to measure the width of a single peak. For *two peak resolution*, the *modified Rayleigh criterion* is used to define the just-resolved separation of two peaks. (Refer to Figure 1 and Lucy (1992a).)

$$R_1 = \frac{2I_2}{I_1 + I_3}, \qquad R_2 = \min\{\frac{I_1}{I_3}, \frac{I_3}{I_1}\}.$$

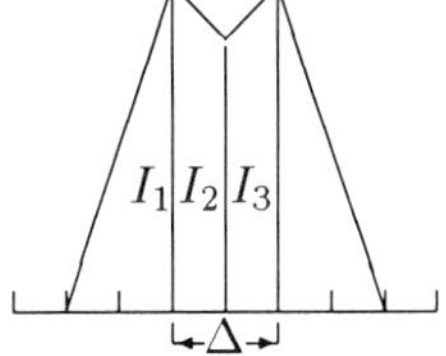

Figure 1. Resolution of two peaks.

Two peaks are said to be *well-resolved* if $R_1 < 0.811$ and $R_2 > 0.667$, and *not-resolved* if $R_1 > 0.811$ or $R_2 < 0.667$. The separation at which two peaks are neither well-resolved nor not-resolved is called the *just-resolved separation*. It is determined with some accuracy by the turning point from well-resolved to not-resolved. A small value of it indicates high resolution.

Each 2-D synthetic image is restored using different methods. However, the single peak width measurement and two peak resolution tests are evaluated using the row containing the peak maximum (maxima).

4. Results

4.1. One Peak Sharpening

In the case of high SNR (peak = 4000), peaks restored by lucy, mem0, MEM/ MemSys5 (historic and classic) and inverse filter are all nearly a δ-function in shape, and their widths are approximately 1.0 pixel. The Wiener filter does not sharpen the peak.

For low SNR (peak = 15), the average peak sharpening factor ($\bar{f}$) for each method is shown in Table 1. The factor f is defined as the ratio between the peak widths before and after restoration.

Table 1. The factor $\bar{f}$ (peak = 15).

	M5 (clas.)	mem0	lucy	M5 (hist.)
$\bar{f}$	3.47	3.09	2.24	2.12

4.2. Two Peak Resolution

The results are shown in Table 2, where Δ is separation in pixels; $\oplus$ means well-resolved while $\ominus$ not-resolved, and they may be prefixed by the numbers of occurrences.

Table 2. Resolution of two point sources.

Δ	No-rest.	lucy	mem0	M5 (hist.)	M5 (clas.)	wiener	inv.
				Peak = 4000			
6	$\oplus$						
5	$\ominus$	$\oplus$	$\oplus$	$\oplus$	$\oplus$	$\ominus$	$\oplus$
4	$\ominus$	$\oplus$	$\oplus$	$\oplus$	$\oplus$		$\oplus$
R_1		0.32	0.30	0.33	~ 0		~ 0
3	$\ominus$	$\ominus$	$\ominus$	$\ominus$	$\oplus$		$\oplus$
2	$\ominus$				$\oplus$		$\oplus$
				Peak = 15			
6	3$\oplus$						
5	3$\ominus$	3$\oplus$	3$\oplus$	3$\oplus$	3$\oplus$		
4	3$\ominus$	3$\ominus$	3$\oplus$	3$\ominus$	3$\oplus$		
3	3$\ominus$		3$\ominus$		3$\oplus$&7$\ominus$		

4.3. Control Experiments

Control experiments are used to determine the rates of resolution (RRs) for two-point sources and the rates of false resolution (RFRs) for extended sources when SNR is low (peak = 15). Based on Table 2, the "critical case" for control experiments is $\Delta = 4$ for mem0 and M5 (clas.), and $\Delta = 5$ for lucy and M5 (hist.).

The results are shown in Table 3. 10 noise seeds are used for each method with each type of source. ($\overline{R_1}$ is the average R_1 for two-point source.)

Table 3. Control experiments (peak = 15).

	M5 (clas.)	mem0	lucy	M5 (hist.)
Δ	4	4	5	5
RR	10/10	9/10	10/10	10/10
$\overline{R_1}$	0.32	0.66	0.28	0.41
RFR	0/10	0/10	0/10	1/10[a]

[a]For the only extented source false-resolved by M5 (hist.), M5 (clas.) also gives false resolution. But mem0 and lucy do not.

4.4. Resolution Improvement Factor

The *resolution improvement factor* r is defined as the ratio between the just-revolved separations (Δ_c) before and after restoration.

Δ_c in the case of low SNR (peak = 15) can be estimated from Table 2. Δ_c is 5.5 (between 5 and 6) before restoration, slightly greater than 3.0 for M5 (clas.), 3.5 (between 3 and 4) for mem0, and 4.5 (between 4 and 5) for lucy and M5 (hist.). The values of r calculated from the above estimation are shown in Table 4. Shown in the table are also the values for high SNR (peak = 4000) calculated in a similar way.

Table 4. The factor r.

	M5 (clas.)	mem0	lucy	M5 (hist.)	inverse
High SNR	2.7	1.5	1.5	1.5	2.7
Low SNR	1.8	1.5	1.2	1.2	

5. Conclusions

MEM/MemSys5 classic, mem0, lucy and MEM/MemSys5 historic are in descending order regarding the effectiveness of resolution enhancement in HST image restoration. The Wiener filter does not enhance resolution. The inverse filter can hardly be used. This conclusion can be used as a guideline in general image restoration.

Resolution may be improved by a factor of $1.2 \sim 1.8$ after restoration. For high SNR, this factor for MEM/MemSys5 classic may be greater than 2.0 (like the inverse filter).

mem0 and lucy have zero rates of false resolution. MEM/MemSys5 (classic and historic) may give false resolution. The reasons might be over-fitting data, and/or the lack of total power constraint (no positivity constraint on the original image).

In the experiments a PSF subsampled by a factor of 8 is used. With a standard (not subsampled) PSF, the factor r would be 1.0. That is, *there is no resolution enhancement.* This confirms the assertion that resolution seems to be improved in a crowded star field because of the elimination of halos but not the sharpening of the core of PSF.

Acknowledgments. This research work was supported by the Space Telescope Science Institute, which is operated by AURA, Inc., for NASA under contract NAS5-26555.

References

Lucy, L.B. 1992a, A&A, 261, 706

Lucy, L.B. 1992b, AJ, 104, 1260

Terebizh, V.Y. 1993, A&A, 270, 543

Astronomical Data Analysis Software and Systems III
ASP Conference Series, Vol. 61, 1994
D. R. Crabtree, R. J. Hanisch, and J. Barnes, eds.

Evaluation of Image Restoration Algorithms Applied to HST Images

I. C. Busko

Space Telescope Science Institute, Baltimore, MD 21218, and Astrophysics Division, INPE, S. J. dos Campos, Brazil.

Abstract. This work reports preliminary results on an on–going intercomparison of publicly–available image restoration algorithms, when used in the specific context of stellar fields imaged by the HST WFPC. Properties as fidelity to the original image and photometric linearity, as well as computation performance, were evaluated.

1. Introduction

The scientific usefulness of HST restored images can be fully assessed only by numerical experiments in which "perfect" images are artificially degraded, next restored by a given technique, and finally the output is compared with the original. A few authors published results on this line (e.g., Cohen 1991), but focussed most of their efforts on the MEMSYS implementation of the Maximum Entropy method.

With the availability of other image restoration techniques in the Space Telescope Science Data Analysis System (STSDAS), as well as a suite of simulated data sets available in public domain on the ST ScI Electronic Information Service (STEIS), more systematic and complete studies are now possible. In this work we report results of an ongoing comparative study between restoration algorithms when aplied to images of stellar fields.

2. Methodology

Data used in this study are simulated WFPC I observations of a "star cluster", similar to the ones available in STEIS. Simulated images were built on a 8X8–times oversampled data grid, later block–averaged to the actual instrument resolution. The PSF was computed by TinyTIM (Krist 1991) at $\lambda = 5500$ Å, at center of CCD #1. The appropriate noise model for WF CCDs was used in all simulated images.

Algorithms studied to date include current STSDAS implementations of the Richardson–Lucy iteration (Richardson 1972, Lucy 1974), Maximum Entropy (Wu 1993), the Wiener filter (Andrews & Hunt 1977), σ–CLEAN (Keel 1991) and an Iterative Least Squares algorithm (e.g., Katsaggelos 1991).

Criteria to evaluate restoration quality include both generic and astronomical–specific ones. An often used goodness–of–fit criterion in image restoration work is a measure of the "distance" between the restored image $\hat{f}(i,j)$ and the "truth" image $f(i,j)$ (the one without any degradation)

$$\eta_2 = 10 \log \frac{\sum_{i,j}(g(i,j) - f(i,j))^2}{\sum_{i,j}(\hat{f}(i,j) - f(i,j))^2}$$

where $g(i,j)$ is the observation at pixel i,j. The measure is expressed in dB, and increases as long as the restored image becomes "closer" to the truth image. We are using also an absolute–value distance

$$\eta_1 = 10 \log \frac{\sum_{i,j}|g(i,j) - f(i,j)|}{\sum_{i,j}|\hat{f}(i,j) - f(i,j)|}$$

which correlates better than η_2 with visual quality evaluation. Both η_1 and η_2 are independent of image content.

Astronomical criteria include photometric linearity, precision, and sky background statistics. Photometric evaluation was performed on star images using standard aperture techniques.

Algorithm sensitivity to PSF errors was evaluated by restoring the test image using both the original PSF, as well as another TinyTIM PSF, computed for a bluer star and situated ~ 200 pixels away from the the CCD center.

3. Results

Table 1 summarizes some distance measure results. These measures were computed both for the full image and for a 50 pixel sq. region centered on the star cluster, dominated by crowded star images. Thus, they are sensitive to different image properties. The full image measures asses the global goodness–of–fit, but in the particular image under study they are strongly sensitive to the sky background fit. The crowded region measures, on the other hand, are more sensitive to how well star images were fitted. The highest peak value is a measure on how close the brightest star's peak approached the truth value, and so it is also a goodness–of–fit measure for star images.

Algorithms are arranged in Table 1 in increasing order of execution speed. All algorithms were iterated up to the point where at least one of η_1 or η_2 just started to decrease. Most often this happened for the full image's η_1, which is sensitive to spurious noise peaks introduced in the sky background. σ–CLEAN was iterated to the 2.6σ level.

Visual inspection alone showed that the three algorithms: R–L, MEM0, and ILS+$\mathcal{P}$ are capable of very similar results. This is confirmed by the similar η_1 values for them. The solution obtained by the ILS+$\mathcal{P}$ algorithm, being a least–squares one, had the highest η_2, as expected. R–L produced a somewhat better solution than MEM0, but at a considerable speed penalty. σ–CLEAN produced not so good η_1 results as the methods above, mostly because of the large PSF residuals left on the CLEAN+residual map. The RILS results show that inclusion of a spatially–adaptive smoothness constraint in the least–squares iteration actually *decreased* restoration quality.

Results suggest that the ILS+$\mathcal{P}$ iteration is the most robust against PSF mismatches, followed by R–L and MEM0, which showed very similar degradations. σ–CLEAN was the most sensitive to PSF errors. Figure 1 depicts some

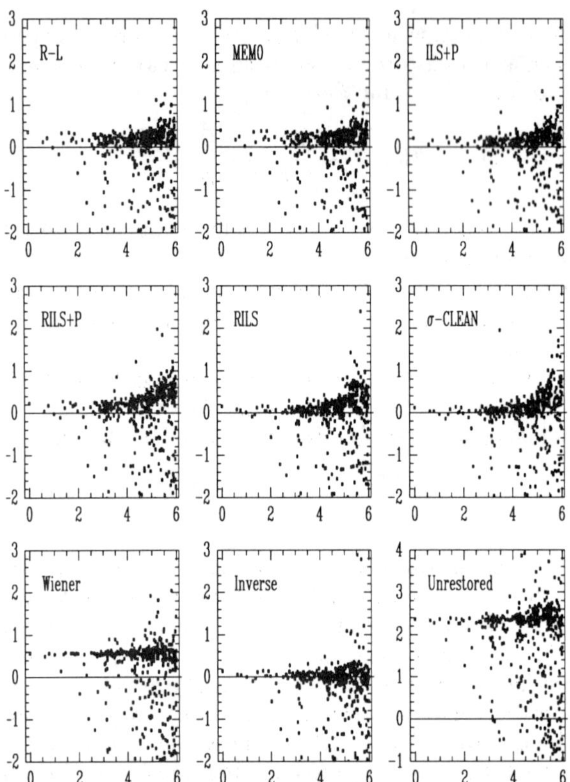

Figure 1. Aperture photometry in restored and unrestored images. A 2.2 pixel diameter aperture was used to measure each one of the 500 stars in the simulation input list. Abcissa is magnitude in truth image, ordinate is residual in the sense restored image minus truth. A negative residual indicates excess light in the aperture, probabily due to crowding. A positive residual means some light was not recovered by the restoration algorithm. Notice that even the unrestored image has missing light in some of the fainter stars.

Table 1. Goodness–of–fit measures. Meaning of symbols is as follows: R–L: Richardson–Lucy; MEM0: Maximum Entropy; ILS: Iterative Least Squares (no regularization); RILS: ILS with Miller regularization and spatial adaptivity; $\mathcal{P}$: projection (positivity).

PSF errors	Algorithm	Number of iterations	CPU time[a] (minutes)	Full image η_2	Full image η_1	Crowded region η_2	Crowded region η_1	Highest peak[b] (%)
NO	R–L	3000	132	6.90	5.08	10.61	5.83	52
	MEM0	300	19	5.98	4.87	9.88	5.68	46
	σ–CLEAN[c]	40000	13	9.32	3.89	10.16	5.01	87
	ILS + $\mathcal{P}$	300	12	9.48	5.43	11.82	5.88	69
	RILS+$\mathcal{P}$[d]	150	12	9.03	4.88	11.94	5.70	65
	RILS[e]	100	3	8.44	0.82	10.31	3.18	78
	Iterative inverse	80	1	7.68	-2.90	9.90	2.61	74
	Wiener		6 sec.	1.30	-0.52	1.39	0.74	29
YES	R–L	1500	65	5.92	4.49	9.64	5.18	46
	MEM0	300	19	4.81	4.21	8.76	4.97	39
	σ–CLEAN	30300	10	7.94	2.84	8.31	3.73	90
	ILS + $\mathcal{P}$	200	8	8.79	5.03	11.06	5.34	66

[a]SPARC10, 256 sq. image.

[b]Ratio of brightest star's highest peak in restored and truth images.

[c]Loop gain = 0.05.

[d]$\alpha = 0.2$, $\mu = 0.5$.

[e]$\alpha = 0.03$, $\mu = 0.5$.

results from aperture photometry in restored images. A trade-off between a particular algorithm's linearity, and its ability to control high-frequency noise buildup, as well as its capability to decrease confusion between neighbor star images is suggested. Tests involving several independent realizations of the same image showed that restoration neither improves nor degrades photometric precision when compared with precision achievable in unrestored images.

References

Andrews, H.C., & Hunt, B.R. 1977, "Digital Image Restoration" Prentice-Hall, New Jersey

Cohen, J.G. 1991, AJ, 101, 734

Katsaggelos, A.K. 1991, "Digital Image Restoration", Springer–Verlag

Keel, W.C. 1991, PASP, 103, 723

Krist, J. 1991, "The Tiny TIM User's Manual", Space Telescope Science Institute

Lucy, L.B. 1974, AJ, 79, 745

Richardson, W.H. 1972, Journal of the Optical Society of America, 62, 55

Wu, N. 1993, Astronomical Data Analysis Software and Systems II, Hanisch, R.J., Brisseden, R.J.V., and Barnes, J., (eds.), ASP Conference Series, 52, 520

Astronomical Data Analysis Software and Systems III
ASP Conference Series, Vol. 61, 1994
D. R. Crabtree, R. J. Hanisch, and J. Barnes, eds.

Morphological Filtering of Infrared Cirrus Emission Using the MasPar

Jeffrey A. Pedelty

Space Data and Computing Division, NASA Goddard Space Flight Center, Greenbelt, MD 20771

Philip N. Appleton

Department of Physics and Astronomy, Iowa State University, Ames, IA 50011

John P. Basart

Department of Electrical and Computer Engineering, Iowa State University, Ames, IA 50011

Abstract. We have implemented a prototype infrared cirrus filter on the MasPar massively parallel computer. The filter uses the techniques of mathematical morphology. The MasPar implementation has reduced the time necessary to filter an IRAS 100μm image from the order of hours to seconds.

We give a brief overview of mathematical morphology and our prototype morphological filter. We discuss our MasPar implementation and plans to refine and extend our filter which are now practical because of the MasPar's tremendously faster speed.

1. Introduction

The Infrared Astronomy Satellite (IRAS) was launched in 1983. It surveyed nearly the entire sky at wavelengths of 12, 25, 60, and 100μm, generating a data archive which has revolutionized our understanding of the infrared universe.

An early discovery from IRAS was the infrared "cirrus" emission. The cirrus appears as wispy, cloud-like structure, hence the name, and is due to thermal radiation from cold dust grains in our galaxy. The cirrus emission dominates IRAS images at 100μm, blocking our view beyond. Our motivation is to remove the cirrus emission and thus reveal the extragalactic far-infrared sky. In the process we hope to also learn something about the cirrus emission itself.

2. Mathematical Morphology

Mathematical morphology (or morphological image processing) is a relatively recent and exciting field. For a recent review see Basart et al. (1992). Current applications of mathematical morphology include robotic vision, medical imaging, remote sensing, and non-destructive evaluation (NDE). The SPIE now holds an annual conference devoted to morphology applications, but astronomy applications have been limited (Lea and Kellar 1989, Huang and Bijaoui 1991, Appleton et al. 1993).

A fundamental feature of mathematical morphology is the use of a structuring element. This is essentially a smaller version of the input image, and contains structure of the type that is to be analyzed in the image. The two basic morphological operations are dilation and erosion. Dilation expands and brightens an image, while erosion shrinks and darkens. Dilation is performed by sliding the structuring element over the image while adding the values of the structuring element to the image, and then recording the maximum value reached at each point. Erosion is similar, except that the structuring element values are subtracted from the image, and the global minimum is taken.

The opening operation is created by first eroding and then dilating an image. Physically, the opening operation extracts all structure brighter than the background whose size is smaller than the structuring element and which conforms to the shape of the structuring element.

3. Morphological Cirrus Filter

A prototype morphological filter for removing infrared cirrus emission was developed by Appleton, Siqueira, and Basart (1993). The filter uses the concept of sieving discussed by Serra and Vincent (1992), using a series of 17 image openings. We first open the image using a truncated Gaussian structuring element that has a FWHM $= w_1 = 5'$. For each successive opening we increase the FWHM by $2'$, up to a maximum of $w_i = 37'$. We create a growth cube with 16 planes by taking the differences between successive openings. This growth cube contains spatial structure on scales $w_i \leq w \leq w_{i+1}$. The growth cube is in essence a series of bandpass filtered images, but *not* in spatial frequency because of the highly nonlinear nature of the opening operator.

Empirically, we find that pixels known to contain cirrus emission have growth curves (i.e., plots down the cube) which are distinct from that of other emission. We subtract a normalized average cirrus growth curve from each pixel of each plane in the growth cube, and then create the final filtered image by summing all 16 growth planes. We find that the residual cirrus is $\approx$ 15x weaker in the M81/M82 field.

4. MasPar Computer System

The results reported in Appleton et al. (1993) are very encouraging, but the workstation implementation of the morphological filter took several hours to run. This makes it very impractical to experiment with the filter or to run it on very many fields in the IRAS database. We expected the MasPar implementation would speed up the filtering.

The MasPar system hardware has 4 major elements (Blank 1990): 1) a Processor Element (PE) array which is a 2-d mesh with fast nearest neighbor and somewhat slower general communications, 2) an Array Control Unit (ACU), 3) a Unix Front End (currently a DECstation 5000/240), and 4) a high-speed I/O Subsystem.

The PE array performs compute intensive tasks under the direction of the ACU. The front end supports interactive use of the PE array in a familiar Unix environment. All coding is done there, including interactive visual debugging. We use the FITSIO package developed at the NASA/GSFC HEASARC (Pence 1992) so that all image I/O can be in the FITS format.

The MP-1 at GSFC has 16K processors, each with 64KB local memory (1 GB total). We also use a 4K processor MP-2 at the ISU Ames Laboratory. The programming model is SIMD (Single Instruction Multiple Data), which means that all PE's perform the same instruction on separate pieces of the image. Two high level languages exist for the MasPar. The MasPar Language (MPL) is based on C, while MasPar Fortran (MPF) is a derivative of Fortran 90.

5. MasPar Implementation of Morphological Cirrus Filter

We begin by assuming that the number of PE's in the MasPar equals the number of pixels in the image (e.g., $128x128 = 16\text{K}$) so each PE processes just one pixel.

Dilation was discussed above in terms of sliding the structuring element across the input image, but it can also be performed by sliding the image instead. We adopt this approach for our MasPar implementation since the size of our largest structuring element is much smaller than the MasPar PE array. We center the structuring element at $(0, 0)$. We shift the input image once for each element of the structuring element, the amount of the shift is given by the (x, y) coordinate of the element. To each of these shifted images we add the value of the structuring element at (x, y). The final value of the image dilation is the pixel by pixel maximum of each of these shifted and offset images. Erosion is procedurally identical, except that the structuring element is subtracted and the global minimum is taken.

Calculating the required image translations is very efficient on the MasPar since nearest neighbor communications are so rapid. In practice, however, the elements in the structuring element are processed beginning with the central value at (0,0), and proceeding outward in a spiral pattern. This minimizes the required number of translations, and means that only nearest neighbor communications are required.

If the image is bigger than the PE array (in our case $512x512$), then more than one pixel is stored in a given PE. We have implemented two ways to map the image to the PE array. The first is known as hierarchical mapping. It divides the input image into $4x4$ pixel subimages which are then assigned to each processor. The other is called cut and stack mapping. This method assigns successive rows of the image to corresponding rows in the PE array. When an image row is longer than the PE array then the row is cut and stacked back on the PE array. Thus, pixels 1, 129, 257, and 385 in row 1 are assigned to the first PE. A similar process is used when the number of rows in the image exceeds the size of the PE array. With hierarchical mapping each processor is assigned pixels which are adjacent in the input image, in contrast to the cut and stack method in which the assigned pixels are uniformly spread throughout the image.

6. Performance Comparisons

The time required to filter a single 512x512 pixel image on a MasPar MP-1 with 16K processors is 15 seconds, translating to a performance of 580 MFLOPS. An MP-2, also with 16K processors cuts this time to 8.5 seconds and reaches about 1 GFLOP. By comparison, a single processor Cray Y-MP takes about 97 seconds, a DECstation AXP requires 730 seconds, and the workstation originally used to develop the filter took over 30000 seconds.

These MasPar performance numbers are for the version of the filter written in MPL. The MPF version is about 15% slower. However, the MPF version is much shorter and is much easier to maintain. The MasPar timings show little dependence on the data mapping used in that hierarchical mapping is only 2% faster than cut and stack. This attests to the high interprocessor communications rate, since this implies that is just as fast to access a datum from the next processor as it is to access one locally.

7. Future Work

Our MasPar implementation now permits us to practically explore: 1) how the filter works on a much larger sample of IRAS $\lambda 100\mu$m images, 2) effects of the size, shape, and amplitude of the structuring element on the final filtered image, and 3) alternate normalizations of the cirrus growth curve prior to subtraction from the growth cube. These topics are now being explored.

Possible topics for future work include: 1) testing using analytical (e.g., fractal) cirrus models as input, 2) adaptive structuring elements (perhaps related to texture analysis and feature classification), 3) morphology neural networks (see Davidson and Hummer 1993), and 4) joint filtering of IRAS $\lambda 60\mu$m and $\lambda 100\mu$m images.

Acknowledgments. This work is funded by the Earth and Space Science Applications Project of the NASA High Performance Computing and Communications Program (HPCCP).

References

Appleton, P.N., Siqueira, P.R., & Basart, J.P. 1993, AJ, 106, 1664

Blank, T. 1990, in Proceedings of IEEE CompCon, IEEE: San Francisco, 20

Basart, J.P., Chackalackal, M.S., & Gonzalez, R.C. 1992, in Advances in Image Analysis, Y. Mahdavieh & R.C. Gonzalez, SPIE: Bellingham, 306

Davidson, J.L., & Hummer, F. 1993, IEEE Transactions on Circuits, Systems, & Signal Processing, 12, 177

Huang, L., & Bijaoui, A. 1991, Exp. Astr., 1, 311

Lea, S.M., & Kellar, L.A. 1989, AJ, 97, 1238

Pence, W.D. 1992, in Astronomical Data Analysis Software & Systems I, A.S.P. Conf. Ser., Vol. 25, eds. D.M. Worrall, C. Biemesderfer & J. Barnes, 22

Serra, J., & Vincent, L. 1992, IEEE Transactions on Circuits, Systems, & Signal Processing, 11, 47

Astronomical Data Analysis Software and Systems III
ASP Conference Series, Vol. 61, 1994
D. R. Crabtree, R. J. Hanisch, and J. Barnes, eds.

Image Quality Assessment for the GONG Project

W. E. Williams, J. Goodrich, R. Toussaint

National Solar Observatory, NOAO[1]*, P.O. Box 26732, Tucson, Arizona 85726*

Abstract. Two methods of image quality assessment are evaluated. The automated versions of these methods allow expeditious identification of acceptable data with a minimum of skilled operator interaction. Data reprocessing due to the inclusion of unacceptable images has been nearly eliminated.

1. Introduction

The GONG (Global Oscillation Network Group) project will observe the sun nearly constantly for three years, from six sites placed around the globe. Approximately 1+ terabyte of image data will be acquired during the course of the project. This massive amount of data — an estimated four million images — must be processed in a small fraction of data acquisition time. A major obstacle to efficient data reduction is the presence of unacceptable images in the data stream. These images must be identified and removed very early on in the data processing.

2. GONG Data Overview

GONG data consists of velocity, modulation amplitude and mean intensity images derived from three amplitude-modulated intensity observations of the sun obtained in the light of Ni I at 677 nm. The relative Doppler velocity is directly proportional to the phase of the modulated signal.

The velocity images are processed by temporal filtering of the time series of images, spatial decomposition into spherical harmonics and temporal Fourier transform of the resulting mode coefficients (Duvall and Harvey 1986, Williams, 1992). The plot of the mode coefficient magnitude as a function of spherical harmonic degree and temporal frequency, the $l - \nu$ diagram, is the ultimate indicator of the presence of unacceptable images. $l - \nu$ diagrams are made at the end of the processing, and require sufficient time to encourage the removal of unacceptable images before the temporal filtering is performed.

---

[1]The National Optical Astronomy Observatories are operated by the Association of Universities for Research in Astronomy, Inc. (AURA) under cooperative agreement with the National Science Foundation.

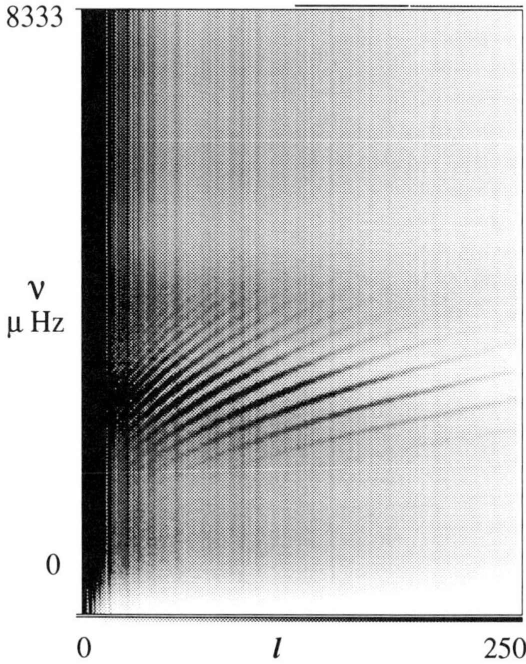

Figure 1. Unacceptable $l - \nu$ diagram for the May 19, 1993 GONG prototype time series data. One unacceptable image was included in the data processing.

3. Image Quality Assessment

Since the helioseismic modes have small amplitudes compared to the mean Doppler velocity, both obviously and subtly unacceptable data can have an equally undesirable impact on the results. Although only the velocity images are processed into mode coefficients, the intensity and modulation images are evaluated as indicators of the velocity image quality. Evaluation of the velocity images alone has been found to be inadequate. Figure 1 shows the impact of one unacceptable image on an $l - \nu$ diagram of a time series of 400+ images.

If any one of the velocity, modulation amplitude or averaged intensity images are obviously or subtly unacceptable, the velocity image is removed from further processing. Obvious unacceptable images may be found in any of the three data types. The modulation amplitude images have been found to be the most sensitive indicator of subtly unacceptable data.

Obvious unacceptable images include partial images, misshapen images, images with gross systematic signal variations on the solar disc, images with anomalous spikes on the solar disk and images with poor definition of the limb of the sun. Subtly unacceptable images may have slight systematic signal variations across the solar disk and/or anomalous signal levels on the solar disk.

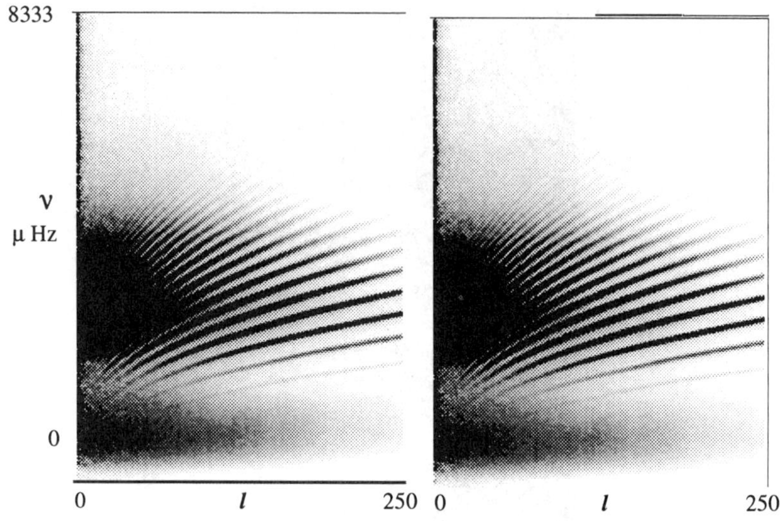

Figure 2. Acceptable $l - \nu$ diagrams for the May 19, 1993 GONG prototype time series data. The images for the left $l - \nu$ diagram were selected by single image evaluation. The images for the right $l - \nu$ diagram were selected by temporal trend evaluation. A few missing time samples have been filled using an autoregressive gap filler on the mode coefficient time series (Brown and Christensen-Dalsgaard 1990, Anderson 1992).

4. Applications

The decision to remove data from further processing is made by the operator. The operator's decision must be made rapidly and accurately. To assist the operator in this decision, software has been developed to automate the identification of possible unacceptable data.

Images are evaluated by comparing single images to a set of standard parameters and examining temporal trends of selected image statistics. Both methods give equivalent results, as shown in Figure 2. The image quality assessment used by the project will make use of the best features of each method, as discussed below.

Evaluation of single images is easily done by software without skilled operator attention. It has the added advantage of being objectively applied to fragmentary data, that is, images recorded between clouds and other interruptions to observation. Fragmentary data is anticipated to be a substantial part of the GONG network data.

The image evaluation software records temporal trends in modulation amplitude, velocity and intensity. Plots of these trends are used to monitor instrument performance and show unusual events. For example, on June 18, 1993, the camera rotator turned on in the middle of the day. The single image evaluation indicated one unacceptable image at the time the camera rotator moved. The temporal trend analysis showed a large step change in velocity, indicating that images before and after the event needed to be temporally filtered as if they were from two different site days.

In the example in Figure 2, the two methods did not agree on the acceptability of every image in the time series. The two methods did agree on the acceptability of most images. The equivalence of the two $l - \nu$ diagrams indicates that the criteria for acceptable images for both methods are probably too conservative.

5. Summary

Use of automated image quality assessment procedures developed over the course of the project have practically eliminated data reprocessing due to unacceptable images. Time required for unacceptable image identification and removal has been reduced to less than ten percent of data acquisition time, that is, 5 seconds or less per time sample. Increasingly sophisticated criteria for automated image quality assessment promises further reduction in time required to identify and remove unacceptable data.

References

Anderson, E. 1993, in GONG 1992: Seismic Investigation of the Sun and Stars, T.M. Brown, ASP Conference Series 42, 445

Brown, T. M., & Christensen-Dalsgaard, J. 1990, ApJ, 349, 667

Duvall, T. L., & Harvey, J. W. 1986, in Seismology of the Sun and the Distant Stars, D. Gough, Dordrecht: Reidel, 1986, 105

Williams, W. E. 1992, "SUNTRANS: Spherical Harmonic Transpose and Time Transpose Package" in GONG Alpha GRASP Version 92.1, J. Pintar, NOAO/NSO GONG project document

Part 4. Data Analysis
Section B. Spectral Analysis

Astronomical Data Analysis Software and Systems III
ASP Conference Series, Vol. 61, 1994
D. R. Crabtree, R. J. Hanisch, and J. Barnes, eds.

Applications of the Hough Transform

Pascal Ballester

European Southern Observatory, Karl-Schwarzschild-Str. 2, D-85748 Garching, Germany

Abstract. The Hough transform is a robust parameter estimator of multi-dimensional features in images. It finds many applications in astronomical data analysis. It enables in particular to develop auto-adaptive, fast algorithms for the detection of echelle orders and automated arc line identification. The Hough Transform provides robustness against discontinuous or missing features. The HT is compared in this paper to other robust fitting techniques in terms of robustness to contamination.

1. Introduction

The Hough Transform has been developed by Paul Hough in 1962 and patented by IBM. It became in the last decade a standard tool in the domain of artificial vision for the recognition of straight lines, circles and ellipses. The Hough Transform is particularly robust to missing and contaminated data. It can also be extended to non-linear characteristic relations and made resistant to noise by use of anti-aliasing techniques.

2. Hough Transform

The Hough Transform has been originally developed to detect analytically representable features in binarized images, such as straight lines, circles or ellipses. The characteristic relation (1) of the sought-for feature is back-projected in the parameter space. Each set pixel (x_i,y_i) defines a relation between the parameters of the characteristic relation which can be represented as a curve in the parameter space.

$$f((x, y), (a_1, ..., a_p)) = 0 \qquad (1)$$

The Figure 1 illustrates the construction of the HT for the detection of straight lines. Each pixel (x_i,y_i) along the lines will generate by back-projection a straight line of equation $b = y_i - a.x_i$. These lines intersect at the locus (a, b) characterizing the line. All points belonging to the curve have been mapped into a single location in the transformed space, allowing an easier detection. The Hough transform involves a peak finding algorithm to detect the features.

A noteworthy characteristic of the HT is that it does not require the connectedness of the co-linear points. Segmented lines will generate a peak in the

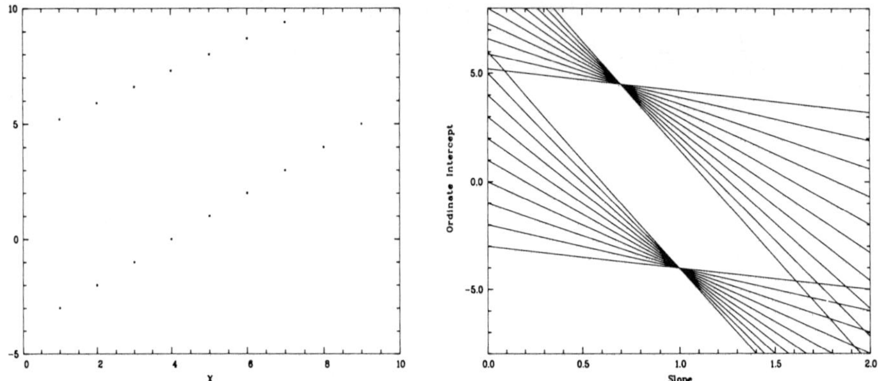

Figure 1. Construction of the HT for detecting straight lines.

parameter space and the lacking segments simply do not contribute to the transform. On the other side, artifact peaks might be generated in presence of noise and high density of features by coincidental intersections in the parameter space. To a certain extent, artifacts can be avoided by using anti-aliasing techniques and adapted peak finding algorithms. Also the HT treats each image point independently, allowing a parallel implementation of the method.

3. Robust Fitting

The sensitivity of least-square (LS) fitting to outliers is a traditional problem in data analysis and a variety of robust techniques have been developed, such as median based techniques, trimming or Winsorization. An interesting technique is the Least Median of Squares (LMS) and the LMS-based Reweighted Least Square (RLS) (Rousseeuw, 1987). The LMS consists of minimizing the term:

$$med(y_i - \alpha x_i - \beta)^2 \quad (2)$$

The RLS corresponds geometrically to finding the narrowest strip covering half of the observations. The breakdown point of the RLS method is 1/2, i.e., it will provide stable estimates of the fit parameters up to 50 % of contamination of the data set. Most of the other techniques do not attain a breakdown point of 30 %.

Although the HT is not a regression method, it provides estimates of regression parameters such as slope or ordinate intercept. It can be compared in terms of breakdown points to robust fitting techniques like the RLS.

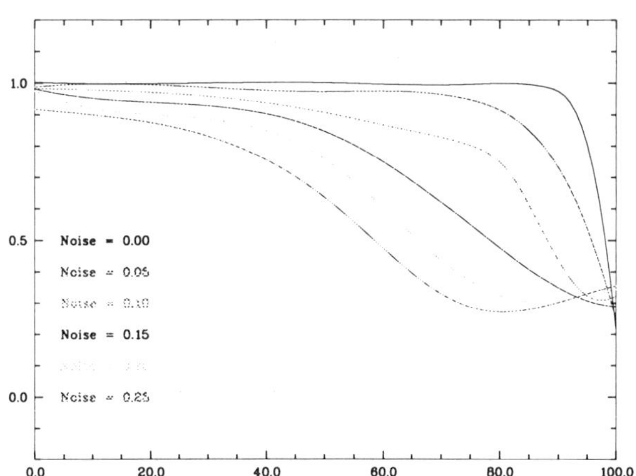

Figure 2. Breakdown points of the HT for different levels of noise.

The Figure 2 shows the breakdown plots of the HT for a case study presented in (Rousseeuw, 1987). For low levels of noise, the HT shows an excellent robustness (the breakdown occurring at 90 % of contamination) resulting from the fact that strictly co-linear points will give rise to a well defined peak brighter than coincidental intersections. As the level of noise increases a greater number of points is required to secure the detection of the features.

4. Detection of echelle orders

The application of the HT to the detection of echelle orders is straightforward, given the linearity of the sought-for echelle orders. The parallelism of the orders allows in principle to detect most of the orders of the spectrum by transforming the pixels along two columns of the image. This property reduces by a large factor the computation cost. In order to detect also the truncated orders at the edges of the spectrum and to improve the signal-to-noise ratio of the peaks in the accumulator, the implementation of the HT involves the transformation of a few tens instead of all columns of the spectrum.

The number of votes accumulated in the peaks corresponding to each order is an integral function of the intensity along the orders. The global maximum of the accumulator gives the parameters of the brightest order. The algorithms performs an iterative peak removal to allow the detection of fainter orders. The detection is stopped either when no new maximum brighter than a predefined fraction of the first maximum can be found, as described in (Risse,1988) or when a user-defined number of orders has been reached.

This method has been implemented in the echelle package of ESO-MIDAS since the version 91NOV (November 1991) in the command DEFINE/HOUGH.

5. Extending the HT

5.1. Three-Dimensional Hough Transform

The application of the Hough transform in parameter space of more than two dimensions has been widely studied. Multi-stage techniques are frequently used to reduce computations. However, these techniques lack of accuracy and stability if the problem is not largely over-determined. The three-dimensional Hough transform has been applied to the problem of automated arc line identification.

The sought-for dispersion relation involves three coefficients of central wavelength (λ_c), average dispersion (α) and non-linearity (β):

$$\lambda = \lambda_c + \alpha x(1 + \beta x) \tag{3}$$

5.2. Anti-Aliasing Techniques

The Hough transform does not account for uncertainties in the positions due to digitization errors or measurement errors. Anti-aliasing techniques consist of defining an influence function: data points will contribute to the number of votes of every candidate depending on this function and of its influence domain.

5.3. Automated Arc Line Identification

The identification of arc lines consists of associating a list of line positions in pixel space to a list of reference wavelengths. The principle of the method is to perform all possible associations in a pixel-wavelength space. The three-dimensional HT allows to detect within a given range of central wavelength and average dispersion, the non-linear dispersion relation maximizing the number of associations. The maximum is searched for in a cube of the Hough space, providing all three parameters (λ_c, α, β) of the equation 3.

The method is by definition robust to contamination in the lists corresponding in that case to unassociable features. Robustness to noise is improved by use of anti-aliasing techniques. This algorithm is presently used for the automated wavelength calibration of long-slit spectra at the ESO New Technology Telescope (NTT). In its current implementation the method allows for an initial uncertainty of $\pm$ 100 pixels in central wavelength and $\pm$ 30 % in average dispersion. Wide ranges of average dispersion (1 to 20 Angstroem per pixel) of central wavelength (5000 to 8000 Angstroems) can be processed with a unique catalog.

References

T. Risse, Hough Transform for Line Recognition, Computer Vision and Image Processing, 1989, 46, 327-345, 1989

P. Ballester, 1991, "Finding Echelle Orders by Hough Transform", Proceedings of the 3rd ESO-ST/ECF Data Analysis Workshop

P.J. Rousseeuw & A.M. Leroy, 1987, "Robust Regression and Outlier Detection", John Wiley & Sons Inc.

Astronomical Data Analysis Software and Systems III
ASP Conference Series, Vol. 61, 1994
D. R. Crabtree, R. J. Hanisch, and J. Barnes, eds.

Determining Wavelength Scales for the EUVE Spectrometers

W. T. Boyd, P. Jelinsky, E. C. Olson, M. J. Abbott, C. A. Christian

Center for EUV Astrophysics, University of California, Berkeley, CA 94720

Abstract. The Extreme Ulraviolet Explorer satellite contains three EUV spectrometers built at UC Berkeley. EUVE Guest Observers must rely on the wavelength correction provided in the EGO software since there is no calibration lamp on board the spacecraft. We discuss the evolution and future status of the spectrometer wavelength scales for EUVE as well as their implementation into the EGO data set.

1. Introduction

Measurement of the resolution and determination of wavelength scales for the EUVE spectrometers has been complicated by several factors. Because the imaged spectral lines are distinctly non-Gaussian, simple Gaussian fits to the lines yield inaccurate results of the line widths, so the actual line width must be measured directly for each line; a custom IDL routine was written to make this process semi-automatic. The optical system has been ray-traced on a computer to produce a theoretical wavelength scale for each of the three spectrometers. The true scales differ from theory because of distortion due to the solid-state detector anodes, among other causes, and so real data must be used to characterize these deviations. Pre-launch spectra of EUV sources were taken but did not cover all three spectrometers completely, hence in-orbit spectra of real astrophysical plasmas have been necessary to fill in the gaps. Line centroids were measured using a custom interactive IDL routine.

2. Evolution of EUVE Wavelength Scales

2.1. Initial Idealized Wavelength Scale

The first extensive EUVE wavelength scale (valid over the entire detector) was a set of polynomial equations fit to artificial data points generated by ray-tracing through a computer model of the Deep Survey telescope and spectrometer optics. This solution was fit to the Long-Wavelength Spectrometer. The Medium- and Short-Wavelength solutions were just scaled versions of the Long-Wavelength fit since the former have dispersions of two and four times that of the latter. This scale did not account for detector distortion or rotation of the detector axes with respect to the spectral dispersion and imaging axes. It did feature separable (no cross terms) functions of detector X and Y and spectral off-axis angle.

2.2. Ground Calibration Spectra: Rotation Correction

Calibration spectra of a variety of laboratory sources were obtained with the spectrometers prior to launch. Line identifications were made using the Kelly (1982) list. Independent determination of a wavelength scale for these spectra confirmed the initial ray-traced scale. When rotation and large-scale distortion were included in the fits, a very good wavelength solution resulted ($\sigma = 0.32$ Å for the LW spectrometer). This fit was only good for off-axis angles less than about 0.2^o , however, and a more general solution was desired.

2.3. Global Scale With Rotation

The next step was to add rotation to the initial ray-traced scale. Five free parameters were added to the fit for each spectrometer: detector rotation angle, X and Y offsets for the rotated coordinate system, plate scale (pixels/cm) and a ratio between the plate scales in the X and Y directions. The best fit values of these parameters came from minimizing $\chi^2 = (X_{measured} - X_{expected})^2 + (Y_{measured} - Y_{expected})^2$, where the expected values of X and Y come from applying the ray-traced fit backwards (using Newton's method), given line wavelengths and off-axis source angles as input. The ground calibration data were used to find the initial wavelength scale until enough in-orbit data were accumulated. Line identifications for the in-orbit spectra were made using the Landini and Fossi (1990) and Mewe et al. (1986) line lists. The current wavelength scale is based on the best fits to the five rotation parameters on in-orbit data. It is best for the SW and MW spectrometers; EUV emission in the LW is highly attenuated by the ISM. These maps typically predict wavelengths to within one resolution element (RMS) which corresponds to about 0.5 Å for the Short-Wavelength Spectrometer (70 - 180 Å), 1.0 Å for the Medium-Wavelength Spectrometer (150 - 370 Å) and 2.0 Å for the Long-Wavelength Spectrometer (290 - 750 Å); the scales are much worse when they are applied near the detector edges.

3. Incorporation of the Solution Into the Reference *EGODATA* Set

Once the best wavelength solution based on in-orbit spectra is obtained, it is fixed in the EUVE Science Baseline Document (a controlled software release). The equations are of the form:

$$\begin{aligned} \lambda &= \lambda(X) + \lambda(Y) + \lambda(\theta_s) \\ \theta_i &= \theta_i(Y) \end{aligned}$$

where θ_s and θ_i are the off-axis incident angles in the spectral and imaging directions, respectively.

Once the numbers are fixed, the *EGODATA* set is built from the Science Baseline Document information. The wavelength maps are stored as 2-D lookup tables in IRAF image format. Both the wavelength and imaging angle equations are now separated into two components: one assuming the source is directly on the telescope boresight and one for the off-axis correction terms.

4. Wavelength Scale Application to Guest Observer Spectra

Application of the wavelength scale to GO data occurs in the Comprehensive Event Pipeline (*cep* in IRAF *cea* package). The following set of operations is performed on each photon in the telemetry stream:

•Convert from Detector X,Y to Initial Wavelength and Imaging Angle Using the Solutions in *EGODATA*, Assuming the Source is on Boresight

$$(X,Y) \mapsto (\lambda_0, \Theta_0)$$

•Compute the Off-Boresight Angles:
i) Calculate Vector from Telescope Aspect to Source RA, Dec;
ii) Apply Several Coordinate Transformations to get Spectral, Imaging Off-Axis Angles for Each Spectrometer

$$(\alpha, \delta) - aspect \mapsto (\theta_s, \theta_i)$$

•Add Off-Axis Corrections to Wavelength, Imaging Angle Using the Solutions in *EGODATA*

$$(\theta_s, \theta_i) \mapsto (\delta\lambda, \delta\Theta)$$
$$(\lambda_0, \Theta_0) + (\delta\lambda, \delta\Theta) \mapsto (\lambda_{final}, \Theta_{final})$$

Note that this produces a two-dimensional spectrum (QPOE file) which is already wavelength dispersion corrected. Extraction of the finished one-dimensional spectrum is then just one step away. The original detector event coordinates are preserved in the QPOE file.

The *cep* routine is part of the IRAF EUVE GO software. When new wavelength calibrations are released, investigators need only copy (via ftp or from the latest EUVE CDROM) the new *EGODATA* set in order to reprocess their spectra. The latest release is available from *cea-ftp.cea.berkeley.edu*; from NCSA *mosaic*, load the file *http://cea-ftp.cea.berkeley.edu/HomePage.html.*

5. Future Scales: Distortion Correction

Recent efforts have focused on measuring and correcting departures from the theoretical maps, such as a curvature of the spectrum in the imaging direction and a compression of the spectrum along the dispersion axis near the detector edges; both of these effects are attributed to the solid-state detector design.

Distortions are being characterized by fitting polynomial equations to the residuals to the best fit wavelength scale. To the extent that these distortions are repeatable, large-scale effects, we are able to correct for them. From several white dwarf spectra that span the full detector, we have a good measurement of the Y distortion for observations near boresight. The X distortion correction is less well constrained, partly because it comes from collected spectral line centroids (which tend to be clumped around common EUV wavelengths), and partly bacause of scatter due to an unknown variable. Errors in line centroid measurements are not large enough to account for all of the scatter. The accuracy of the final wavelength solution is limited by this scatter.

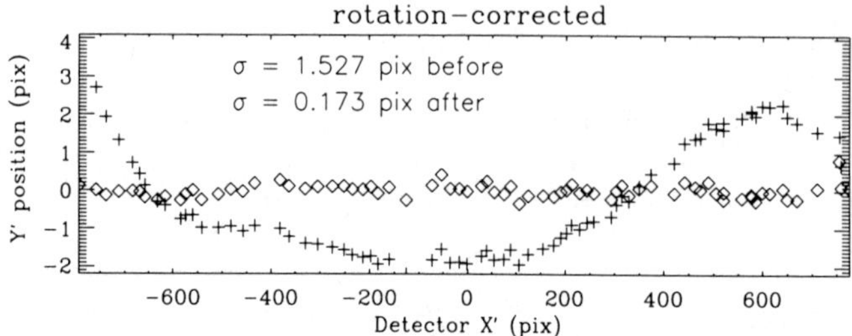

Figure 1. LW Spectrum trace before and after Y distortion correction.

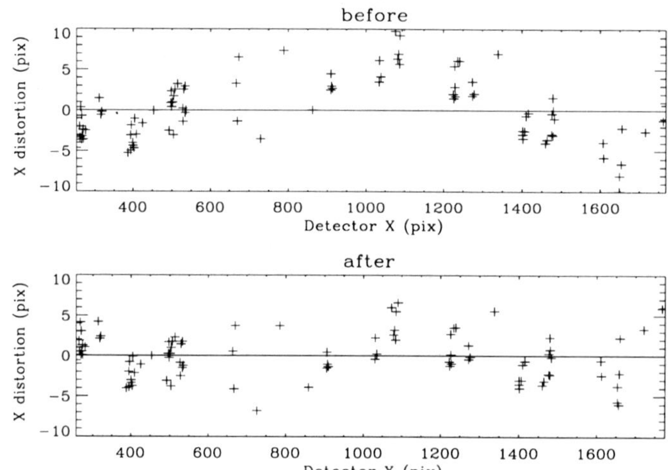

Figure 2. Effect of X distortion correction on LW line centroids.

The effect of this distortion correction as applied to 94 spectral line centroids from several targets in the LW spectrometer is a 37% improvement in the scatter. Before correction, the wavelength scatter is $\sigma = 1.2$ Å and after correction it is $\sigma = 0.75$ Å . Not only will wavelength scale accuracy be improved (especially near detector edges) with distortions accounted for, but the spectra will curve less on the detector, greatly facilitating extraction. This correction will have the largest effect on Long-Wavelength spectra, which have the lowest dispersion. Examples of distortion correction are seen in Figures 1 and 2.

Acknowledgments. This work was completed under NASA contract NAS5-29298.

References

Kelly, R.L. 1982, "Atomic and Ionic Spectrum Lines Below 2000 Angstroms", Oak Ridge National Laboratory Publication ORNL-5922

Landini, M., & Monsignori Fossi, B. C. 1990, A&AS, 82, 229

Mewe, R., Lemen, J. R., & Van den Oord, G. H. J. 1986, A&AS, 65, 511

Astronomical Data Analysis Software and Systems III
ASP Conference Series, Vol. 61, 1994
D. R. Crabtree, R. J. Hanisch, and J. Barnes, eds.

The FIVEL Nebular Analysis Tasks in STSDAS

Richard A. Shaw

ST ScI, 3700 San Martin Dr., Baltimore, MD 21218

Reginald J. Dufour

Dept. of Space Physics & Astronomy, Rice Univ., Houston, TX 77251

Abstract. A package of STSDAS tasks has been developed to derive the physical conditions in a low-density (nebular) gas given appropriate diagnostic emission line ratios; and line emissivities given appropriate emission line fluxes, the electron temperature (T_e) and density (N_e). The tasks in this package are based on the FIVEL program developed by De Robertis, Dufour & Hunt (1987). Two of these tasks make use of a 3-zone nebular model to derive T_e and N_e simultaneously in separate zones of low-, intermediate-, and high-ionization. They also make use of STSDAS binary tables for efficient storage of the input fluxes and the output diagnostics and ionic abundances. These tasks have been extended somewhat beyond the original FIVEL program to provide diagnostics from a greater set of emission lines, most particularly those in the vacuum ultraviolet that are now available from the *IUE* and *HST* archives.

1. Introduction

The interpretation of emission line radiation from an ionized gas is important in a wide variety of astrophysical contexts, such as H II regions, planetary nebulae, active galactic nuclei, and nova and supernova remnants. The physical basis for line emission from a photoionized nebula has been well understood for decades, and is discussed in many excellent references (see, e.g., Osterbrock 1989). Briefly, most of the common ions that dominate the nebular cooling rate have either p^2, p^3, or p^4 ground-state electron configurations, which have five low-lying levels. To fair approximation, only these five levels are usually relevant to calculating the observed emission line spectrum. Transitions between these levels span the range from the satellite ultraviolet to the infrared, and all are now observable with a combination of ground-based and space-based observing facilities.

A package of STSDAS tasks has been developed to derive the physical conditions in a low-density (nebular) gas given appropriate diagnostic emission line ratios; and line emissivities given appropriate emission line fluxes, the electron temperature (T_e) and density (N_e). The tasks in this package are based on the FIVEL program developed by De Robertis, Dufour & Hunt (1987), who describe the equations to be solved and their method of solution. These tasks extend the functionality of the original FIVEL program, and also provide a very simple model within which to derive the nebular ionic abundances. These tasks are most useful in the fairly common instances where one has somewhat incomplete information about a complicated physical system, such as a narrow-line region in an active galactic nucleus, or somewhat more information about a physically simple system, such as a fairly evolved planetary nebula. In these cases it is useful to calculate nebular densities or temperatures from the traditional diagnostic line ratios, either to provide some reasonable input parameters for a more complicated physical model, or to calculate ionic abundances (or other quantities) within some simplifying assumptions.

2. Nebular Diagnostics from Single Ions

Two new tasks provide a simple IRAF-style parameter interface for calculating nebular diagnostics for a single ion. The **temden** task will calculate N_e given T_e, or T_e given N_e, for one of several ions given the associated diagnostic flux ratio. The result is displayed and also stored in a task parameter. The available diagnostic line ratios, the ionization potential of the associated ion (in eV), and the nebular ionization zone to which they are attributed, are listed below. The line ratios are given as $I(\lambda_1)/I(\lambda_2)$, where λ_1 and λ_2 are in Angstroms; ratios involving sums of line strengths are given as $I(\lambda_1+\lambda_2)/I(\lambda_3+\lambda_4)$.

Electron Density Diagnostics

Ion	Spectrum	Line Ratio	I.P.	Zone
N^0	[N I]	$I(5200)$ / $I(5198)$	0.0	Low
S^+	[S II]	$I(6716)$ / $I(6731)$	10.4	Low
O^+	[O II]	$I(3726)$ / $I(3729)$	13.6	Low
Cl^{+2}	[Cl III]	$I(5517)$ / $I(5537)$	23.8	Med
Ar^{+3}	[Ar IV]	$I(4711)$ / $I(4740)$	40.9	Med
C^{+2}	C III]	$I(1907)$ / $I(1909)$	47.9	Med
Ne^{+3}	[Ne IV]	$I(2423)$ / $I(2425)$	63.5	High

Electron Temperature Diagnostics

Ion	Spectrum	Line Ratio	I.P.	Zone
O^0	[O I]	$I(6300+6363)$ / $I(5577)$	0.0	Low
S^+	[S II]	$I(6716+6731)$ / $I(4068+4076)$	10.4	Low
O^+	[O II]	$I(3726+3729)$ / $I(7320+7330)$	13.6	Low
N^+	[N II]	$I(6548+6583)$ / $I(5755)$	14.5	Low
S^{+2}	[S III]	$I(9069+9532)$ / $I(6312)$	23.4	Med
Ar^{+2}	[Ar III]	$I(7136+7751)$ / $I(5192)$	27.6	Med
O^{+2}	[O III]	$I(4959+5007)$ / $I(4363)$	35.1	Med
Cl^{+3}	[Cl IV]	$I(7530+8045)$ / $I(5323)$	39.9	Med
Ar^{+3}	[Ar IV]	$I(4711+4740)$ / $I(2854+2868)$	40.9	Med
Ne^{+2}	[Ne III]	$I(3869+3969)$ / $I(3342)$	41.1	Med
Ar^{+4}	[Ar V]	$I(6435+7006)$ / $I(4626)$	59.8	High
Ne^{+4}	[Ne V]	$I(3426+3346)$ / $I(2975)$	97.0	High

The atomic parameters are largely taken from the compilation by Mendoza (1984), with more recent published values for some ions, as compiled by Osterbrock (1989). The collision strengths, which are continuous functions of temperature, are tabulated in the literature for only a few selected temperatures. The appropriate collision strengths are computed at run-time for any particular T_e by evaluating polynomial fits to the tabulated values for each transition.

The **ionic** task will calculate the level populations, critical densities, and line emissivities for a specified ion, given N_e and T_e. It will also calculate the ionic abundance relative to H^+ if the wavelength and relative flux (on the scale $I[H\beta] = 100$) of one of the emission lines are also specified; the result is stored in a task parameter. The ionic abundance can be derived from the flux of any of the possible transitions between the fivel levels, not just those lines typically used for diagnostics. In addition, ionic abundances can be derived for Cl^+, Cl^{+3}, K^{+3}, and K^{+4}. The set of ions and diagnostics for these tasks is being extended beyond that of the original FIVEL program in order to make use of those lines in the vacuum ultraviolet (such as C III] 1907, 1909 ÅÅ, and [Si III] 1892, 1884 ÅÅ) that are now available from the *IUE* and *HST* archives. While C III] and [Si III] do not have a 5-level ground-state electron configuration, the same general method applies for deriving the level populations and line emissivities.

3. 3-Zone Nebular Model

In order to calculate ionic abundances in a real nebula, it is necessary to know the electron temperature and density where the various ionic emissions are produced. In some physical contexts it makes sense to view the structure of a nebula as an "onion skin," where the ionization drops off radially from some central source of ionizing radiation, and T_e drops somewhat as N_e increases (on average) with distance. Different ions are found in spherical shells of various radii, depending on the ionization potential of the ion.

Two tasks in this package were designed to model nebulae in just this way, with 3 zones of low-, intermediate-, and high-ionization. The nebular physical parameters are derived within each zone by making simultaneous use of temperature- and density-sensitive line ratios from different ions with similar ionization potentials. The small dependence of the temperature indicators upon N_e, and of the density indicators upon T_e, is removed with an iterative technique and ultimately results in an average T_e and N_e within each zone.

The modelling tasks make use of the TABLES external package in order to provide a simple and powerful data structure and ancillary tools for accessing the observed fluxes and the derived results. The input tables may contain line fluxes for many nebulae and/or many regions within nebulae, one object/region per row. The flux(es) for a given emission line (usually, but not necessarily, given relative to $I[H\beta]=100$) are placed in separate columns. The tasks locate particular emission line fluxes and temperatures/densities via names of specific columns in the input table(s). These columns have suggestive default names, but are entirely user-definable.

Since it is often difficult to provide a complete set of diagnostic line ratios (owing to limited signal-to-noise ratio, spectral resolution, or wavelength coverage of the observed spectra) these tasks were designed to make use of whatever information is available, and to use reasonable defaults (e.g., T_e = 10,000 K, or N_e = 1000 cm^{-3}) when necessary. In particular, any emission line flux that is unavailable (unless it can be computed: e.g., the [Ar III] $I(7751)$ flux from $I(7136)$) is excluded from the calculations. If there are insufficient valid diagnostic line fluxes available for a given ion, the result stored for that diagnostic is "INDEF".

The diagnostic line ratios and ionic abundances are derived from the input line fluxes, corrected for interstellar reddening. The reddening corrected line flux I is derived from the input line flux F by:

$$I_\lambda = F_\lambda * 10^{-c*f_\lambda}$$

where c is the extinction constant (i.e., the logarithmic extinction at Hβ, 4861 Å), and f_λ is derived from the Whitford (1958) and Seaton (1979) extinction functions. The value of c must be given in the input table if a correction for reddening is desired. However, the correction may be disabled if a correction flag (stored in another table column), is set to "yes".

The abundance calculations in **abund** are based upon the diagnostics from the zone most appropriate for each ion. The **abund** task uses only those lines that are typically strongest in real nebulae, some of which are really sums from closely spaced line pairs, such as [O II] 3726+29 ÅÅ. (Refer to the on-line help for more details.) The calculated ionic abundance is a weighted average of that derived from each of the observed emission lines for that ion, where the weights are roughly proportional to the relative line intensities.

4. Future Development

These tasks will be available soon in STSDAS, in a new package called **nebular**. Additional tasks are planned for this package which will fit simple functions to the run of T_e, and of N_e, with ionization potential, based upon the observed line strengths. Other tasks will derive the interstellar extinction coefficient from (among other methods) the hydrogen Balmer decrement, as well as compute ionic abundances (relative to H^+) from recombination lines.

Acknowledgments. Support for this software development was provided through grant no. NAG5–1432 from the NASA Astrophysics Data Program.

References

De Robertis, M. M., Dufour, R. J. & Hunt, R. W. 1987, JRASC, 81, No. 6, 195

Mendoza, C. 1984, in Planetary Nebulae, I.A.U. Symposium No. 103, ed. D. R. Flower, (Dordrecht:Reidel), 143

Osterbrock, D. 1989, "Astrophysics of Gaseous Nebulae and Active Galactic Nuclei" (Mill Valley:University Science Books)

Seaton, M. 1979, MNRAS, 187, 75P

Whitford, A. 1958, AJ, 63, 201

Astronomical Data Analysis Software and Systems III
ASP Conference Series, Vol. 61, 1994
D. R. Crabtree, R. J. Hanisch, and J. Barnes, eds.

Band Selection Procedure for Reduction of High Resolution Spectra

L. Pásztor

MTA TAKI, Budapest Herman Ottó út 15., H-1022, Hungary

F. Csillag

Centre for Surveying Science, Univ. of Toronto, Erindale College 3359 Mississauga Road N, Mississauga, ONT, L5L 1C6, Canada

Abstract. In this paper we present a technique for reduction of spectra based on the methods of multivariate statistical analysis. The procedure was developed for general processing of digital, high resolution spectra. The recursive band selection method can be applied in studies for weighting spectral bands according to their sensitivity to a predefined classification scheme. Additionally, definition of medium and broad band systems is possible, which can efficiently substitute the original spectrum. According to the characteristics of the method resulted from a remote sensing application, it is suggested for use in different astronomical studies, too.

1. Introduction

Recent astronomical techniques make the recording of detailed spectra possible in almost all ranges of the electromagnetic spectrum. Radio telescopes, spectrographs of IUE etc., covers large wavelength ranges with more or less high dispersion. Such observations lead to very large dimensional data sets having tens or even hundreds of narrow spectral bands. A major challenge for such technological advancement is to increase the efficiency of information processing. In a general sense, it primarily means reduction of data set, data dimensionality with constraints on keeping high proportion of information, while in particular cases it requires specific, calibration-type preprocessing.

A general approach to the problem is the introduction of stochastic models and the application of the various methods of multivariate statistical analysis. There are some evidences in recent astronomical works that the automatizable multivariate descriptive data processing methods can be efficiently applied to the analysis and reduction of high resolution spectra (Murtagh & Heck 1987; Tóth et al. 1992). One of the most noteworthy features of multivariate statistical methodology is the flexible ability to describe and analyze large data matrices.

Another recent area of research, where (imaging) spectrometry plays an important role, is remote sensing. Originally the authors developed and applied their technique first in this context (Csillag et al. 1993). A slightly modified and improved version together with the astronomical aspects is discussed in details in Pásztor (1993).

2. The Problem

In many cases the variables characterizing objects of a study may be ranked into two categories. The relation of these variable groups may be analyzed by various multivariate techniques. An important case is when one of the two variable groups is of qualitative nature, either because of inherent or technical reasons. In these cases the observations can be categorized according to the qualitative data. Thus the relationship between the variable groups can be described in a way when the multivariate methods are focussed on the recognition of a predefined classification scheme.

Another important characteristic of multivariate data sets is their inclination to contain redundant information. Generally, the inherent information content does not require usage of all the descriptive variables, since they are greatly correlated. In this case a few number of certain, artificial variables may provide the same informativity.

A very good example of data sets with the above described features is provided by observations with very high resolution spectra which take over the role from previously used variables in the characterization of various astronomical objects. The qualitative ex-variables are generally based on morphological, locational etc., features or on spectral characteristics registered in other spectral ranges.

3. The Method

Our recursive band selection procedure assigns weights to the original spectral bands, whose weights characterize their sensitivity to a predefined classification scheme defined independently of the spectral features. The method eliminates, step by step, the less informative band of the original high resolution spectrum as compared to the recognition of the classification scheme. Every step is combined as follows (Figure 1).

3.1. Principal Component Analysis on the Spectral Bands

Principal Component Analysis is quite widely applied in multiwavelength image processing for reducing dimensionality. It has been successfully used to identify a linear combination of spectral measurements, leading to a smaller number of uncorrelated axes in the measurement space. This approach determines the importance of each original band in each resultant feature and finds optimum features in the sense that the number of features can be much less than the number of original bands without significant loss of information in terms of proportion of variance. Our band selection procedure keeps all the resulted principal components for the next stage of the step.

3.2. Discriminant Analysis on the Principal Components

Once principal components are resulted discriminant analysis can be performed to test the performance of PCA-derived features on the predefined classification scheme. Formally, it is quite similar to the equations used in PCA, because it also leads to an eigenvalue/eigenvector problem. Here the eigenvectors provide the discriminant factors, while the eigenvalues characterize their merits. The number of discriminant factors is determined by the number of groups (g) and variables (m). If $g - 1 < m$, the space of discriminant factors is of $d = g - 1$ dimension, otherwise of $d = m$.

3.3. Determination of Discrimination Efficiency, and Removal of the Least Informative ("Noisiest") Band

The discriminant functions are obtained with another set of weights applied on features, which are themselves linear combinations of the original bands. Therefore, the final role of the lth original band in the jth discriminant function can be summarized in a matrix element $w_{jl} = \sum_{k=1}^{m} c_{kj} b_{lk}$. Here c_{kj} is the weight of the kth principal component in the jth discriminant factor and b_{lk} is the weight of the lth original variable in the jth principal component. For selection of bands a final weight assignment is proposed. The resulted weights for the original bands in a discriminant function are normalized and then multiplied by the total between-groups variability attributable to the given function. These values are then summarized for each original band. The resulted values (called "goodness") well characterize the overall role of an original band in the recognition of classification. The original band having the lowest "goodness" value will be considered as noisiest and removed. When this band is dropped from further processing, it is one iterative step in our procedure.

3.4. Stopping the Procedure

Effectiveness of a set of discriminant functions can be measured by the reclassification accuracy resulted from test data. According to the general behavior of multivariate data sets and based on our experiences, the efficiency of reclassification shows saturation versus the number of bands. That is beyond a threshold, the increase in the number of bands does not provide improvement in the goodness of reclassification. The recursion needs to be stopped at this threshold.

The output of the method consists of a subset of the original bands providing almost the same classification accuracy as the whole spectrum; the "goodness" values of the bands in the subset and the $w_{jl} = \sum_{k=1}^{m} c_{kj} b_{lk}$ matrix elements for the evaluation of subsequent measurements.

4. Conclusion

The presented band selection procedure provides an optimal subset of original spectral bands in a sense that it is composed of the least number of spectral bands, which, however, contains all the relevant information as for recognition of the predefined classification scheme. If the resulted number of bands is even more than what is acceptable, the usage of a smaller subset is proposed, which consists of the bands with the best "goodness" values (although such truncation of data sets may already result in much more significant loss of information).

In the real remote sensing problem the resulted bands clustered along the spectrum, which fact made the definition of broader bands meaningful. Usage of these synthetic bands resulted in much better recognition of the classification scheme, than that of other widespread broad band systems. Similar distribution

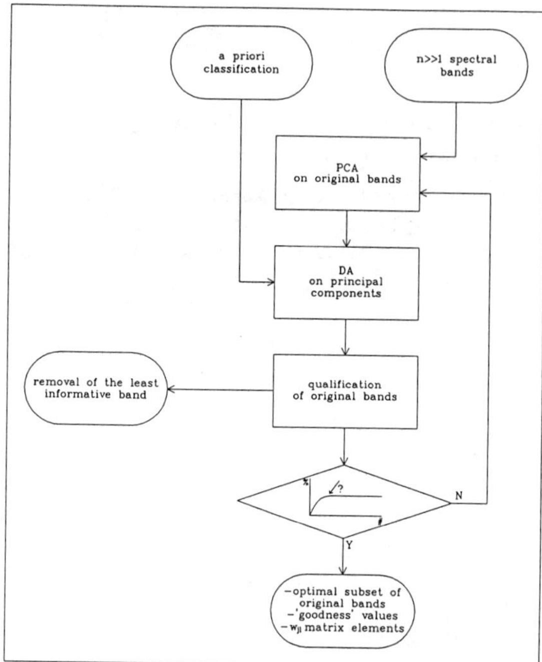

Figure 1. Scheme of the procedure.

of the "optimal" narrow bands along a spectrum may provide similar improvement in the definition of classification-sensitive broad band systems.

It is important to note here that, although recent reports on optimum band selection using PCA suggest a certain combination on "purely" statistical grounds, this procedure leads to an application-sensitive measure of the potential of the data set in terms of classification accuracy.

Acknowledgments. The financial support of the Hungarian Academy of Sciences, the National Science Foundation (Grant No. INT-8721949) are gratefully acknowledged. L. Pásztor is thankful to SAO and the Hungarian State Research Fund (Grant No. OTKA-F 4239) for the travel grants.

References

Csillag F., Pásztor L., & Biehl L.L. 1993, Remote Sens. Environ. 43, 231

Murtagh F. & Heck A., 1987, "Multivariate Data Analysis". Reidel, Dordrecht

Pásztor, L. 1993, "Statistical Methods in Investigation of Formation and Early Evolution of Stellar Systems", Ph.D. thesis Eötvös University, Budapest

Tóth L.V., Ábrahám P. & Balázs L.G. 1992, "Multivariate Statistics on HI and IRAS Images", in Astronomical Data Analysis Software and Systems I, A.S.P. Conf. Ser., Vol. 25, eds. D.M. Worrall, C. Biemesderfer & J. Barnes, 251 responding class number. In Figure 3, the final results of object shown, on the training image *(a)* and on a different one *(b)*. Each correspond to a spectrum "head" (i.e., a *class 2* pattern).

sions and Future Developments

work, we noticed that an ANN used to detect specific patterns allow sults similar to those obtainable by visual inspection, and in general those obtainable by means of the algorithmic techniques known to ANN algorithms, anyhow, are definitely superior in terms of elapsed east an order of magnitude. Furthermore, the scientist's control on re is possible: the possibility of interacting with the algorithm to o correct detections allow a great degree of flexibility, and combines ges of a visual inspection with computing speed.

developments of this work will go in the direction of the refinement od to allow the detection of objective prism spectra in more complex such as badly overlapped spectra (e.g., globular clusters) or steeply kground, possibly using unsupervised learning methods. Scientific n of the spectra extracted from the original image, matching the 2 b-classes accuracy reachable by the rule-based classification method y Hantzios et al. (1993), will be attempted.

wledgments. The software used for the purposes of this work is ttgart Neural Network Simulator), which is (Copyright) 1990-93 Uni-Stuttgart, and is distributed as 'Free Software' by IPVR, University rt. The project for stellar classification of stars in clusters and asso- the Magellanic Clouds using objective prism data is carried out in ion with A.Dapergolas, E.Kontizas and M.Kontizas, of the National ry and University of Athens, and has been partially supported by A and by the Italian Ministry of Foreign Affairs.

es

A., Micol A., Pasian F., Pucillo M., Santin P., Sedmak G., & Smareglia , 1990, in: Proc. 2nd ESO/ST-ECF Data Analysis Workshop, D.Baade, J.Grosbøl eds., ESO Conferences and Workshop Proceedings, 34, 31

& Hart P., 1973, "Pattern Classification and Scene Analysis", Wiley Sons

P., Kontizas E., Pasian F., Dapergolas A., Kontizas M., & Smareglia R., 93, in: Astronomy from Wide-Field Imaging, Proc. IAU Symposium 1, in press

nn J.,& Krësel U., 1992, Technical Report, Daimler-Benz AG, Institut r Informatik, Forschungszentrum Ulm

Mache N., Hübner R., Mamier G., Vogt M., Herrmann K.U., Schmalzl ., Sommer T., Hatzigeorgiou A., Döring S., Posselt D., and Reczko .,& Riedmiller M., 1993, SNNS User Manual, Report 3/93, University f Stuttgart

Astronomical Data Analysis Software and Systems III
ASP Conference Series, Vol. 61, 1994
D. R. Crabtree, R. J. Hanisch, and J. Barnes, eds.

Detection of Spectra in Objective Prism Images Using Neural Networks

R. Smareglia, F. Pasian

Osservatorio Astronomico, Via G.B.Tiepolo 11, 34131 - Trieste, Italy

Abstract. Some methods for the automatic detection of spectra in objective prism images have already been developed, but none of them is completely satisfactory, and an interactive session including a visual check is always necessary. In this paper, a neural network approach is proposed to solve this problem. A multi–layer feed forward neural network has been developed to distinguish single, multiple or overlapped spectra, and some encouraging results are shown.

1. Rationale

In the framework of a scientific collaboration between the Astronomical Observatory of Trieste and the National Observatory of Athens, Greece, a system for the processing of objective prism data has been developed (Balestra et al. 1990). One of the weak points of the package is the automatic detection of spectra on the original image. All of the methods available in the package for this purpose (binarization + thinning, Sobel filtering, *Logical Feature* filtering) are fairly costly in terms of computing resources, and require a quite extensive interactive correction step to guarantee acceptable results.

To overcome these problems, the use of an Artificial Neural Network (ANN) approach has been considered. If a supervised learning approach is chosen, the ANN just needs a certain number of selected patterns (sub-images, as seen by a human operator) to be assigned to a *training set*, for an initial learning phase. In subsequent phases of the learning process, human intervention is required to correct or refine the choices and selections made by the ANN until a satisfactory outcome of the learning phase is obtained, and the real *"production"* phase can start. Using this approach, the flexibility of the object detection process is therefore guaranteed, while keeping a high degree of efficiency.

2. Methods and Results

The ANN approach used for the purposes of this work is the so-called Dynamic Learning Vector Quantization (DLVQ) method. The idea behind this algorithm is to find a natural grouping in a set of data (see Duda & Hart 1973 and Schürmann & Krësel 1992 for reference). DLVQ actually extends the original LVQ learning rules adding hidden ANN layers dynamically during the learning phase only when they are needed. The DLVQ algorithm has been used as implemented in the SNNS package software (Zell et al. 1993). In Figure 1, the

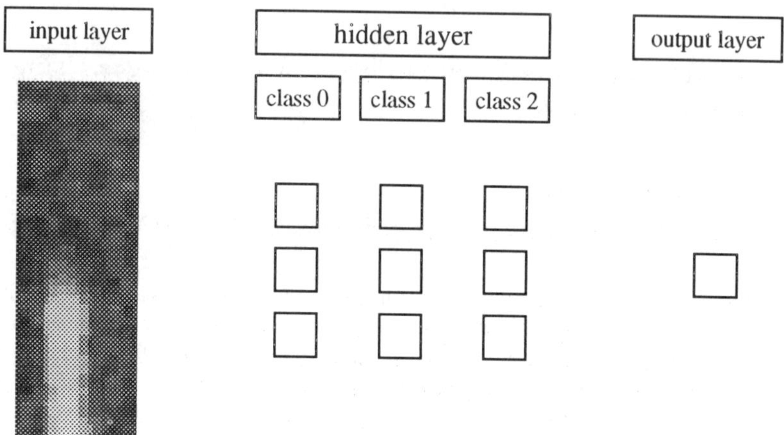

Figure 1. Topology of a network trained with DLQV.

topology of a network trained with DLVQ is shown. The output layer must consist of only one unit. At the start of the learning phase it does not matter whether the output layer and input layer are connected, since the links, including the ones to possibly existing hidden units will be created during the learning phase. The output pattern contains the information on which class the input pattern belongs to.

In our case, 3 classes of patterns have been chosen to train the network: background or non-relevant objects (scratches or dust grains on the digitized plate, etc.) have been labeled as **class 0**, "tails" of spectra as **class 1**, "heads" of spectra (single and overlapped) as **class 2**. Since spectra are aligned along the Y–coordinate axis patterns have been chosen to be sub-images of size 10x30 pixels, since they contain the whole width of a spectrum, and enough length to contain enough features relevant to spectrum detection.

The initial training set included 9 patterns chosen as the most significant ones; during the learning phase, detections (i.e., classification of patterns as *class 2* objects) were interactively corrected or confirmed. Due to the statistical method used for the learning phase by ANNs, it is important that the various classes of patterns used are balanced in number.

At the end of the learning phase, 40 patterns have been used in order to reach a 90% efficiency in the detection of spectra "heads". A single "head" may be detected more than once, but this effect is due to the size of the mask used to extract the input pattern (10x30 pixels sub-images, with a 60% overlap in the X direction and a 50% overlap in Y). This is not a problem: when the correct "head" position is determined, if two position values match within a certain error box, one is rejected. The efficiency in the detection of spectra can be furtherly improved if the learning phase is expanded during the "production" phase of the algorithm. In this case, the human operator is allowed to refine the network scheme by interactively confirming or correcting the classification made by the algorithm.

In Figure 2, the patterns chosen at the end of the learning phase of the DLVQ algorithm are shown. Each pattern is a 10x30 pixels sub-image labeled

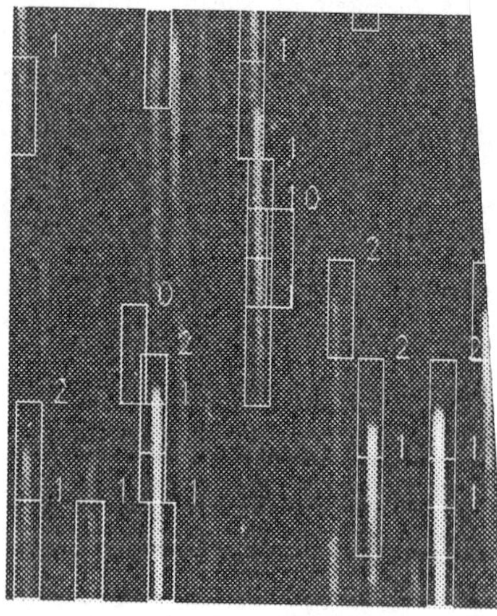

Figure 2. Topology patterns at the end
pattern is a 10x30 pixels sub-image label
class number.

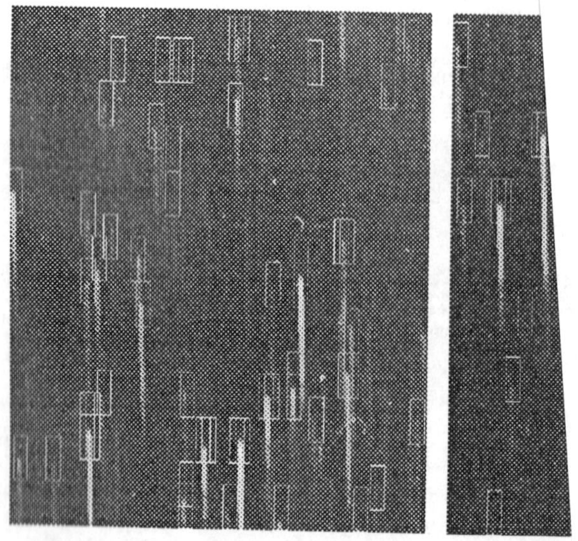

Figure 3. Final results of object detection on
on a different one *(b)*. Each of the 10x30 pixels
to a spectrum "head" (*class 2* pattern).

with the co
detection ar
of the boxes

3. Concl

During this
obtaining r
superior to
the authors
time by at
the proced
confirm or
the advant

Future
of the met
situations,
varying b
classificat
spectral s
proposed

Ackno
SNNS (St
versity of
of Stuttga
ciations o
collabora
Observat
CNR/GN

Referen

Balestra
R.
P.

Duda R.
&

Hantzios
1
1

Schürm
f

Zell A.,

Astronomical Data Analysis Software and Systems III
ASP Conference Series, Vol. 61, 1994
D. R. Crabtree, R. J. Hanisch, and J. Barnes, eds.

The IRAF/STSDAS Synthetic Photometry Package

H. Bushouse and B. Simon
Space Telescope Science Institute, 3700 San Martin Drive, Baltimore, MD 21218

Abstract. The *STSDAS* synthetic photometry package provides capabilities for generating and manipulating instrumental passbands and spectral data and for performing synthetic photometric calculations using these data. The primary purpose of the package is to provide cross-calibration of multiple HST instrument modes and to provide observers with tools for simulating HST observations of real astrophysical targets. The completely data-driven structure of the package allows it to be applied to other telescopes and instruments simply by supplying the appropriate instrumental throughput data.

1. Introduction

The Space Telescope Science Data Analysis System (*STSDAS*) synthetic photometry (`synphot`) package is an IRAF-based suite of tasks designed to simulate photometric and spectrophotometric data resulting from Hubble Space Telescope (HST) observations of astrophysical targets. Tasks in the `synphot` package can be used to make plots of HST instrument sensitivity curves and stellar library spectra, to predict count rates for observations in any available mode of the HST science instruments, and to examine photometric transformation relationships amongst the various HST observing modes, as well as conventional photometric systems such as *UBVRI* and *uvby*. Synphot is available to assist guest observers in preparing HST observing proposals and has proven useful in planning and optimizing HST observing programs due to its cross-instrument simulation capabilities.

2. How It Works

Throughput curves for all HST instrument components—such as mirrors, apertures, filters, gratings, and detectors—are stored in *STSDAS*-format tables. In addition to HST instruments, throughput data for several conventional photometric systems, such as Johnson-Cousins *UBVRI*, Stromgren *uvby*, and Walraven *VBLUW*, are also included. The individual throughput data tables are referenced via master instrument graph and component lookup tables. The instrument graph table essentially provides a map of all of the HST instruments and describes all allowed combinations of the instrument components. The `synphot` passband calculator utilizes user-supplied keywords that describe a desired instrument observing mode in order to trace a path through the instrument graph table, load the necessary throughput tables, and then calculate the composite passband from the product of the throughputs of the individual optical components that make up that mode. The fact that all pertinent instrument data resides in external tables means that any telescope and instrument can be simulated simply by providing the appropriate throughput data and instrument graph table.

3. The Synphot Tasks

There are three main categories of tasks in the `synphot` package:

- Tasks that create, manipulate, and plot passbands and spectra.
- Tasks that fit model passbands and spectra to photometric and spectrophotometric data.
- General utility tasks that convert IRAF/STSDAS image data to and from table format and tasks that check or display information from the instrument graph and component lookup tables.

The following sections describe the basic functions of the tasks that create and manipulate passbands and spectra and perform synthetic photometry on these data.

3.1. Generating and Manipulating Passbands

The tasks `calcband` and `plband` calculate and plot passband (throughput) data. They can either read existing throughput curves from data tables or synthesize passbands from gaussian, box, and polynomial functional forms where the user specifies the central wavelength, width, or polynomial coefficients. For example, the following IRAF command will produce a plot of total throughput for the Faint Object Spectrograph (FOS), using its blue-side detector, its 4.3 arcsecond entrance aperture, and its G270H grating:

```
sy> plband fos,blue,4.3,g270h
```

The string "fos,blue,4.3,g270h" specifies the desired HST instrument observing mode and simply consists of keywords associated with the particular instrument components that make up the observing mode. Throughput data for the HST Optical Telescope Assembly (OTA) is included by default in calculations of any HST instrument mode.

3.2. Generating and Manipulating Spectra

The tasks `calcspec` and `plspec` calculate and plot photometric and spectrophotometric data. They can read or synthesize spectral data and convolve it with selected passbands or HST observing modes to simulate observed data. Spectral data can be read from tables of your own observational data or from available spectral libraries. Synthetic spectra can be generated through the use of built-in blackbody, powerlaw, and HI emission functions, where the user specifies the desired temperature, slope, or column density. Spectra can also be modified to add or remove extinction effects and can be renormalized to a chosen absolute flux level within a given passband. The spectrum calculator supports the use of data in a variety of physical units, including F_λ, F_ν, counts, AB_ν and ST_λ magnitudes, Jy, and mJy. Unit conversion is performed automatically when necessary to combine data of different forms or to produce results in a chosen form that differs from that of the original data.

The following IRAF command demonstrates how one might synthesize an 8000 K blackbody spectrum that is normalized to an AB magnitude of 15.5 in the Johnson V passband and includes the effects of interstellar extinction at a level of 0.2 $E(B-V)$:

```
sy> calcspec "bb 8000 rn v abmag 15.5 ebmv 0.2" bb.tab flam
```

The substring "bb 8000" specifies the desired blackbody function, "rn v abmag 15.5" specifies that the spectrum is to be renormalized to a level of 15.5 AB mags in the V passband, and "ebmv 0.2" specifies the desired level of extinction. The resulting spectrum will be stored in the table "bb.tab" and the data will be in units of F_λ ("flam").

3.3. Estimating HST Countrates

The tasks `calcphot` and `countrate` are especially useful for simulating HST observations and estimating detected count rates. `Calcphot` provides the ability to compute photometric quantities for spectra within selected passbands. `Countrate` computes count rates for spectra within selected passbands and will also compute count rate spectra for spectrographic instruments. The `countrate` input parameter list has been designed to mimic the specifications on an HST proposal exposure log sheet (see Table 1). For non-spectroscopic instrument modes, such as any of the HST Wide Field/Planetary Camera (WFPC) or Faint Object Camera (FOC) modes, `countrate` computes the total detected counts within the specified filter passband that one would expect to receive from a target having a given spectrum. For example, Table 1 shows the `countrate` parameter settings one would use to estimate the total counts from a source having a powerlaw spectrum in a 100 second exposure using the WF/PC detector 3 and the F439W filter.

Table 1. `Countrate` Task Parameter List.

Parameter	Value	Description
output =	"wfpl.tab"	Output table name
instrument =	"wfpc"	Science instrument
detector =	"3"	Detector used
spec_elem =	"f439w"	Spectral elements used
aperture =	" "	Aperture / field of view
userspec =	" "	User supplied input spectrum
synspec =	"pl 5000 1.5"	Synthetic spectrum
synmag =	"17.5 V"	Magnitude of synthetic spectrum
reddening =	0.15	Interstellar reddening E(B-V)
exptime =	100.	Exposure time in seconds

The form of the powerlaw spectrum is specified in the "synspec" parameter, and we've asked to have the spectrum normalized to a V magnitude of 17.5 via

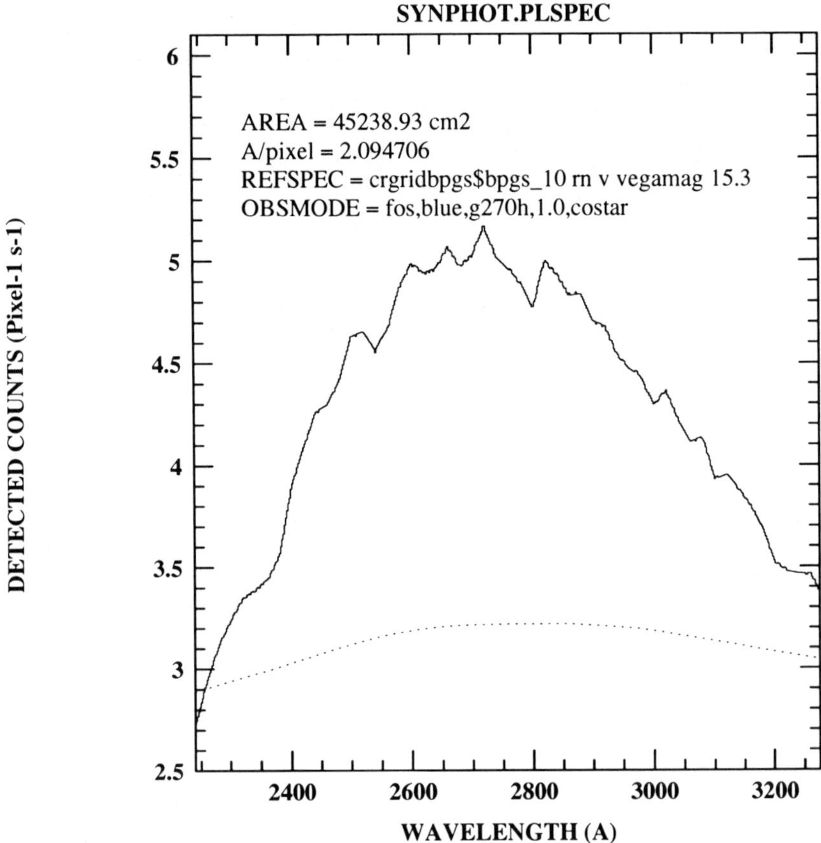

Figure 1. FOS countrate spectrum produced using `plspec`.

the "synmag" parameter. We've also asked to have the spectrum modified for effects of reddening at a level of 0.15 $E(B-V)$. For this example, the integrated number of counts that we expect is 10202.

Finally, we show an example using `plspec` to simulate an FOS observation of a star that we'll renormalize to a V magnitude of 15.3. The stellar spectrum is read from the table "crgridbpgs$bpgs_10.tab". The FOS instrument mode that we'll use is the blue-side detector, the 1.0 arcsecond aperture, the G270H grating and we'll include the effects of the COSTAR package:

```
sy> plspec fos,blue,g270h,1.0,costar \
>>> "crgridbpgs$bpgs_10.tab rn v vegamag 15.3" counts
```

The resulting countrate spectrum is shown in Figure 1.

Astronomical Data Analysis Software and Systems III
ASP Conference Series, Vol. 61, 1994
D. R. Crabtree, R. J. Hanisch, and J. Barnes, eds.

Faint Object Spectrograph Polarimetry Data Analysis

H. Bushouse

Space Telescope Science Institute, 3700 San Martin Drive, Baltimore, MD 21218

Abstract. Software tools are now available in the *IRAF/STSDAS* environment for the analysis of HST Faint Object Spectrograph (FOS) spectropolarimetry datasets. Tools exist that allow an observer to examine, compare, and combine multiple polarimetric datasets, to rebin polarimetry spectra in order to increase signal-to-noise ratio, and to produce stacked plots of flux and polarization spectra.

1. Introduction

Because of the special nature of spectropolarimetric data, normal spectral reduction and analysis tools usually do not provide an efficient or effective means of handling polarimetry spectra. A suite of software tools has been created within the *IRAF/STSDAS* environment that is specifically tailored to handle HST Faint Object Spectrograph (FOS) polarimetry data. These tools are contained within the `fos/spec_polar` package of the version 1.3 release of *STSDAS*. FOS polarimetry datasets normally consist of orthogonal pairs of source spectra obtained at either 4, 8, or 16 different polarimeter waveplate position angle settings. Stokes IQUV parameter spectra and polarization spectra—including linear and circular polarization and polarization position angle—are computed both from each of the two orthogonal sets ("pass directions") of spectra, as well as from the combined data for both pass directions.

The *STSDAS* polarimetry tasks allow an observer to:

- Intercompare individual sets of flux-calibrated spectra in order to identify spurious data;
- Flag spurious data within individual sets of flux-calibrated spectra and recompute Stokes and polarization spectra using only the unflagged data;
- Combine individual sets of either flux-calibrated or Stokes IQUV spectra for a given target and recompute polarization spectra from the combined data;
- Rebin flux and polarization spectra to increase the signal-to-noise ratio;
- Compute normalized Stokes Q/I, U/I, and V/I spectra from IQUV; and
- Produce stacked plots of flux, Stokes parameter, and polarization spectra.

2. The Polarimetry Tasks

The `spec_polar` package currently contains the eight tasks listed in Table 1. The tasks are designed to perform various operations on three different stages of FOS polarimetry datasets. These three stages are 1) flux-calibrated spectra obtained at different position angles of the instrument's waveplate, 2) Stokes IQUV spectra that are computed from the flux-calibrated spectra, and 3) polarimetry spectra (linear, circular, and polarization position angle) that are computed from the Stokes IQUV spectra. All of the tasks are designed to make use of and correctly propagate (where appropriate) statistical error and data quality flag spectra that accompany the flux and polarimetry spectra.

Table 1. *STSDAS* FOS SPEC_POLAR Package Menu.

Task name	Description
`calpolar`	Perform polarimetry processing of calibrated data
`comparesets`	Compare spectra from different datasets
`pcombine`	Combine multiple flux-calibrated datasets
`polave`	Average Stokes parameters for multiple datasets
`polbin`	Rebin polarization spectra
`polcalc`	Calculate polarization spectra from Stokes IQUV spectra
`polnorm`	Normalize Stokes Q, U, and V by Stokes I
`polplot`	Stacked plot of flux, Stokes, and polarization spectra

2.1. Inspecting Datasets

The task `comparesets` produces a stacked plot of flux-calibrated spectra from either different waveplate positions within an individual dataset, or individual waveplate positions from different datasets. This allows an observer to identify potentially spurious data from individual datasets or waveplate positions. The task also computes statistics over a user-specified wavelength region within each spectrum so that quantitative comparisons can be made. The `calpolar` task will then allow an observer to recompute Stokes IQUV and polarimetry spectra for a given dataset while ignoring spectra from waveplate positions that have been determined to be bad.

2.2. Combining Datasets

Because FOS polarimetry observations are usually limited in duration to one orbital period, it is often necessary to acquire several individual observations of faint sources in order to obtain sufficient signal. Polarization spectra for individual datasets cannot be averaged reliably, therefore either the flux or Stokes parameter spectra from the individual datasets must be averaged, and then the polarization spectra can be recomputed from the combined data. The task `pcombine` can be used to combine multiple sets of flux-calibrated spectra. Following that, the task `polcalc` can be used to recompute Stokes parameter and polarization spectra from the combined flux spectra. Alternatively, `polave` can be used to directly average the Stokes parameter spectra for multiple datasets

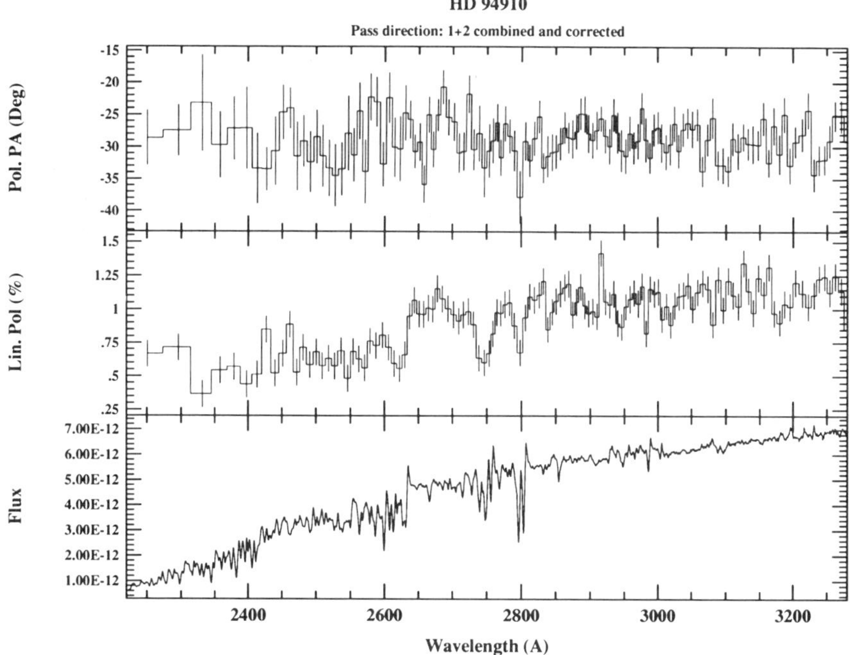

Figure 1. Binned polarization spectra of HD 94910.

and recompute the polarization spectra from the averaged Stokes parameter data. Each routine propagates the statistical errors associated with the input spectra and produces a properly averaged output error spectrum.

2.3. Rebinning Polarization Spectra

The task `polbin` will rebin Stokes IQUV and polarization (linear, circular, position angle) spectra in order to increase the signal-to-noise ratio of the spectra. The rebinning may be performed in one of two modes: 1) equal-sized bins containing a constant number of input pixels per output bin; or 2) variable-sized output bins containing a constant level of linear polarization error per bin. The `polbin` task works by computing the mean Stokes IQUV values in each output bin, and then recomputes the linear, circular, and polarization position angle values for each bin from the mean IQUV. In the case where the user has selected the equal-errors per bin mode, a running mean of Stokes IQUV is calculated as it moves through the input pixel values, and when the linear polarization error calculated from the mean Q and U reaches the desired level, the mean data are placed in an output bin and the process is restarted for the next bin.

2.4. Plotting Polarization Spectra

Flux, Stokes, and polarization spectra can be displayed in various combinations using the `polplot` task. The total flux (Stokes I) spectrum is included by default in all plots, while the user has the option to plot either the Stokes Q and U or the linear polarization and polarization position angle spectra along with the flux spectrum. There is also an option to include Stokes V or circular polarization along with the other Stokes or polarization spectra, respectively. The user can also choose the desired plot style: continuous line, histogram, or point mode. Error bars may be included with any of the available plot styles. Figure 1 shows an example of a polarimetry dataset that has been rebinned to equal linear polarization error per bin using `polbin` and plotted using `polplot` in histogram mode.

Astronomical Data Analysis Software and Systems III
ASP Conference Series, Vol. 61, 1994
D. R. Crabtree, R. J. Hanisch, and J. Barnes, eds.

Deconvolving Spatially Undersampled Infrared Spectroscopic Images

A. Bridger, G. S. Wright, W. R. F. Dent, P. N. Daly

Joint Astronomy Centre, University Park, 660 N. Aohoku Pl., Hilo, HI, 96720

Abstract. We summarize the results of attempts to deconvolve spatially undersampled long slit infra-red spectra of an extended source. The recovery of such data can be important in the infra-red as sources are not always known to be extended *a priori* and as substantial increases in efficiency could be realised.

1. Introduction

The technique of deconvolving dual-beam maps of spatially extended sources is well known in the radio and sub-millimetre wavebands. In the infra-red similar methods are used, *chopping* the secondary mirror and/or *nodding* the telescope, because there is a strong spatially and temporally variable background. However, the chopping or nodding has usually simply been to an off-source patch of sky; an inefficient method.

The infra-red array detectors used in both cameras and spectrometers still have relatively few pixels, so that infra-red data are frequently undersampled and/or the field of view is small. For small sources a common technique with array spectrometers is to nod the source along the slit, on the array, achieving a background measurement nearby in space and time, while also always observing the source. However, this is not done for large sources, to avoid confusing the two beams. If deconvolving dual beam observations could be applied in the infra-red then the efficiency of observing extended sources could be doubled.

The technique used in the sub-millimeter (Emerson, Klein and Haslam 1979, hereafter EKH) works in one dimension on maps, so applying it to the spatial dimension of spectral images presents no real difficulty.

2. The Experimental Data

To test the value of the deconvolution technique in the infra-red, observations of a standard star and of a planetary nebula, NGC7662, were obtained with the Cooled Grating Spectrometer 4 (CGS4, Mountain et al. 1990) on the United Kingdom Infra Red Telescope (UKIRT). The planetary was observed by nodding off to blank sky for background subtraction, and then also by nodding the source along the slit by two different separations, ensuring overlapping images. The acquisition and reduction system is described elsewhere (Daly et al. 1993). The spectra are spatially undersampled: the plate scale in this experiment was

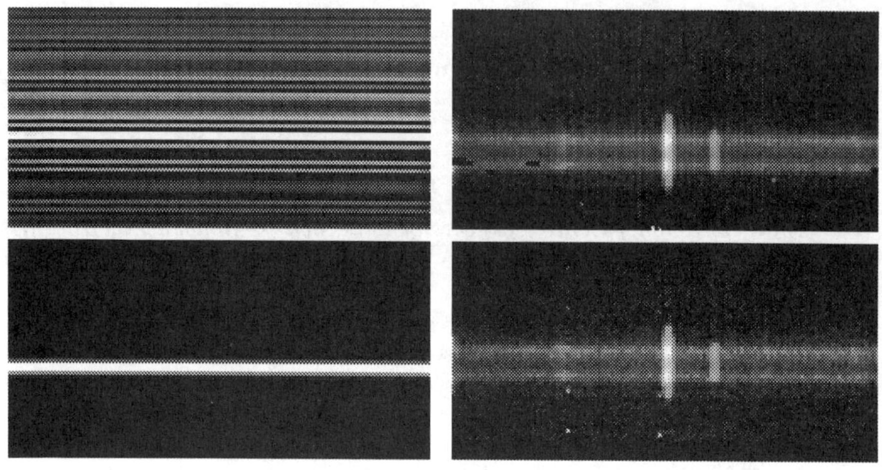

Figure 1. Left: Restored frame A - see text. Right: Traditional (top) and restored (bottom) images of the planetary.

1."51/pixel, and the seeing was ≤1."5. The observations are summarized in the table below.

Table 1. The observational data.

Frame Reference	Object	Total exp. time (secs)	On source exp. time (secs)	Approx. object size (arcsecs)	Beam separation (arcsecs)
A	SAO52194	64	64	1.5	7.7
B	NGC7662	160	160	30	15.4
C	NGC7662	160	160	30	7.7
D	NGC7662	320	160	30	100

3. The Restored Spectra

After basic reduction by the CGS4 software system (Daly and Beard 1992) each frame consisted of a noisy background with a positive and negative view of each image, with the exception of frame D. For frames B and C these images overlapped. Bad pixels were replaced with interpolated values, as the deconvolution algorithm is sensitive to such features. Frames A, B and C were then restored using the RESTORE program of the JCMTDR package (Lightfoot and Harrison 1992), which is based on the algorithm described in EKH, also used in the NOD2 package (Emerson 1986).

The left half of Figure 1 shows frame A, the star, restored in two different

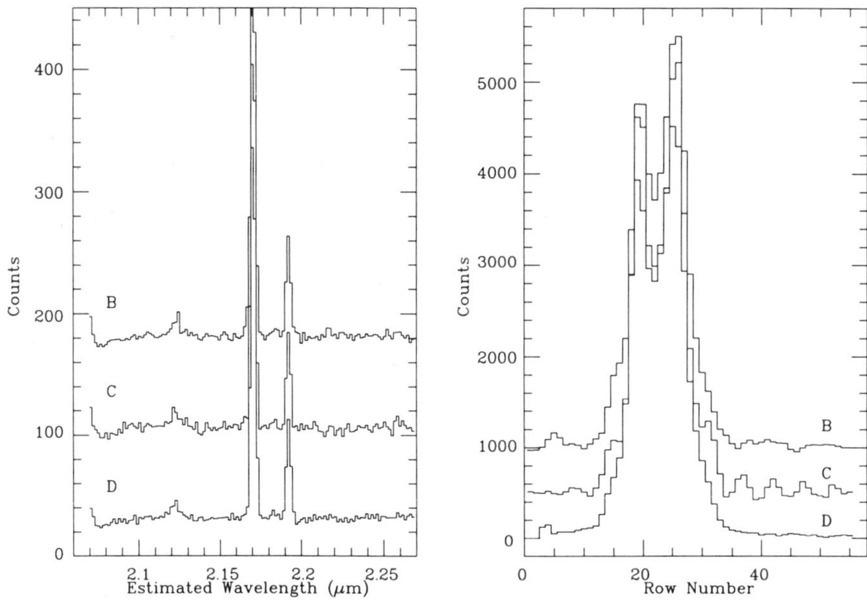

Figure 2. Details from the NGC7662 images: Left, example extracted spectra, right, integrated spatial profiles.

ways, because a simple restore was found to be insufficient (see discussion). The right half of Figure 1 shows a comparison between the restoration of frame B and the reduced frame D. Figure 2 shows

extracted spectra from each of the three frames of the planetary (from the top bright edge of the image) on the left and spatial profiles from the images on the right, each graph shifted for clarity.

4. Discussion

The technique worked well on the spectra of the planetary, but failed to restore the unextended spectrum of the star. Stellar spectra extracted "traditionally" and using RESTORE were identical in shape, but with different flux levels. The reason is clear in the top left image of Figure 1; "ripples" or echoes either side of the restored spectrum. Although such echoes may be seen in the restoration of bright sub-millimetre sources when the edges of the map do not go close enough to zero, it transpires that the effect seen here is caused by a different problem; the stellar image is spatially undersampled. To check this the beams were artificially balanced and then the spectral image was smoothed in the spatial direction by convolution with a gaussian of half-width 1 pixel, simulating fully sampling the image. The contribution of the slight beam imbalance was found to be small. The restored image of this modified stellar spectrum, the lower left image in Figure 1, shows that the restoration worked reasonably well this time.

Spectra taken from various parts of the extended planetary images differ only at a level well below the noise. The only exception being a spectrum from the extreme faint edge of the planetary extracted from frame C. In frame C small echoes of the planetary spectrum do appear, however, the echoes are faint, about 3 times the residual background noise. In frame B echoes are barely detectable. The cuts across the spatial direction of the planetary spectral images shown on the right in Figure 2 demonstrate that the deconvolved images, especially frame B, match frame D well. Spectra of the bright top edge extracted from frames B, C and D and plotted on the left in Figure 2 are almost indistinguishable. The residual background level (sky minus sky) is very similar in all three images, about 1% of the peak flux, but the noise in this background in the two deconvolved images is about 50% higher than in the traditional image. This may be due to having spent less time integrating on the sky background in these frames. The signal to noise measured in the continuum of the spectra at the bright top edge is about 6–7 and is identical in all three cases.

5. Conclusions and Further Work

In conclusion it seems safe to say that the differences between the spectrum of the planetary taken in the traditional infrared manner and those taken using an overlapped dual-beam technique are small, probably insignificant. It is clear that the larger beam separation produces a significantly better result.

However, the technique fails when applied to point sources that are spatially undersampled. Modification of the algorithm to handle undersampled cases would be desirable. In addition, when used with an array it is clear that a clean detector, with very few bad pixels, is required, as the necessary "invention" of data for the bad pixels should be minimised.

This technique should be investigated further, particularly in the thermal infrared, where the field is likely to be smaller and the problems of background subtraction are greater, and also for larger objects, where a technique of scanning the array across the object to ensure that the background is seen on either side of it might be well employed. In addition this technique might become essential when the UKIRT, as with many other telescopes, begins to employ an active secondary which has (for mechanical reasons) a limited chop throw.

References

Daly, P. N., & Beard, S. M. 1992, "CGS4DR Users' Guide", Starlink User Note 27, Starlink Project, RAL

Daly, P. N., Bridger, A., & Krisciunas, K. 1993, "Automated Observing at UKIRT", this volume

Emerson, D. T. 1986, "NOD2 Commands and Programs", Institut de Radio Astronomie Millimetrique

Emerson, D. T., Klein, U., & Haslam, C. G. T. 1979, A&A, 76, 92

Lightfoot, J. F. & Harrison, P. A. 1992, "JCMTDR - Figaro Applications for Reducing JCMT data", Starlink User Note 132, Starlink Project, RAL

Mountain, C. M., Robertson, D. J., Lee, T. J., & Wade, R. 1990, in Instrumentation in Astronomy VII, D. L. Crawford, SPIE, 1235, 25

Part 4. Data Analysis
Section C. Time Series Analysis

Astronomical Data Analysis Software and Systems III
ASP Conference Series, Vol. 61, 1994
D. R. Crabtree, R. J. Hanisch, and J. Barnes, eds.

Simulation of Aperiodic and Periodic Variabilities in X-RAY Sources

L. Burderi, M. Guainazzi, N. R. Robba

Istituto di Fisica, Universitá di Palermo, Via Archirafi 36, I- 90133 Palermo, Italy

Abstract. We have developed a simulation program of the time behavior of a flickering X ray source. Taking into account some physical models of the variability, this program generates a time series of sampled data suitable to analysis through the Fast Fourier Transform techniques. Using slightly different models, corresponding to different physical conditions in the source, we have been able to generate power spectra of different shapes. A semiquantitative interpretation of the results is performed and physical considerations for each of the models considered are outlined and a comparison with observed spectra is performed.

1. Introduction

The Power Spectrum Density (PSD) has been extensively used to study the periodic (Harmonic Lines, HL) and the aperiodic (Red Noise, RN) features of the temporal variability of the High Mass X Ray Binaries (HMXRB). Some authors have proposed a very simple Shot Noise model (SN) in order to explain the RN feature (Terrel and Olsen 1970). In this scenario the plasma instabilities that can occur during the accretion process (Rayleigh-Taylor, Kelvin-Helmoltz, Alfvén zone instabilities above the accretion column) could clump the matter in blobs (Elsner and Lamb 1977,1984; Arons and Lea 1976); the shot emission associated with the accretion of these blobs is the cause of the aperiodic variability. The timescale associated with the instabilities mentioned above cover a very wide range (roughly from milliseconds to tens of seconds). Due to the casual occurrence of the blobs, as a consequence of the instability phenomena, these emissions are supposed to occur randomly (stochastically) in time.

Having this picture in mind, we try to study the PSDs of HMXRB in order to obtain information and constraints on some parameters describing the physics of these systems. In particular we considered the PSDs of HMXRB as built up by two distinct main components:

- the broad band RN feature as the signature of the inhomogeneities (blobs) in the plasma flow during the accretion process;
- the harmonic lines as the signature of the lighthouse effect associated with the neutron star spin.

From the shape of the PSD at the base of the harmonic lines some information on the degree of coupling between the shot noise component and the periodic modulation can be extracted; this means that is in principle possible to infer on the degree of funneling of the clumped matter onto the polar caps. In fact, in case of coupling between the periodic modulation and the shot noise, we expect a broadening of the base of PSD's harmonic lines due to the superposition of a RN-shape function (deriving from the PSD shot noise broad band continuum) on each of the lines of the harmonic system (Burderi et al. 1992).

An exact analytical computation of this effect on PSD of finite length is extremely heavy; in order to investigate this kind of phenomenon, taking into account all the effects involved we have performed simulations of this kind of signal, using a Monte Carlo code, and compared the resulting PSD with that obtained from the real data.

2. Results

We suppose that the simulated signal is due to the incoherent sum of a shot and a non shot (uniform) component. Each component is, in principle, partly modulated by the "lighthouse effect" due to the spinning neutron star. β_{un} is the fraction of modulated over the total uniform intensity, while β_{sh} is the fraction of modulated over total shot intensity. The total signal can be written as:

$$I_{un} + I_{sh}(t) = [\beta_{un} I_{un} + \beta_{sh} I_{sh}(t)]M(t) + (1 - \beta_{un})I_{un} + (1 - \beta_{sh})I_{sh}(t). \quad (1)$$

$I_{sh}(t)$ is built up by the superposition of exponential shots of total area S and mean occurence rate λ. M(t) is the periodic modulation function:

$M(t) = C + \sum_{k=1}^{N_H} c_k \sin(k\omega_0 + \phi_k)$ and C is a normalization constant.

It is possible to prove that the PSD of such a signal observed for a time T (normalized adopting the Lehay factor $F = 2/[\bar{I}_{tot}T]$ can be approximately written as the sum of several terms, of which one takes into account the broadening at the base of the harmonic lines in the case of coupling between the SN and the HL:

$$\Sigma(\nu) = F\beta_{sh}^2 \lambda S^2 \sum_k \frac{c_k^2}{4} N(\nu - \nu_k) \quad (2)$$

where $N(\nu)$ depends on the particular shape and timescale associated with the emission of each single blob.

The HMXRB Cen X-3 has been observed by ME detector on board the EXOSAT satellite on 14 May 1985. Its PSD is showed in Figure 1. In order to reproduce this PSD, we have performed simulations of this source using a different "blend" of SN signal and uniform signal, each signal partially or totally coupled with the periodic modulation function. To reproduce the observed PSD shape we have used in our simulations shots of exponential shape characterized by different decay times τ with a power law distribution (spectrum index $\alpha = 0.3$; $0.1(s) \leq \tau \leq 40(s)$). This kind of distribution well reproduce the observed shape of the RN feature, in particular the observed slope of -1.3 in the PSD of real data. With these assumptions the function $N(\nu)$ becomes (Letho 1989; Burderi et al. 1993): $N(\nu) \simeq [1 + (2\pi\tau_{max}\nu)^{1+\alpha}]^{-1}$.

In Figure 2 are shown the PSDs corresponding to three different physical scenarios simulated, and the differences (residuals) between each of these PSDs and the PSD of the real data.

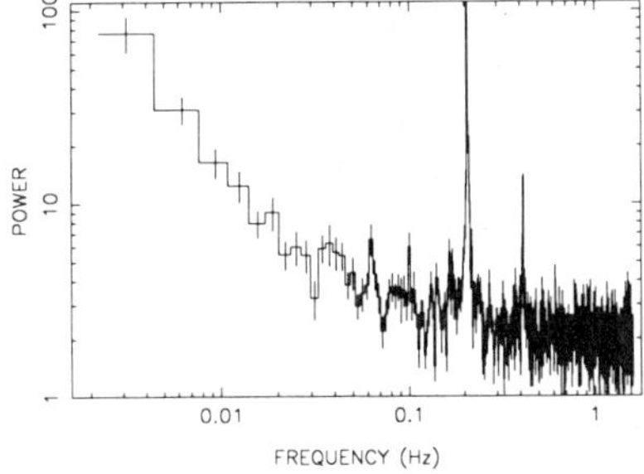

Figure 1. PSD of CenX-3 EXOSAT-ME 134/85 observation, with $T_{start} = 1594.20673$ and $T_{end} = 1594.26094$ (MJD).

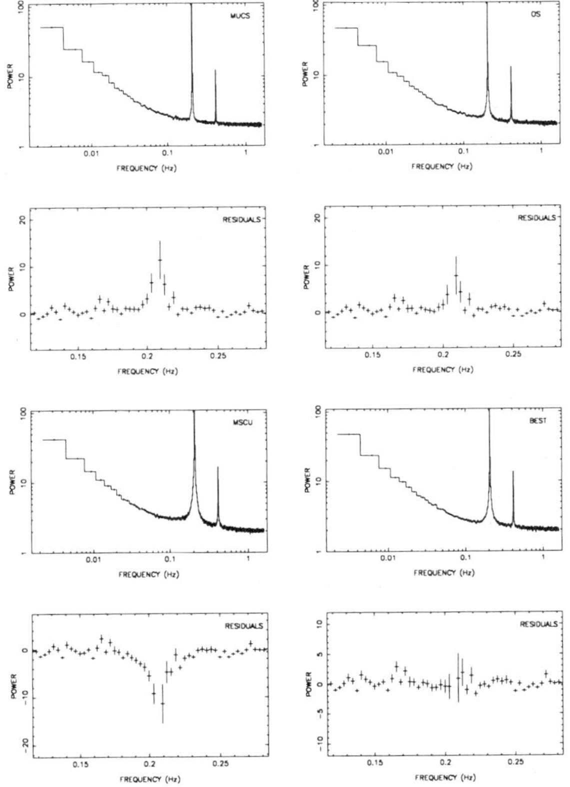

Figure 2. Simulated PSD and residuals in four different physical scenarios: *upper left panels* - totally modulated uniform and continuum shot (MUCS); *upper right panels* - only shot emission (OS); *bottom left panels* - totally modulated shot and continuum uniform (MSCU); *bottom right panels* - Best simulation with 40% of continuum emission consituted of shots and 60% of uniform.

- MUCS: signal obtained with $\beta_{un} = 1$; $\beta_{sh} = 0$ in formula (2) (with the minimum value of the modulation function exactly zero). The underlying physical scenario is that the component funneled by the magnetic field lines (i.e., the modulated one) is composed of not clumped matter, while the shot emission originates from matter not funneled onto the polar caps (i.e., is not modulated);

- OS: signal with $I_{un} = 0$; $\beta_{sh} = 1$. In this scenario the accreting matter is totally clumped in blobs and funneled onto the polar caps, giving the observable periodic modulation.

- MSCU: signal with $\beta_{un} = 0$; $\beta_{sh} = 1$ (with the minimum value of the modulation function exactly zero). In this scenario the clumped matter is funneled, while a not modulated uniform component is still present.

If we assume the reasonable hypothesis (which doesn't imply a qualitative change in our conclusions) that the two components, shot and uniform, are not mixed in building up the modulated part of the signal we can draw the following considerations, from an analysis of the residuals.

In the models MUCS and OS the base of the first harmonic line in the simulated PSD is narrower than the real one. This broader feature of the real data seems to indicate that some degree of coupling between the shots and the periodic modulation is present and that not all the emission is due to shots. A reasonable possibility is then that a fraction of the emission comes from not clumped matter not accreting onto the polar caps, in fact this is the only way in which the base of the harmonic line can be broadened. In model MSCU, in which the shot component is totally modulated - although the minimum value is not zero - while a not funneled uniform component is still present, the broadening of the base of the line is too enhanced. The exact fraction of the uniform component can be determined looking for the data that show no residuals around the first (and the second) harmonic line. The best PSD is showed in the bottom panel of Figure 2, together with the corresponding residuals.

Therefore we can conclude that: i) the aperiodic variability observed in this source seems associated with the matter funneled onto the polar caps; ii) some uniform (non shot) emission also seems to be present which comes from zones not directly illuminated by the "lighthouse effect".

References

Arons, J., & Lea, S. M. 1976, ApJ, 207, 914

Arons, J., & Lea, S. M. 1976, ApJ, 210, 792

Burderi, L., Robba, N. R., & Cusumano, G. 1992, Proceedings of the COSPAR Symposium, Washington D.C., in press

Burderi, L., et al., 1993, in preparation

Elsner, R.F., & Lamb, F. K. 1977, ApJ, 215, 897

Elsner, R.F., & Lamb, F. K. 1984, ApJ, 278, 326

Letho, H. J. 1989, ESASP, 296, 499

Terrel, J., & Olsen, K. H. 1970, ApJ, 161, 399

Astronomical Data Analysis Software and Systems III
ASP Conference Series, Vol. 61, 1994
D. R. Crabtree, R. J. Hanisch, and J. Barnes, eds.

Some IDL Routines for Time-Series Analysis

Raymond E. Rusk

Defence Research Establishment Pacific, Victoria, BC, V0S 1B0

Abstract. An IDL function to provide a time averaged power spectral density estimate, with confidence limits, for one or two data sequences, and the cross spectrum between two data sequences is described. The user can select the data windowing function and the degree of overlap processing. Routines to efficiently plot the original time-series and to display the moving power spectral density estimate are discussed.

1. Introduction

Over the last decade the Electromagnetics Section at DREP has developed an in-house interactive data analysis and graphics software package tailored towards the specific needs of our research group. This software is well suited to most of our data reduction requirements but it is not as flexible nor as extensible, by the end-user, as many of the commercial data analysis and visualization packages which have since become available.

Two commercial packages, IDL and MATLAB, have become widely used at DREP because they provide the basic tools needed for numerical simulation, algorithm prototyping and data analysis in one computing environment. Flexible file I/O, convenient graphics and the ability to work with large data sets have attracted the author to IDL.

While IDL provides many ready-to-use tools for image processing it has less support than MATLAB for time-series analysis. In the author's work, it is often necessary to estimate instrumental and environmental noise from very long data sequences. An IDL function, patterned after a similar MATLAB routine, to provide a time averaged power spectral density estimate is reported here as are some other basic tools for time-series analysis.

2. Power Spectrum Estimation

IDL lacks a routine for spectral estimation.[1] This is inconvenient since one often wishes to characterize the frequency content of a measured signal. Fortunately, it is not difficult to write new signal processing routines in IDL.

There are many ways to estimate the spectral content of a given finite data set (cf., Kay, 1988 or Marple, 1987). Significant differences can exist between

[1]This may change soon. RSI have announced that Numerical Recipes published by Cambridge University Press will soon be incorporated into IDL.

these estimates because of different assumptions made about the signal and noise content of the data. For larger data sets such as those considered here, classical estimation methods based on the periodogram are adequate.

An unsmoothed periodogram does not produce a consistent estimate of the spectrum because the variance of the estimate does not approach zero as the number of data points used to calculate the periodogram increases. Instead, the resolution of the spectral estimate increases. A consistent estimate of the spectrum can be obtained by trading off some of this increased resolution for decreased variance. This can be done by sectioning the data set, Fourier transforming each section to obtain a spectrum estimate, and then averaging these estimates. Alternately, the periodogram of the entire data set can be smoothed by convolution with an appropriate spectrum window.

Welch (1967) discusses a method of power spectrum estimation which combines averaging periodograms and windowing. The data record is sectioned and a window is applied to the data segments before computation of the periodograms. This method is widely used because it can be efficiently implemented using the fast Fourier transform algorithm. This and other methods which section the data in the time domain have the useful property that they facilitate testing for and measuring nonstationarity.

An implementation of the Welch method is presented in §11.6.1 of Oppenheim and Schafer (1975). Rather than code this implementation directly into IDL, or modify FORTRAN or C code from Rabiner et al. (1979) or Press et al. (1988), for instance, the code for SPECTRUM, a MATLAB function which computes the power spectrum estimate of one or two data sequences and the cross spectrum between two data sequences, was translated into IDL. This was possible because DREP has a site licence for MATLAB and MATLAB routines are written in a high level "language" similar to IDL. A working IDL routine for spectrum estimation was created within a couple of hours of discovering the need for one.[2] The IDL code was verified by comparing its results to output from MATLAB and from in-house software. MATLAB has received a lot of attention and support from authorities in the field of signal processing and numerical analysis so there can be reasonable confidence in its algorithms.

The MATLAB implementation of Welch's method contained in SPECTRUM uses linear trend removal in each data segment. The final spectral estimate has the same total power as the "detrended" data sequences. Removing the mean from each segment of data and then adding it back as a DC component in the spectrum of that data segment preserves Parseval's relation between the total power in the original data and the final spectrum. It also allows a step-wise correction for linear or polynomial trends in the data. The translated MATLAB algorithm was modified to handle trends in this way.[3]

---

[2]The author has since learned that John Hopkins University/Applied Physics Laboratory maintains a library of IDL routines which include functions for power spectral estimation (rspec, cspec and xspec). This library is available by anonymous ftp from fermi.jhuapl.edu.

[3]The advantages and disadvantages of removing large means (DC offsets), linear and higher-order polynomial trends are discussed in more detail by Marple (1987). Large sample means lead to corrupted low-frequency spectral estimates because of spectral leakage from the zero frequency bin.

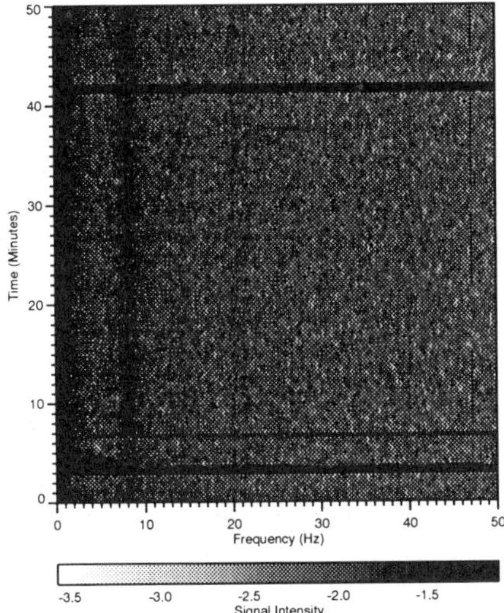

Figure 1. An image composed of time ordered modified periodograms.

The variance in the spectral estimate can be determined by accumulating the square of the power density as well as the power density for each modified periodogram included in the spectral estimate. With 0% overlapping the variance is inversely proportional to the number of segments included in the final averaged spectral estimate. A further reduction in variance is possible when data segments are overlapped but the reduction is no longer proportional to the number of averaged segments because the segments are not independent. Welch (1967) shows that a nearly optimal reduction in variance occurs with 50% overlapping of data segments and that the variance with 50% overlapping is about 11/18 of the variance without overlapping.

3. An Example

The PLOT command in IDL uses the NSUM keyword to display long time-series more rapidly. However, replacing every NSUM points with an average value masks "glitches" in the data. An IDL routine which divides the time-series into a sequence of arrays similar in number to the resolution of the display and plots only the MIN and MAX for each array speeds up display of long time-series while highlighting bad data. When examined in this way, spurious values are apparent at the 3.5, 7 and 41 minute points in the time series used in this example.

Figure 1 was created using the TVSCL command from an array of (unnormalized) modified periodograms calculated from the sample time-series. Sections of bad data, intermittent or transient signals, and nonstationarity are sometimes more apparent in such an image than in the original time series. An intermittent signal is seen near 47 Hz. The broad feature at 8 Hz is the first Schumann res-

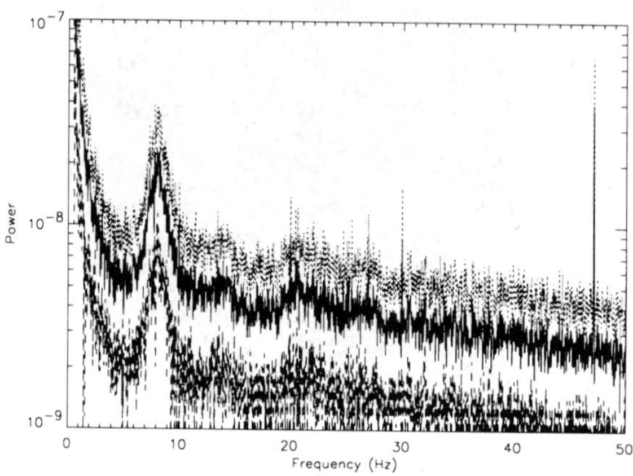

Figure 2. A sample power spectrum (solid line) estimated using time averaging over short, modified periodograms. The dotted and dashed lines are four sigma error bounds.

onance in the ELF radio band. Figure 2 shows a spectral estimate obtained by averaging periodograms of data collected between the 20 and 40 minute points in the data record.

The IDL routines described in this paper will be contributed to the IDL Astronomy User's Library maintained at the ftp site idlastro.gsfc.nasa.gov. The code can also be obtained directly from the author.

References

Kay, S. M., 1988, "Modern Spectral Estimation", Prentice-Hall

Marple, S. L., 1987, "Digital Spectral Analysis with Applications", Prentice-Hall

Oppenheim, A. V. & Schafer, R. W., 1975, "Digital Signal Processing", Prentice-Hall

Press, W. H., Flannery, B. P., Teukolsky, S. A. & Vetterling, S. A., 1988, "Numerical Recipes in C", Cambridge University Press

Rabiner, L. R., Schafer, R. W., & Dlugos, D., 1979, "Periodogram Method for Power Spectrum Estimation", in Programs for Digital Signal Processing, IEEE Press

Welch, P. D., 1967, IEEE Trans. Audio & Electroacoust., AU-15, 70

Part 4. Data Analysis
Section D. High Energy Data Analysis

Astronomical Data Analysis Software and Systems III
ASP Conference Series, Vol. 61, 1994
D. R. Crabtree, R. J. Hanisch, and J. Barnes, eds.

QPTOOLS: Tools for Creating and Manipulating IRAF/QPOE Data

M. A. Conroy

Smithsonian Astrophysical Observatory, Cambridge, MA 02138

Abstract. The QPOE data format was developed by IRAF to support photon event-list data. We are developing several tasks that allow users to generate QPOE files of their own description and to modify existing data files. For instance, users can create their own QPOE file by replacing columns in an existing file, or by importing *ASCII* lists of events.

1. Introduction

The Quick Position Ordered Event (QPOE) data format was developed by the Image Reduction and Analysis Facility (IRAF) to support photon event-list data (Tody 1986). The filtering mechanisms that IRAF provides with this data structure make it a very powerful tool for data analysis. One of the key features of this data format is the automatic conversion from event-list format to *image* format. However, it is often desirable to view and manipulate the QPOE file in tabular form. Unlike the *table* data format developed by ST ScI, the QPOE format does not have an accompanying package of manipulation tasks. We have started work on developing a few of these basic utilities, modeled after existing IRAF *image* or TABLE tasks.

2. What is QPOE?

The IRAF/QPOE file structure was designed specifically to provide storage and access of event-lists (photon-lists) generated by photon-counting detectors, such as those common in High Energy Astrophysics. The representation of X-ray data in QPOE data structures is the heart of the PROS package (Worrall et al. 1992). The interface provides several powerful capabilities:

- Dynamic conversion of event-list data to *image* arrays
- Rich set of built-in dynamic filtering capabilities, including run-time calculation of EXPOSURE time
- Support for World Coordinate System representations
- Support for dynamic spatial masking
- Support for application specific extensions

Since this data structure is so fundamental to ROSAT and *Einstein* analysis, we have been developing tools to extend the IRAF built-in capabilities. Since EUV is another project that uses the QPOE data format, they are developing QPOE tools as well.

3. Sample QPOE tasks

3.1. QPCREATE

Qpcreate is a script that executes a sequence of IRAF tasks to produce a QPOE file from an input *ASCII* list.

```
<ASCII list>
<HEADER list>   ---->  FITS/BINTABLE    ---->  QPOE file
<GTI list>
```

The only requirement for the *ASCII* list is that it include 'x' and 'y' columns. Optionally, the user may generate a set of HEADER parameters by modifying the supplied template. The other optional file is an input list of "good-time" intervals (GTI) that is used to define the correct exposure intervals.

3.2. QPCALC

Qpcalc is a general task (modeled on the TABLES **tcalc** task) that performs algebraic operations on QPOE event-list attributes. This task evaluates an arbitrary expression that includes event-attribute names, constants, and operators, and either creates the specified event-list attribute in the QPOE file or overwrites an existing attribute. The usual arithmetic and logical operations are supported, as well as the most common arithmetic functions.

3.3. QPLINTRAN

Qplintran is a task (modeled on IRAF **imlintran** task) that performs a general linear transformation on the X/Y axes of a a QPOE event-list. The task implements the following transformation equation, to transform (X1,Y1) into (X2,Y2):

```
X2 = Xout + (X1-Xin)*Xmag*cos(Angle) - (Y1-Yin)*Ymag*sin(Angle)
Y2 = Yout + (Y1-Yin)*Ymag*cos(Angle) + (X1-Xin)*Xmag*sin(Angle)

    where  (Xin, Yin)       - input reference pixel
           (Xout,Yout)      - output reference pixel
           (Xmag,Ymag)      - relative plate scale in each axis
             Angle          - the rotation angle
```

Two common examples of this transformation are 'shift' and 'rotate'. These two utilities are useful with ROSAT observations, to correctly align separate observing 'segments'. The **qpappend** task will then 'concatenate' the aligned segments. (See Figure 1)

```
xr> qpshift shockb.qp -13.0 10.0
xr> qpshift shockc.qp -10.0 5.0
xr> qpappend @shock.lst "" "cyga.qp" ""
```

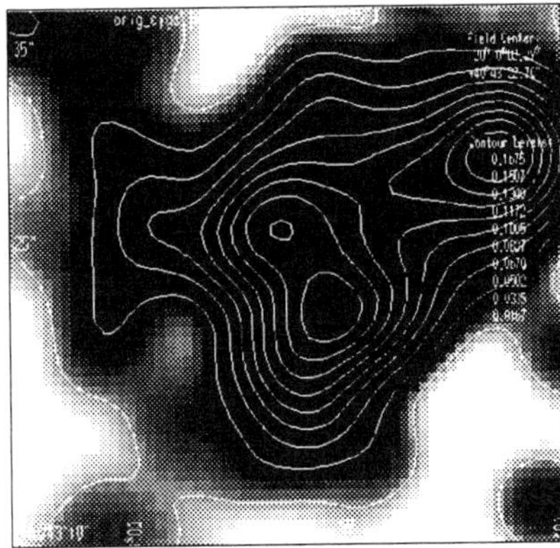

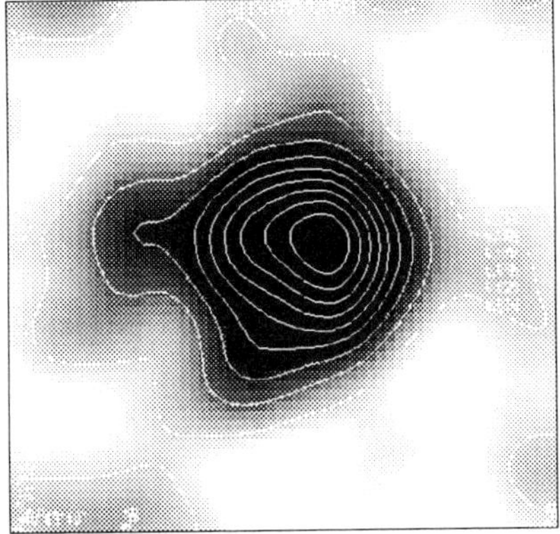

Figure 1. On the left is the original ROSAT observation of Cygnus A, combining 5 observing segments. On the right is the same observation, in which the segments have been separated, aligned to correct for aspect errors and re-combined.

3.4. QPGAPMAP

Qpgapmap is a special purpose task to generate corrected detector pixel coordinates from raw instrument coordinates. This process for the ROSAT/HRI instrument is referred to as 'de-gapping'. The detector coordinate system is the system used for all the HRI calibration maps. Once the detector coordinates are available in the QPOE file, it is possible to use them to 'look-up' the appropriate pixels of the calibration map.

3.5. QPPHASE

Qpphase is a task that calculates a period phase from the event-list times and produces a new event-attribute, 'phase'. IRAF allows user filtering on any event attribute, thus allowing 'on' and 'off' images of a pulsar to be separated (see Figure 2). The 'phase' attribute is calculated as:

```
phase = time / period + 0.5 * (-1.0*pdot/period**2) * time**2
phi   = mod(phase,1)

     where   time    is the photon arrival time
                     (optimally barycenter-corrected)
             period  user input period in seconds
             pdot    is the user specified rate-of-change
                     in the period.

     xr> qpphase crab.qp "" 33.4081E-3 4.22E-13 crabphase.qp
     xr> xdisp crabphase.qp[phase=(0.5:0.6)]     'on'  PHASE
     xr> xdisp crabphase.qp[phase=(0.7:0.8)]     'off' PHASE
```

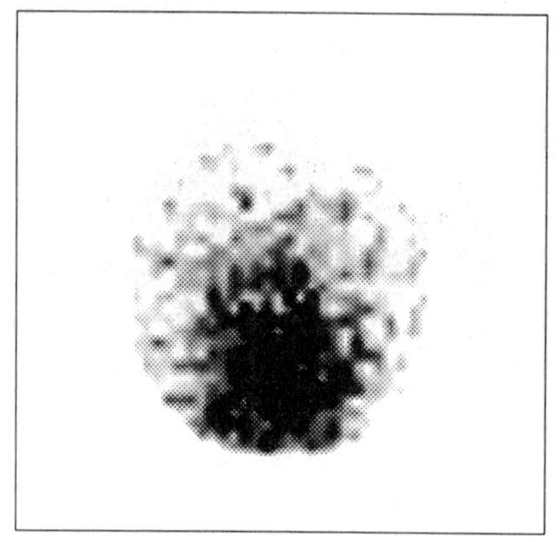

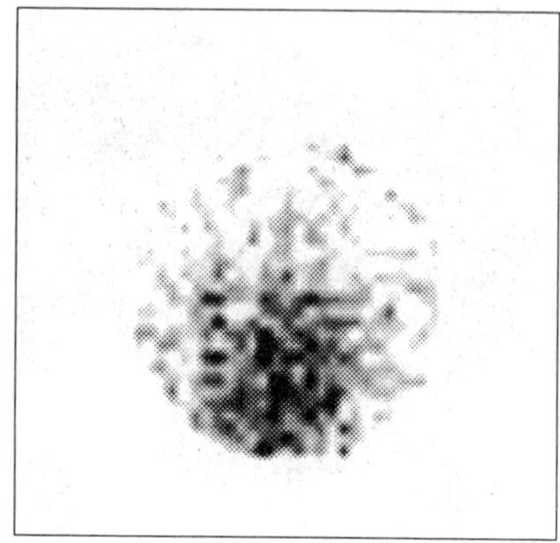

Figure 2. On the left is the ROSAT observation of the CRAB pulsar during the 'on' phase and on the right the same observation during the 'off' phase.

4. Future Directions

The existing tools for manipulating QPOE files are a loose collection of tasks scattered throughout the IRAF layered packages such as PROS and EUV. As the existing support tools for QPOE mature and expand it might be reasonable to consider the following:

- Consolidate QPTOOLS into a single IRAF layered package
- Add a GUI interface to these tools
- Integrate the ST *table* file structure into QPOE
- Expand the user input options for QPCREATE
- Add more utilities to access QPLINTRAN and QPCALC for common tasks, e.g., QPFLIP

Acknowledgments. The PROS project is partially supported by NASA contracts NAS5-30934 (RSDC) and NAS5-30751 (*Einstein*). PROS is available at no charge by contacting rsdc@cfa.harvard.edu, (6699::RSDC, rsdc@cfa).

References

Tody, D. 1986, in Instrumentation in Astronomy VI, SPIE, 627, part 2

Worrall, D.M., Conroy, M., DePonte, J.,Harnden,Jr.,F.R.,Mandel,E., Murray, S.S, Trinchieri, G. , VanHilst, M., & Wilkes, B.J., 1992, in Data Analysis in Astronomy IV, eds. V. Di Gesu et al., (Plenum Press), 145

ent Filters in Use

ers will find that a combination of QPOE-based and TABLE-based time ill be needed to achieve optimal results. The following is an example OSAT HRI data:

un RFITS2PROS to make an 'unscreened' event QPOE file.

un MKHKSCR and build a 'standard' attribute filter; edit the file and move the entry for HIBACK.

un HKFILTER to generate a filter of accepted times.

un TABFILTER on the event-rate file (_evr.tab)

- Display and edit high background (BACKLEV) column
- Generate a temporal filter for BACKLEV

un QPCOPY to filter the QPOE event-list through the time filters gen- rated by HKFILTER and TABFILTER to create a new QPOE file with he specified screens.

knowledgments. This work is partially supported by NASA contracts ROSAT Science Data Center (NAS5-30934) and *Einstein* (NAS8-30751).

ences

y, M.A., Garcia, M.R., Mandel, E.G., Roll, J., & Worrall, D.M. 1992, in Astronomical Data Analysis Software and Systems I, A.S.P. Conf. Ser., Vol. 25, eds. D.M. Worrall, C. Biemesderfer & J. Barnes, 17

ran, M.F., White, R. & Conroy, M. 1993, in Astronomical Data Analysis Software and Systems II, A.S.P. Conf. Ser., Vol. 52, eds. R.J. Hanisch, R.J.V. Brissenden & J. Barnes, 549

, D. 1986, in Instrumentation in Astronomy VI, SPIE, 627, part 2

all, D.M., Conroy, M., DePonte, J. ,Harnden, Jr.,F.R., Mandel,E., Murray, S.S, Trinchieri, G., VanHilst, M. & Wilkes, B.J. 1992, in Data Analysis in Astronomy IV, eds. V. Di Gesu et al., (Plenum Press), 145

Astronomical Data Analysis Software and Systems III
ASP Conference Series, Vol. 61, 1994
D. R. Crabtree, R. J. Hanisch, and J. Barnes, eds.

Temporal Data Screening in PROS

Janet DePonte and Maureen A. Conroy

Smithsonian Astrophysical Observatory, Cambridge MA 02138

Abstract. We present several tasks available in the PROS/IRAF environment to assist the user in screening and filtering event data. The technique employed translates all user filters into a common time-based filter. The time intervals are then used to filter events in the QPOE file. The rationalized data format (RDF) for ROSAT make both the unscreened photons and time-tagged ancillary data available to the user. Thus, with the temporal screening tools presented below the user has the ability to completely specify the criteria for acceptance or rejection of photons.

1. Introduction

Filtering and selection of photons is often a user's first step in the scientific analysis of astronomical data. Satellite based X-ray observations are acquired under a set of constantly changing conditions. For instance, the position of the target on the detector is not constant, and observing conditions such as background rate, aspect quality, and bright object contamination change as a function of satellite position and telemetry quality. These conditions are recorded and saved in the data files provided to the observer, as well as the event data that are stored with spatial, temporal and energy attributes.

ROSAT data distributed in the rationalized data format (RDF) provide the community with the complete (unscreened) events in the *Basic* FITS file (Corcoran et. al 1993). In addition to the primary event list there are several additional extensions. The status of instrument parameters are stored in the *temporal status interval (TSI)* FITS bintable extension. The limits of acceptable data quality for the screened events are stored in the *standard quality limits (STDQLM)* extension. The limits of acceptable data quality for the unscreened events are stored in the *all quality limits (ALLQLM)* extension. The start and stop times during which all observing conditions meet the *standard quality limits* are stored in the *standard good time intervals (STDGTI)* extension.

The RDF Basic FITS file is self-defining and supported in PROS by the IRAF/QPOE data structure (Conroy et. al 1992). Ancillary data such as orbit, aspect, and housekeeping data are provided in separate FITS bintable files. The ancillary data is supported in IRAF by the TABLES data structure.

Several tasks have been developed in PROS/IRAF to implement the extended filtering capabilities that these data make available. The design is to build temporal filters based on screens applied to time-based data. The temporal filters are then used to screen events in the QPOE file.

2. Event Filters Derived from the *BASIC* FITS File

The *BASIC* FITS file provide user access to all parameters used by the standard processing system to screen photons. The tasks are modeled after standard processing procedures. The standard (screened) event-list of standard processing or the complete (unscreened) event-list can be defined using these tasks. More importantly, the user may edit one or more of the attribute filters to create a time filter that defines an event-list with user-specified conditions.

MKHKSCR - This task creates an *ASCII* file with housekeeping limits in IRAF filter format. Many instrument parameters are recorded in the *TSI* extension, either as bit-encoded flags or binned data levels. The task utilizes a lookup table that matches quality parameters (either *ALLQLM* or *STDQLM*) with *TSI*s and provides the mapping between the two QPOE data structures. The output attribute filter is a list of temporal status identifiers with their quality limit range.

Example Data Quality Filter for 'Unscreened' HRI Events

```
hiback=1:100, hvlevel=12:16, viewgeom=1:5, aspstat=1:5,
asperr=1:14, hqual=0:0, saadind=1:3, saada=1:4, saadb=1:4,
temp1=1:8, temp2=1:8, temp3=1:8, logicals=%44X
```

Example Data Quality Filter for 'Screened' HRI Events

```
hiback=1:8, hvlevel=12:16, viewgeom=1:3, aspstat=3:5,
asperr=2:4, hqual=0:0, saadind=1:3, saada=1:4, saadb=1:4,
temp1=1:8, temp2=1:8, temp3=1:8, logicals=%4X
```

HKFILTER - This task creates an *ASCII* file with time limits in IRAF filter format. The task reads the *TSI* attribute filter generated by MKHKSCR and applies it to the *TSI*s. The result is a list of time intervals during which all the limits expressed in the input screen are true.

Example Time Filter for 'Unscreened' Events

```
time=(2585242.00:2585666.00,2585668.00:2588182.00, ...
      2684948.00:2686271.00,2707077.00:2708791.00)
```

Example Time Filter for 'Screened' Events

```
time=(2585242.00:2585666.00,2585668.00:2586464.00, ...
      2684948.00:2686271.00,2707077.00:2708790.00)
```

3. Event Filters Derived from the *Ancillary* FITS Files

Users will often find that the quality of their data is dependent on *Ancillary* data not used by the standard processing system, and therefore not recorded in

the *Basic* FITS file. Thus, we present tools that use generate event-list filters.

Any TABLE with a time attribute column is a ca QPOE time filter only space-craft units are compatib QPOE file. Some data in the *Ancillary* TABLES are ra levels in the *TSI* (i.e., high background in _evr.tab). are not available in the *TSI* (i.e., master veto rate in _

TABFILTER - This task is a data quality selection tas interactively display time dependent data stored in can be displayed and graphically edited to select time filter that can be written to the QPOE as a r definitions are time filters that are stored in the Q of this macro task is the EUV task DQSELECT.

TABFILTER Session

```
xp> tabfilter rp800020_evr.tab

 Loading table: rp800020_evr.tab

 Valid Commands:
        display <monitor>                   hide
        list <table>                        mode
        edit <monitor>                      delet
        limload <file_name>                 limsa
        showlim <monitor>                   hidel
        gtgen                               gtnew
        gtload <file_name>                  gtsave
        redraw                              gap <w
        filter <qpoe_file> <filter_name>    window
        q,quit                              help <

Command> list rp800020_evr.tab    ... lists table
Command> disp mv_aco              ... tex window di
Command> window                   ... manipulate gr
Command> edit mv_aco              ... choose min/ma
Command> gtnew                    ... init time fil
Command> gtgen                    ... generate time
                                      from selectio
Command> filter rp800020.qp mvbk  ... add time filt
                                      as macro defi
Command> q                        ... exit

xp> imhead rp800020.qp long+      ... to see the mac

xp> qpcopy 'rp800020.qp[time=(mvbk)]'
```

Tasks such as QPCOPY accept either *ASCII* filter file macros to screen the events according to the time intervals ex

4. Ev

Most us filters using R

- R
- R r
- F
- F
- I

Ac to the

Refe

Conr

Corc

Tody Wor

Astronomical Data Analysis Software and Systems III
ASP Conference Series, Vol. 61, 1994
D. R. Crabtree, R. J. Hanisch, and J. Barnes, eds.

An IRAF-Based Pipeline for Reduction and Analysis of Archived ROSAT Data

Katherine L. Rhode, Giuseppina Fabbiano, and Glen Mackie

Smithsonian Astrophysical Observatory, 60 Garden Street, Cambridge, MA 02138

Abstract. A pipeline is being developed for the reduction and analysis of ROSAT pointed observations. The pipeline is part of a NASA Long Term Space Astrophysics project entitled "An X-ray Perspective on the Components and Structure of Galaxies" (Fabbiano et al. 1990), and is designed for the purpose of determining structural and spectral characteristics of a large sample of galaxies.

1. Introduction

The study of the X-ray spectral and structural properties of galaxies, though still in its infancy, has already proven to be an important area for scientific exploration. During the 1980's, the *Einstein Observatory* provided, for the first time, detailed information about the soft X-ray emission from galaxies. *Einstein* observations led to the discovery of a hot interstellar medium in bright early-type galaxies. They also showed that the X-ray emission from normal spiral galaxies originates in the evolved component of their stellar populations (e.g., supernova remnants, neutron stars, and black holes). In addition, *Einstein* provided data for studies of powerful active galactic nuclei (AGN), and revealed that some otherwise normal galaxies may harbor small-scale, low-luminosity AGN.

The intriguing results from *Einstein* can be substantially expanded using observations from this decade's ROSAT Observatory. The U.S. ROSAT Data Archive at NASA's Goddard Space Flight Center contains data for all non-proprietary ROSAT Position Sensitive Proportional Counter (PSPC) and High Resolution Imager (HRI) pointings. The archive will eventually contain several thousand data sets. To exploit this large volume of data, we are developing a pipeline to systematically reduce and analyze PSPC and HRI observations.

2. Details of the Pipeline

In order to utilize the many existing IRAF/PROS tools designed for analysis of ROSAT data, we chose IRAF as the basis of our pipeline. The pipeline's main calling routine is an IRAF script which reads a list of ROSAT observation sequences. The script operates on each sequence in turn, performing step-by-step reduction and analysis of the data. Most of the steps are accomplished with calls to IRAF tasks from the *xray* and *stsdas* packages, but the script also invokes

AWK and Unix commands to perform functions such as output formatting and data conversion.

Because of the differing characteristics of the PSPC and the HRI, a separate pipeline script was written for each instrument. Except for a spectral reduction section in the PSPC script, they perform the same basic operations:

- **File Manipulation**

  An output directory is created for the sequence; the associated ROSAT archive FITS files are moved there, and used to create IRAF/PROS-format files.

- **Tabulation of Possible Counterparts**

  Sky maps and lists of objects in the field are created using the HST Guide Star Catalog (GSC) (Lasker et al. 1990) and the SIMBAD source list from the archive.

- **Image Reduction**

  Images are corrected for exposure time and vignetting, and smoothed.

- **Source Detection**

  Existence and location of X-ray sources are determined from the IRAF/QPOE event list, using the *ldetect* task (from *xray.xspatial*). (Presently, the pipeline runs *ldetect* once, with a detect cell size of 120".)

- **Flux Determination**

  X-ray counts are calculated for SIMBAD and *ldetect* sources.

- **Surface Brightness Profiles, and PSF Derivation**

  Surface brightness profiles are derived for sources with $\geq$ 300 net counts, via the *imcnts* task (from *xray.xspatial*).

  Appropriate point spread functions are calculated for each *ldetect* source.

- **Spectral Reduction (PSPC version only)**

  Spectra are extracted, and hardness ratios calculated, for *ldetect* sources.

  For *ldetect* sources with $\geq$ 300 net counts, files are created for input into the XANADU/XSPEC spectral fitting program.

3. Pipeline Results and Outputs

The steps described in the previous section yield a set of output files for each ROSAT observation that includes:

- **Image Analysis**

  Smoothed images and contour maps

  Surface brightness profiles for selected *ldetect* sources

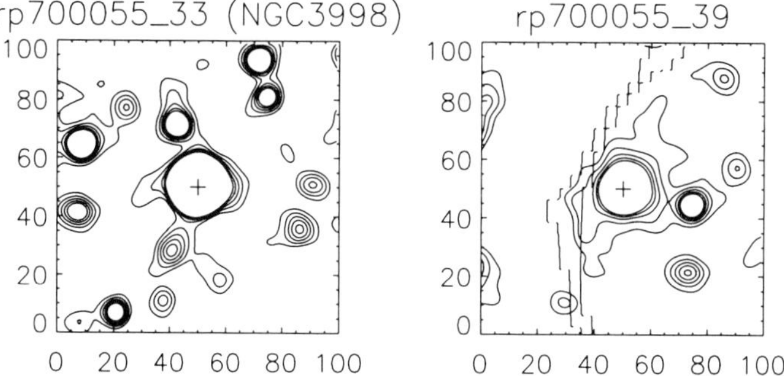

Figure 1. X-ray contour maps, made by passing pipeline output to IDL routines. Each square is a 100x100-pixel (15"x15") sub-image of the NGC 3998 field, centered on an *ldetect* source. 59 sources were detected in this field using a detect cell size of 120"; source 33 (the field target) and 36 are shown. The solid lines are X-ray contours at 2, 4, 6, 8 and 10σ above background; broken lines indicate portions of the field obscured by the PSPC support structure.

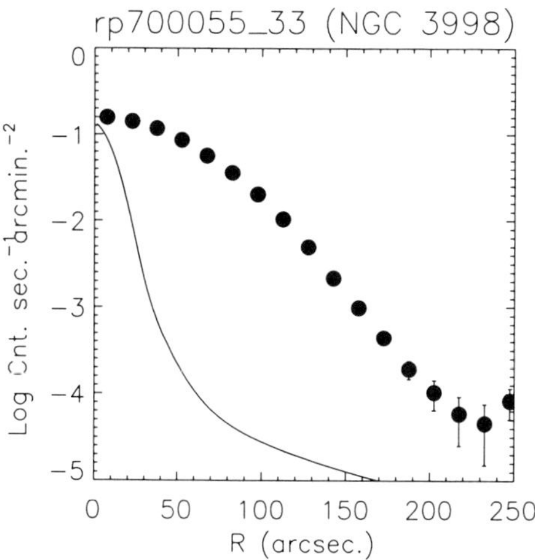

Figure 2. The X-ray surface brightness profile produced by the pipeline for *ldetect* source 33 (NGC 3998). The profile is plotted as filled circles; the line is the 1.4 keV PSPC point spread function.

- **Spectral Analysis (PSPC data only)**

  Extracted spectra for *ldetect* sources, in IRAF/PROS and XSPEC format

- **Files for Plotting and Graphics**

  Plots of GSC objects (stellar and non-stellar) in the field

  Input files for specialized IDL graphics routines

- **Other**

  A log of each IRAF command, with the date and time of execution

  Final results, tabulated in Unix **/rdb** databases

Example pipeline output for ROSAT sequence rp700055, a 58.5 ksec PSPC observation centered on the S0 galaxy NGC 3998, is shown in the figures.

4. Planned Improvements to the Pipeline

The first version of the pipeline was completed in September 1993. As we begin processing data from the archive, we are also "fine-tuning" the pipeline, and expect to complete a second version by early 1994. Along with other improvements, future versions will permit:

- Preliminary spectral fitting on some sources, performed automatically with XANADU/XSPEC (PSPC version only)
- Calculation of X-ray color ratios (PSPC version only)
- Incorporation of overlays of the CD-ROM ST ScI Guide Star (Digitized Schmidt Plate) Survey, when it becomes available
- More accurate background subtraction, using templates from long-exposure PSPC and HRI frames
- Cross-correlation of X-ray source positions with positions in catalogs such as the Third Reference Catalogue of Bright Galaxies (de Vaucouleurs et al. 1991), and the Catalogue of Principal Galaxies (Paturel et al. 1989)

Acknowledgments. This work was supported in part under NASA grant NAGW-2681(LTSA).

References

Fabbiano, G., Canizares, C. R., Kim, D.-W., Schechter, P., & Trinchieri, G. 1990, NASA Long-Term Space Astrophysics Research Program Proposal P2309-10-90

Lasker, B. M., Sturch, C. R., McLean, B. J., Russell, J. L., Jenkner, H., & Shara, M. M. 1990, AJ, 99, 2019

Paturel, G., Fouque, P., Bottinelli, L., & Gouguenheim, L. 1989, A&AS, 80, 299

de Vaucouleurs, G., de Vaucouleurs, A., Corwin, H. G., Buta, R. J., Paturel, G., & Fouqué, P. 1991 Springer-Verlag: New York

Astronomical Data Analysis Software and Systems III
ASP Conference Series, Vol. 61, 1994
D. R. Crabtree, R. J. Hanisch, and J. Barnes, eds.

Spatial Modelling Techniques Applied to ROSAT HRI Observations of Cygnus A

D. E. Harris

CfA, Cambridge, MA 02138

R. A. Perley

NRAO, Socorro, NM 87801

C. L. Carilli

Sterrewacht te Leiden, The Netherlands

Abstract. The X-ray image of the powerful radio galaxy Cygnus A contains several types of morphological structures. We describe image processing steps used to (a) partially correct the image for poor registration; (b) subtract a modified King distribution for the hot cluster gas; and (c) model the residual features associated with the radio core source and hot spots. For the hot spots, we have used previous knowledge concerning the high brightness areas of the radio emission since we believe the hot spot X-ray emission is non-thermal.

1. Registration

Observations of Cygnus A were taken with the High Resolution Imager (HRI) of the ROSAT satellite over a period extending from 25 Apr 1991 to 23 May 1992, with 27 separate observing intervals. The effective exposure time was 42,848 sec.

To realize the potential angular resolution of the HRI for our observation it was necessary to register the various time segments of the data. There is a problem generally attributed to the star trackers which often produces an aspect solution which changes by 10 to 15 arcsec between 'observing seasons' (separated by 6 months). Therefore we divided our data into 5 segments such that each segment was comprised of data taken over no more than a few days. We then measured the apparent position of a serendipitous point source situated 6.′4 east of Cyg A, shifted each map so as to align this source (Figure 1), and then summed the shifted maps. The required shift for each segment is given in Table 1. This procedure improved the image quality by sharpening the features associated with the radio core and hot spots (Figure 2). Absolute position uncertainties are believed to be 2", since the peak of the central X-ray emission is aligned with the center of the optical galaxy to that accuracy.

Table 1. IRAF/IMSHIFT.

	Date	Obstime (ksec)	RAshift (arcsec)	DECshift (arcsec)
A	91apr25	2.2	0	+4.5
B	91nov8/9	11.6	-8.75	-5.0
C	91dec6-8	10.1	-9.0	-4.5
D	92apr21	1.9	+8.5	-2.0
E	92may22/23	17.4	0	0

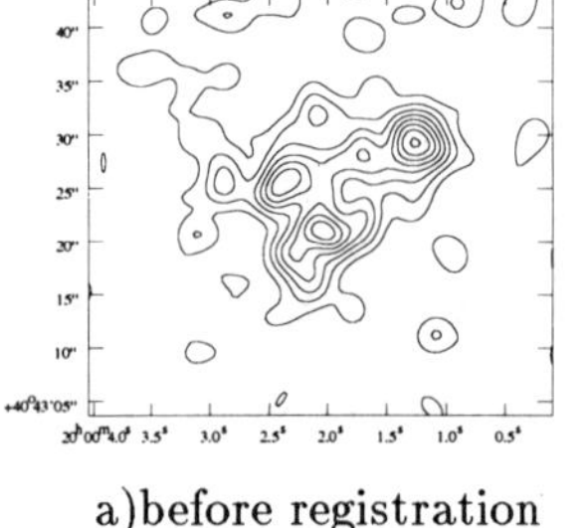

a)before registration

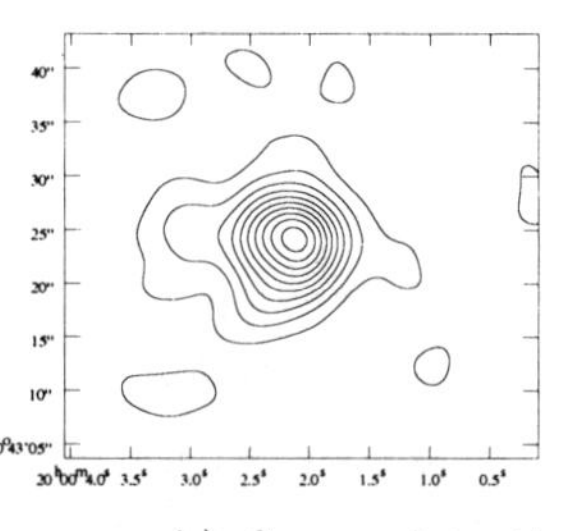

b)after registration

Figure 1. The source 6.4' east of Cygnus A used for registration. The data have been smoothed with a Gaussian of FWHM=3". Contour levels are linear: 0.03, 0.06, 0.09....0.30 cts/pixel.

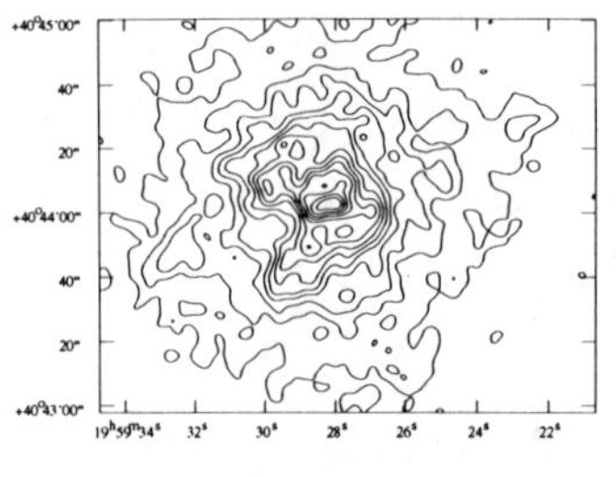

a) before registration

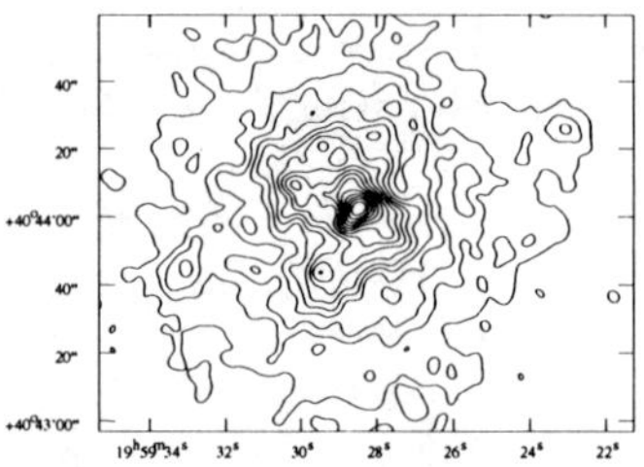

b)after registration

Figure 2. The HRI image of Cygnus A. The data have been smoothed with a Gaussian of FWHM=5". Contour levels are linear: 0.07, 0.132, 0.194,...1.0 cts/pixel.

2. Cluster Gas — Subtraction of King Model

The dominant feature of the X-ray map is the centrally condensed cluster emission. To better investigate the effects of the radio source on the X-ray morphology, we fitted a 'King model' Surface Brightness $\propto a/[1+(\frac{r}{a})^2]^{3\beta-0.5}$ to the radial profiles of brightness in 'pie sections' away from the radio source. Values of $\beta = 0.75$ and core radius, $a = 35''$ (35 kpc for H=75 km/s/Mpc) were derived from the radial profiles. The amplitude of the King model was set to match the net number of photons from the smooth distribution within a circle of radius 125". After subtraction, the residual map (Figure 3) shows remarkable signatures of the presence of the radio source on the X-ray morphology: (1) a central source coincident with the radio nucleus in the optical galaxy, (2) excess X-ray emission co-spatial with the radio emitting hotspots, (3) a decrement in emission located in the inner regions of the radio lobes, and (4) two regions of enhanced emission (east and southeast of the core) which may be associated with the bowshock. The results on the core component will appear in the proceedings of IAU Symposium 159, "AGN Across the Electromagnetic Spectrum" (Harris, Perley, & Carilli). A paper on the hotspots has been submitted (Harris, Carilli, & Perley) and work on the interaction of the radio source with the cluster gas is in preparation (Carilli, Clarke, Perley, and Harris).

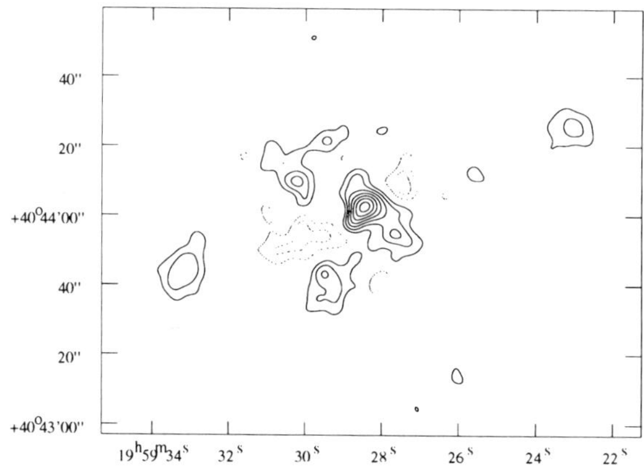

Figure 3. The residual image after subtraction of a modified King Model with core radius = $35''$ and $\beta = 3/4$. Positive contour levels are the same as in Figure 2; negative contours have been added at -0.132 and -0.07.

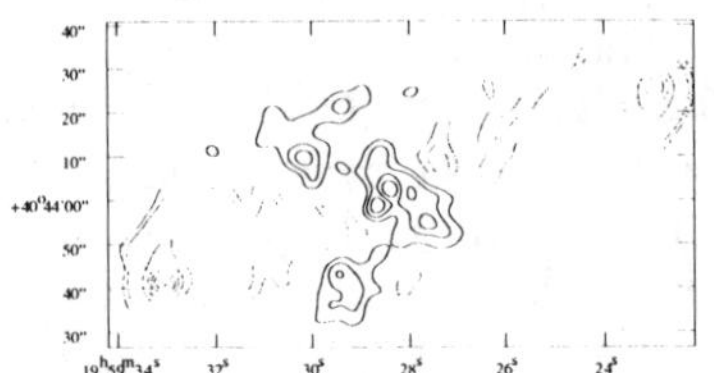

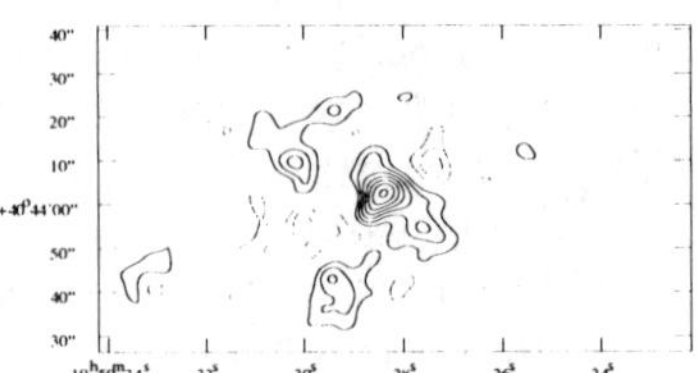

Figure 4. Residuals after subtraction of the 15GHz radio image convolved with the HRI PRF. Contour levels are the same as those in Figure 3. (a) The amplitude of the radio model was chosen so that the peak brightness at the X-ray core would be reduced to the background level. Since the X-ray core is comparatively bright compared to the hotspots, there are large negative areas representing the radio lobes. These are shown by the extra contours -6, -5, ... -1. (b) The model amplitude was chosen to match the peak surface brightness of hotspot A.

3. Modeling the Residual Features — The Core and Radio Hotspots

Since we are convinced that the X-ray emission from the radio hotspots is non-thermal (i.e., synchrotron self-Compton, SSC, emission), we generated a model by convolving the ROSAT HRI point response function (PRF) with the 15 GHz radio map of Cygnus A. The resulting model represents an image that might have been observed if the HRI were sensitive to photons of 0.000000062 keV. The SSC emissivity will not necessarily mimic the synchrotron emissivity, but the brighter parts of the 15 GHz map approximately isolate regions with large numbers of relativistic electrons and high photon energy densities. This model was then subtracted from the residual of the King subtract. Normalization was chosen to match the peak brightness of the each feature in turn.

Figure 4a shows the results when the core intensity is chosen for normalization. The strong radio emission of the lobes is seen as pronounced negative features — thus displaying the model as a byproduct. One can also see weak excess emission near the position of the core source.

In Figure 4b we see that hotspot A (west) is completely removed when the model is normalized to match the peak intensity of A. Note that the core is basically unaffected, meaning that the ratio of X-ray to radio emission is quite different for the core and hotspot. This difference is quantified in the normalizing factors used: Hotspot A required only 1/23 of that used for the core.

4. Acknowledgements

This work was partially supported by NASA grant NAG5-1536 and NASA contract NAS5-30934.

Astronomical Data Analysis Software and Systems III
ASP Conference Series, Vol. 61, 1994
D. R. Crabtree, R. J. Hanisch, and J. Barnes, eds.

Einstein Hardness Ratios from a User-Developed IRAF Script

K. L. Rhode and F. R. Harnden, Jr.

Smithsonian Astrophysical Observatory, 60 Garden Street, Cambridge, MA 02138

Abstract. In 1992 a "bug" was discovered in the software that had been used to process *Einstein* Imaging Proportional Counter data. One implication of this bug was that source hardness ratios calculated for inclusion in the *Einstein* **Catalog of IPC X-ray Sources** were invalidated. IRAF/PROS was used to generate a corrected table of hardness ratios in time for the catalog's publication.

1. Introduction

1.1. The *Einstein* Mission

The *Einstein* Observatory (HEAO-2) was launched in November 1978 and operated until April 1981, making more than 5,000 pointed observations. *Einstein*'s grazing-incidence focusing optics provided vastly improved sensitivity over previous X-ray astronomy missions (Bradt 1992): it could detect up to 1,000 times fainter sources than could previous missions, and had sufficient positional accuracy to make it possible to locate optical counterparts (Harris et al. 1993). In addition to detecting most of its targeted objects, *Einstein* discovered several thousand "serendipitous" sources, increasing by an order of magnitude the number of X-ray sources known (Harris et al. 1993).

Two of *Einstein*'s four focal plane instruments had imaging capabilities: a high-resolution imager (HRI), and an imaging proportional counter (IPC). The HRI had a field-of-view 24 arc minutes across, and spatial resolution of ~4 arc seconds, the best of the *Einstein* instruments (Bradt 1992). The IPC had angular resolution of only ~1 arc minute over its 1-degree-square field-of-view, but was equipped with modest spectral resolution (Bradt 1992). It was by far the most frequently used of *Einstein*'s instruments, performing nearly seventy-five percent of the targeted observations (Harris et al. 1993).

1.2. The Source Catalog

The *Einstein* **Observatory Catalog of IPC X-Ray Sources** (EOSCAT) was completed several years after the end of the mission. The IPC was chosen for the catalog effort because it provided more sky coverage, higher sensitivity, and spectral capabilities. A machine-readable version of the catalog was released in January 1990, and a seven-volume printed version was published in 1993.

The catalog comprises most of the relevant IPC data from the post-mission data reprocessing, called "Rev1". Volume 1 contains the primary documentation, a complete source list (with 4,806 unique sources), and several appendices containing results. Volumes 2 through 7 contain contour maps of each IPC observation, ordered by right ascension, with one map per page. Below each contour map is information about the observation, such as the start and stop times, live time, and roll angle. There is also a Source Table, which lists information about the sources detected in the field. One of the items in the Source Table is a "source flag", which can indicate that more information about the source exists elsewhere in the catalog. For example, a source flag of "H" means the source appears in the table of hardness ratios in Appendix D.

1.3. EOSCAT Hardness Ratios, and the PI-binning Bug

Appendix D lists hardness ratios for nearly half of the detections in the EOSCAT master source list. The hardness ratio of a source can provide a rough indication of its temperature, spectral index, or associated column density. For EOSCAT, the hardness ratio is defined as the number of counts in the higher energy band minus those in the lower energy band, divided by the sum of the counts in both bands. The IPC lower energy (or "soft") band covers 0.16 to 0.81 keV, and the higher energy ("hard") band covers 0.81 to 3.5 keV.

In 1992, it was discovered that the conversion from pulse-height to "energy" (or "PI") binning in the Rev1 processing of IPC data had used a gain map which was not rolled to match the observation (Prestwich et al. 1992a). Because of this, the hardness ratios that had been calculated for EOSCAT Appendix D were invalidated. Since the catalog had not yet gone to press, we were able to generate a table of corrected hardness ratios in time for its printing.

2. Producing Corrected Hardness Ratios

2.1. Requirements of the Project

Generating a table of corrected hardness ratios would require that we perform spectral data extraction and calculate source counts and background counts for almost 3,000 IPC sources. Our objectives were to:

- process a large volume of data quickly, and in an organized and automated fashion
- use the PI-corrected data, which were readily available on the *Einstein* IPC Event List CDROMs (Prestwich et al. 1992b)
- minimize the software effort, by using an existing IRAF/PROS task (*qpspec*) to do the necessary spectral extraction.

We decided that we could accomplish these objectives by writing an IRAF script task. Given a list of sources, the IRAF task would access the PI-corrected data on CDROM, and perform spectral extraction for each source. The results could then be input into the maximum-likelihood algorithm that had been used to produce the original hardness ratio values for EOSCAT Table D. We named the script task *pre_mlhr*, because it was to be executed prior to the maximum-likelihood program, called *mlhr*.

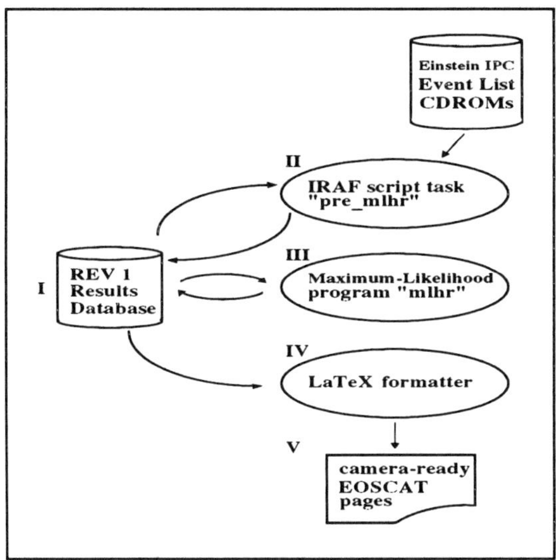

Figure 1. The steps required to generate the corrected EOSCAT hardness ratios.

2.2. Details of the Process

Running the script was one of a series of steps required to generate a corrected hardness ratio table:

- **Step I — Creating the input source list**

  INGRES databases of Rev1 processing results were used to make a list of every EOSCAT source. Sources appearing in the same observation sequence were listed together, and sources for which a hardness ratio would be calculated were flagged.

- **Step II — Running the script task** ***pre_mlhr***

  The source list was used as input for the *pre_mlhr* task. For each observation sequence in the list, *pre_mlhr* created an output directory, accessed the PI-corrected data for that observation on CDROM, and ran *qpspec* for those sources for which a hardness ratio was to be calculated. Results were written to an output table.

- **Step III — Calculating source hardness ratios**

  The *pre_mlhr* output was used to make a database. The maximum-likelihood program, *mlhr*, operated on the database, calculating a hardness ratio and errors for each source and updating the database with these values.

- **Step IV — Creating camera-ready catalog pages**

  An ASCII dump from the database was input into a LaTeX formatting script, to produce the camera-ready pages of corrected hardness ratios — the final version of EOSCAT Table D.

Because we were able to execute these steps quickly and successfully, EOSCAT went to press with 29 pages of corrected hardness ratio values, rather than with a single page stating that Table D had been omitted.

3. Conclusions

The fact that we used IRAF/PROS to accomplish arguably the most important part of our task — the spectral extraction for ~2,800 sources in ~1,700 IPC fields — showed how effective IRAF/PROS is as a tool for astronomical data analysis. When we began this project, we had little experience with writing IRAF script tasks; still, it took only about 2 weeks for us to develop and use an IRAF script that would carry out our objective of processing a large amount of X-ray data quickly and conveniently.

We did encounter some apparent limitations to IRAF's capabilities, however, in the areas of file manipulation and producing formatted output. In those cases, we found it useful to invoke external tools from IRAF. Specifically, we called Unix and AWK scripts from our IRAF script, to perform certain file manipulation tasks, formatted reads and writes, and ASCII table manipulation.

This solution is not entirely satisfactory, because invoking AWK or Unix commands from within IRAF can itself produce difficulties. Complications arise mainly because IRAF and the host environment (Unix, in our case) are not able to communicate fully with each other. We could not pass an IRAF variable (such as a sequence number or a file name) to a Unix command as an argument, because Unix did not have access to the IRAF variables we had declared. This situation is frustrating to users, who, while inside a software environment loaded with useful tasks for astronomical data analysis, are at the same time isolated from the software tools they are accustomed to using in their host environments.

To conclude, although at times we were frustrated by some of IRAF's limitations — and perhaps more so by the difficulties we had trying to overcome those limitations — overall we were satisfied with how much we were able to accomplish with IRAF in a short period of time.

Acknowledgments. This work was supported in part under NASA contracts NAS8-30751 and NAS5-30934.

References

Bradt, H. V. D., Ohashi, T., & Pounds, K. A. 1992, ARA&A, 30, 391

Harris, D. E., Forman, W., Gioia, I. M., Hale, J. A., Harnden, Jr., F. R., Jones, C., Karakashian, T., Maccacaro, T., McSweeney, J. D., Primini, F. A., Schwarz, J., Tananbaum, H. D., & Thurman, J. 1993, NASA Technical Memorandum 108401

Prestwich, A., McDowell, J., Garcia, M., & Harnden, R. 1992, HEAO Newsletter, Vol. 1, No. 7, 4

Prestwich, A., McDowell, J., Plummer, D., Manning, K., & Garcia, M. 1992, "The Einstein Observatory Database of IPC Images in Event List Format", CDROM Volumes 1–4, SAO, Cambridge, USA

Astronomical Data Analysis Software and Systems III
ASP Conference Series, Vol. 61, 1994
D. R. Crabtree, R. J. Hanisch, and J. Barnes, eds.

ASCA Initial Data Processing and the ODB (Observation Database)

M. Itoh, K. Mitsuda, A. J. Antunes, R. Fujimoto, H. Honda, E. Matsuba

Institute of Space and Astronautical Science, Yoshinodai 3-1-1, Sagamihara, Kanagawa, 229, Japan

J. Butcher, J. Osborne, J. Ashley

Department of Physics and Astronomy, University of Leicester, University Road, Leicester LE1 7RH, England

T. Takeshima

Institute of Physical and Chemical Research, Hirosawa 2-1, Wako, Saitama, 351-01, Japan

Abstract. This paper describes the initial data processing of the X-ray astronomy satellite ASCA and the Observation Database (ODB) installed at the Institute of Space and Astronautical Science. In the initial data processing, the raw telemetry data downlinked from the satellite are reformatted to FITS binary table format. Attitude solutions and the orbital elements obtained on the ground are also compiled as FITS files. The set of telemetry, attitude and orbital files at this stage is called FRFs (First Reduction Files). Information such as a unique observation ID, the target and the observer's name is written in the header of the telemetry FRF at the reformatting. Such information is maintained and provided by the ODB. The ODB is a relational database that contains information on accepted proposals, scheduled time lines, and observation history. The ODB is also used for mission operation as well as to provide information to general observers through the Internet.

1. Introduction

ASCA (ASTRO-D) is the fourth Japanese X-ray astronomy satellite, it was launched in February, 1993 (Tanaka et al. 1994). The payload consists of high throughput X-ray telescopes covering a wide energy range with moderate spatial resolution. On the focal plane, there are X-ray CCD cameras with high spectral resolution and imaging scintillation proportional counters. With these instruments, ASCA is one of the most powerful X-ray observatories ever flown. The observational instruments and data processing/analysis software have been developed under collaboration between US and Japan. This paper describes the initial data processing of ASCA and the Observation Database (ODB) developed by a team at the University of Leicester and installed at the Institute of Space and Astronautical Science (ISAS).

2. Initial Data Processing

A flow chart of the initial data processing is shown in Figure 1. The goal of the initial data processing at ISAS is to produce a set of files called FRFs (First Reduction Files). FRFs consist of (1) telemetry, (2) attitude and (3) orbit files. All these files are in FITS binary table format. While the telemetry and attitude files are created for each observation, the orbit file contains all the orbit information throughout the mission with new information added sequentially.

Telemetry data from ASCA are downlinked at ground stations including KSC (Kagoshima Space Center), and NASA's Deep Space Network Stations. Those data are transmitted to ISAS through networks and stored in a database called SIRIUS on a FUJITSU main frame computer. Some cleaning is done before the storage in SIRIUS by throwing away any part of the data that suffers from apparent telemetry errors.

Telemetry data to be processed into FRFs are extracted from the SIRIUS database into a Unix environment, and minor reformatting is performed to the FITS binary table format. Each record of the table corresponds to the information unit of the telemetry format called a superframe (8192 Bytes).

One of the important tasks at this stage is time assignment to the data. The telemetry data contain time information based on the satellite clock. Comparison of the satellite clock value and a reference clock on the ground is made during the real-time contacts at KSC, and the absolute time is assigned to the data. Corrections are made for the radio propagation time between the satellite and the ground station, delays in the on-board telemetry data formatter and the ground system, and temperature-dependent drift of the satellite clock frequency. Correction for the time lag in the on-board X-ray detectors and the front-end data processing are done in the downstream software because it is dependent on the detector and the data processing mode.

Using the telemetry FRF, the satellite's attitude is determined. The result is compiled as the attitude FRF. The orbit of the satellite is determined weekly by the Japanese NASDA agency. The updated orbital elements are sent to ISAS every week and compiled as the orbit FRF.

When the FRF production software produces the final telemetry/attitude FRF, it obtains such information as the unique ID for the observation (the sequence number), the target and PI's name from the ODB, and puts this information in the header. To Japanese observers, data are distributed in the form of FRFs. For US observers, telemetry FRFs are converted to photon event files called science FITS files and HK FITS files containing the housekeeping information.

3. Observation Database (ODB)

The ODB is a relational database running under ULTRIX/SQL. It contains information on the accepted ASCA observation proposals, time lines, and the observations as carried out.

Following is a list of "tables" in the database with brief descriptions;

- Proposals : summary of information from accepted proposals.

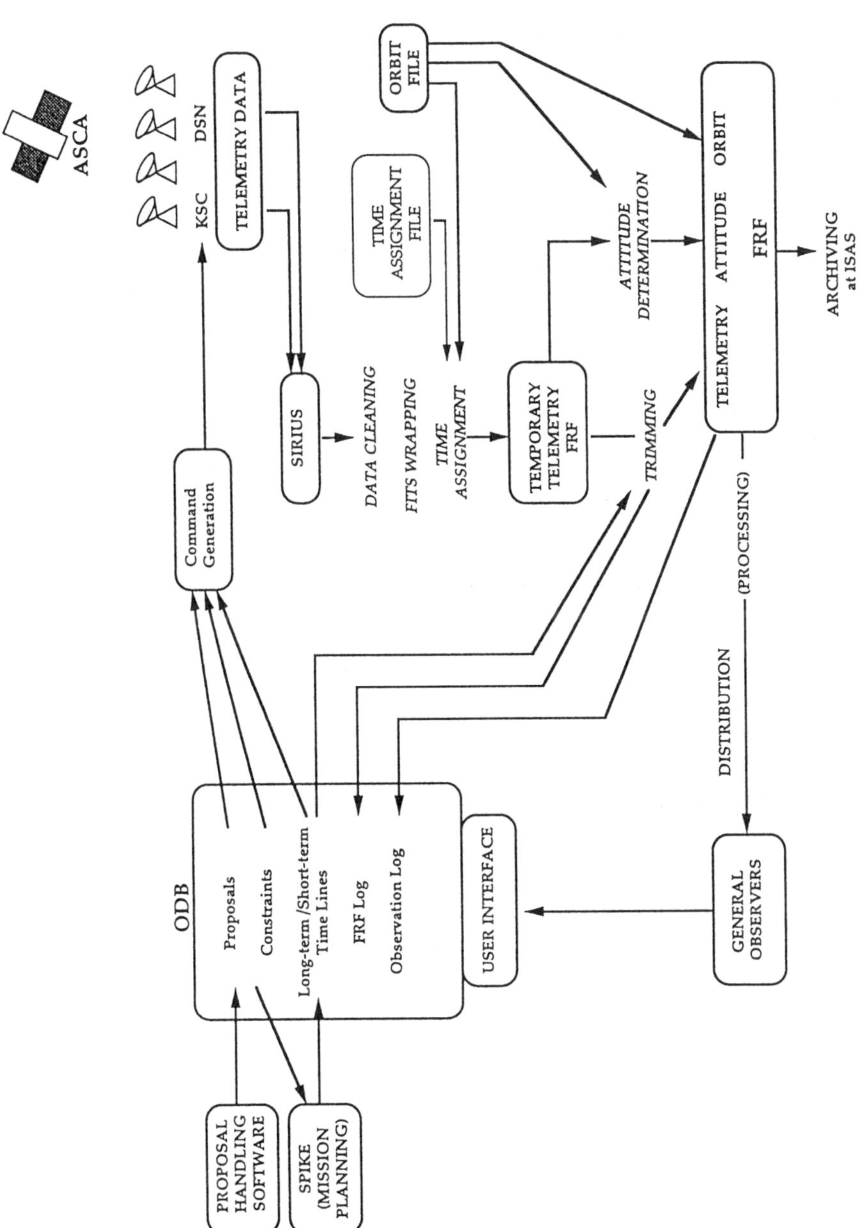

Figure 1. Data processing at ISAS and the ODB.

- Remarks : proposer's remarks.
- Observation constraint : constraints of the time-critical observations.
- Long term timeline : schedule for the whole AO interval.
- Short term timeline : detailed timeline appended to week by week.
- FRF log : status of the FRFs for each observation.
- Observation log : includes actual pointing position, summary of HK values, instrument modes and settings, etc.

The ODB not only receives and stores information, but also provides other mission planning/operation software with essential information. This exchange of information is shown schematically in Figure 1. The proposal handling software at NASA/GSFC provides information on accepted proposals to ODB. The mission planning software SPIKE takes the proposal information to produce observation time lines (Antunes et al. 1994). For the daily mission operation, the ODB provides the time lines and the proposal information such as target position, observation modes, and other requests from the proposer. For the FRF generation, the ODB provides time line information and the sequence number etc. It also receives information on the FRFs' status. After the FRF generation, software to generate a complete observation log is executed.

The ODB has a screen-based user interface. Guest Observers can access the ODB through the Internet to get information on the ASCA observations. There is a captive account named "odb", of which the password is "ascaasca". Opening a session in "xray.astro.isas.ac.jp" (ip address 133.74.7.1) with this account automatically invokes the user interface of the ODB. Because the ODB user interface is based on curses, an X-terminal is not required to access the ODB. The user interface to the ODB includes on-line help and a facility to mail query results back to the user.

Acknowledgments. The authors thank all the ASCA team members.

References

Antunes, A., Nagase, F., Isobe & T. 1994, this volume

Tanaka, Y., et al., 1994, PASJ, submitted

Astronomical Data Analysis Software and Systems III
ASP Conference Series, Vol. 61, 1994
D. R. Crabtree, R. J. Hanisch, and J. Barnes, eds.

PROS Support for PSPCs on the SRG Mission

A. Hornstrup, N. J. Westergaard, C. Budtz-Jørgensen

Danish Space Research Institute, Gl. Lundtoftevej 7, DK-2800 Lyngby Denmark

M. A. Conroy, J. S. Orszak

Center for Astrophysics, 60 Garden Street, Cambridge, MA 02138, USA

Abstract. In 1995, the Russian Spectrum Röntgen Gamma (SRG) satellite will be launched. The satellite contains several X-ray telescopes, among which are the SODART telescopes, developed in collaboration between the Danish Space Research Institute (DSRI), the Russian Space Research Institute (IKI, Moscow), Helsinki Observatory (Finland), NASA and others.

DSRI provides 4 Position Sensitive Proportional Counters (PSPC) detectors (called HEPC and LEPC) in the focal planes of the SODART mirror modules. PROS, the X-ray analysis package under IRAF, developed by SAO initially to reduce ROSAT and Einstein data, has now been extended to support reduction of data from the HEPC/ LEPC detectors on-board SRG.

This paper presents the results of the implementation of the HEPC/ LEPC instruments in the PROS package and examples on reduction of simulated HEPC observations.

1. Introduction

The Danish Space Research Institute is currently producing two X-ray telescopes (called SODART) to be launched on the Russian Spectrum Röntgen Gamma (SRG) satellite in December 1995. The telescopes consist of two mirror modules, two sledges with focal plane instruments and an Objective Crystal Spectrometer (OXS) in front of one of the mirror-modules.

The focal plane instruments for the SODART telescopes are: two Russian PSPC's (KFRD), a US produced polarimeter (SXRP), a Finnish solid state detector (SIXA) and two sets of PSPC's from DSRI. The DSRI PSPCs are one HEPC (High Energy Proportional Counter (2-30keV)) and one LEPC (Low Energy Proportional Counter (0.2-8keV)) in each focal plane.

DSRI has decided to primarily use the X-ray analysis package PROS (a package under the image analysis system IRAF) as reduction and analysis software. PROS is developed by SAO initially to reduce EINSTEIN and ROSAT data.

We here describe briefly the implementation of the Danish PSPCs into the PROS xspectral package. A few simulated observations are also presented. The "xray.xspectral" part will be released by November 1, 1993. Implementation of HEPC and LEPC into the spatial and into the timing package of PROS is less instrument specific, and will be ready by December 1993, to be released with the spring 1994 release of PROS.

The HEPC and LEPC instruments will be continuously supported in future SAO-releases of PROS.

The special parts for analysis of observations with the OXS have not yet been implemented in PROS.

2. Implementation

The PROS spectral package needs definitions of instrument specific parts. Since HEPC and LEPC are PSPC's, the basic idea has been to use the current version of ROSAT PSPC as a template. In the general "xray" package of PROS, the SRG_HEPC1 (-2) and SRG_LEPC1 (-2) instruments are defined as valid telescope / instruments.

In the "xray.xspectral" package of PROS, all instrument specific routines have been rewritten to implement the HEPC and LEPC detectors. This work has introduced several new parameters in the parameter files. Besides the (default) setting of these parameters, the implementation of HEPC/LEPC is transparent to the user.

Input files are QPOE files, which are made using "fits2qp" from FITS files defined close to the RDF (Rationalized Data Files) FITS standard to be released soon (Corcoran et al. 1992). All scientific data from the HEPC/LEPC detectors will eventually be distributed in RDF FITS format, directly readable by "fits2qp" in "xray.xdataio".

Response matrices and definitions of energy-grid follow closely the principles from the ROSAT PSPC implementation. Size and format for the final HEPC/LEPC response matrices are currently being defined. The response function is in the PROS version split in three parts:

1) The `*.rmdat`-files - the response matrices, with the response from the (Xenon) gas in the detector sensor box given in a 729(energies) x 128(channels) matrix. The energies range from 0.1 to 30 keV in both cases, and the energy grid is defined in the `*.egr` files.

2) The `*.window`-files window transmission function, which gives the transmission of the window in each of the 729 energies. Note, that there are no filters in connection with the HEPC/ LEPC detectors.

3) The `*.ofa`-files, giving the off-axis dependence. In analogy with the ROSAT approach the off-axis angle range is divided in 14 parts. The dependence is described in a 729(energies) x 14(angles) matrix. The first row in the file defines the off-axis angle (in arcmin) in question.

The response matrices for the HEPC and LEPC detectors are calculated based on the detector design. These matrices will be updated following the on-ground and in-flight calibration. The matrices will be part of the future PROS releases and also available via anonymous ftp from `dsri.dk:/pub/srg/rmdat`. The off-axis behaviour of the telescope is calculated using the ray-tracing program shortly described below.

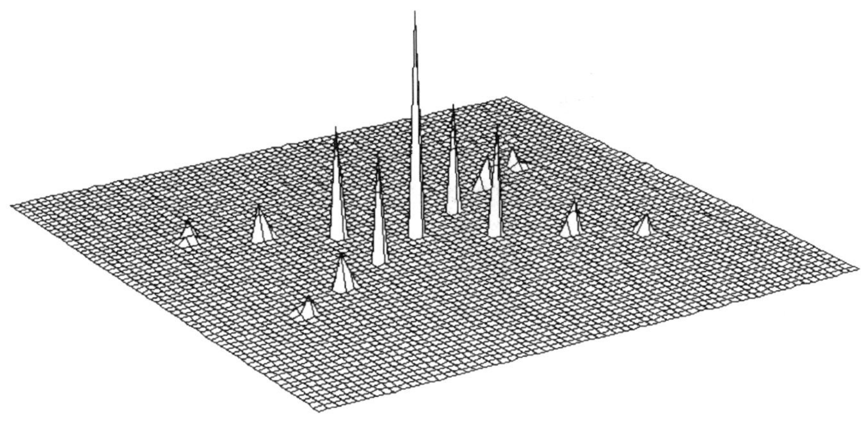

Figure 1. 13 Crab sources in a surface plot to illustrate the vignetting of the telescope.

3. Simulation

To test the PROS support of HEPC/LEPC, a simulated observation of the Crab source is performed, with a powerlaw model, where energy index is 1.1 and absorption is defined by $N_H = 3\times10^{21}\ cm^{-2}$. The simulation is done with local (DSRI) ray-tracing software. In the telescope simulation part, scattering from the mirrors and figure errors are not (yet) considered. Therefore, the resolution is only geometry dependent (and is around 15 arcsec), while the design goal of the telescope (including scattering and figure errors) is a 2 arcmin resolution (HPW).

The telescope has a strong vignetting, illustrated in Figure 1, which shows 13 simulated Crab sources positioned in different off-axis angles. The vignetting has effect on the fitting accuracy. In Table 1, the spectral fits to the simulated

Table 1. Table showing fitting parameters for the simulation of the Crab source in the HEPC1 detector of the SODART telescope. The input spectrum has energy index 1.1 and log(N_H)=21.48.

Distance (arcmin)	Counts/sec	Energy Index	log(N_H)
0	3800±70	1.13±0.05	21.70±0.32
8	1916±104	1.01±0.12	20.85±0.97
16	880±79	0.97±0.16	20.70±0.84
22	598±70	1.07±0.12	20.77±0.96

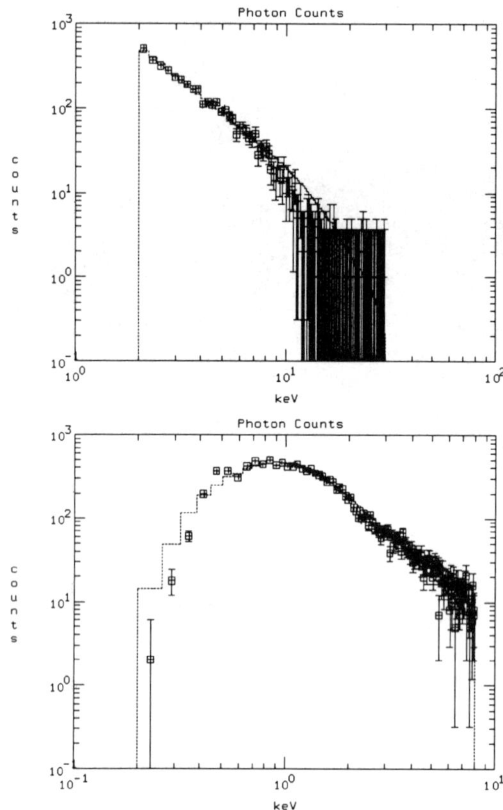

Figure 2. Simulated spectra of a Crab source, for the HEPC and the LEPC detectors. Fits are shown as solid lines.

data are shown. Figure 2 shows the spectra and the fits for the simulation of the Crab source in the on-axis case for both the HEPC and the LEPC detectors.

References

Corcoran, M.F., Pence,W., White, R. & Conroy, M.A. 1992, in Astronomical Data Analysis Software and Systems, A.S.P. Conf. Ser., Vol. 52, eds. R.J. Hanisch, R.J.V. Brissenden & J. Barnes, 549

Astronomical Data Analysis Software and Systems III
ASP Conference Series, Vol. 61, 1994
D. R. Crabtree, R. J. Hanisch, and J. Barnes, eds.

Gazing at X-ray Sources: The SAX Mission

M. B. Negri

Agenzia Spaziale Italiana, Via R. Margherita 202, I-00198 Roma, Italy

L. Piro

I.A.S. / C.N.R., Via Enrico Fermi 21, I-00044, Frascati, Italy

L. Salotti

Telespazio SpA, Via Corcolle 19, I-00131, Roma, Italy

Abstract. The SAX, "Satellite per Astronomia X", is a program of the Italian Space Agency (ASI) and the Netherlands Agency for Aerospace Programs (NIVR) devoted to systematic, integrated and comprehensive studies of X-ray sources in the energy band 0.1–200 keV. Scientific objectives are: imaging and broad band spectroscopy (0.1–10 keV), spectroscopy and timing (3–200 keV) and all sky monitoring (2–30 keV). SAX will be launched at the end of 1995 in to a circular orbit at 600 km. The minimum mission lifetime will be two years. This paper outlines the SAX mission focusing on the SAX ground segment, which is composed of an Operational Control Centre (OCC) and of a Scientific Data Centre (SDC). Emphases shall be devoted to the data flow from the spacecraft to the final end user and to the data treatment facilities which are available to duty scientists.

1. Introduction

SAX (Butler and Scarsi 1990) will be the first X-ray mission with the capability of observing more than three decades of energy with a relatively large area, a good energy resolution and imaging capabilities (resolution of ~1') in the range of 0.1–10 keV. The SAX mission is supported by a Consortium of institutes in Italy together with institutes in the Netherlands and in the SSD/ESA.

The scientific payload on-board SAX comprises:
A package of co-aligned narrow field of view instruments (NFIs) consisting of:
- a Low Energy Concentrator Spectrometer (LECS): 1 unit (0.1–10 keV).
- a Medium Energy Concentrator Spectrometer (MECS): 3 units (1–10 keV).
- a High Pressure Gas Scintillation Proportional Counter (HPGSPC): (3–120 keV).
- a Phoswich Detector System (PDS): (15–200 keV).

A set of wide field instruments (WFIs) consisting of:
- two Wide Field Cameras (2–30 keV) pointing in opposed directions.

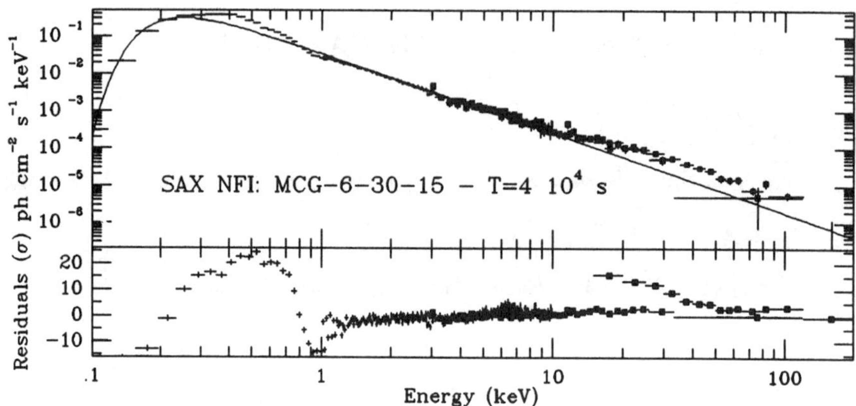

Figure 1. The complex spectrum of the Seyfert galaxy MGC–6–30–15 as would be obtained by SAX–NFI (LECS, MECS, HPGSPC = filled boxes, PDS = empty boxes) in 40000 s.

SAX will be able to perform more than 2000 pointings. While the NFIs will be the prime instruments most of the time, the WFIs will be periodically used to scan the galactic plane in order to monitor the temporal behaviour of sources above 1 milliCrab and to detect transient phenomena. Pointings will be organized in a Core Program and in a Guest Observer Program disposing of ~ 20% of the first year time, this fraction to be increased the following years. Participation in the Core Program is open to members of the national communities involved in the mission and will be ruled, along with application for Guest Observer time, on a peer review basis.

2. Wide Band Capability of SAX

The sensitivity of the NFIs will allow the exploitation of the full band of SAX for sources as weak as 1/20 of 3C273, opening new perspectives in the study of spectral shape and variability of several classes of objects. In this respect, the example shown in Figure 1 is very enlightening. It shows the complex spectrum of a Seyfert 1 galaxy, namely MCG–6–30–15, observed by SAX-NFIs in 40000 s with several spectral features detected by previous X-ray satellites: a soft excess EXOSAT: Pounds et al. 1986), an OVII edge around 0.8 keV (ROSAT: Nandra & Pounds 1992), an iron line at 6.4 keV and a high energy bump above 10 keV (Ginga: Matsuoka et al. 1990). All these components can be measured with good accuracy with SAX in a single shot for the first time.

3. SAX Data Flow and Processing

The two main structures of the SAX ground segment are the OCC and the SDC. They have been designed bearing in mind the goal of keeping fast reaction times to interesting on board science events (such as transient detection) as well as of making shrewd the data delivery to the astronomical community. Both of them are located inside the new Telespazio's Rome 2 operation building with a LAN connecting the OCC to the SDC. Furthermore, the OCC is relayed, via an Intelsat-V international channel, with the Singapore ground station antenna in order to up-link telecommands and to retrieve telemetry from the SAX spacecraft while the SDC is electronically reachable from the exterior through DECNet/SPAN and Internet.

The OCC pursues the SAX spacecraft orbital management activities such as planning of operations, telecommand sequence generation and transmission, attitude and orbit reconstruction, telemetry retrieving from the remote antenna. In addition it hosts the Science Operation Centre (SOC) which is the spot where the SAX telemetry is firstly archived and treated. SAX data (science and housekeeping) are foreseen to reach the SOC within a one orbit time span from their collection on ground. At SOC the orbit-by-orbit telemetry supply flow is subject to a "cut and paste" process in order to group together data from the same scientific pointing which are transmitted during different links. This process, mostly automatic, outputs a pointing oriented data archive which is immediately available to quick-look processing.

A Quick-Look (QL) facility is provided to the SOC duty scientists in order to inspect the SAX raw telemetry data. The supported functions are mainly devoted to scientific data reductions for a first assessment of the instrument in orbit performances and results. Basically, QL functions belongs to three classes: accumulation, display and treatment. Accumulation functions ingest raw telemetry archive data files in order to build working files, i.e., files containing images, spectra or count rate profiles, which are in turn the inputs of display and treatment functions. Obviously, treatments may give birth to second generation working files. The QL facility runs on two DEC RISC DECstations 5000/240, for what it concerns display and computation, and on the SOC DEC VAX 6000/510 host, for what it concerns data retrieving and accumulation. A DECNet link connects the computers. The idea that lies behind accumulations is that for each raw telemetry format type a routine is provided which can extract all the working structures admitted by the concerned format. Before execution the duty scientist can choose which type of structures, among those admitted, are to be produced and which accumulation parameter ranges are to be applied in the course of the accumulation. Display functions implement all the state-of-the-art Workstation graphics capabilities on data presentation including 3-D. For sky images with known pointing coordinates a function has been foreseen which marks the already known X-ray source positions; these are extracted from a SAX X-ray source input catalogue obtained merging the Ariel V, HEAO-I, Einstein, EXOSAT and, possibly, ROSAT source catalogues. Treatments include basic statistics, smoothing and line fitting for calibration source line parameter determination and excess evaluation on sky images. A particular effort has been devoted to WFC sky field reconstruction with the additional capability of focusing on a significative excess in order to extract an energy histogram or a count rate time profile during the pointing. The WFC detector image deconvolution is accomplished by means of a balanced correlation algorithm. No use of calibration or correction matrices is foreseen in the QL context.

Profitting by such a tool case the SOC scientific personnel should be in measure to detect evident celestial events or payload failures with reaction times of a few orbit time spans. This should prove particularly useful in case of WFC transient detection when an acceptable source position localization starts a rescheduling of the payload activities with the goal of an NFI pointing to the transient within one/two days from the first reaction. As soon as data have reached their final pointing structure in the archive they are ready to be transmitted to the SDC. In a nominal operation context data belonging to a pointing could be theoretically transferred to SDC within a 2-3 orbit time span from the on-board completion of the concerned pointing, however, during real-life operations, this delay should be somehow stretched but not exceeding the one day limit. SOC is able to store a shadowed data volume peaking to a 7 day SAX telemetry output.

The SDC is the real scientific core of the whole SAX observing activity assuring such tasks as data analysis, data dispatching, support to Guest Observers and planning. Functionally it has been divided in two components: the Mission Support Component (MSC) and the Scientific Analysis Component (SAC).

SDC-MSC is the terminal for the SAX raw telemetry data which is received from the OCC-SOC and stored into the SAX Mission Archive. This archive, which is supported on optical disk media, has the same structure as the SOC archive requiring no telemetry rearrangement before data insertion. As the mission is expected to output about one Gbyte per day, the archive shall contain not less than 750 Gbytes on completion of the two year nominal life time. 6.5 Gbyte single optical disk are used. The fact that the SDC archive structure is not different from SOC's enables to install at SDC all the QL functions that run at SOC.

Besides archiving, the SDC-MSC is in charge of collecting observation proposals from the astronomical community, of supporting the proposal preparation by Guest Observers and of making plans for the instrument use over the long range period (six months). A special attention has been deserved to the Final Observation Tape (FOTs) production and dispatching with the prime objectives of making easy their analysis by different computer systems and of a fast deliver to the end user as well. In this respect, the coding of a FOT production software package is in progress at SDC-MSC which could automatically ingest the raw telemetry data pertaining to a Guest Observer as soon as they reach optical disks in order to reformat and store them on tape media. The package interfaces with an Observation Proposal Database where all informations about Guest Observers and their proposals are stored. FOT are dispatched via normal mail. In absence of problems it is foreseen to accomplish the FOT production procedure relative to a SAX pointing within two/three days from the on-board data collection.

References

Butler, C., & Scarsi L. 1990, SPIE Proceedings, 1344, 464

Matsuoka, M., et al. 1990, ApJ, 361, 440

Nandra, K., & Pounds, K.A. 1992, Nature, 359, 215

Pounds, K.A., et al. 1986, MNRAS, 221, 7P

Astronomical Data Analysis Software and Systems III
ASP Conference Series, Vol. 61, 1994
D. R. Crabtree, R. J. Hanisch, and J. Barnes, eds.

The Calibration Data Archive and Analysis System for PDS, the High Energy Instrument on board the SAX Satellite

Daniele Dal Fiume , Filippo Frontera [1], Mauro Orlandini and Massimo Trifoglio

Istituto TeSRE — Consiglio Nazionale delle Ricerche — Bologna - Italy

Abstract. In this report we describe the archival and data analysis system for PDS under development at TeSRE. We foresee that the complete system will begin to be operative in the first part of 1994. Some of its parts are already operating. The procurement of the HW and of the commercial SW is almost complete.

1. Introduction

The high energy (15–300 keV) experiment Phoswich Detection System (PDS) on board the Italian (with important Dutch collaboration) SAX satellite is an array of four phoswich scintillation detectors (3mm of NaI(Tl)/5 cm of CsI(Na)) with a total geometric area of 800 cm^2 and a field of view of 1.4°. Detailed description of the instrument is already available in literature (Frontera et al. 1991, 1992). The experiment is built by LABEN S.p.A. (Vimodrone, Italy), under scientific supervision of the PDS Scientific Hardware Group (at TeSRE — Bologna, and IAS — Frascati, institutes of CNR, the National Research Council of Italy) and it is funded by ASI, the Italian space agency.

The PDS Scientific Hardware Group will be responsible for the timely analysis of data during the testing phases, preliminary to the acceptance of the flight and spare models, and during the calibrations of the instrument. Some data concerning preliminary results of tests (in particular tests of the on-board electronics) were already analyzed and the observed performances were published (Frontera et al. 1993). It is expected that the bulk of data coming from tests will arrive at TeSRE starting during the last quarter of 1993 and in 1994. The on-ground calibrations will take place approximately during fall 1994. Other tests when the instrument will be integrated in the satellite are also foreseen up to few months before the launch. The system for the analysis of the PDS calibration data is based on a network of 4 medium-sized workstations, in which data will be both stored and analyzed. The final configuration of the system will include 3 HP9000 RISC workstations, 1 DEC ULTRIX workstation, at least 5 GB of magnetic disks, magnetooptical disks, CD-ROM writers and readers, DAT drives. Most of the hardware is already available. It is foreseen that the operative configuration will be available in 1994.

---

[1] Dipartimento di Fisica — Università di Ferrara — Ferrara — Italy

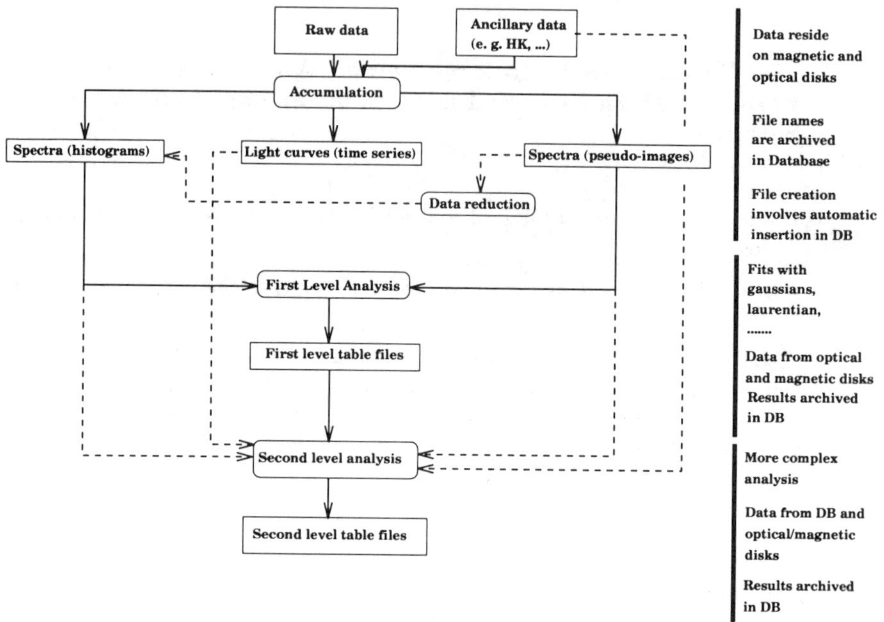

Figure 1. PDS calibration data processing.

Data coming from the PDS experiment during most of the tests and calibrations performed on ground will be handled by the electronic digital processor (Frontera et al. 1991,1992) that formats data for transmission to ground. Currently the following set of operative modes is available for PDS:
in direct modes each qualified event is transmitted with a programmable set of fields, selected between those available for each event from the analog processor; each photon is tagged with its arrival time with maximum time resolution of 15 μs, and the other available information is PHA, pulse shape (also called "rise time"), unit id, coincidence flags;
in indirect modes the digital processor accumulates histograms of counts versus ADC PHA channel, selecting events on the basis of on-board programmable thresholds on pulse shape; the integration time for these on-board accumulated spectra is between 1 and 16 seconds. Coupled with these modes, one can in parallel start high time resolution modes, that produce time series of counts integrated on times between 1 and 125 ms, in 1 to 4 integrated energy channels.

2. Processing PDS data

A very sketchy scheme for PDS data processing is shown in Figure 1. From calibrations we have raw data, containing the output from direct or indirect PDS modes, and ancillary data, including HK. Raw data files contain actually raw "telemetry" packets, similar to those formatted by on-board processors to be transmitted to ground during operative life. The main difference from in-flight data is the lack of on-ground post-processing that involves field expansion to integer number of bytes, stripping of telemetry frames and reordering. The maximum data rate from the PDS experiment is 50 kbit/s.

These raw data are processed, and the results are in form of reordered or accumulated data: Photon lists: only from direct modes — each photon is one record, each input field is mapped to one output field or is filtered — this format can produce the minimum loss of information in the output files; Pseudo-images: only from direct modes — a matrix of counts versus PHA ADC channel versus PSA ADC channel is produced — typical dimensions are 1024×512 pixels; Spectra: histograms of counts versus PHA ADC channel or versus PSA ADC channel; Light curves: time series — histograms of counts versus time.

The format of PDS files is compliant with the specifications of SAX data file formats as defined by the SAX Data Analysis Working Group (Maccarone et al. 1992, Chiappetti 1991, 1992). The SAX data files format was designed having in mind an easy portability to FITS format, via translator (and actually we already have translators from-to the FITS format).

In the PDS implementation of this format, the keywords are divided in mandatory and "private". Mandatory keywords are and will be always present in PDS data files, also in those produced for the scientific analysis of the observations of celestial sources. "Private" keywords are intended for use mainly during calibrations.

3. The PDS Calibrations Archival and Database

In order to correctly size the analysis system, we have to estimate the amount of data kept on line to be processed. As a "worst" case for the ground tests we can assume a week of uninterrupted data production at the PDS calibration facility. If we have data coming from PDS for not more than 50% of the time this results in 1.9 GB of data per week. These data of course do not need to be all processed in almost real-time, but need to be periodically checked to control the quality of the tests/calibrations, and to be archived tidily. However a full processing must be possible in almost real time at least for part of the data, if needed.

The heart of our archival is a relational database based on the INGRES (t.m. of Ask Corp.) commercial database. In this database we have tables at different levels:
(a) tables containing information on files: these tables contain all what is needed by the user in order to correctly select the data set of interest; these tables are automatically updated each time a new file is archived; for off-line files, also information on the archival location is available;
(b) tables of data from first level analysis: in this case the results of data analysis are directly inserted in the database;
(c) tables of data from second level analysis: as in previous case we insert in the database the results of the analysis.

Interactive transactions (queries) with the DB will be handled using GUIs created by 4GL scripts. Programmatic transactions with the DB will be performed using embedded SQL calls in Fortran and C programs.

4. Analysis of PDS Calibration Data

A first level, very simple, analysis involves the processing of spectra and pseudo images (e.g., fits with gaussians or laurentians to photopeaks or to rise-time peaks). This analysis is quite simple but, to be performed effectively, needs a high level of interaction with the visualization of data and model. The results of this analysis level are obtained as table files and are then directly archived in the PDS database.

The second level analysis is more complex. It uses data produced by first level analysis, original spectra, HK and ancillary files. In this level we foresee to perform trend analysis, to calculate the conversion from PHA ADC channel to equivalent energy of input photons, to optimize the background rejection using pulse shape amplitude, to perform analysis of the stability of the detector using series of counts versus time or counts versus other parameters, to estimate the performance of the automatic gain control. Also in this case the results of this analysis are obtained mainly as table files and are then directly archived in the PDS database.

Most of the functions on data display and on data manipulations will be implemented using IDL (t.m. of RSI Inc.). More scientific analysis will be performed using tools developed by us: C and Fortran programs (written conforming to the POSIX standard) and IDL procedures. In the development of our tools, we will strictly conform to the guidelines for SW development issued by the SAX Data Analysis Working Group. These choices should grant to the system a life time long enough to cover the pre-operational and the full operational phases of PDS.

Acknowledgments. This work was funded by dedicated ASI grants. We like to acknowledge invaluable discussions with all the members of the SAX Data Analysis Working Group. In particular we are deeply indebted to Lucio Chiappetti for many and many excellent hints, suggestions, criticisms, ideas that undoubtedly improved the quality of our work.

References

Chiappetti, L. 1991, SAX DAWG Report 15/91

Chiappetti, L. 1992, SAX DAWG Report 20/92

Frontera, F., Dal Fiume, D., Pamini, M., Poulsen, J. M., Basili, A., Franceschini, T., Landini, G., Silvestri, S., Costa, E., Cardini, D., Emanuele, A., & Rubini, A. 1991, Adv. Space Res., 11, 281

Frontera, F., Dal Fiume, D., Pamini, M., Zhang, C. M., Basili, A., Franceschini, T., Landini, G., Silvestri, S., Costa, E., Cardini, D., Emanuele, A., & Rubini, A. 1992 Il Nuovo Cimento, 15C, 86

Frontera, F., Dal Fiume, D., Landini, G., Artina, E., Biserni, M., Chiaverini, V., Monzani, F., Costa, E., & Butler, R. C. 1993, IEEE TNS, in press

Maccarone, M. C., Sacco, B., Chiappetti, L., Dal Fiume, D., Trifoglio, M., Piro, L., Ubertini, P., Perola, G. C., Heise, J., & Parmar, A. 1992, in Astronomical Data Analysis Software and Systems I, A.S.P. Conf. Ser., Vol. 25, eds. D.M. Worrall, C. Biemesderfer, & J. Barnes, 151

Astronomical Data Analysis Software and Systems III
ASP Conference Series, Vol. 61, 1994
D. R. Crabtree, R. J. Hanisch, and J. Barnes, eds.

The COMPTEL Processing and Analysis Software System (COMPASS)

C. P. de Vries[1]

SRON, P.O. Box 9504, 2300RA Leiden, The Netherlands

Abstract. The data analysis system COMPASS of the γ-ray Compton telescope COMPTEL aboard the Compton-GRO spacecraft is described. A continuous stream of data of the order of 1 kbytes per second is generated by the instrument. The data processing and analysis software is build around a relational database management system (RDBMS) in order to be able to trace processing history and current status of all COMPTEL data. Four institutes cooperate in this effort, requiring procedures to keep local RDBMS contents identical among the sites and efficient exchange of data using network facilities. Lately there has been a gradual migration of the system from central processing facilities towards clusters of workstations.

1. Introduction

The imaging compton telescope COMPTEL is one of the four γ-ray detectors aboard the Compton Gamma-ray Observatory CGRO. Detailed descriptions have been presented by: Schönfelder et al. (1993) and Diehl et al. (1992). COMPTEL is sensitive to γ-rays from 750 keV to 30 MeV with a field of view of approximately 1 sr. The instrument consists of two detector layers. The upper one has 7 liquid scintillator modules (D1) each module surrounded by 8 photomultiplier tubes (PMTs). The bottom layer has 14 crystal scintillator modules (D2) each with 7 PMTs attached. Each detector layer is surrounded by anticoincidence shields to reject charged-particle events.

A γ-ray event is normally registered after an incident photon is Comp-ton-scattered in a D1 module, and then totally absorbed in a D2 module. Relative PMT signals define the scatter and absorption location in the instrument. Total signals define the energy. These parameters allow computation of scatter direction and scatter angle of the photon.

A maximum of 20 events per second can be registered. For each event all relevant PMT signals, anticoincidence signals, time of flight between D1 and D2 and a few other parameters are recorded, which results in a continuous stream of data of about 1 kbytes per second, about 30 Gbyte per year. Routine instrument calibration- and response processing (see Diehl et al. (1992), den Herder et al.

[1]On behalf of the COMPTEL collaboration: Max-Planck-Institut for Extraterrestische Physik, Garching, Germany; SRON-Leiden, Leiden, The Netherlands; Space Science Department of ESA/ESTEC, Noordwijk, The Netherlands; University of New Hampshire, Durham, USA

(1992)), expands the initial data stream to about 150 Gbyte per year. These data are exported to all sites of the collaboration for further analysis.

2. Software Requirements

Four sites at widely separated places having a great variety of computer hardware had to work closely together. This led to the following set of requirements initially defined to design the COMPTEL Processing and Analysis Software System (COMPASS):

- The system should be functionally equivalent at all sites including a uniform user interface. It must be possible to develop data-reduction and analysis codes at each of the four sites and install at all sites without modifications, requiring a uniform systems interface layer.
- All original data and processing products should have uniform descriptions and should be properly catalogued. The descriptions should be available at all sites.
- The heritage of data within COMPASS should be traceable. This includes the possibility of tracing all processing done, starting from the raw data including the task input parameters in all processing programs to arrive at a particular result. This immediately implies proper software configuration and version control as well.
- There should be separate test environments where new code or updates can be tested without interfering in any way with the usual production processing.
- Easy exchange of data, software and their respective descriptors between the various sites allowing for differences in file systems, binary formats and operating systems.

Within this system the data analysis tasks are developed. This includes software to separate the various data streams, calibration software, and software for instrument-response determination and scientific analysis.

3. Implementation

User interaction is through a user interface program, or user shell (USH). A database keeps a description of which tasks the user may run and to which datafiles the user has access. The USH sets up (a) panel(s) for task parameter input based on the description in the database. Processing is usually done in background. The executable is built based on the current software version description in the database. Actual task parameters plus the descriptors of the newly generated files are stored in the database, thus ensuring the possibility of proper heritage tracing of all data files.

To keep all task I/O separated from the operating system a separate data access layer (DAL) is available. The DAL interface takes care of differences between operating systems. In addition, the DAL permits for changing binary data formats (e.g., to permit more efficient data storage) for a given type of dataset without affecting the science programs source codes.

After task execution has taken place the user can assign a quality to an output dataset descriptor depending on scientific arguments, thus making future selections of good data easier.

The ORACLE RDBMS has been selected because of the availability on all hardware platforms used and its built-in mechanisms for easy exchange of database contents via so-called dump files. Weekly database dump files are exchanged between sites to keep the RDBMS contents identical. Oracle forms and report procedures are used for more direct interaction with the database. Separate administrative database tables exist to keep track of software changes and status of exchange of database contents between sites. The update of relevant information is done automatically whenever possible. This all ensures a proper history tracing of the complete system configuration at all times.

4. Experience and Operations

Using an RDBMS has proved to make quite a flexible system. It allows not only for easy software and task parameter changes, but e.g., also to administer changes in dataset binary formats which has proved to be critical.

The large volumes of data involved deserve special attention. Disk space is limited and consequently data required for a job are not always readily available. The fact that task input dataset identifiers are available in the database prior to task execution, permits easy automatic retrieval of the required datafiles from archive. Many input dataset files may be combined in lists having a single descriptor. The DAL takes care of switching between one file in the list to the next one, upon getting a read request from a task. For lists combining huge volumes of data not all physical files have to be on disk simultaneously, but may be retrieved when actually needed and deleted when used.

Most smaller items (source codes, database dumps etc.) which need exchange between sites are transferred via e-mail or FTP. For large datafiles transport takes place on tapes (more recently DAT tapes). Each type of file sent is made uniquely identifiable allowing for completely automatic processing.

For data files a special encoding scheme has been designed which all operating systems can process and convert into their own binary datafile format. This encoding scheme adds the complete datafile descriptor, normally stored in the database, to the export format file. This allows the receiving site to restore the descriptor of the file received in the database, independent of the actual (weekly) database update. Descriptors sent via database dump files normally get a special flag at the receiving site to indicate the actual file is not yet present at the site. Once the actual file has been received the flag is set accordingly.

Smooth site communication and exchange of data and software is critical in order to allow distributed analysis of data. The bandwidth of international networks at this moment is sufficient to meet the current COMPASS requirements. A continuous effort is being made to automate the routine maintenance and updates to the COMPASS system between the COMPTEL sites.

5. Current Status and Future

Originally all COMPASS sites made use of central processing facilities, in most cases consisting of large mainframe computers. Designing the COMPASS software to work on a great diversity of systems initially seemed quite a burden, however, when sufficiently powerful and inexpensive Unix workstations came on the market the COMPASS system could be easily ported.

The COMPASS code structure is such that all scientific source code is system independent. Only a limited set of host-specific routines in the USH and DAL needs to be modified to adapt to a different operating system environment. Gradually now, sites move away from their mainframe computers towards the workstation concept. At two of the sites (Leiden, and New Hampshire) the conversion is essentially complete. At Leiden for example, COMPASS is now run on a cluster of (20) Sun workstations. Background job execution can be evenly divided over the workstations, either by system choice or user selection. The database configuration is based on the client-server architecture using one particular workstation as a database server. Another workstation acts as a data server keeping the COMPASS data files being used on its disks. A problem so far has been to keep all COMPASS data in an online archive accessible by the workstations. Traditionally mainframe computers did have a 24-hour possibility for archival-tape setup. Now however, inexpensive reliable tape "stackers" have become available on the market. In the Leiden system, an Exabyte tape stacker, capable of holding 300 Gbyte of data has been installed to replace the mainframe based data archive.

6. Concluding Remarks

The COMPASS system design has proved to be a flexible system capable of dealing with changing requirements, software changes, data formats etc., in an easy way. An RDBMS as heart of the system has fulfilled its promises. To guarantee good quality of scientific processing, the possibility for heritage tracing of data has proved to be invaluable as has the possibility of proper overviews of data and processing status.

Hardware has now become available to enable online data archival on workstations in sufficient quantities. This allows one to make full use of the cheap computing power of workstations.

References

Schönfelder V. et al., 1993, ApJS, 86, 657

Diehl R. et al., Data Analysis in Astronomy IV., ed Di Gesù et al (Plenum), 201

Den Herder J.W. et al., Data Analysis in Astronomy IV., ed Di Gesù et al (Plenum), 217

Astronomical Data Analysis Software and Systems III
ASP Conference Series, Vol. 61, 1994
D. R. Crabtree, R. J. Hanisch, and J. Barnes, eds.

Tools for Use with Low Signal/Noise Data

Herman L. Marshall
Massachusetts Institute of Technology, Center for Space Research, Rm 37-667a, Cambridge, MA 02139

Abstract. Tools that may be used for examining low signal/noise data are described. The first example is sensitive detection of point sources even when the background is so low that Gaussian statistics may not be applicable, such as the case in high energy astronomy. Second, an algorithm has been developed for estimating and fitting distributions of source samples when there are marginal (e.g., $< 3\sigma$) detections without resorting to the assignation of upper limits. Finally, a spectrum fitting routine has been developed that can be used to estimate spectral parameters in very faint spectra where χ^2 statistics cannot be applied. An application to ROSAT data is shown.

1. Introduction

I'd like to report here on progress made on three tools that can be used for handling astronomical data with low signal to noise ratio (SNR). Some of these results draw directly from earlier work reported in 1991 (Marshall 1992, hereafter Paper I). Paper I gives more details setting up the problems of interest.

2. Source Detection Threshold

In Paper I, an outline of an efficient source detection method was outlined and investigated in detail. The basic method relied upon use of a "matched filter" to look for point sources in a large data set. Most practical problems in the method were solved there except one: determining the threshold for detecting a source when the data have a low average rate, obeying Poissonian statistics.

In the 2D case, the detection statistic is formed by convolving the data, C_{kl}, with the point spread function, ϕ_{ij}:

$$S_{kl} = \sum_i \sum_j \phi_{ij} C_{k-i,l-j} \tag{1}$$

and a source is declared if the value of S is larger than some threshold. When the background, B, is large then Gaussian statistics hold and the Normal distribution can be used. In the case of small B, then the $P(S)$ is not easy to compute. After extensive simulations, however, I have found that there is a simple approximation for determining the threshold for small B:

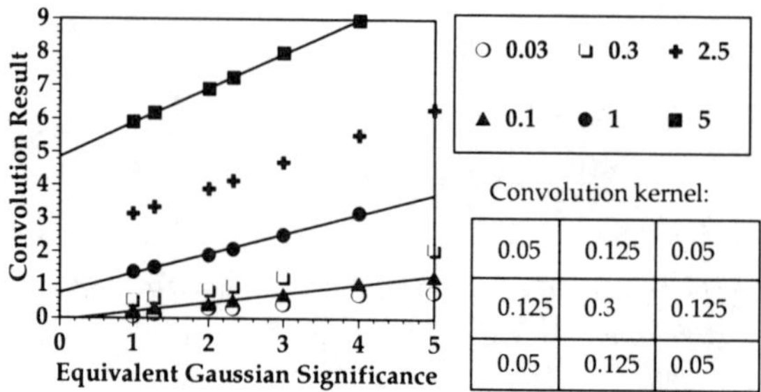

Figure 1. Results of convolving a 2D kernel with simulated images at various background levels.

$$T = K[(\sum_i \sum_j \phi_{ij}^2)^{1/2}] + \Delta] + B - \Delta \quad (2)$$

$$\Delta = 0.7 \sum_i \sum_j \phi_{ij}^2 \quad (3)$$

This formula holds for a variety of PSF shapes and over a large range of significance levels, approaches the Gaussian case for high background, and works well down to very small background values. Figure 1 shows results of one simulation run. The formula also works in one dimension, so it can be applied to running sums and, when the PSF consists of only one pixel, approximates the Poisson probability distribution.

3. Generating Distributions from Samples with Marginal Detections

The basic methods were outlined in Paper I. In the first method, one may derive an *a posteriori* probability distribution for a quantity (say, a flux) given the observed value (observed counts) and the uncertainty in the measurement. Using these distributions, one may use the second method to derive a non-parametric estimate of the distribution of the intrinsic quantity x, when given a population with the measured values and their uncertainties. I have performed extensive Monte Carlo simulations of distributions derived from random data samples using this method and compared them to the distributions that would be derived using the Kaplan-Meier (K-M) method from survival statistics. Figure 2 shows one such sample generated using an exponential probability density (with unit scale) and uncorrelated uncertainties which are uniformly distributed on the interval [0,1].

The new method comes closer to the intrinsic distribution than a Kaplan-Meier (K-M) method based on forming "detections" and "bounds" (cf. Avni et al. 1980). This conclusion is quantified in the right side of Figure 2, which shows the distribution of maximum differences between the derived and intrinsic distributions (a measure which is analogous to the Kolmogorov-Smirnov test). For a sample of 60 objects (which is heavily censored when 3σ limits are applied),

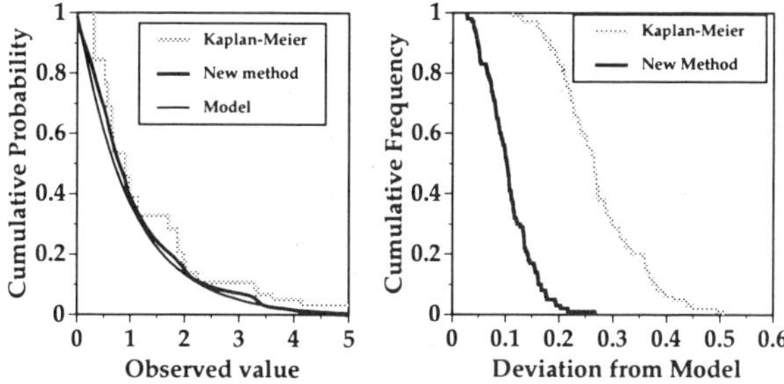

Figure 2. Illustration of two low signal methods. *Left* Comparison of estimates of a distribution to the model; 60 objects were simulated. *Right* Distribution of maximum differences between the model and the estimated distribuiton for 100 independent simulation sets such as used for the Figure at left.

distributions derived by the new method deviate by no more than 0.2 in 5% of the random samples while deviations at least this large occur more than 80% of the time using the K-M distribution estimator. Thus, the new method provides a better approximation to the intrinsic distribution and can be used to compare models to data with low signal/noise.

4. Spectral Modelling of Low Signal/Noise Data

The current version of the IRAF/PROS package employs a spectral fitting routine based on minimizing the value of χ^2 computed from the residuals between the observed and expected pulse height distributions. For the brightest sources, this procedure is acceptable but it breaks down when the number of events per channel is less than about 25, as will be the case near the endpoints of the spectrum and for the entire spectra of many sources. Figure 3 shows spectral fitting results from the IRAF/xray/spectral package for 3C 263, a quasar observed because of its low intrinsic N_H.

We extracted source and background regions to external tables and then developed a C program to test various spectral models. In order to optimize the analysis, the ROSAT PSF is used in the likelihood minimization so that the source parameters and the background rate per pulse height bin are estimated simultaneously (for a total of 69 parameters). The likelihood equation that was minimized was

$$\begin{aligned} S \;=\; & -2\sum_{ijk} C_{ijk}\log(\phi_i n_{jk} + A_{jk}b_{jk}) - 2\sum_{jk} B_{jk}\,\log(b_{jk}) \;+ \\ & 2\sum_j \tau_k \sum_k n_{jk} + 2\left(\sum_i \omega_i\right)\left(\sum_{jk} b_{jk}\right) + 2\omega_B \sum_j \tau_k \sum_k b_{jk} \qquad (4) \end{aligned}$$

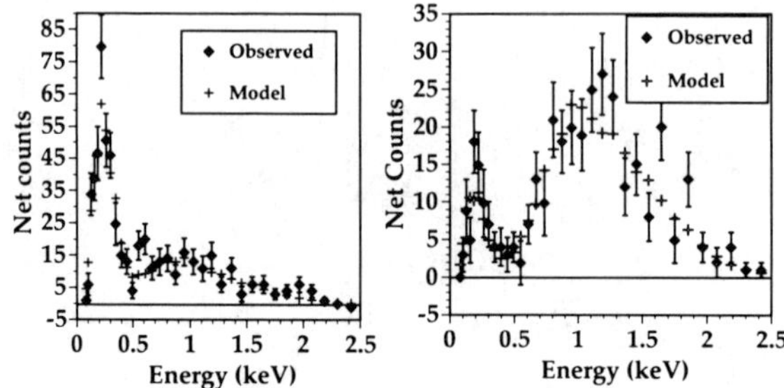

Figure 3. Observed net counts and model for ROSAT PSPC observations of 3C 263. Error bars come from the IRAF `xray/spectral` routine. *Left* Data for the PSPC observation without a filter, *right* PSPC data with Boron filter.

where B_{jk} is the observed background counts in pulse height (PH) j and filter k (for Boron filter or none), b_{jk} are the model background values in these bins, C_{ijk} are the counts in angular annulus bin i from the center of the source and in bin jk, n_{jk} is the expected counts in these bins (formed as an integral over the PH effective area *vs.* energy), A_{jk} is the effective area in bin jk, τ_k is the exposure time in filter k, ω_i is the solid angular area of the ith annulus, and ω_B is the size of the background region.

Using this approach, we have obtained the best possible uncertainties on the model parameters. This analysis results in a very different set of spectral parameters: instead of accepting the Galactic value of the neutral hydrogen column, N_H, the likelihood analysis indicates $N_H < N_{H,gal}$, so that a steep spectral component below 1 keV must be added to the model.

Acknowledgments. I would like to thank William T. Boyd for his work on the ROSAT data analysis. This work was supported in part by MIT subcontract SV1-61010 to SAO and NASA contract NAS5-30180 to U.C. Berkeley.

References

Avni, Y., Soltan, A., Tananbaum, H., and Zamorani, G. 1980, ApJ, 238, 800

Marshall, H.L. 1992, in proceedings of Statistical Challenges in Modern Astronomy, E. D. Feigelson and G. Babu, eds., p. 247

Part 4. Data Analysis
Section E. Radio Astronomy

Astronomical Data Analysis Software and Systems III
ASP Conference Series, Vol. 61, 1994
D. R. Crabtree, R. J. Hanisch, and J. Barnes, eds.

Radio Synthesis Imaging — A High Performance Computing and Communications Project

Richard M. Crutcher

National Center for Supercomputing Applications and Astronomy Department, University of Illinois, Urbana, IL 61801

Abstract. This is a report on a major project in High Performance Computing and Communications at the National Center for Supercomputing Applications. The project has as its goal the support of the imaging, data storage, visualization and analysis needs of radio astronomy aperture-synthesis arrays in the highest performance computing and communications environments.

1. Introduction

Aperture synthesis array telescopes such as the VLA and BIMA make it possible to produce maps with high angular resolution at radio wavelengths. Because the images produced by a synthesis array are formed in a digital computer by Fourier transformation of the observed visibilities, the computer system is an integral part of the synthesis array telescope — its image-forming element. The computational requirements of synthesis arrays are driven by algorithms developed to overcome two factors which greatly degrade performance — incomplete coverage of the aperture plane and time varying systematic errors due to instrumental and atmospheric instabilities. Deconvolution and self-calibration can largely compensate for these factors and lead to the production of very significantly improved images. However, both techniques are very computationally intensive. Radio astronomy synthesis imaging is an area which has been severely handicapped by lack of sufficient computing power. Many important types of problems in astrophysics simply cannot be attacked today — not because radio arrays cannot obtain the data, but because high performance computing facilities with appropriate software to properly process the data are not available!

The National Science Foundation has funded a five-year High Performance Computing and Communications project at the National Center for Supercomputing Applications (NCSA) to address the problem of inadequate computing support for radio synthesis arrays. The project has four components: 1) Implementation of synthesis-array software on a very high performance computer system. 2) Development and prototyping of real-time processing of synthesis array data from a remote telescope facility, and development of distributed computing technology. 3) Establishment of a data archive for both unprocessed visibility data and completed images, so that astronomers may obtain data or images over the national networks. 4) Development of advanced tools for visualization and analysis of three-dimensional scientific data sets. We discuss these components below.

2. High Performance Computing

The Thinking Machines Corporation CM-5 at NCSA is a 512 node massively parallel processor (MPP) computer with 16 gigabytes of RAM; each node has approximately the processing power of the original Cray 1 supercomputer. Our plan is to port the BIMA software system MIRIAD to the CM-5. The computationally intensive routines are being rewritten and optimized to take advantage of the MPP architecture of the CM-5. This effort is being done with assistance from on-site Thinking Machines Corporation personnel at NCSA. We are developing a distributed computing environment. The bulk of the MIRIAD code, including the user interface and non-parallel portions, will run on workstations with high-speed connections to the supercomputer. The project has acquired a SGI Onyx supercomputing workstation with HiPPI (800 Mbit/sec) connection to the CM-5. The NCSA Data Transfer Mechanism (DTM) will be used for interprocess communications between the CM-5 and the Onyx. The CM-5 will appear (transparently) to be an extremely fast co-processor in the workstation. This system in production mode will allow astronomers throughout the nation transparent access to sufficient computing power to process any synthesis array data which the BIMA or NRAO telescopes can produce, resulting in significantly increased scientific productivity. After implementing and optimizing existing algorithms, we will explore the development of new algorithms specifically aimed at efficient data reduction of large scale problems using MPP architectures. Finally, the experience gained will be used to implement the computationally intensive synthesis array tasks in the AIPS++ environment as the AIPS++ software system matures.

3. High Performance Communications

Astronomers have two communications problems in using radio synthesis arrays. One is access by the telescope to the required high performance computing system; the second is access by the astronomer to the system.

Arrays are located in remote sites, and it is often not practical to go to the telescope for the experiment. Even when it is feasible, the telescopes do not have on-site high performance computing systems for data processing. Data are usually recorded on magnetic tape for off-site processing. There is generally no practical means for observers to interact with their data until many hours or days after the observing run is completed. This unfortunate limitation means that all observing files must be prepared well ahead of the observing time, with no opportunity to react to changes in observing conditions or errors in the observing setup. Moreover, there are time-variable astronomical objects — such as solar flares and comets — which demand real-time observing. Now, the data processing for such an experiment may take place weeks or months later, after the variable emission has changed. This problem could be solved if telescope data could be transmitted in real time to a high performance computing system. The BIMA array will be a prototype of such a system. This project will install a T1 data link between Hat Creek and NCSA so that the BIMA full sustained data rate can be transmitted in real time. BIMA astronomers will be able to access their data from their offices as the observations are in progress and to rapidly and fully process the data on the NCSA system.

Astronomers who use U.S. radio synthesis arrays are located throughout the country. Although workstations will be used extensively for the less computationally intensive data processing and analysis tasks, use of the highest performance computers is essential for parts of the research. Efficient use of the high performance computing system requires that it be accessible with high bandwidths to astronomers' individual institutes or universities. Because so much of the computational work is interactive image processing, the most efficient strategy is a "distributed metacomputer" mode in which the user interface and smaller tasks are performed on a local workstation while the computationally intensive tasks are off-loaded to the supercomputer. Therefore, in addition to the development of a local distributed computer environment, we plan to develop a remote distributed computing system. We will use the 600 Mbit/sec BLANCA testbed network to connect an Onyx workstation at the University of Wisconsin to the NCSA HiPPI network. Wisconsin astronomers will use the system from Madison to monitor processing of their data at NCSA, interrupt the system as required in order to change processing parameters, and visualize and analyze their data remotely.

4. On-Line Archive and Digital Library

The BIMA synthesis array will produce about 300 Gbytes of data per year, with an approximately equal volume of fully processed image cubes. Management of such large data sets is a major problem, particularly since scientists who want to use the data are located throughout the country. We will address this problem by developing a nationally accessible system in which we will store the data and supporting information in an on-line archive. For the data archive hardware we expect to use an optical WORM jukebox with a capacity of at least 1 terabyte. We have begun development of a software system which allows a user anywhere on the network to access the catalog computer with an X Window client interface, set criteria for search of the database, and select among the data sets which meet the criteria. The user can then transfer the data set to the supercomputer system using a special protocol designed for rapid transfer of the large datasets across high speed networks. Alternatively, the data can be transferred to the user's home workstation for processing there. Another paper in this volume more fully describes the archive and database work of the project.

In addition to serving as an archive for BIMA visibility data and images, we intend the system to function as a digital library. We will solicit contributions of relevant images produced by other telescopes and at other wavelengths. For example, astronomers studying star formation in Orion might be able to call up BIMA data cubes in several molecular lines, VLA images in the neutral hydrogen line and the continuum, images at infrared wavelengths obtained with IRAS, and optical images produced with CCD cameras on ground-based telescopes. The goal will be to enable astronomers interested in a particular object to obtain images and data, placed in the digital library by other scientists, for planning new experiments, comparison with new data, and testing of theoretical results.

5. Visualization and Image Analysis

Visualization of the very large data cubes obtained with modern spectral-line synthesis arrays such as BIMA is an essential tool for the analysis of the data. This project will focus on developing strategies for utilizing interactive visualization in a high-performance, heterogeneous computing environment. The visualization software development will be within the framework of the AIPS++ system. Astronomers will be able to use the real-time visualization to edit the data, change processing parameters, try multiple processing options, and very quickly produce optimal images which can be analyzed with the visualization software. There will be two phases of the visualization development.

The first phase will extend the capabilities of Nicer-Slicer-Dicer, a volumetric visualization tool written at NCSA for high-end SGI workstations. Nicer-Slicer-Dicer is a distributed tool using a client-server model for data analysis and visualization. Nicer-Slicer-Dicer's visualization capabilities include isosurface extraction and a hardware algorithm for direct volume visualization. In the current version, a data server process runs on the NCSA Convex 3880 and can be used for data management and analysis functions. A client process runs on a SGI graphics workstation and provides visualization and user interaction capabilities. Nicer-Slicer-Dicer was designed to handle the very large data sets that are increasingly common in radio astronomy. The user can request that data be sub-sampled, or that only sub-regions of the data be retrieved from the server and sent to the client for visualization.

The second phase will involve the development and implementation of parallel visualization algorithms for the CM-5. While using the graphics machine to perform the visual mapping allows for exploiting the graphics hardware to support some aspects of the operation, this requires transferring a considerable amount of data from the computational machine to the graphics machine. The alternative is to use the parallel machine for the mapping from data to polygonal representation or image. Implementing visualization algorithms on the parallel machine can also produce higher quality images in shorter time. Algorithms for direct volume visualization can be particularly compute intensive. Implementing volume visualization techniques on the CM-5 will allow this to be done with a fast enough turn-around to support interactive visualization. Performing visualization operations on the parallel machine can also make three-dimensional visualization techniques, such as direct volume rendering, accessible to the segment of the research community that might not have access to high-performance desktop graphics machines. If only a two-dimensional image needs to be displayed, then a machine limited to X Windows is adequate. Finally, if a volume visualization algorithm is applied to the data on the same machine used for computation, only the resultant two-dimensional image must be transferred from the computational machine to the visualization workstation.

Acknowledgments. This project is funded by the National Science Foundation under grant NSF ASC 92-17384 and by the National Center for Supercomputing Applications.

Astronomical Data Analysis Software and Systems III
ASP Conference Series, Vol. 61, 1994
D. R. Crabtree, R. J. Hanisch, and J. Barnes, eds.

The AIPS++ Array and Image Classes

B. E. Glendenning

National Radio Astronomy Observatory, 520 Edgemont Road, Charlottesville, VA 22903-2475

Abstract. An Object-Oriented library of powerful and convenient C++ classes is being developed as a part of the AIPS++ project. Two sets of classes which provide Array and Image functionality are described.

The Array classes (which have been available for some time) provide arbitrary N-dimensional arrays, as well as 1-, 2- and 3-dimensional specializations. Convenient "whole-array" expressions may be written, and decimated array sections are available as first class objects.

The Image classes (which are presently under construction) offer the same sorts of features as arrays. Additionally, they must provide services such as binding coordinates to pixels, providing error estimates, and mapping sections of large multi-dimensional images to memory efficiently.

1. Introduction

AIPS++ is a successor to the popular AIPS package which is being constructed by a consortium of seven radio observatories. Some general information about AIPS++ can be found in Croes (1992) and in Norris (1993).

AIPS++ is being written in an object-oriented style using the C++ programming language. Object-oriented programming is a style of programming in which implementation details are encapsulated in a *class* (which corresponds to a "type") so that those implementation details may change without forcing many other changes in the program. Object-oriented programming also allows subsequent classes to extend or redefine the behavior of existing classes through a mechanism known as *inheritance.*

C++ is an object-oriented extension to the C language. A reasonable introduction to C++ with some discussion of object-oriented programming may be found in Lippman (1991).

Much of the work to date has concentrated on various foundation classes, such as the Array classes described here, and the Table data system described elsewhere in these proceedings (van Diepen 1993). With foundation classes in place, work is proceeding on higher level ("astronomical") classes; these classes can be said to be layered on top of the foundation classes. The Image classes (presently under construction) are described later in this paper.

An initial public release of the library (which contains the Array classes, but not the preliminary image classes) was made in November, 1993. Fuller public releases are anticipated in 1994 and beyond. Contact *aips2-request@nrao.edu*

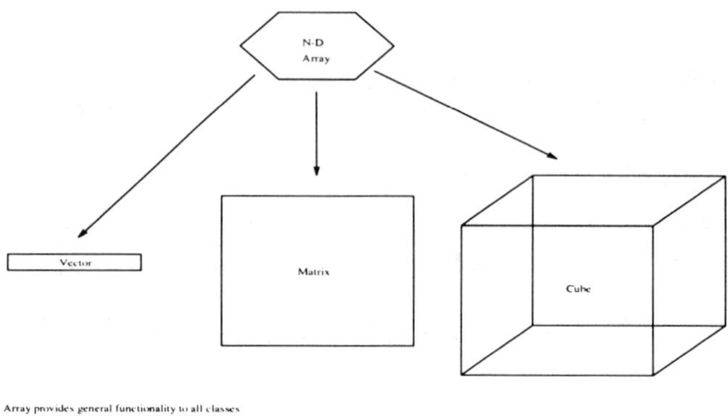

Figure 1. Basic Array Classes.

via Internet e-mail, or *The AIPS++ Project Office* at the author's address for current release, or any other, information.

2. Array Classes

The basic Array classes (Figure 1) are used to represent a multidimensional lattice of regularly spaced values. An *Array* object can be of any dimensionality. There are specialized Vector, Matrix, and Cube classes to represent one-, two-, and three-dimensional arrays. These specializations are provided both for convenience (in general an n-dimensional array must be indexed with a length-n integral vector, known as an *IPosition*) and efficiency. Features of the array classes include:

Templating The array classes are able to hold values of essentially any type, so for example you could have a Matrix of strings, or a Vector of images. The ability to be able to use containers (arrays, linked lists, associative arrays, ...) is very powerful, although it stresses the compilation system.

Variable Origin While zero is the default origin, it may be changed to any value.

Whole-array expressions Rather than having to write multiple loops to step through an array, one can write expressions that work on whole-arrays much as in the languages FORTRAN 90 and IDL. While the default is to use "element-by-element" arithmetic, the linear algebra products are also available. All the usual arithmetic, logical, transcendental, statistical and other functions are available. (One particularly useful such function applies an arbitrary user function to every element of the array).

Other Mathematics There are various higher level mathematical classes and functions that use the Array class. For example there are classes that:

- Grid irregularly sampled data onto a grid (including convolution corrections).

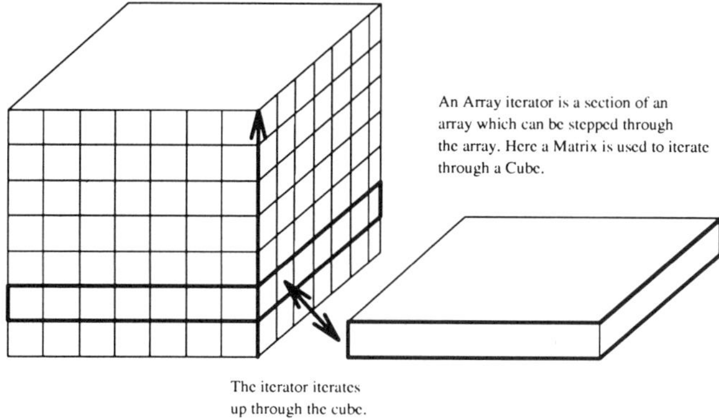

Figure 2. Iterating Through an Array.

- Perform FFT's and DFT's.
- Perform various Matrix decompositions (LU and QR)

Array references It is often useful for an array to reference storage inside another array. For example, in the following line of C++:

```
matrix.row(5) = 1;
```

the result of "matrix.row(5)" is a Vector whose storage corresponds to storage in the variable "matrix." Assignment to the resulting vector changes the values along the appropriate row of the matrix. Another way of creating a reference is to take an array section ("slice"). An Array section is a regular subvolume of the array, possibly decimated ("every third column") along one or more axes. The above "row" function is a special case of this.

Iteration When you are given an array of potentially arbitrary dimensionality, for example an argument to a function you are writing, it is often convenient or efficient to iterate through the array using sub-arrays of a known dimensionality. For example, Figure 2 illustrates iterating through a Cube array with a cursor which is a Matrix. Note that the cursor array is a reference; changing values through the cursor will change the underlying storage in the Cube.

More information on the array classes is available in Willis *et. al.* (1993) and in internal project documentation. For the latter contact the project office.

3. Image Classes

The image classes need to provide all the services the Array classes do (*access* to pixels, *whole-image* manipulations, *iteration*, etc.) as well as some others.

Coordinates Images have astronomical coordinates. Besides simply converting between pixel and image coordinates, one must be able to perform operations like registering images based on their coordinates, changing coordinate systems, and expressing uncertainties in positions.

Pixel Types and Errors Pixels have both a type (including units) and errors which must be accounted for and propagated in image operations. Propagation of image errors is a research topic. Initial versions will have very simple statistical models.

I/O Images can still be much larger than typical memories. For example, radio-interferometric cubes could conceivably be many gigabytes. The approach that is being used to encapsulate image I/O is to use an image iterator. In particular, one can choose to take a default cursor, which maps to an efficient shape (a tile) for the I/O system, or one can choose a shape that is convenient for the algorithm (i.e., a sliding window). One can also define the cursor to be as large as the entire image to lock it all into memory so you can ignore I/O (memory and swap space permitting).

Other Image Types Besides the straight storage of pixels, we also need to be able to deal with masked images (i.e., blanks), and possibly with images that are formed from analytical functions (e.g., simulations).

4. Status

The Array classes described here have been available for some time and were part of the November 93 AIPS++ Library release. The Image classes are currently under construction. They should be available internally to the project in early 1994.

References

Croes, G.A. 1992, "On AIPS++, A New Astronomical Information Processing System", in Astronomical Data Analysis Software and Systems II, A.S.P. Conf. Ser., Vol. 52, eds. R.J. Hanisch, R.J.V. Brissenden & J. Barnes, 156

Diepen, G. van 1993, AIPS++ Table Data System, this volume

Lippman, Stanley B. 1991, "C++ Primer — Second Edition", Addison Wesley

Norris, R.P. 1993, "AIPS++: A New Astronomical Imaging Package", in IAU Symposium #158: very High Angular Resolution Imaging, Ed. W. Tango, Kluwer, in press.

Willis, A.G., Healey, M.P., & Glendenning B.E. 1993, "The AIPS++ N-Dimensional Array Classes", in OON-SKI '93 — Proceedings of the First Annual Object-Oriented Numerics Conference

Astronomical Data Analysis Software and Systems III
ASP Conference Series, Vol. 61, 1994
D. R. Crabtree, R. J. Hanisch, and J. Barnes, eds.

AIPS++ Table Data System

G. van Diepen

Netherlands Foundation for Research in Astronomy, P.O.Box 2, 7990 AA Dwingeloo, The Netherlands

Abstract. The AIPS++ data system is based on tables, which can be mapped to different file formats. It allows for a uniform interface to all data files. Each column or row in a table can be handled as an ordinary vector. An X-based browser allows a user to view and possibly edit all data items in the table. Selection and sorting on rows and columns is possible using an SQL-like language. The system is written in C++. Two coding examples are given to show the ease of use of the table classes and the C++ language.

1. Introduction

AIPS++ is currently being developed by an international consortium of mainly radio astronomical institutes. It will almost entirely be written in C++. It was decided in an early stage that the data system should be based on tables. They offer great flexibility and can be mapped to different file formats (e.g., FITS). It also allows that common tools can be used for all data files. For example, an X-based table browser has been developed. This tool enables a user to view and possibly edit all data in any table.

In November 1993 a beta release of the available AIPS++ code will be announced, which contains the first version of the table system. This version already has a rich functionality, but the (well encapsulated) underlying IO-system is rather simple minded. In the next months a more sophisticated IO-system will be developed The TexInfo file Database.texi in the beta release contains a manual describing how to use the table classes.

2. Table Model

The table model is based on the FITS binary tables. It has, however, some extensions to increase the flexibility. An AIPS++ table consists of:

- A set of keywords (similar to FITS headers).
- A set of columns.
- Each table keyword or column can have a set of keywords attached to it. These keywords are useful to store, for example, a unit or FITS items crval, crpix, etc.

- Each keyword or cell in a column can hold a scalar, an N-dimensional array or a table. The latter allows for a hierarchy of tables.
- All built-in data types (Bool, int, float, etc.) are supported as well as strings and complex numbers (classes String and (D)Complex).
- A keyword or column containing arrays or tables can be direct or indirect. The data of a direct array or table is directly stored in the table cell. The shape of direct arrays and tables must be the same in all cells of a column, because each row in a table must have the same length. In case of indirect arrays or tables, the table cells only contain a reference to the array or table. Henceforth, their shape (and dimensionality) may vary per cell.

3. Kinds of Tables

The ordinary kind of table is the so-called filled table. All data in it is mapped to files using a storage manager. Multiple storage managers may exist to support multiple file formats. Currently only the simple AipsIO file format is used, but in the near future other formats like FITS or Karma (developed by Richard Gooch at CSIRO) will be used. All data is stored in a canonical format to allow easy sharing between different machines.

Virtual tables, on the contrary, will perform some kind of mapping between the table data and the possibly underlying files. Examples of virtual tables are:

- A reference table contains the result of a selection, projection or sort. It only contains references to the appropriate rows and columns in the actual table.
- A simulated observation can be calculated on the fly. Only its parameters needs to be stored in a file.

The application programmer and end user will in principle see no difference between the various kinds of tables. The differences are hidden in the lower level classes. The top level classes, which form the interface to the application programmer, offer the same consistent interface to all kinds of tables.

There can be a wide variety of virtual tables. A new kind of virtual table has to be derived from the base tables classes and implemented by the programmer.

4. Table Interface

The entire table system consists of many classes, but only a few of them are of interest to the application programmer. They are described in detail in the manual mentioned in the introduction. They make use of the many classes in the AIPS++ toolkit and of many features of the C++ language (like templates and overloaded operators).

The main operations a programmer can perform are:

- Accessing values in an arbitrary keyword, column and row.
- Accessing slices of an array.

- Getting and putting an entire column.
- Selecting and sorting rows
- Handling columns and rows as vectors for mathematical operations on them.
- Iterating through a table in any order.

Below two examples using the table classes are given to show the power of the table classes and the C++ language.

```
// Select and sort the data from an existing table.

#include <aips/Table.h>
#include <aips/TabExprNode.h>  // for select expression

main()
{
// Open an existing table.
// Select the rows for which U*U + V*V < 1.
// It uses the overloaded operators *, + and <.
// Sort the result on time (default is ascending).
//
    Table tab("table.name");
    Table seltab = tab(tab.col("U") * tab.col("U") +
                       tab.col("V") * tab.col("V") < 1);
    Table sortab = seltab.sort ("time");
}
```

```
// Example of use of table vectors.

#include <aips/Table.h>
#include <aips/TableVector.h>
#include <aips/TabVecMath.h>   // for math on TableVector
main()
{
// Open the table; allow for updates.
// Create a table vector for column time.
//
    Table tab("table.name", Table::Update);
    TableVector<float> timetv(tab,"time");
//
// Update the vector.
// This also updates the values in column time in the table.
//
    timetv = 2*timetv + 10;
}
```

So far, most attention has been paid to the programmer's interface. A few tools have been developed for the end user, but in the near future more attention will be paid to this.

- An SQL-like prototype language has been developed (using the GNU-tools bison and flex), which allows to perform a selection or sort on a table.
- A X-based table browser is written to interactively view the contents of a table. If possible, one can also change the contents. It uses the above mentioned SQL-like language to make it possible to browse only through a subset of a table.
- A simple table filler has been written to create a table from an ASCII file. This is mainly useful to create some prototype tables.

5. Conclusions

Now that the astronomical classes are being designed and developed, it appears that the AIPS++ table system is indeed very useful. At the lower class level a lot of work still has to be done, but the high level table classes are stable now. However, they are not used enough yet to decide if they are sufficiently complete.

C++ is a powerful language, albeit with its syntactical peculiarities. Especially the template mechanism is a very powerful feature. However, it is at the moment also a weakness of the language, because the available compilers cannot handle them efficiently yet. A link of a program can be a very time-consuming, and therefore sometimes frustrating process, when many templates have to be instantiated.

Acknowledgments. Allen Farris (ST ScI, Baltimore) made the original design of the table model.

Working with Brian Glendenning and Darrell Schiebel at NRAO in Charlottesville meant a giant step forwards on the path to mastering C++.

Rick Copeland (a summer student) designed and implemented the table browser using the Interviews package.

Bob Duquet (NRAO, Socorro) wrote some very useful functions to create a table from an ASCII file.

Several other people gave good advice to make the table classes and documentation more easily accessible.

Part 4. Data Analysis
Section F. Multi-Wavelength Analysis

Astronomical Data Analysis Software and Systems III
ASP Conference Series, Vol. 61, 1994
D. R. Crabtree, R. J. Hanisch, and J. Barnes, eds.

Radio to X-ray Observation of Quasars

Belinda Wilkes and Jonathan McDowell
Smithsonian Astrophysical Observatory, 60 Garden St., Cambridge, MA 02138

Abstract. Quasars are multi-wavelength emitters, with roughly equal energy output per decade from far-infrared through to X-ray frequencies. In recent years observations of individual quasars over this full frequency range have become increasingly possible with a corresponding increase in the amount of multi-wavelength data. By nature the data are diverse, being observed with a variety of telescopes, satellites and instruments. There is no standard software available that can combine and analyze such complex datasets. This paper describes and demonstrates the major software requirements of an investigation of quasar spectral energy distributions (SEDs) with reference to our own software (TIGER) as a useful example.

1. Introduction

Quasars are the brightest, most powerful objects known in the universe today. They are generally believed to be embedded in the central regions of a galaxy and to be powered by a super-massive black hole. With the exception of the extended radio emission, their vast energy output is concentrated into a small region so that in general they look like point sources rather than extended ones. It is only for those that are relatively weak and nearby that we can see the surrounding fuzz of a galaxy and be certain that it exists. Other defining characteristics are the presence of strong, broad, redshifted emission lines in the optical/ultra-violet and a flat, often blue, optical continuum and roughly equal energy output per decade in frequency from far-infrared (IR) to X-ray. About 10% are also strong radio sources.

This paper is concerned with the software requirements for an investigation of the characteristics of and mechanisms for producing the large amount of energy emitted by a quasar. Clearly a pre-requisite for attempting such an investigation is the availability of complete, multi-wavelength observations of a sample of quasars in order to delineate their spectral energy distributions (SEDs). Most quasars are at large distances from us and thus relatively faint. Over the past decade, advances in instrument, satellite and telescope technology have resulted in more and more quasars being accessible to multi-wavelength observation. During this time we have generated a database of $\gtrsim$ 100 quasar SEDs covering a full range of redshift and relative continuum properties (Elvis et al. 1993). We are currently in the process of analyzing these SEDs with the aim of determining the mechanisms involved. We will describe the types of data

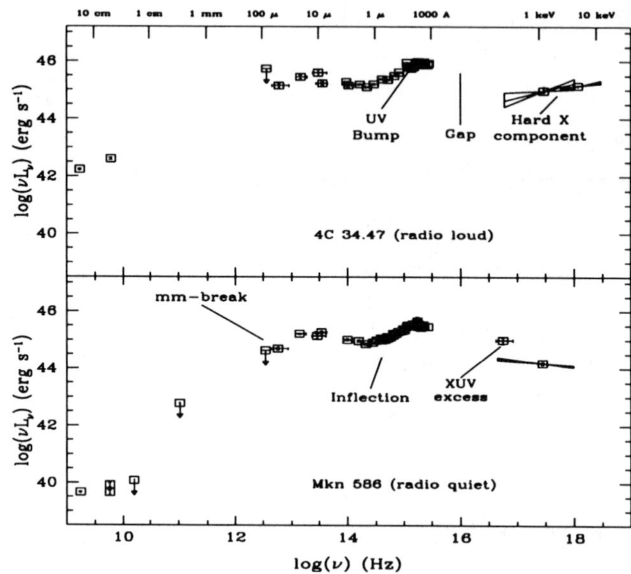

Figure 1. Examples of the radio–X-ray SED for a radio-loud and a radio-quiet quasar illustrating the main continuum features. The SEDs show logarithm of energy per unit logarithmic frequency interval in the rest frame (from Elvis et al. 1993 Fig 1).

with which we are dealing, software requirements in order to optimize data input and output and tools required for analysis and modeling the data.

2. Why Study Quasar SEDs?

The motivation for our study is to understand the energy generation mechanisms responsible for a quasar's energy output in the various spectral bands. Figure 1 shows the radio–X-ray SEDs of two quasars, one radio-loud and the other radio-quiet. It is clear that, although the energy output is roughly comparable throughout the whole far-IR–X-ray region of the SED, there is a characteristic structure which is key to our task. The IR–optical/UV region appears to consist of two bumps with a dip at $\sim 1\mu$m in between. The optical/UV bump does not peak in the observed range, but must do so in the so far unobserved extreme UV (EUV) region in order to meet the lower X-ray flux. The main differences between the two well-established classes of quasars: radio-loud and radio-quiet, are the slope of the X-ray spectrum and the strength of the mm break, with the radio-loud objects having ~ 3 orders of magnitude stronger radio emission and flatter X-ray slopes. In the IR–optical/UV however, the two classes look very similar and it is to this part of the spectrum that we will refer in discussing our software requirements.

Given the shape and different characteristics described above, a number of competing models have been suggested as contributors to the far-IR–UV quasar SED. These include: the host galaxy, thermal emission peaking around $\sim 1\mu$m; dust with a range of temperatures, some of it heated by the central continuum source, providing some/all of the IR emission (Sanders et al. 1989); thermal emission from an accretion disk with a range of temperatures, providing some/all of the optical/UV bump (Malkan and Sargent 1982); direct synchrotron emission, providing an IR–X-ray power law (Green et al. 1992) optically thin free-free emission dominating the optical/UV emission (Barvainis 1993). All these possible mechanisms would contribute over a wide range of frequencies so that the final SED would be a convolution of their various forms, emphasizing the need for complete observations in our investigation.

3. Observational Data

A single quasar SED consists of data from a large variety of sources and in a variety of forms. We summarize the facilities and software needed to obtain data in each wavelength range and the final form in which the data are generally presented.

X-ray data are obtained with satellites, e.g., *Einstein*, EXOSAT, *Ginga*, ROSAT, ASCA. In each case the details of energy range and resolution are different. The energy resolution of X-ray detectors is low so that a spectral study involves an iterative procedure of fitting a model folded through the instrument response to the data. Standard software to perform this analysis is available (e.g., IRAF/PROS, MIDAS/XSAS). The astrophysical result is then a flux at a specific energy combined with constraints on the parameters of the fitted spectral form, e.g., a power law slope with associated errors.

Spectral data are obtained either via satellite observations for ultra-violet (UV) data (e.g., IUE, HST) or ground-based telescopes for optical and infrared. Due to the faintness of most quasars, large, 4-metre-class ground-based telescopes are generally necessary. The resulting data are specified as flux as a function of wavelength, and can be reduced and analysed by IRAF, MIDAS and other, similar software packages.

Photometric data in the optical or infrared are obtained using small 1-metre-class ground-based telescopes. Again the data can be reduced using standard software and the results are numbers (fluxes or magnitudes) in each band/filter, e.g., UBVRIJHK color photometry.

Far-infrared data are available from the IRAS database as flux in each band/filter. The data are analyzed and the numbers generated using IRAS-specific software.

Radio and milli-metre data, fluxes or brightness temperatures in specific frequency bands, are obtained with ground-based radio telescopes. The data are generally reduced using the AIPS software.

While IRAF, MIDAS and similar software packages allow reduction of data in many different wavebands, there is no facility to combine data of different kinds into a single dataset. In principle it is possible to combine different spectral data. However this would require that all the data be binned to the same dispersion (angstroms per channel). This is not generally appropriate given the very different resolutions in optical, IR, and UV spectra. Other combinations, even simple ones such as optical photometry and spectra, are not currently possible in existing, standard software packages.

4. Software Requirements

There are four main areas with which software to handle multi-wavelength data must deal. These will be described in turn in the following sections.

4.1. Data Input: Constructing a Spectral Energy Distribution

The software must be able to accept input of data in all the diverse forms outlined above and be able to convert between each set of units and all data types allowing all the data for a given object to be placed in a single SED. In order to minimize operational error, it is advisable that the data input to the software be the data from the source, i.e., the numbers or datasets output from the reduction software or the actual numbers given in a paper and that the input file be readable, i.e., ascii, so that numbers can be easily checked and modified. All conversions are then performed by the software, ensuring uniformity and minimizing error. Information contained in the individual datasets should not be lost during this procedure, i.e., the resolution of each dataset should be preserved. Finally the software should be sufficiently robust that a change in a single flux point of an input dataset can easily be propagated into the final SED with no risk of additional errors being incurred.

An example of a data input file from our own software (TIGER) for the quasar 3C212 containing data from many different wavebands and in many different forms is given below:

```
# This file is in TIL (TIGER Input Language)
# TIL is designed to let you enter data from the literature in as
# close as possible to its original form. The TIGER program will
# compile the file into SED format, converting fluxes to nu f(nu)
# values in Jy Hz.
#
# First we select the object to which the data apply
obj 3C212
#
# FAR-INFRARED
# Specify information for these observations
ref Impey,Neugebauer.1988; inst IRAS; beam 400
date 830000
# Define the 4 IRAS bands using the FILTER command
# giving center, low and high frequencies
FILTER IRAS1 12 mu;15 mu;8.5 mu
FILTER IRAS2 25 mu;30 mu;19 mu
FILTER IRAS3 60 mu;80 mu;40 mu
FILTER IRAS4 100 mu;120 mu;83 mu
# Specify that fluxes are linear F(nu) values in mJy
lin fnu mJy
err1   # Tell the F command to expect one error value
```

```
F(IRAS1)  <94     # No error value required with limit
F(IRAS2)  <146   #
F(IRAS3)  <140    #
F(IRAS4)  599 105 # Flux and error for IRAS4 band
#
# OPTICAL
#
ref Smith,Spinrad.1980; inst Hale; beam 5; date 781023
lin fnu mJy err1
F(3400A) 0.021 0.005
F(3700A) 0.017 0.005
F(4000A) 0.021 0.005
F(4300A) 0.025 0.005
F(4600A) 0.025 0.005
F(4900A) 0.041 0.005
F(5200A) 0.054 0.005
F(5500A) 0.062 0.005
F(5800A) 0.095 0.005
F(6100A) 0.090 0.005
F(6400A) 0.103 0.005
F(6700A) 0.11 0.01
F(7000A) 0.13 0.03
F(7300A) 0.15 0.04
# Specify a magnitude; program knows the bands and
# will calculate B and V from V and B-V. User can
# specify an alternate set of magnitude zero points if desired.
V 19.1 0  B-V 0.4 0
#
# RADIO
#
ref Bregman,etal.1985; inst VLA; beam 5; date 830000
lin fnu Jy
F(5 GHz) 0.89 0
F(0.17GHz) 15.1 0
#
# X-RAY
# Define the IPC bandpass
ref This Paper
inst IPC; beam 120; date 810213
filter IPC  1 keV;0.5 keV;3 keV
# Specify that fluxes are integrated over band
# in units of 1E-13 cgs units. Software will
# convert to a flux at band center (1 keV)
lin nufnu int cgs dlf -13
# Specify a power law slope and errors to go with next
# flux value
alpha 1 0 0
F(IPC) 14.0 2.5
filter PSPC 1 keV;0.1 keV; 2.4keV
```

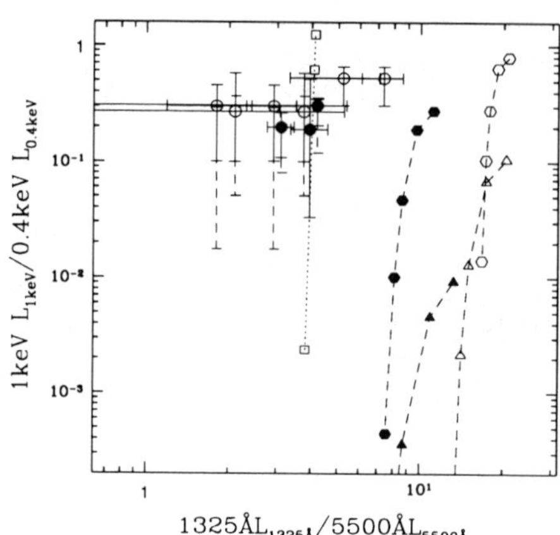

Figure 2. The SED of the radio-loud quasar 3C273.

```
alpha 1.3 0.8 0.5
F(PSPC) 17.0 0.2
```

In addition to the data themselves, the software must keep track of salient information on each object and on each dataset. Object information for quasars includes: name(s), celestial coordinates, redshift, Galactic absorption column density. Essential information on datasets includes: date of observation, telescope, beamsize, source (reference/telescope/instrument/satellite).

Each dataset will have different parameters of which the software must keep track. For example X-ray spectra have: a flux at a specific energy and associated errors, a spectral form (power law slope) with associated errors, and a total energy range. Optical spectra include flux and associated error at a series of equally-spaced wavelengths. As well as reading in data in such diverse formats, data storage formats must also allow such diversity while conforming to well-defined standards. To this end a specialized SED storage format was designed in TIGER. This will soon be converted to a FITS BINTABLE format but is currently specific to TIGER itself.

4.2. Transformations

The scope of required transformations is large including both simple unit conversions, e.g., erg cm^{-2} s^{-1} Hz^{-1} to Jansky, and astrophysical transformations such as those from magnitude to flux or from flux to luminosity. For these latter cases, the software must be sufficiently flexible that assumptions such as cosmological parameters can be modified and results rederived when required.

4.3. Data Analysis

Plotting and Interrogating Datasets As well as the usual plotting and zooming capabilities required in any software package, in an SED it is necessary to distinguish the various datasets, include error and waveband information and be able to plot spectral forms in place of data points. Figure 2 shows the radio–X-ray SED of 3C273, the brightest and most completely observed quasar in our sample. The style of data point is different for each dataset. TIGER provides commands

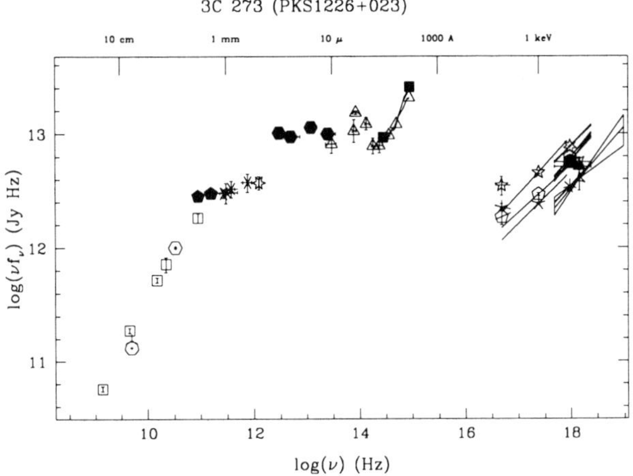

Figure 3. The Soft X-ray color ($L_{1keV}/L_{0.4keV}$) plotted against the optical-UV color ($L_{1325\mathring{A}}/L_{5500\mathring{A}}$). Circles identify the six quasars. The dotted line identifies the prediction of a pure free-free model as a function of temperature. Dashed lines identify accretion disk models in a Kerr geometry as a function of disk inclination and accretion rate (Fiore et al. 1993).

which allow the user to modify the data display, either using the cursor or by specific commands, or edit the data in the buffer, removing erroneous points or unreliable datasets. It is also possible to interrogate the data, displaying the values of specific points and/or their source.

Astrophysical Corrections In a study of the SEDs of quasars, we wish to determine the SED as it was emitted by the quasar rather than as it appears to us. This requires corrections for reprocessing of the light along the line-of-sight to the quasar. Perhaps the most important of these is interstellar reddening. This correction is complex and is based upon a combination of models and published measurements. It is essential that it be possible to modify the correction and rederive the results robustly. This allows for example an investigation of the effects of various assumptions as to the reddening law on the results of a specific set of measurements.

Quasars are situated at significant redshifts. In order to view the SED as it was emitted, the observed SED must be shifted to the rest frame of the quasar and transformed from flux to luminosity. As mentioned earlier, the software should be sufficiently flexible to allow different assumptions of the cosmological parameters and also the use of various cosmologies. In general it should be possible to set a default but to override that default when necessary and, as with reddening corrections, it should be easy to rederive results with a revised cosmological model.

The most characteristic feature of a quasar are its strong, broad, optical/UV emission lines. However for a continuum study, these lines are of no interest and add uncertainty and noise to the continuum measurement and fitting procedure. The software can indicate the positions of the prominent emission lines and remove data points which are affected by these lines.

Measuring the SED There are two complementary approaches to the analysis of a large, complex dataset such as this one. The first is to view the whole sample in terms of simple measurements which characterize a particular component of the SED, e.g., the optical/UV bump. This allows a statistical study of the diversity of the SEDs and a general comparison of the observations with the predictions of various theoretical models. We have used this approach in a number of cases, most recently in a study of the optical/UV–soft-X-ray bump in comparison with accretion disk and free-free models. The comparison is made of colors, or flux ratios, in a number of broad bands chosen to characterize the shape of the bump, with those predicted by the various models. Figure 3 shows the results for six quasars with high quality ROSAT X-ray spectra in comparison with free-free and pure accretion disk models (Fiore et al. 1993, their Fig. 1). The figure demonstrates that neither model can reproduce the full range of both optical/UV and soft X-ray data.

This kind of analysis requires that measurements be made of the SEDs. In order to measure the flux or luminosity in a specific waveband, it is necessary either to fit a model to the data and measure that model or to interpolate between the various data points and measure the interpolation. For simple color measurements such as those used here either a simple power law fit or an interpolation is best, the choice depending on the nature and quality of the data. Thus the software should be able to interpolate, fit and integrate over specified ranges of the dataset. In addition, the use of such measurements to characterize the quasar SED, also requires that the contribution of the host galaxy, with which we are not concerned in our data/model comparison, be subtracted first. Figure 4 shows the SED for the radio-quiet quasar, NAB0205+024 with a simple galaxy model superposed. The model is stronger than is appropriate for this particular quasar in order to make its form clearly visible. This again is an astrophysical correction with built in assumptions for which the software should be sufficiently flexible for a modification to be made and results re-derived with ease and robustness. In addition the subtraction introduces errors in the output fluxes which must then be included in the errors on individual data points ready for use during later analysis of the galaxy-subtracted SED.

The second, complementary approach to this study is to perform detailed modeling of individual SEDs. Figure 5 shows the same IR–X-ray SED of the radio-quiet quasar PG1211+143 with a series of possible contributing components superposed. In this example, the components include thermal dust emission with a range of temperatures, the host galaxy (which in this case can and should be specified as part of the model), and optically thin free-free emission with a temperature high enough that it peaks in the EUV and extends into the soft X-ray region (Barvainis 1993). The software should allow generation of each model, variation of its parameters and strength relative to the other components and a determination of the goodness of fit between the data and the combined models taking into account the errors as well as upper limits in the data.

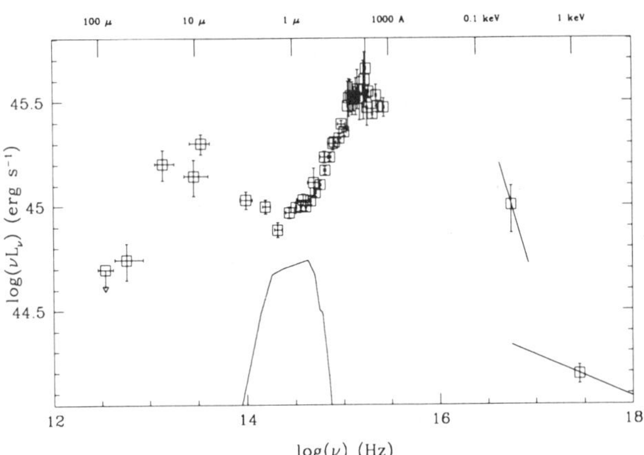

Figure 4. The far-IR–X-ray SED of the radio-quiet quasar nab0204+024 with a simple galaxy template superposed. Note that this particular template is stronger than that present in this particular SED.

4.4. Database Function

For any large collection of data such as this, it is frequently necessary to sort or select sub-samples of objects based upon various basic data such as coordinates redshift, luminosity, radio strength, etc. The properties of any specific dataset should be easily displayed, references for specific data points displayed and details (e.g., band width, beamsize) made available as well as providing a list of all objects for which a particular reference or data source is relevant.

Finally, in order to facilitate efficient and reproducible analysis of a large dataset, it must be possible to run scripts which will perform a series of tasks on a well-defined subset of the sample. An example includes:

- deredden
- shift to rest frame
- subtract the Galaxy contribution
- interpolate between the data points
- determine the luminosities within specific energy bands
- write the result in tabular form

The software should be designed so that modification of individual data points, of the cosmological model, the magnitude scale for example can be made and the results rederived, from beginning to end, robustly without the need for additional intervention. This is essential in order to minimize errors and ensure uniformity in the complex analysis of a large sample.

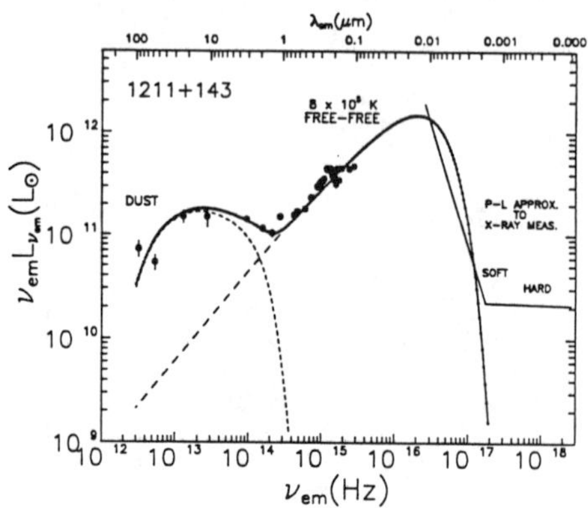

Figure 5. The IR–X-ray SED of the radio-quiet quasar PG1211+143 with a model superposed consisting of a galaxy disk containing dust with a range of temperatures and an optically thin, free-free component dominating the optical/UV and X-ray emission (reproduced from Barvainis 1993).

5. Summary

We have reviewed the major software requirements of our study of quasar energy distributions. We expect that similar studies of other astronomical objects will have similar general requirements though the details of astrophysical conversions and corrections may be different. As multi-wavelength data become available for more and fainter objects, we expect a growing demand for software which fulfills these requirements.

Acknowledgments. This research was funded in part by NASA contracts NAS5-30934 and NAS8-30751 and NASA grants NAG88-689 (ADP) and NAGW-2201(LTSA).

References

Barvainis, R. 1993 ApJ 412, 513

Elvis, M., Wilkes, B. J., McDowell, J. C., Green, R. F., Bechtold, J., Willner, S. P., Cutri, R., Oey, M, S., & Polomski, E. 1993 ApJS, submitted

Fiore, F., Martin Elvis, Aneta Siemiginowska, Belinda J. Wilkes, Jonathan C. McDowell, & Smita Mathur 1993, in preparation

Green, P. J., Anderson, S. F., & Ward, M. J. 1992 MNRAS 254, 30

Laor, A. 1990 MNRAS 246, 369

Malkan, M. A., & Sargent, W. L. W. 1982 ApJ 254, 22

Sanders, D., Phinney, E. S., Neugebauer, G., Soifer, B. T., & Matthews, K. 1989 ApJ 347, 29

Astronomical Data Analysis Software and Systems III
ASP Conference Series, Vol. 61, 1994
D. R. Crabtree, R. J. Hanisch, and J. Barnes, eds.

Analysis Techniques for a Multiwavelength Study of Radio Galaxies

D. M. Worrall and M. Birkinshaw

Harvard-Smithsonian Center for Astrophysics, Cambridge, MA 02138

Abstract. Our multiwavelength study of radio galaxies requires the use of a mixture of home-grown software and widely-used analysis packages. We outline current procedures and provide an assessment directed at software developers.

1. Introduction

Our study of radio galaxies combines X-ray, radio, and optical data to address scientific objectives including: (a) Are the radio jets in pressure balance with an external hot medium? (b) What is the rate of fuel supply to the active nuclei? (c) What physical mechanisms produce the nuclear and jet emission?

Our current data-reduction, display, and analysis requires the use of IRAF, AIPS, SM, MONGO, and home-grown FORTRAN programs, and simplifications are made in the analysis due to the current limitations of the tools and procedures. Work on the radio galaxy NGC 6251 is used to illustrate our current methods, in the interest of stimulating discussion of improved analysis tools and procedures. Scientific results can be found in Birkinshaw & Worrall (1993) and Worrall & Birkinshaw (1994). Details concerning the software packages can be found elsewhere in this book and in volumes I and II of the series.

2. Comparison of X-ray, Radio, and Optical Images

We performed the basic reduction of our VLA radio data and ROSAT PSPC X-ray data for NGC 6251 using the AIPS and IRAF/PROS systems, respectively. We then transferred the X-ray image to AIPS (via FITS-format conversions) and used the 'regrd' and 'hgeom' tasks to re-grid the radio image to match the header of the X-ray image; this included precession between epochs B1950 and J2000. The re-gridded radio image was transferred to IRAF and an overlay contour plot (produced by the IRAF/PROS/imcontour task) of the radio image was drawn on the X-ray image using IRAF. An SAOimage display of the result was converted into a PostScript file using a Unix X-windows to PostScript conversion procedure ('xwd' followed by 'xwd2ps'); the resulting plot is shown as Figure 1a.

The X-rays are only slightly resolved (see §3) and radially symmetric. No X-ray emission is detected associated with the 4-arcmin radio jet. IRAF/PROS tasks and home-grown IRAF scripts were run to quantify these statements.

Our KPNO CCD red image was analyzed using IRAF (Figure 1b). By chance, the scale is similar to that of Figure 1a, and north points up on both

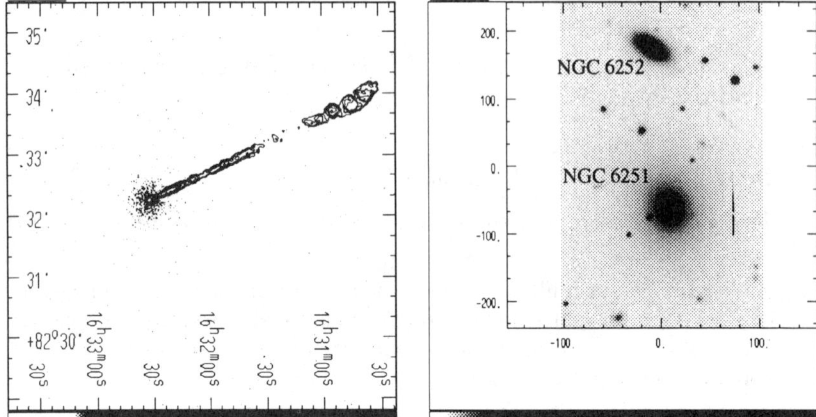

Figure 1. (a) Contour plot of 330 MHz VLA A-array reduced data overlayed on ROSAT PSPC X-ray image. The X-rays are concentrated at NGC 6251's core. (b) KPNO CCD red image of similar, but not identical, spatial scale (0.59 arcsec/pixel).

figures; celestial-coordinate-system parameters need adding to the optical image before overlays are possible. Display and plotting used the same method as for Figure 1a. X-rays are not detected from the companion galaxy NGC 6252.

3. X-ray Extent and Multiwavelength Emission from the Radio Jet

The energy distribution and radial profile of the X-ray photons (Figure 1a) were extracted using IRAF/PROS. Home-grown FORTRAN software was used to fit the radial profile to source models convolved with the energy-dependent radially-symmetric ROSAT PSPC point response function (PRF; Figure 2a), using the technique described by Birkinshaw (1994). The best fit is to a point-source (narrow curve) plus hot gas described by a hydrostatic β-model (broad curve). The plot was made using the FORTRAN interface to the SM package which in turn allows PostScript output easily.

The most prominent feature in the radio jet was modeled as synchrotron emission, and model parameters were found by comparing the calculated multi-wavelength spectrum with observations (including the X-ray upper limit). The model fitting was achieved using home-grown FORTRAN software and the MONGO plotting package (Figure 2b); a similar fit for M 87 is shown for comparison.

4. Line of Sight Absorption and Galaxy Environment

An independent measure of the amount of X-ray absorbing gas along the line of sight helps us to constrain our X-ray spectral-model fits. We mapped NGC 6251 with the VLA over a range of frequencies sensitive to HI absorption at velocities associated with the radio galaxy and intervening material. The spectral channel

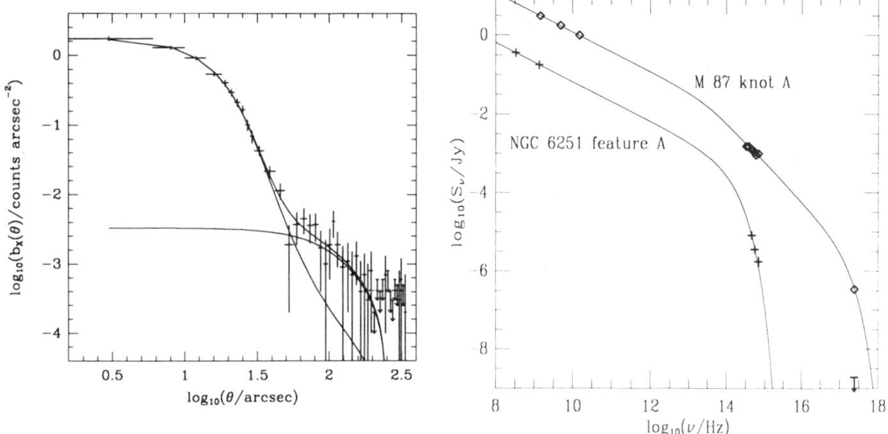

Figure 2. (a) X-ray radial profile and best-fit PRF-convolved two-component model. (b) Comparison of multiwavelength synchrotron spectrum with observations for the most prominent feature in NGC 6251's jet and for knot A of another radio galaxy, M 87.

maps were analyzed using AIPS. It was then necessary for several hundred numbers to be typed in by hand and averaged for overlapping bins to produce an HI spectrum (Figure3a). The absence of strong features in this spectrum limits the amount of X-ray absorption and helps us to constrain our X-ray spectral-model fits (not shown) which are accomplished using IRAF/PROS.

A catalog was searched for bright galaxies in the vicinity of NGC 6251. These galaxies were plotted and labeled with velocities using SM (Figure 3b); this is a reference plot for us, not intended to be of publication quality. We have MMT spectra (not shown) which we are analyzing with IRAF to measure velocities for several galaxies in the field. The dispersion of these velocities constrains the temperature of hot gas which can be associated with any galaxy groups, and further helps us to determine the correct X-ray spectral model.

5. Assessment

- Major packages are good at providing standard data reduction and calibration routines. Transfer of images between these packages (e.g., IRAF and AIPS) is easy and convenient via the FITS format.

- All ground-based telescope systems need to make their best attempts at attaching the celestial coordinate system to their images, as is already true for radio interferometer and space-based systems. Re-gridding, mosaicing, and correcting distortions are difficult problems affecting our ability to overlay images; more convenient software would help.

- The electronic preparation of papers and posters requires analysis packages to produce publication-quality output graphics in encapsulated PostScript.

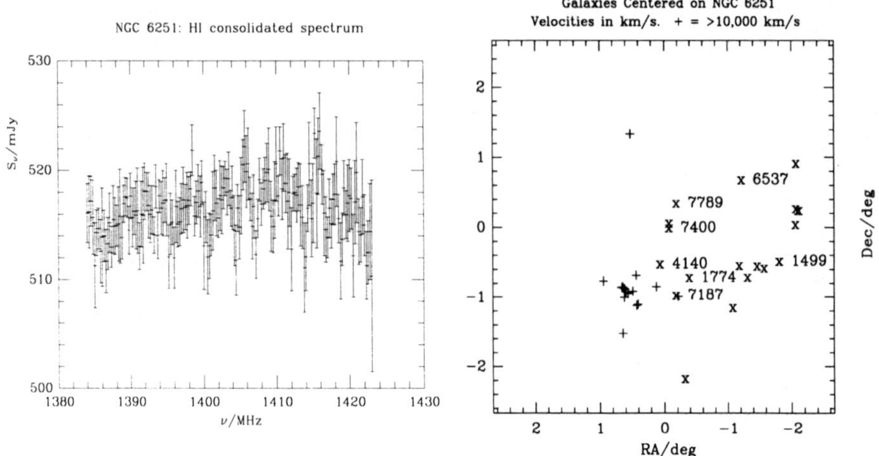

Figure 3. (a) No strong HI absorption features are seen in the radio spectrum. (b) Positions and velocities (where available) of bright nearby galaxies extracted from a catalog.

Switches should be provided for turning on and off information text within or surrounding the plot (as in, e.g., the 'imcontour' task in IRAF/PROS).

- Our scientific objectives require home-grown specialized science programs which use results from the major packages.
  - Since coding inside major packages requires substantial experience, it is more convenient for us to write specialized code as external home-grown routines; more provision of dummy tasks with places for user-supplied analysis code (e.g., 'taffy' in AIPS) would encourage us to develop code within major packages.
  - Transfer of *results* (cf images) is inconvenient; we would like major packages to provide results in ASCII files of user-definable format for transfer to our specialized science code.

Acknowledgments. Support from NASA grant NAG5-1882 and contract NAS8-39073 is gratefully acknowledged.

References

Birkinshaw, M. 1994, this volume

Birkinshaw, M., & Worrall, D.M. 1993, ApJ, 412, 568

Worrall, D.M., & Birkinshaw, M. 1994, ApJ, in press

Astronomical Data Analysis Software and Systems III
ASP Conference Series, Vol. 61, 1994
D. R. Crabtree, R. J. Hanisch, and J. Barnes, eds.

Fitting Models to UV and Optical Spectra: Using SPECFIT in IRAF

Gerard A. Kriss

Department of Physics & Astronomy, Johns Hopkins University, Baltimore, MD 21218

Abstract. Fast workstations make the fitting of complex, multi-parameter models to UV and optical spectra practical for the average user. I describe an interactive tool called `specfit` that runs in the IRAF environment which can fit data spanning a large range in wavelength from a variety of instruments using a non-linear χ^2-minimization technique. Data can be input as IRAF images or as ASCII files containing triplet entries of wavelength, flux, and errors. A variety of functional forms can be used to describe the continuum, emission lines and absorption lines, including user-defined functions that are input as ASCII files. The user has a choice of a simplex algorithm or a Levenburg-Marquardt procedure for minimizing χ^2. `specfit` can run interactively, in which case the user can modify parameters and display various representations of the data and the current fit, or as a background task. Output includes a log file summarizing the quality of the fit, the best-fit parameter values with error bars and optional integrations over user-defined line and continuum intervals, and an optional ASCII file suitable for plotting the data, the fit, and the individual fit components.

1. The General Problem of Spectral Fitting

The ultimate goal of the analysis of the spectrum of a celestial object is to obtain some insight into the physical processes at work. Inferring physical information from the observed spectrum is intimately coupled to the properties of the instrument used to obtain the data. In starting with a raw, uncalibrated spectrum, the observer generally has two objectives: 1) converting the observed data into a representation of the true incident flux through a process of instrumental calibration, and 2) characterizing the spectrum in terms of astrophysical processes. Before attempting to take the second step by fitting the spectrum, it is a good idea to understand how data are calibrated for a given instrument and the limitations that procedure might impose on any subsequent analysis.

If we start with the photon spectrum incident on our instrument,

$$N_\lambda = \frac{F_\lambda \, \Delta t}{(hc/\lambda)},$$

convolution with the effective area of the instrument and its response function (which includes the resolution of the instrument and scattered light properties) gives the counts detected as a function of position on the detector as

$$C_x = \int N_\lambda \, R(\lambda, x) \, d\lambda \ , \text{ or}$$

$$C_{x_j} = \sum_i D_{ij} \, N_i \ ,$$

where D_{ij} is the "detector matrix", and $N_i = \int_{\lambda_i}^{\lambda_{i+1}} A_\lambda \, N_\lambda \, d\lambda$. (Real observations will also include both internal and external background which I am neglecting in this simplified treatment.)

Mathematically inverting this process is generally highly unstable, especially in the presence of noise, and two different approaches are commonly used to fit the spectrum and obtain the true incident flux. The first approach, used mostly by X-ray and γ-ray astronomers, is to model the data in natural detector units. Representations of the incident source flux are propagated through the instrument model and compared to the data. The incident source flux is inferred in a model-dependent way via parametric fits.

The second approach is appropriate to instruments with moderate to high spectral resolution and low amounts of scattered light, as is common for dispersive UV and optical spectrographs. For such instruments the detector response function can be approximated as a delta function, and the corresponding detector matrix is the identity matrix. Direct inversion of the problem is now stable. With the additional approximation $N_i \sim A_\lambda N_\lambda \Delta\lambda_i$, we obtain

$$F_{\lambda_j} = \frac{(hc/\lambda_j)}{A_\lambda \, \Delta\lambda_j} \, \frac{C_{x_j}}{\Delta t} .$$

The first term on the right hand side is frequently called the "inverse sensitivity function".

After obtaining a flux-calibrated spectrum using this method, the observer is now ready to make fits to measure line fluxes, model line profiles and the continuum shape, and to obtain physical diagnostics. Fitting models to the spectrum in this approximation benefits from a significant computational savings since the incident spectrum need not be convolved with the instrument response function.

While the flux-calibrated spectrum is a model-independent approximation to the true incident flux, one must keep in mind its limitations. Narrow or unresolved emission lines will have profiles dominated by the instrument response function. Likewise, spectra with high contrast between the emission lines and the continuum will be affected by the scattered-light properties of the instrument at the lowest signal levels in a spectrum with a large dynamic range.

2. Desirable Characteristics of a Spectral Fitting Program

The prime requirement for any spectral fitting program is that it provide a statistical evaluation of the goodness of fit and of the error bars for the fitted parameters. After this, general ease of use and interactive modes that provide a graphical display of the data, the fit and the residuals are highly desirable. Since complex, multi-parameter fits can require a lot of computer time, background or batch modes of operation are also useful. In spite of the advances in data portability made with the FITS format and widely distributed software systems such as IRAF and IDL-based reduction systems, there are still a wide variety of instrument-dependent data storage formats, and a flexible means of data input for a general use fitting program is useful. Since the primary aim of spectral fitting is to infer physical characteristics from the data, the fitting program should also incorporate a wide range of physically meaningful fitting functions, and it should be easy to add additional functions as needed. Finally, a fitting program of general utility should be easily ported to a wide variety of systems. The easiest way to achieve this last goal is to design a task that runs within a widely distributed environment such as IRAF so that portability is virtually automatic.

While spectral fitting programs that are generally available have had a long history in high-energy astronomy, similar tools have not been widespread in the optical and UV astronomical communities. The reason for this may be largely historical. Digital electronic detector systems with easily characterized noise properties have only come into widespread use in UV and optical astronomy during the past decade. Since well characterized errors for the data are essential for a proper statistical analysis, complex fitting programs were not terribly apt in earlier days. Another, and perhaps more important reason, is the sheer amount of data in a UV or optical spectrum and the complexity of any model that must account for each independent resolution element. Typical UV or optical spectra contain of order 500 resolution elements spread over 1000 or more pixels. This can be compared to the few dozen channels in an X-ray spectrum that are usually fit with models containing fewer than 10 free parameters. Until recently, the computer time required to fit a model with the hundred or more free parameters required by a 1000 point UV or optical spectrum would have been unreasonable. Most UV or optical fitting tools therefore were simply "deblending" tasks for fitting a few unresolved or barely resolved emission or absorption lines.

Currently available spectral fitting programs that are widely distributed include `xray.xspectral.fit` in PROS and `xspec` in the XANADU X-ray package, both for X-ray spectra, `noao.images.fit1d` in IRAF, the "deblend" option in `noao.onedspec.splot`, and the various routines in `stsdas.fitting` in IRAF/STSDAS: `gfit1d`, `nfit1d`, and `ngaussfit`. The latter routines suitable for fitting UV and optical spectra provide some capability for fitting emission lines or smooth simple continua, but they lack a wide range of physical models, and they are generally intended only for spectra from single instrument configurations. There may well be more sophisticated tools available in the UV/optical astronomical community at individual sites.

In this paper I describe an interactive tool called `specfit` that runs in the IRAF environment and is currently being distributed as part of the `contrib` package in `stsdas`. `specfit` can fit complex physical models to data from several instruments spanning a wide range in wavelength, and it goes a long ways toward meeting most of the desirable characteristics for a general purpose spectral fitting program.

3. Using SPECFIT

3.1. Data Input

The ease of entering data into a fitting package is a key element in facilitating its use. Special binary formats have frequently hindered the portability of data and also the portability of software. While the FITS format has increased the transportability of data, programs that access data through simple ASCII files widen their applicability to a variety of instruments. The primary means of data input for `specfit` is a simple ASCII file with two header lines followed by the data. The first of these is a descriptive name, and the second gives the number of data points, the integration time, and the Julian date of the observation. The data are organized in three columns per line giving the wavelength, flux, and 1σ error for each data point.

`specfit` will also accept one-dimensional IRAF images. In this case the wavelengths are assumed to follow a linear dispersion relation, and the transformation from pixel number to wavelength is read from the standard FITS parameters in the image header (CRPIX1, CRVAL1, and CD1_1 or CDELT1). For this input format, since errors are not explicitly given, they are approximated by having the user specify a linear relation between the data and the variance for each point. This permits an acceptable statistical evaluation for Poisson-distributed data that are in raw counts or for uniformly weighted data such as one might encounter in a high signal-to-noise ratio spectrum dominated by systematic rather than statistical errors.

A final data format recognized by `specfit` is one developed for spectra from the Hopkins Ultraviolet Telescope which flew aboard the space shuttle *Columbia* in December 1990 as part of the Astro-1 mission (see the review by Davidsen [1993]). These one dimensional spectra are stored as two-dimensional IRAF images with the spectrum in the first line, the 1σ error bars in the second line, and the linear wavelength scale stored in the header using the standard FITS parameters for a linear dispersion relation.

3.2. Model Fitting Functions

`specfit` has a wide variety of internally defined fitting functions which are summarized in Table 1. Additional functions can be added to `specfit` in two different ways. Simple functions can be incorporated into the existing code by modifying two routines which call the fitting functions, plus one initialization routine. Re-installation in IRAF is facilitated by the `mkpkg` utility.

A simpler technique for including complex models is to use the functions `usercont`, `userline`, or `userabs`. These functions are specified by the user as ASCII files containing paired entries of wavelength and flux that are evenly spaced in wavelength. They are a powerful method of using grids of pre-calculated, computationally expensive models to fit complex spectra. Examples of this use are fitting the high-order Lyman line absorption in the far-UV spectrum of NGC 4151 (Kriss et al. 1992), fitting the far-UV spectra of elliptical galaxies using stellar population and stellar atmosphere models (Ferguson et al. 1991), and fitting accretion disk models to the far-UV spectra of cataclysmic variables (Long et al. 1991). A linear grid of models is specified by listing the file name of the model along with the key parameter which varies among the

Table 1. Internally Defined Functions in `specfit`.

Function	Description
linear	linear continuum in F_λ
powerlaw	power law continuum in F_λ
bpl	broken power law continuum in F_λ
blackbody	blackbody continuum in F_λ
recomb	optically thin recombination continuum
gaussian	Gaussian line profile
lorentz	modified Lorentzian line profile
logarith	power-law line profile $F = I_0 * (\lambda/\lambda_0)^\alpha$
labsorp	Gaussian profile absorption line
dampabs	damped absorption line
logabs	absorption line with $\tau \sim \|\lambda - \lambda_0\|^\alpha$
extinc	mean galactic extinction, Seaton law
eabsorp	absorption edge, $\tau \sim (\lambda/\lambda_0)^3$
usercont	user-supplied continuum
userline	user-supplied emission line profile
userabs	user-supplied absorption line profile

listed models in an ASCII file. The example shown in Table 2 is for a grid of Comptonized accretion disk models (Lee, Kriss and Davidsen 1992) in which the accretion rate onto the central black hole varies from 0.20 to 0.35 $M_\odot yr^{-1}$. This list of file names must be stored in the file `continuum.ls` in the running directory when using `usercont`. The file names given in `continuum.ls` are for the ASCII files containing the wavelength/flux pairs for each of the specified models. For `userline` and `userabs` the model file names must be listed in `profile.ls` and `absorption.ls`, respectively.

Table 2. File Format for Input of User-defined Models to `specfit`.

Filename	Key Value
`cdisk/cbm0750r0020T874t10.flx`	0.20
`cdisk/cbm0750r0025T874t10.flx`	0.25
`cdisk/cbm0750r0030T874t10.flx`	0.30
`cdisk/cbm0750r0035T874t10.flx`	0.35

3.3. Parameter Input

Since `specfit` is an IRAF task, general parameter input is controlled by the IRAF parameter file mechanism. Modifications are made using the `eparam` task. These general parameters govern things like the input file name, whether the task should be run interactively or not, names of log files and plot files, and the names of the database files used for the parameters in the model.

```
begin   halpha_n2
task    specfit
components      4
                powerlaw
                gaussian
                gaussian
                gaussian
                powerlaw1       2
                        1.1e-14  0.  2.  0.1e-14  0.001 0
                        1.0  -5.  5.  0.05  0.001  0
                gaussian2       4
                        3.0e-13 0. 10. 0.2e-13  0.001 0
                        6548.1 6545.  6551.  0.2  0.001 0
                        800.  200.  1500.  50.  0.001 0
                        1. 1. 1. 0.1 0.001 -1
                gaussian3       4
                        1.2e-12  0. 10. 0.5e-13  0.001 0
                        6562.9  6560.  6566.  0.2  0.001 0
                        800.  200.  1500.  50.  0.001 0
                        1. 1. 1. 0.1 0.001 -1
                gaussian4       4
                        9.0e-13 0. 10. 3.0  0.001 2
                        6583.4  6580.  6590.  1.00539  0.001  2
                        800.  200.  1500.  1.0  0.001  2
                        1. 1. 1. 0.1 0.001 -1
```

Figure 1. specfit uses the IRAF ASCII database format for specifying functional forms and initial guesses for parameters. The database is free-format and keyword-oriented.

Specifying a multi-component model with many parameters requires a great deal more information from the user than can be easily specified in an IRAF parameter file. Besides describing the model itself, one must provide initial guesses for the values of the parameters, upper and lower bounds within which they are permitted to vary, step sizes to be used by the minimization algorithm, and a tolerance for judging convergence of the fit. specfit uses the IRAF text database format for the files which describe the models and input all the necessary information. A sample database file suitable for deblending Hα+[N II] line emission in a spectrum is shown in Figure 1.

The first line in the database file gives the name of the database model, and it must match the name of the file itself, minus a prefix of "sf". The lines following the keyword "components" describe the model itself. All model components are assumed to be additive, except for absorption components that are assumed to be multiplicative. To allow for overlying emission that is not absorbed by the absorption components (e.g., night sky lines or airglow in the spectrum), specfit assumes that absorption applies only to components appearing previously in the list. After listing the functional ingredients of the model, the parameters for each component appear following keywords consisting of the component name with the component number appended. The number of parameters for the component follows the keyword, and then the entries for each parameter follow on successive lines. Each line gives, in order, the parameter value, the minimum allowable value, the maximum allowable value, the step size for the minimization search, the tolerance for judging convergence of the fit, and a flag indicating whether the parameter is free to vary (0), fixed at the given value (−1) or linked to another parameter (a positive integer).

The ability to link one parameter with another is a powerful feature for reducing the number of free parameters in a fit and for testing realistic physical constraints on a spectrum. When a parameter is linked to the corresponding parameter of another component, the step size is used as a scale factor. In the example shown in Figure 1, the widths of the [N II] components are forced to be the same, the intensity of the red component is fixed at 3.0 times the value of the blue line, and the wavelength of the red line is forced to be in the same ratio to the blue line as observed in the laboratory. Since the Hα line may be formed in a different region with different physical conditions, none of its parameters are linked to those of the [N II] lines.

Table 3. Interactive Keystroke Commands in `specfit`.

Keystroke	Definition
a	Change the tolerance on a parameter
c	Change the value of a parameter
e	Evaluate the current χ^2
f	Minimize using the Simplex algorithm
m	Minimize using the Marquardt algorithm
i	Print information on a parameter
l	Change the lower limit
p	Overplot the fit on the data
q	Exit from `specfit`
r	Re-plot the data
s	Select new sample ranges
t	Change the maximum number of iterations
u	Change the upper limit
x	Change the fix or free status
z	Change the step size
d	Plot the distribution of χ^2
−	Plot the residual spectrum
+	Plot the model alone
w	Window the graph
?	Print this help menu
:	Issue an IRAF colon command
	(space) Print current cursor position

3.4. Operational Modes and Performance

specfit has both an interactive mode and a batch mode. The interactive mode permits display of the data, the fit, residuals, χ^2 distributions, manual updates to parameter values and their constraints, and other functions as summarized in Table 3. This is the most useful method of operation for simple fits and for debugging complex fits. The batch mode is appropriate for complex fits that will require much computer time to reach convergence, or for the pipeline processing of many similar data sets. It is relatively straightforward to set up a series of complex fits using IRAF cl scripts that use the same template fitting function to analyze a whole series of spectra as was done for the International AGN Watch campaign on NGC 5548 (Clavel et al. 1991).

Two minimization techniques are available in specfit — a simplex algorithm written in SPP adapted from the Pascal version published in *Byte* magazine (1984), and a modified version written in SPP of the Levenburg-Marquardt algorithm from Bevington's (1969) CURFIT. The simplex algorithm provides fast initial convergence, even for parameters with poor initial guesses, but the Marquardt algorithm converges more effectively as the fit nears the minimum. Very close to the minimum both methods tend to slow down, but alternating between the two tends to prevent either from getting stuck on false minima. This "alternate" method is an option for running specfit in the non-interactive background mode.

Iterations halt either when their number matches the maximum value specified by the user, or when the convergence criteria have been satisfied. For a fit to converge, the change in χ^2 from the previous iteration must be less than the tolerance chosen in the parameter file, *and* the change in each parameter must be less than the tolerance given in the database file. Simple fits to a few hundred data points with only a handful of free parameters (typical for deblending a few nearby emission lines) converge interactively on a SPARCstation 2 in seconds to a few minutes. At the opposite extreme, single iterations of the Marquardt algorithm can take many hours. As an example, one iteration on the FOS spectrum of NGC 5548 from the spring 1993 International AGN Watch monitoring campaign (3648 points, 161 free parameters) requires ~12 hours on a dedicated SPARCstation 2. Since the computation time scales roughly linearly with the number of data points and quadratically with the number of free parameters, for such complex fits it pays to "tweak up" the initial guesses used for the final model by initially fitting smaller sections of the spectrum containing only a few components before iterating on the full fit to the whole spectrum.

3.5. Interpretive Output

Besides making a new database entry for the final fit, specfit also produces a log file and an optional plot file to aid the user in evaluating the results and displaying them for publication. Error messages during execution are directed to stderr, but the log file can be directed to wherever the user has specified in the parameter file. This may be stdout during debugging, or a file in the current directory. As shown in Figure 2, the log file summarizes the input data, the number of data points and free parameters, χ^2 for the fit, and the sample ranges specified. The fitted model is then listed in the same order as given in

```
JD-2440000= 49113.151
** OBJECT: NGC5548 H19 A-1 4902  FILE: ../fosfluxes/mean_40.lis **

Chisquare = 10000.41 for 3648 points and 161 freely varying parameters.

Regions of data included in the fit.
Sample  1: 1155 to 1205.
Sample  2: 1225 to 2320.

Param Value     +/-        Low Limit   Up Limit   Step Size Tolerance Fix?
powerlaw1        2
  6.1676E-14  2.9765E-17          0. 1.0000E-12  2.0000E-15     0.001    0
      1.1492  6.1441E-4          -5.          5.         0.1     0.001    0
gaussian2        4
  8.7853E-13  2.0273E-14          0.          1.  1.0000E-14     0.001    0
      1236.5     0.04792       1230.       1245.         0.4     0.001    0
       1280.      25.958        300.       9000.        200.     0.001    0
          1.          1.          1.          1.         0.1     0.001   -1
```

Figure 2. Sample portions from a `specfit` log file.

the database file, but in a more easily read format that includes the 1σ errors on the data points.

The error bars in `specfit` are obtained from the error matrix, which is the inverse of the curvature matrix. This method assumes that χ^2 contours in parameter space can be approximated by parabolic surfaces. The error bars correspond to $\Delta\chi^2 = 1$, which is appropriate for a 1σ confidence interval on a single interesting parameter. For multi-parameter fits with more than one interesting parameter, however, the correct $\Delta\chi^2$ for the 1σ confidence interval is larger. (See Avni [1976] for appropriate values of $\Delta\chi^2$ for more than one interesting parameter.) The rigorous method for determining confidence intervals for more than one interesting parameter is to actually solve for the appropriate contours of $\Delta\chi^2$ in parameter space, but this is impractical for the many parameters often necessary for UV and optical spectra. A straightforward approximation is to assume that the surface of $\Delta\chi^2$ continues to follow a parabola, and then to scale the $\Delta\chi^2 = 1$ error bars from the error matrix by the square root of the desired $\Delta\chi^2$ obtained from Avni (1976). This appears to work reasonably well, as long as the parameters are not strongly coupled. A typical situation that *does not* work in this approximation are the error bars for the power-law continuum normalization and the E_{B-V} in a fit that includes an extinction correction. Whenever in doubt, it is a good idea to compute a grid in parameter space to trace the contours in χ^2. Unfortunately, `specfit` does not have a mode that facilitates computing such a grid.

In interpreting the error bars from the fit, another caveat one must always keep in mind is that `specfit` assumes Gaussian statistics. For low signal-to-noise ratio spectra obtained with photon-counting instruments this assumption can lead to biased or erroneous results. In these situations binning the data more coarsely to get into the Gaussian regime before using `specfit` is one recourse.

Users can optionally have `specfit` perform integrations of selected wavelength intervals in the input spectrum to obtain tabulations of emission line or continuum fluxes. For emission lines both the fitted model and the data are directly integrated above the fitted continuum within the wavelength intervals specified by the user. For continuum intervals, the flux in the data and in the fitted model are directly integrated. Errors for the integrated quantities are propagated using the error bars from the input data and the error bars computed for the fitted parameters in the model.

Another optional output from `specfit` is an ASCII file that can be used to plot the data, the total fit, or individual components of the fit. The file has a format that is similar to the ASCII input file format — two header lines are followed by individual lines containing entries for the wavelength, input flux, input errors, the model flux, and the fitted values at each wavelength for the individual fit components. This file is easily read and plotted with most plotting programs, e.g., Mongo, SuperMongo, or `igi` in the `stsdas.stplot` package.

4. Future Work

While `specfit` provides a powerful tool to the astronomical community for fitting physical models to UV and optical spectra, there are clear places to improve the interaction with the user. The new Xwindow graphical user interface tools that will soon be available in IRAF will facilitate this. Future improvements will permit the use of alternative statistics to χ^2 for evaluating the goodness of fit. This will improve the ability to fit low signal-to-noise ratio spectra with non-Gaussian errors. A fitting option that would allow the user to compute automatically grids in parameter space for evaluating contours of χ^2 would also be useful. Finally, many complex models such as stellar atmosphere and accretion disk spectra require more than a single key parameter. Future versions of `specfit` may allow two or more dimensions for the user-specified continuum, line, and absorption models.

Acknowledgments. This work was supported by NASA contract NAS 5-27000 to the Johns Hopkins University for the Hopkins Ultraviolet Telescope project, and NASA-GSFC grant NAG 5-1630 to the FOS team.

References

Avni, Y. 1976, ApJ, 210, 642

Bevington, P. R. 1969, "Data Reduction and Error Analysis for the Physical Sciences", (McGraw-Hill: New York), 232-245

Clavel, J., et al. 1991, ApJ, 366, 64

Davidsen, A. F. 1993, Science, 259, 327

Ferguson, H. C., et al. 1991, ApJ, 382, L69

Kriss, G. A., et al. 1992, ApJ, 392, 485

Lee, G, Kriss, G. A. & Davidsen, A. F. 1992, in Testing the AGN Paradigm, ed. S. S. Holt, S. G. Neff, and C. M. Urry, (New York: AIP), 159

Long, K. S., et al. 1991, ApJ, 381, L25

Astronomical Data Analysis Software and Systems III
ASP Conference Series, Vol. 61, 1994
D. R. Crabtree, R. J. Hanisch, and J. Barnes, eds.

The ESIS Spectral Package

S. G. Ansari, P. Giommi, H. Berle

European Space Information System, Information Systems Division of ESA, via Galileo Galilei, I-00044 Frascati, Italy

A. Ulla

Niels Bohr Institute, Blegdamsvej 17, 2100 Copenhagen Ø, Denmark and LAEFF-INTA, Apdo.50727, 28080 Madrid, Spain

Abstract. The ESIS Spectral package provides access to a number of multiwavelength data through a graphical user interface, which contains a variety of tools for pre-analysis of spectra. Currently, ESIS provides access to EXOSAT ME, IUE Low Dispersion, and IRAS Low Resolution spectra, which can be homogeneously accessed for any given object. The package features chemical element identification of spectral lines, automatic redshift correction, de-reddening functions, and a radio to X-ray energy distribution of single objects. The spectral package provides links to native mode analysis packages, such as XSPEC. In the future links to other packages, such as MIDAS and IRAF will be available.

In this paper, we discuss the ESIS spectral package tool and give a number of examples to demonstrate its advantages.

1. Introduction

The ESIS system is an ESA service designed to access, retrieve and manipulate data from heterogeneous remote databases. A brief description of ESIS can be found in this volume (Giommi & Ansari 1993). In this paper we describe in some detail the part of the system that retrieves and manipulates energy spectra of astronomical objects.

The Spectral Package of ESIS is a graphical user interface based on OSF-Motif and on the Athena Widget Plotter (Klingebiel 1992). It interfaces to the ESIS catalogues through a client-server architecture.

The application allows the user to search for objects using the major search capabilities found in all other ESIS packages:

- Search by name of an object known to the SIMBAD database, where all the names of the given object are searched for in the selected catalogues.
- Search by coordinates. Coordinates may, either be directly entered manually, or the name of the object may be used to extract the coordinates from the SIMBAD database.

- Search by an overview of all spectra available for a given object. The name of the object is given by the user. ESIS then extracts its coordinates from SIMBAD and searches in the spectral catalogues by using internal default radii for each individual catalogue before building an interactive graphical user interface from which spectra can be selected by the click of a mouse button.

- Search by parameters through the use of strict mathematical expressions.

Catalogues containing information on spectral products are then searched and displayed in tables to the user, except in the case of the overview, where the user graphically chooses spectra from an interactive frequency vs. observation time plot.

The spectral application currently supports IRAS, IUE, HST and EXOSAT data. As other data sources become available, they may be easily incorporated.

2. Spectral Functionalities

The spectral application allows the user to compare spectra from various missions and wavelength regions. One such example is the capability of extracting flux values from various catalogues from the Radio to the X-ray and to automatically plot the energy distribution of any object both in the form of a $f(\nu)$ vs. ν and $\nu f(\nu)$ vs. ν.

Spectra of the same wavelength region are automatically superimposed, or the scale is automatically readjusted, depending on the spectral wavelength region and resolution. Other available functionalities range from zooming in on a selected region, customizing plots of each displayed spectrum by choosing styles, line widths, colors, etc.

2.1. Spectral Line Identification

In order to help the user in identifying spectral features, the ESIS spectral application provides the semi-empirical values of Kurucz and Peytremann (1975) as a reference. To determine central wavelengths, the user may fit the spectral profile with a parabolic, a Gaussian or rotational profile. When the central depth is determined, the spectral application then searches in the Kurucz Peytremann database for all lines falling within a given tolerance.

For practical reference, the spectral application also provides a periodic table of chemical elements, providing basic information of astronomical interest for each element (Allen 1973):

- Atomic Number
- Atomic Weight
- Solar Abundance
- The first five levels of ionization potential

Furthermore, if the redshift of a given object is known, the values are extracted from the ESIS catalogues and applied to the measured central wavelength, before a spectral line identification is made.

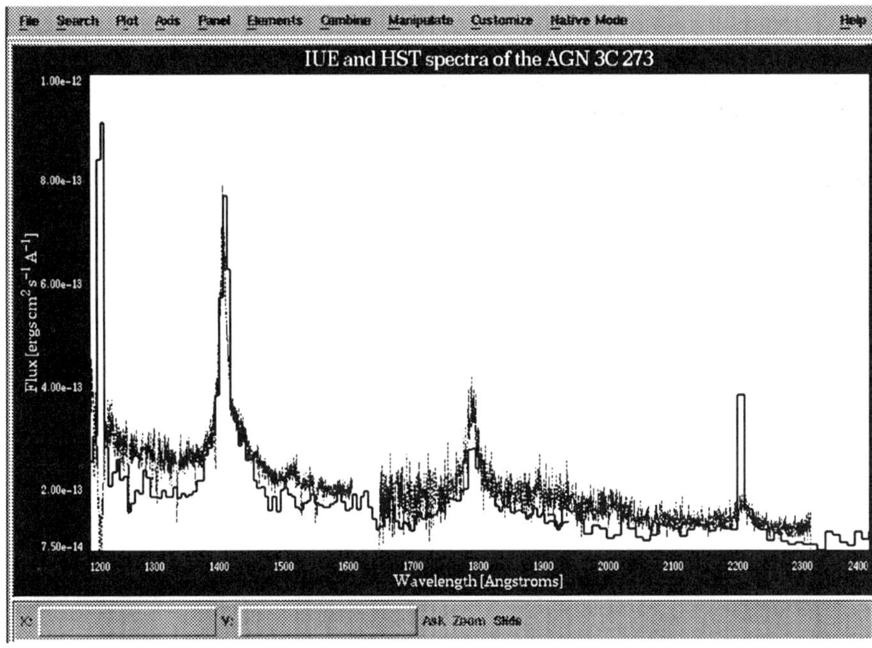

Figure 1. Short wave and long wave IUE spectra of the AGN 3C 273 are plotted. Superimposed are a number of HST FOS spectra.

2.2. Correlation Functionalities

A number of correlation functionalities are currently under development, which will enhance the spectral manipulation functionalities to allow the user to combine spectra in various ways, including addition, subtraction, division and multiplication of spectra. The user will also be able to integrate continua and to extract individual flux values.

3. Data Origins

Where do all the data come from ? ESIS data are distributed and are available from various connected databases. The IUE spectra, mainly those from the Uniform Low Dispersion Archive at the IUE Tracking Station near Madrid, Spain, but also high resolution spectra may be ordered off-line. IRAS spectra originate from the Low Resolution Spectra Catalog. The HST spectra are made available by the Space Telescope - European Coordinating Facility near Munich, Germany. The EXOSAT Medium Energy spectra are accessed from ESTEC, Noordwijk in the Netherlands.

Plans to include a number of other spectral sources are underway. In the near future spectra from the EUVE data archive and Einstein SSS spectra will be accessible through ESIS, in order to continue to establish a complete coverage of the electromagnetic spectrum for a large number of objects.

4. Acknowledgement

A. Ulla is at present an *European Economic Community* postdoctoral fellow at the Niels Bohr Institute in Copenhagen.

References

Allen, C.W. 1973, "Astrophysical Quantities", The Athlone Press, London, U.K.

Giommi, P., & Ansari, S.G. 1993, this volume

Klingebiel, Peter 1992, "Using The AthenaTools Plotter Widget Set", Univ. of Paderborn, Germany

Kurucz, R.L., & Peytremann, E. 1975, SAO Spec. Rep. 309

Astronomical Data Analysis Software and Systems III
ASP Conference Series, Vol. 61, 1994
D. R. Crabtree, R. J. Hanisch, and J. Barnes, eds.

The Prediction of Stellar Ultraviolet Colours

A. Shemi, G. Mersov, N. Brosch and E. Almoznino

Wise Observatory & School of Physics and Astronomy Raymond and Beverly Sackler Faculty of Exact Sciences, Tel Aviv University, Tel Aviv, 69978, Israel.

Abstract. A comprehensive database was compiled from several IUE spectral data sets, optical star catalogues, and a selected UV photometric data set, of more than 10,000 stars measured by the TD-1 satellite. The database is useful for the prediction of stellar UV colours. A method to derive an intrinsic colour-colour relation $(UV - B)_0$ vs. $(B - V)_0$ is described. [UV - Blue] colour - colour diagrams for the four TD-1 bandpasses, for different stellar luminosity classes, are presented. We obtained a list of selected 333 stars that can be used as "standard candles" for the determination of $(UV - B)_0$ in any arbitrary UV band. The benefit of the method is demonstrated by a comparison of TD-1 observations and images of sky areas from FAUST against predicted UV sky maps.

1. Derivation of the Ultraviolet Colour

Colour - colour diagrams provide a basic tool for understanding photospheres and coronae of stars, as well as stellar systems. Such diagrams usually rely on colours in the optical range. Unlike photometry in this range, the ultraviolet has no standard filters with universally recognized response curves, and the monochromatic UV magnitude of stars is usually determined for specific filters of UV space missions. The prediction of the UV magnitude of stars is of great importance for any efficient UV astronomy space mission. Particularly, prior to such a mission, one can derive sky maps of the expected UV sources and use them to plan the observations, and afterwards, to compare them with the results. Such a preliminary study was carried out by Brosch[1], who approximated the $(UV - V)$ versus $(B - V)$ relations for monochromatic bands at 1500, 2000 and 2500 Å .

The present work is based on four spectral data sets and two star catalogues[2]. All the spectral data were obtained from IUE. The intrinsic ultraviolet colour of a star $u \equiv (UV - B)_0$ can be written in the form

$$u = UV - V + (B - V) - E(UV - B) \quad (1)$$

where $E(UV - B) \equiv UV - B - (UV - B)_0 = \kappa E(B - V)$ is the ultraviolet colour excess. The values of UV, V and $B - V$ were derived synthetically from IUE spectra and compared to photometric data from the TD-1 satellite. Here κ is defined by $\kappa \equiv [A(UV) - A(B)]/[A(B) - A(V)]$, where $A(UV)$, $A(B)$ and $A(V)$

are the values of the interstellar extinctions in the UV, blue and visual bands [3]

$$\kappa_{2740} = 2.08, \quad \kappa_{2365} = 4.14, \quad \kappa_{1965} = 4.27, \quad \kappa_{1565} = 4.00 \ . \tag{2}$$

The TD-*Catalogue of Stellar Ultraviolet Fluxes*[4] contains 31,215 stars with apparent monochromatic magnitudes in four bands: 2585-2895 Å ($\bar{\lambda}$ = 2740 Å), 2200-2530 Å ($\bar{\lambda}$ = 2365 Å), 1800-2130 Å ($\bar{\lambda}$ = 1965 Å), and 1400-1730 Å ($\bar{\lambda}$ = 1565 Å). The Hipparcos input catalogue[5] contains 118,209 stars with measured V, $(B-V)$ colour, spectral type and luminosity class. We identified stars in both TD-1 catalogue and HIC by their coordinates, and obtained a very large UV + optical data set of $\sim 23,330$ stars. Among these we selected $10,617$ stars where their luminosity class and spectral type are given. Also, we obtained a list of selected 333 stars that can be used as "standard candles" for the determination of $(UV-B)_0$ in any arbitrary UV band[2].

Table 1 gives the coefficients a_i and their accuracies Δa_i for prediction of $(UV-B)_0$:,

$$(UV-B)_0 = a_0 + a_1 b + a_2 b^2 + a_3 b^3 \tag{3}$$

for the four TD-1 bandpasses. Here $b \equiv (B-V)_0$, $(B-V)_m$ is the upper limit for the prediction, $\bar{\sigma}$ is an averaged value of the standard deviation, and EV and Σ relate to test of the predictions vs. the TD-1 measurements.

Table 1.

LC	a_0	Δa_0	a_1	Δa_1	a_2	Δa_2	a_3	Δa_3	$(B-V)_m$	$\bar{\sigma}$	EV	Σ
2740Å												
I	-0.373	0.104	5.364	0.311	-4.079	0.793	1.619	0.418	1.7	0.164	-0.202	0.555
II	0.032	0.103	8.200	0.695	0.542	1.245	-7.497	2.273	0.8	0.204	-0.347	2.570
III	-0.676	0.065	5.227	0.192	-0.477	0.418	-0.247	0.200	1.7	0.067	-0.233	0.663
IV	-0.179	0.120	4.703	0.479	-8.702	2.200	8.408	1.897	1.7	0.109	-0.180	0.330
V	-0.291	0.021	4.483	0.048	-5.106	0.205	4.327	0.232	1.05	0.021	-0.157	0.300
2365Å												
I	-0.433	0.107	7.263	0.317	-4.645	0.673	1.521	0.312	1.7	0.131	-0.000	0.952
II	-0.080	0.074	9.141	0.452	-0.482	0.738	-4.283	1.395	0.85	0.136	-0.226	1.084
III	-0.630	0.074	7.160	0.237	-0.280	0.492	-0.868	0.233	1.7	0.088	0.030	0.922
IV	-0.237	0.131	6.181	0.527	-7.991	2.373	7.227	2.040	0.98	0.127	0.092	0.457
V	-0.387	0.019	6.175	0.046	-4.397	0.191	4.121	0.206	1.0	0.021	0.082	0.433
1965Å												
I	-0.338	0.097	8.018	0.296	-6.387	0.651	2.077	0.338	1.7	0.145	0.395	1.018
II	-0.181	0.107	11.555	0.751	3.732	1.325	-11.539	2.441	0.7	0.219	0.327	2.580
III	-0.748	0.069	8.368	0.188	-1.432	0.426	-0.634	0.206	1.7	0.087	0.358	1.170
IV	-0.712	0.136	7.983	0.300	-2.262	1.625	-0.547	1.559	0.96	0.116	0.264	0.621
V	-0.647	0.029	8.202	0.072	-2.132	0.295	0.247	0.318	1.0	0.031	0.332	0.616
1565Å												
I	-0.003	0.094	16.593	0.559	6.217	1.111	-20.893	2.342	0.6	0.160	-0.144	4.329
II	-0.682	0.120	9.254	0.721	-1.232	1.880	0.621	0.767	0.25	0.112	0.359	1.438
III	0.180	0.114	18.833	0.785	3.540	1.733	-41.964	7.815	0.4	0.141	-0.620	6.053
IV	-0.409	0.143	12.922	0.708	1.297	2.116	-6.440	5.010	0.6	0.179	0.205	0.715
V	-0.465	0.032	14.934	0.132	7.314	0.429	-14.240	0.598	0.9	0.061	0.210	0.619

Figure 1 shows Colour - colour diagrams of $(2740\mathring{A} - B)_0$ vs. $(B-V)_0$ (1.a), $(2365\mathring{A} - B)_0$ vs. $(B-V)_0$ (1.b), $(1965\mathring{A} - B)_0$ vs. $(B-V)_0$ (1.c), and $(1565\mathring{A} - B)_0$ vs. $(B-V)_0$ (1.d) for LC Ia+Ib. The solid line represents the third-degree polynomial fit for the selected stars (full squares). The dotted curves confine the range where the prediction accuracy is within the 3σ level.

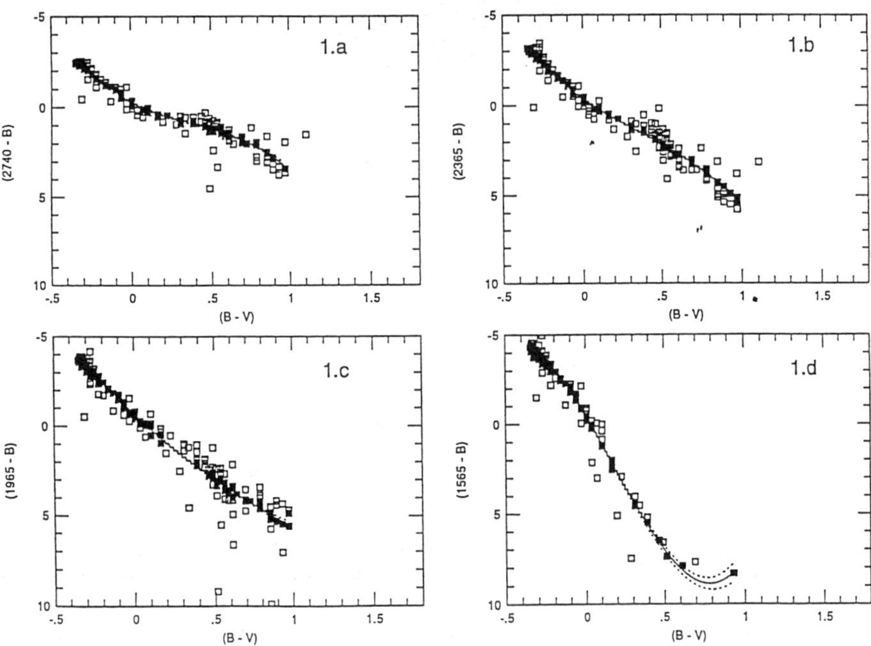

Figure 1. (UV-B) vs. (V-B) for luminosity class V

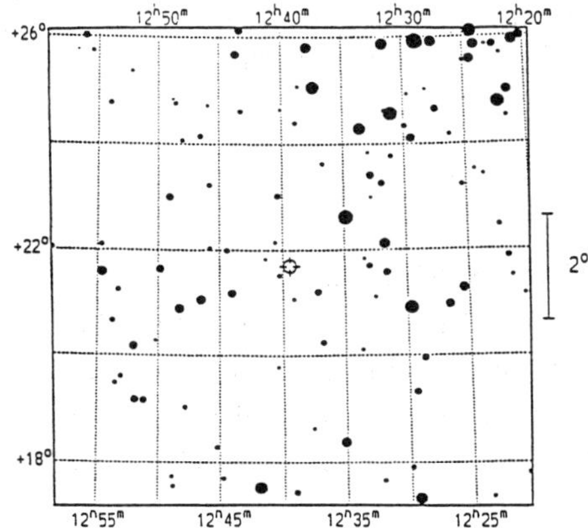

Figure 2. North galactic pole, a predicted UV map

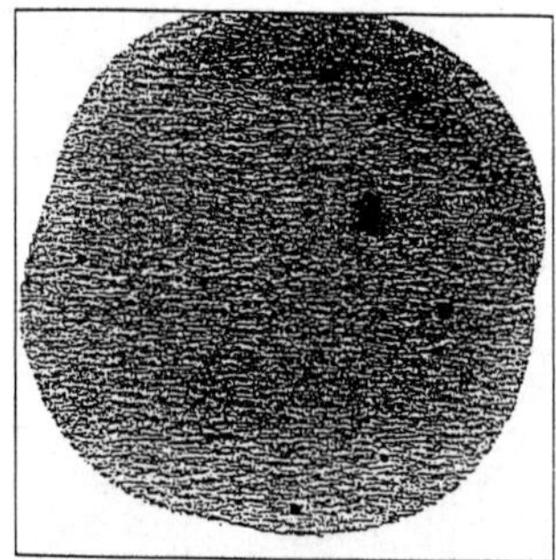

Figure 3. North galactic pole, the FAUST map

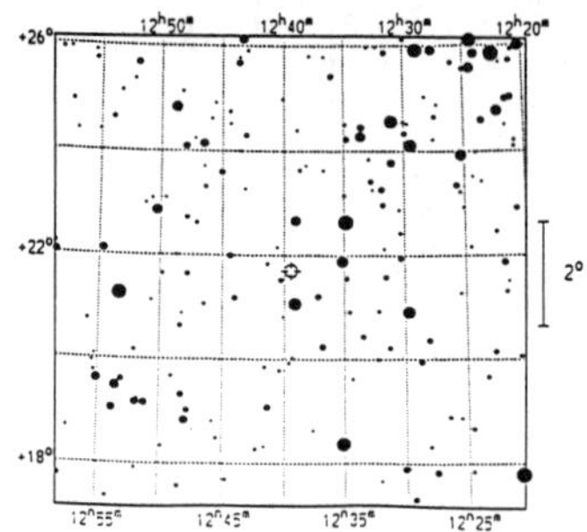

Figure 4. North galactic pole, an optical (V) map

The empty squares refer to spectra appearing in IUE spectral catalogues but excluded by the 3σ criterion. To demonstrate our method we show the prediction of the sky in the region of the north galactic pole, for the FAUST bandpass (Figure 2), against the actual aboard image (Figure 3). The star magnitudes in Figure 2, symbolized with full circles, are proportional to -2.5×log(UV flux of the star). Note that the fainter the star, the larger its magnitude. The spread in the stellar images in figure 3 also indicates its intensity, but it should not be compared quantitatively with that of Figure 2. For comparison we enclose the optical (V) map of the same region (Figure 4).

References

Brosch, N., 1991, MNRAS, 250, 780

Mersov, G., Shemi, A. and Brosch, N., 1993, Preprint

Savage, B.D. and Mathis, J.S. 1979, ARA&A, 17, 73

Thompson, G.I. et. al. 1978, Catalogue of Stellar Ultraviolet Flaxes, SRC

Turon, C. *et al.* 1992, The Hipparcos Input Catalogue, ESA. (HIC)

Part 5. Real-Time Software, Control Systems, and Scheduling

Astronomical Data Analysis Software and Systems III
ASP Conference Series, Vol. 61, 1994
D. R. Crabtree, R. J. Hanisch, and J. Barnes, eds.

Automated Observing on UKIRT

P. N. Daly, A. Bridger and K. Krisciunas

Joint Astronomy Centre, 660 N. A'ohōkū Place, University Park, Hilo HI 96720, U S A.

Abstract. We describe the four elements of the UKIRT observing system used for both the infra-red camera, IRCAM, and the long slit spectrometer, CGS4.

1. Introduction

The United Kingdom Infra-Red Telescope (UKIRT) is an equatorially mounted, 3.8m diameter primary Cassegrain telescope situated on the 4200m summit of *Mauna Kea* on the island of Hawaii. Some 85% of the available dark time is scheduled for observations with either the infra-red camera, IRCAM, or the long-slit spectrometer, CGS4. At the heart of each instrument lies an InSb 62 $\times$ 58 SBRC array which is held at 35K by a closed cycle cooler although plans are afoot to upgrade both instruments to 256 $\times$ 256 arrays during 1994.

IRCAM (McLean et al. 1986) was commissioned in September 1986 and is used for near infra-red (1 – 5 μm) imaging. Presently, four different scales are available from $\sim 0\farcs3$ to $\sim 2\farcs4$ pixel^{-1}. After the array upgrade, IRCAM will yield plate scales of $\sim 0\farcs05$ to $\sim 0\farcs3$ pixel^{-1}. As well as conventional imaging, IRCAM may be used with a rotating waveplate for polarimetry and with external Fabry-Perot etalons for narrow band work.

CGS4 (Mountain et al. 1990) also operates in the near infra-red window at resolving powers of $300 < \frac{\lambda}{\Delta\lambda} < 20000$. It was commissioned in February 1991 and was then 16000 times faster than previous detectors. It is currently scheduled for more than 200 nights per year and can generate ~100 MB of high quality data each night. A choice of gratings (75 l/mm, 150 l/mm and an Echelle) combined with either a 150 mm (f/1.35) or 300 mm (f/2.7) focal length camera provide scales of $\sim 1\farcs5$ or $\sim 3\farcs0$ pixel^{-1}. The slit length is $\sim 90\farcs0$ on the sky and the slit width can be matched to a single row on the detector. The resulting spectra are fully sampled by moving the array by fractions of a pixel over one or two pixels. A spectro-polarimetry facility has recently been commissioned.

Because of the complexity of CGS4, and the necessity of reducing data in near real-time (to allow assessment of data quality against a background many orders of magnitude brighter than the source), a software system was developed (Stewart et al. 1992) that automated the observing and reduction procedure as much as possible, while providing the observer with great flexibility. It became clear that extending this software to run IRCAM, too, would bring several benefits: enhanced efficiency when using IRCAM; reduced requirements for support of the software; and easier installation of the new electronics that will arrive with the larger arrays. This new comprehensive system, backwards compatible with its original task of acquiring and reducing CGS4 data, is now undergoing commissioning on the telescope and is proving to be successful with an 'uptime' of over 99.5 %. All software has been implemented under the ADAM environment (Lawden and Hartley 1992) and, where necessary, writes Figaro compatible data files. The user interface is Starlink's SMS layered over ICL.

2. The Data Acquisition System

To the observer, the *data acquisition system* is centred around the use of EXECs. These ASCII text files contain high-level commands to control the telescope and instrument configurations. A BREAK command allows the EXEC to be interrupted, to peak-up on a source for example, before resuming astronomical observations. EXECs may call other EXECs as shown in Table 1 which lists a sequence of commands for nodding the telescope between two positions on the CGS4 slit. It is also possible to create, automatically, EXECs which will take mosaics (maps) of large areas complete with sky observations at random locations at regular intervals. Observers may prepare their EXEC and configuration files in advance from templates provided.

The only difference between the systems used for IRCAM and CGS4 are in the user interfaces reflecting the different qualities of each instrument and in the underlying tasks that control motors. The other tasks in the system are identical and, once the data has been transferred safely to file store, the data acquisition system interacts with the data reduction system by writing a suitable command into the data reduction queue.

Table 1. Some Example Data Acquisition Commands.

! BS2421.EXEC	! Comment for naming the EXEC
CONFIG BS2421	! Recall acquisition configuration from file BS2421.CD4
SET OBJECT	! Set the instrument but do not start observation
BREAK Now Peak-up	! Break with comment message 'Now Peak-up'
NAME BS2421	! Name the object
TEL MAIN	! Set the telescope to main beam
STARTGROUP	! Tell data reduction to start a new group
DO 2 QUAD_SLIDE	! Do 2 repeats of EXEC called QUAD_SLIDE
ENDGROUP	! Tell data reduction that this group is complete
! QUAD_SLIDE.EXEC	! Comment for naming the EXEC
OBJECT	! Take an OBJECT observation
SLIDE SLIT -21.56	! Slew along the slit by 21.56 arcseconds
SKY	! Take a SKY observation
SKY	! Take a SKY observation
SLIDE SLIT 0	! Slew back to the original slit position
OBJECT	! Take an OBJECT observation

3. The Data Reduction System

The *data reduction system* monitors the reduction queue every two seconds by re-scheduling. A variety of commands are available and a common sequence is shown in Table 2. The software 'knows' all about the different observation types such as BIAS, DARK, FLAT, ARC, OBJ, SKY, CALIB and STANDARD and will reduce such frames according to the configuration selected by the observer. Observations may be combined into GROUPs of sky-subtracted data automatically. An index is also maintained of reduced calibration frames should these be required again.

The software is organized as a suite of co-operating tasks commanded by a control task. During the reduction sequence, the software may take all or some of the actions described in Table 3 depending upon the instrument in use and the observer's preferred configuration. For CGS4, there are further manual operations available to extract the spectra, remove the seeing ripple and apply a flux calibration. For IRCAM, the data can be read directly and manipulated by the existing IRCAM data reduction package or other Starlink packages.

The output is produced in real-time at the telescope. Indeed, automated on-line reduction is *vital* to many observing programmes and several weak spectral features have been identified during observing nights which might otherwise have been overlooked.

Table 2. Some Example Data Reduction Commands.

Command	Comment
DRCONFIG BS2421	! Recall reduction configuration from file BS2421.CRED4
REDUCE O931015_12	! Reduce observation 12 from UT date 15 October, 1993
END O931015_12	! File observation 12 in the reduced observations index file
REDUCE O931015_13	! Reduce observation 13 from UT date 15 October, 1993
END O931015_13	! File observation 13 in the reduced observations index file
REDUCE O931015_14	! Reduce observation 14 from UT date 15 October, 1993
END O931015_14	! File observation 14 in the reduced observations index file
REDUCE O931015_15	! Reduce observation 15 from UT date 15 October, 1993
END O931015_15	! File observation 15 in the reduced observations index file
ENDGROUP RG931015_12	! End the current group and perform housekeeping functions

4. The Automatic Archiving System

Because of the large amount of data produced by the instruments, it is essential that a reliable method of archiving the data exists. The *automatic archiving system* copies each raw data file produced by the data acquisition system to the base facility in Hilo within a few seconds of it being closed at the summit. This provides completely transparent, secure data handling. At the end of each night reduced data files are copied to Hilo, and both the raw and reduced data are copied to DAT tape and optical disk. Thus very soon after the completion of a night's observing, *four copies* of the data-set exist. Further, older data are removed, if they have been successfully copied to the archive media, to keep disk space free. The entire system operates without user input but can send re-assuring message to named personnel or warnings if a problem has occurred.

Table 3. Actions in a Data Reduction Sequence.

Sequence	Action	Mode
1	Apply a bad pixel mask	IRCAM, CGS4
2	Subtract a BIAS frame	IRCAM, CGS4
3	Linearise the signal	Not usually enabled
4	Subtract a DARK frame	IRCAM, CGS4
5	Divide by a FLAT field	IRCAM, CGS4
6	Interlace integrations (for over-sampling)	CGS4
7	Calibrate X-axis into wavelength	CGS4
8	Co-add observations and subtract SKY frames	IRCAM, CGS4
9	Remove residual OH and thermal emission	CGS4
10	Divide by a STANDARD source	CGS4

5. The Automatic Peak-up System

Placing the maximum amount of signal from the source on the desired row of the array proved to be a major inefficiency with CGS4. The instrument has no direct imaging facility, so 'peak-up' was achieved by using a 'movie' facility and manually scanning the telescope across the source. With the small slits used by the CGS4 long camera, this could take as much as 20 minutes even with experienced telescope operators.

To address this problem an *automatic peak-up system* has been designed and commissioned. This software scans the telescope across and up and down the slit while controlling CGS4's data taking thus achieving much greater efficiency. It also produces a more quantitative location of the peak signal than an equivalent manual procedure. Temporal variations in the background can be removed by utilizing sky rows in a data frame. The routine may be used to peak-up the source on any desired row and is able to peak-up from selected emission lines. For a typical, moderately faint point source the time taken to peak-up using this routine is 1–2 minutes.

Although peak-up is most important for spectroscopy, the routine is nevertheless designed to be used with other instruments such as IRCAM where, for instance, it is desirable to re-position a source in the same location on the array.

References

McLean, I. S., Chuter, T. C., McCaughrean, M. J., & Rayner, J. T. 1986, in Instrumentation in Astronomy VI, D. L. Crawford, SPIE, 627, 430

Mountain, C. M., Robertson, D. J., Lee, T. J., & Wade, R. 1990, in Instrumentation in Astronomy VII, D. L. Crawford, SPIE, 1235, 25

Stewart, J. M., Beard, S. M., Mountain, C. M., Pickup, D. A., & Bridger, A. 1992, in Astronomical Data Analysis Software and Systems I, A.S.P. Conf. Ser., Vol. 25, eds. D.M. Worrall, C. Biemesderfer & J. Barnes, 479

Lawden, M. D., & Hartley, K. F. 1992, "ADAM — The Starlink Software Environment", SG/4.2, Starlink Project, RAL

Astronomical Data Analysis Software and Systems III
ASP Conference Series, Vol. 61, 1994
D. R. Crabtree, R. J. Hanisch, and J. Barnes, eds.

Keck Autoguider and Camera Server Architecture

W. F. Lupton

W. M. Keck Observatory, P. O. Box 220, Kamuela, HI 96743, USA

Abstract. Keck autoguiders are not tightly coupled to the telescope control system but instead are just instruments which happen to have been asked to send guide star positions to the telescope. In addition, a camera-independent "camera server" protocol allows generic software to control several types of physical camera. Advantages and disadvantages of this approach are discussed.

1. Introduction

The Keck telescope currently has five autoguider cameras. All are supplied by Photometrics Ltd. and all use the same Thomson TH--7863 frame transfer CCD. One is permanently available for engineering tests; one is shared by the three infra-red instruments; two are used in the Caltech low resolution imaging spectrograph; the final one is used in the Lick high resolution echelle spectrograph.

2. Software Environment

These cameras are all controlled by the software system whose major components are described below.

xguide (the autoguider Motif user interface) is a KTL (Keck Task Library; Lupton 1992) application written by Al Conrad and runs under SunOS. It uses the Lick Observatory MUSIC (Stover 1989) message system to communicate with ...

autoguider server is also a KTL application running under SunOS. It provides various high-level functions and manages the interfaces with the telescope and ...

camera server is an RPC server and controls the cameras. It runs under VMS but is portable (its GPIB access routines need some changes in order to use a SunOS-specific library).

display server is also an RPC server and runs under SunOS. It works in tandem with a Motif display manager which manages X events and uses X server resources to communicate window ids, scaling details and so on to the display server and the user interface.

telescope server represents the telescope control system, which is split between VMS and VxWorks. Tasks are structured in such a way that the planned migration to KTL will be straightforward. Telescope user interface commands enter at the VMS level. Autoguider server communication is with the VxWorks level.

3. Camera Server Protocol

The idea of having a standard camera interface emerged early during the autoguider software design process. The motivation then, as now, was to make it easy to use several types of cameras as the source of guide star positions. Initially we were just going to define a subroutine interface but the idea of a separate server process emerged as being more flexible in our multi-processor environment.

The protocol is defined in terms of the RPC data description language. The big attractions of an RPC implementation are

1. ready-made design encompassing rules about low-level message formats, program lookup on remote hosts and authentication, and
2. widely available implementations across multiple host types.

3.1. Camera Server Example

This example shows a session with a camera client whose commands map closely on to the camera server RPC procedures.

```
Command: servstat
Servstat ...succeeded
cameras=(hiresn,9) (hires,8) (lris50m,7) (lris50s,5) (ssc,3) (f15,2)
(f25,1) (null,0)
```

This executes the `SERVSTAT` procedure to return a structure describing camera server status. The `cameras` list indicates the available cameras and their ids.

```
Command: select hires
Select, name=hires ...succeeded
Current camera has id 8
```

This executes the `LOOKUP` procedure to determine the id of a given camera. This id is a parameter to most of the camera server procedures.

```
Command: camstat
Camstat, id=8 ...succeeded
name=hires, type=Photo
disp=1, store=1, arith=1, stats=1, frame=1, little=0, signed=1
chip=Thomson---TH7863, chipid=, adcbits=0, xsize=384, ysize=288
xpixsize=-0.153000, ypixsize=0.153000
state=0, buffs=
```

This executes the `CAMSTAT` procedure to return status of a specified camera. `Photo` means Photometrics. The various `0/1` values on the next line are camera capabilities (does it have a display? can it store images? etc).

;, interrupt, and clock driven record processors, and sequencer (state- programs.

EPICS Host Applications

workstation or OPI (OPerator Interface platform) is the actual de- t area. Cross development is done with the GNU toolkit. Besides ard VxWorks development environment a number of EPICS tools are under SunOS.

Channel Access Tools for real time applications include three major . ALH (Alarm Handler) is a general purpose alarm handler driven by ı configuration file. AR (Archiver) is a tool to gather and save data IOC. DM (Display Manager) takes display list files created by EDD all monitors and variables to create a visual display of the run-time ent.

Building Tools for setting up and editing are another group of three . DCT (Database Configuration Tool) creates and modifies run-time ;. EDD (Display Editor) creates a display list file for the DM. SNC ɔtation Compiler) is a compiler that generates a C program which rep- ate sets for the IOC Sequencer Tool.

EPICS Database

ry in the database has an associated record type that can be purely related to a specific hardware channel, or linked via input, output, d processing sequence to another database record. A sample database can be seen in Figure 1.

e are a number of Input record types. These are Analog, Binary, Multi- y, Digitized Analog, and State.

database also supports a variety of Algorithm record types. These are ons, PID, Signal Select, Data Compression, Subroutine, and Fanout.

e are also a number of Output records. These are Analog, Binary, Binary, Stepper Motors, Timing Signals, and Permissive.

Gemini Control System Interface

ini Control Ssytem will use the native EPICS interface as the only con- to the real-time systems (see Figure 2). The EPICS system provides for flexible, and full-featured Application Programmer's Interface (API) istributed database system. This API is know as Channel Access and ented as a C-callable library that works under SunOS, VAX/VMS, orks. Channel Access provides a uniform interface across all of the control work packages.

Channel Access API can be divided into three classes of functions: Database Access, and Event.

General functions are *ca_task_initialize()* to initialize a Channel Access n, *ca_task_exit()* to exit a Channel Access application, *ca_search()* to


```
Command: expose
Expose, id=8, etime=0.02, dest=1, frame=0, stime=1, sopen=1, sclose=1,
cread=1, clear=1, simpwind=(0,0,384,288,1,1,0.00) ...succeeded
```

This executes the `EXPOSE` procedure to take an exposure of duration 0.02s. The other parameters are all Boolean flags indicating things like whether the shutter should be used or a readout performed.

```
Command: recv
Receive, id=8, src=1, destlist=(,,) ...succeeded
```

This invokes the `RECV` procedure to receive the specified camera-resident image buffer from the camera and transfer it to the supplied list of destinations. Here the list has a single entry whose fields are blank, which means "return the data in the RPC response".

```
Command: rpc 1
RPC=1
Command: recv
Receive, id=8, src=1, destlist=(rpc,pololu,0x40000000) ...succeeded
```

This is similar, but the `RPC` command has changed the destination list to specify that the camera server should instead send the data to a third party by invoking an RPC on a machine called `pololu` with RPC program number `0x40000000`. This RPC server can do anything it likes with the data — it may for example display it (this is how the Keck autoguider system implements its display). The camera client finds out the program number by negotiating directly with the server in question (in this case it uses X server properties).

4. Supporting Other Cameras

When it is controlling a conventional CCD camera, there is a more-or-less 1–1 correspondence between camera server procedures and operations on the physical camera ... but this is not necessary.

Consider a thermal infra-red camera running at a high frame rate of, say, 1kHz. These frames will be being co-added and sky-subtracted down at a low level but occasionally, say at 1Hz, the resulting frames will be made available for display. The camera server could grab these frames. A request to "take a 1s picture" could be interpreted as "if possible, ask the infra-red camera control system to send back pictures corresponding to 1s total integration time and then grab the next such picture". The camera server protocol is sufficiently general that this should work with no change to the existing autoguider software.

As another example, Keck is purchasing a wavefront sensor camera from Georgia Tech. This is a fast low-noise CCD camera and will be read out at perhaps 100Hz. Centroids will be calculated at this rate and will be sent to a fast steering mechanism (initially our chopping secondary). Of course only the AC component of the variation should be sent to the fast steering mechanism. The DC component must be sent to the autoguider. It could be sent at, say, 1Hz and, provided that the fast centroiding is able to ignore the relatively slow response of the telescope to the autoguider corrections, everything should be fine. Given that the camera server calculates guide star positions (using the `STATS` procedure), it should be a simple matter to implement a camera server that will work with the wavefront sensor.

5. Conclusions

What are the advantages of the Keck approach?

1. Keeping the autoguider separate from the telescope has allowed autoguider software to be developed independently from telescope software, and the well-defined interface between the two means that it is easy to change either without affecting the other.
2. The existence of the camera server interface allowed independent development and testing of the low-level camera control software. Since first being released in May 1992, this software has been extremely reliable.
3. The existence of the camera server also provides an obvious way to support new sources of guide star positions, as already discussed.

Of course, there are also some disadvantages.

1. The many components and interfaces, plus the fact of the relatively unsophisticated command set supported by the Photometrics CC200 camera controller, mean that the Keck architecture is not suitable for guiding at faster than about 1Hz (which happens to be the spec). Really fast guiding is always going to have to make use of direct dedicated connections.
2. The camera server sends the full 16-bit images to the display server. If the data were compressed and reduced to 8 bits, the display would be much faster, especially when running over a network.
3. The home-grown display server offers very few features. We would like to use SAOimage. Unfortunately, SAOimage does not offer quite all the features which we require, in particular the means of placing a binned sub-image (the guide box) in the correct place on the display. Wright and Conrad (1993) discuss this further.

References

Lupton, W. F. 1992, Keck Software Document 28, "KTL Programming Manual"

Stover, R. J. 1989, UCO/Lick Technical Report 54, "MUSIC — a Multi-User System for Instrument Control"

Wright, J. & Conrad, A. R. 1993, "The SAO-IIS Communication Package", this volume

Astronomical Data Analysis Software and Systems III
ASP Conference Series, Vol. 61, 1994
D. R. Crabtree, R. J. Hanisch, and J. Barnes, eds.

The Real-Time Environment for the Telescopes

P. McGehee

Gemini 8-m Telescopes Project, Tucson, AZ

Abstract. The real-time control system fo project will be formed from the integration of responsible for operation of separate physical created at distinct development sites, it is n interface to the core observatory control syst

Since the work packages cover the entire issues, including primary mirror support, e ment controllers it is expected that the re can be quite different and present different solve this integration problem it was decid package that was flexible and powerful enc ini real-time control work packages. The pr real-time control system is EPICS (Experim Control System).

The essence of the EPICS environme database that is built upon the VxWorks re TCP/IP-based networking. It is this data as the interface between the diverse real-ti Observatory Control System.

1. Introduction to EPICS

EPICS stands for Experiment Physics and Industr inal development was started in the late 1980's by Controls and Automation Group (AT-8) at Los and by the Advanced Photon Source Controls a gonne National Laboratory.

EPICS is currently in use at over 30 sites ranging from linear accelerator control to large formation about EPICS is now available via the *http://epics.aps.anl.gov/controls.html* at the Adva

2. The EPICS Real-Time Tasks

The VxWorks target or IOC (Input-Output Co EPICS system. The tasking environment includ

the eve machine

3. Th

The ho velopme the star availabl Th program the alar from th and link environ The program databas (State N resents s

4. Th

Each en software or forwa fragmen The Bit Bina The Calculat The Multi-Bi

5. Th

The Ger nection i a simple into the is imple and Vx real-time The General, The applicat

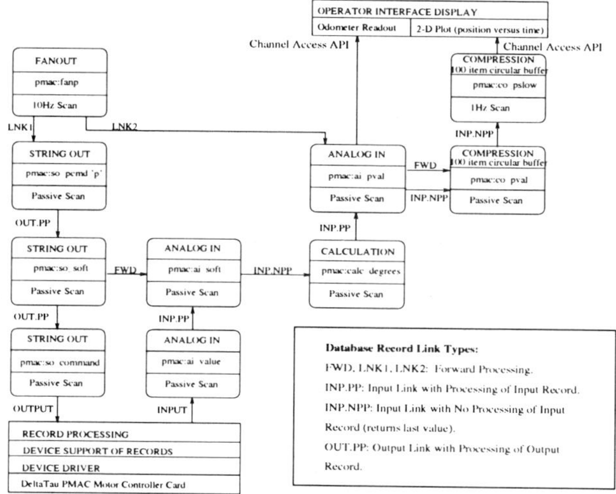

Figure 1. Sample EPICS database layout showing the readback and display of encoder values from a DeltaTau PMAC motor controller VMEbus card. The encoder values are obtained from the PMAC card at a 10Hz rate by issuing a "p" command and reading back the response. The returned encoder counts are then converted into degrees and copied into a 100 element circular buffer for slow readout by the SunOS host.

query system for database entry connection data, and *ca_clear_channel()* to close a Channel Access channel opened by *ca_search()*.

The Database Access functions are *ca_get()* to read a value from a channel, *ca_get_callback()* to read a value from a channel and run a callback when it is ready, *ca_put()* to write a value to a channel, *ca_pend_io()* to flush the send buffer and wait until outstanding I/O completes or timeout, and *ca_flush_io()* to send outstanding I/O requests.

The Event functions are *ca_test_event()* which is a canned event handler for debugging purposes, *ca_add_event()* which specifies a subroutine to be run whenever significant changes occur to a channel, *ca_pend_event()* which will flush the send buffer and wait for asynchronous events or timeout, and *ca_clear_event()* which will remove a subroutine specified by *ca_add_event()*.

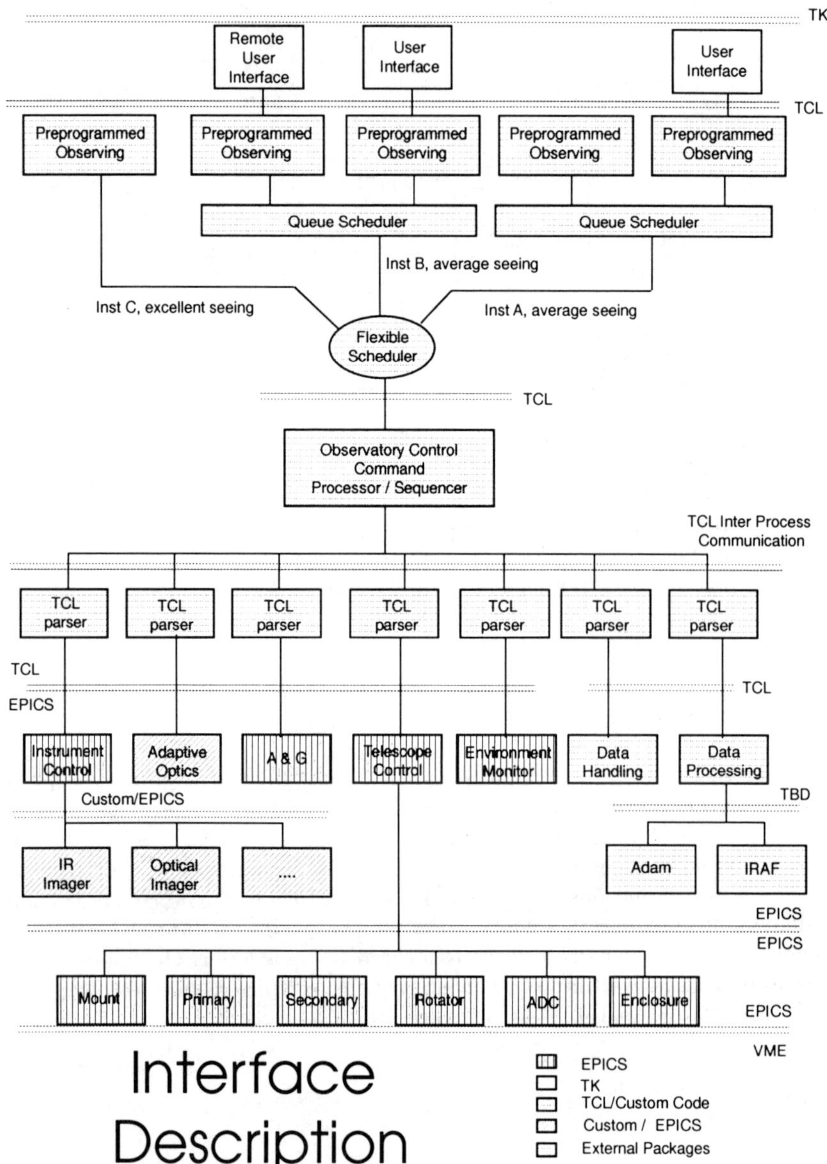

Figure 2. Baseline Gemini control system architecture. Each of the boxes in the lower EPICS layers are handled by separate work packages. The Observatory Control System will use the Channel Access API to interact with the real-time control systems.

Astronomical Data Analysis Software and Systems III
ASP Conference Series, Vol. 61, 1994
D. R. Crabtree, R. J. Hanisch, and J. Barnes, eds.

The Software for the LRIS on the Keck 10-Meter Telescope

J. G. Cohen, J. L. Cromer, S. Southard Jr. and D. Clowe[1]

Astronomy Department, California Institute of Technology. Pasadena, Ca. 91125

Abstract. We discuss the software for the Low Resolution and Imaging Spectrograph, one of the first light instruments built for the Keck 10−meter telescope. Details of the CCD detector readout scheme, the motor control system, the user interface, the astrometric preparatory software, etc., are given.

1. Introduction

The Low Resolution and Imaging Spectrograph (LRIS) for the Keck telescope is an effort under the direction of Bev Oke, principal investigator, (now retired from the California Institute of Technology), Judy Cohen, deputy principal investigator, and a team of engineers and technicians at Caltech. The software, computer interfacing, etc., was the responsibility of J.Cohen. The spectrograph provides direct imaging and multi-slit and long slit low and moderate resolution spectroscopy over a field of view 6 x 8 arc-minutes in size at the Cassegrain focus of the ten−meter Keck telescope.

2. Functions of the LRIS Software (as Specified in 1988)

To read out the detector using multiple amplifiers possibly on multiple CCDs (i.e., to allow for the possibility of a mosaic of CCDs) as rapidly as possible (The formal requirement is to read out and store in the disks of the main instrument computer a 2048 x 2048 image in less than 2 minutes.); to control the remotely operated mechanical mechanisms in a safe yet efficient manner; to support quick look analysis, and to provide the astrometric capabilities needed to make multi-slit masks under various circumstances (i.e., via externally generated coordinate lists or generating a coordinate list from a LRIS image).

All this is to be done through code which is easy to use, with a simple yet powerful user interface, easy to maintain, easy to modify and upgrade (i.e., adding a new motorized stage), and well documented.

The detector used in LRIS is a 2048 x 2048 pixel thinned Tektronix CCD. The only other optical instrument planned for first light at the Keck telescope is HIRES (a high resolution echelle spectrograph located at one of the Nasmyth foci), built under the direction of Steve Vogt of Lick Observatory, with Bob Kibrick assuming responsibility for the software, computer interfacing, and related matters. HIRES uses the same CCD and the same CCD electronics as does

LRIS. It was therefore deemed highly desirable to maintain common software and computer hardware, maintain identical user interfaces (in style, although not in the details of names of motor stages or optical components), and given the large amount of work to be done and very limited available funds, to share the software development between the two institutions.

3. The Two LRIS VME Crates

As standards for the first light Keck instruments, Unix was chosen as the operating system, and Sun machines were chosen as the hardware platform. Because of concern at the use of Unix as a real time system, time critical functions were divorced from the main Sun instrument computer. LRIS thus has two separate VME crates, the CCD detector crate and the LRIS motor control crate, each attached to the Sun workstation by a private ethernet. (The Sun workstation has two ethernet boards. One is used to talk to other workstations on the Keck network. The second is reserved for instrument/detector control.) The Sun 1E CPUs on the VME crates are running VxWorks, a real time operating system developed by Wind River Systems. In addition, the CCD crate contains a CCD controller (CCD computer interface card, designed by Fred Harris of Caltech and Terry Ricketts of UCSC), 32 Mbytes of Chrislin memory, and an Ikon 10089 DMA controller. The motor control crate contains a serial ports card as well.

The desire to have minimum cables going though the cable wrap to the Cassegrain focus, and to minimize heat input into the dome, meant that both of these crates are located in the Keck computer room, while the Sun workstation is located in the control room. There are 2 fiber optic cables which communicate between the CCD computer interface card and the CCD saddle bag electronics. That chassis contains cards designed by Bob Leach of San Diego State University, namely a timing board, a utility board, and an analog board for each amplifier in use. Some of these boards use a Motorola DSP 560001 chip. The fiber optic cable that transmits the data down to the VME crate is running at 40 Mbaud. The second fiber cable which receives instructions from the CCD crate runs at 4 Mbaud.

For the motor control VME crate, there are too many motors to assign each to its own serial cable running through the cable wrap. Instead the signals are multiplexed onto a single serial cable, demultiplexed at the instrument, and distributed to the various motors. Additionally the 22 motors are multiplexed to 4 motor controllers

The partition of work with Lick left Caltech with the responsibility for the code to read out the CCD into the workstation's memory, for the LRIS motor control software, for the responsibility of developing a fast X windows display server, and for the necessary astrometry codes specific to LRIS.

The CCD readout process, designed and coded by S. Southard, Jr., is rather complicated due to the two-stage transfer (CCD computer interface card to the VME crate via DMA, VME crate to the Sun workstation via Ethernet) required by the hardware design. Our analog to digital amplifiers can operate at 10 micro-sec/pixel, or 200 Kbytes/sec. We are actually running at 22 micro-sec/pixel, with 2 amplifiers interleaved, for a total data rate of 182 Kbytes/sec.

It is the number of working amplifiers on the CCD and the pixel transfer time which are limiting our readout rate. However, when the data transfer software was designed, the system design called for up to 16 interleaved amplifiers, for a maximum total data rate of 3.2 MBytes/sec, well exceeding the ethernet capacity. For this reason, the UDP protocol, which is faster but does not guarantee packet delivery, was chosen. Packets which are lost in ethernet transmission are re-sent at the end of the image.

Windowing and binning of CCD frames are also supported.

We are routinely reading out and storing to disk in the Sun instrument computer 2048 x 2048 16 bit deep (no sign bit, 16 bits of data) images with 2 amplifiers interleaved in under 90 seconds. We are routinely manipulating such images with our quick look software, as well as with standard large astronomical packages (IRAF and FIGARO).

4. LRIS Mechanisms

The LRIS motor control VME crate, whose software was designed and written by J.Cromer, runs the motorized mechanisms listed below (plus their associated brakes, absolute or incremental encoders, etc.). We are using API (American Precision Instruments) controllers, Compumotor AC stepping motors, Canon rotary incremental encoders, and Compumotor rotary absolute encoders.

Table 1. LRIS Mechanisms.

Mechanism	No. of motors	Type of Position Feedback
red filter juke	2	switch-encoder, limit switches
slitmask juke	2	switch-encoder, limit switches
red camera focus	1	absolute rotary encoder
offset guider	2	absolute rotary encoders
guider filter wheels	2	none
grating turret	1	switch-encoder
grating turret detent	1	limit switches
grating cells	10	incremental rotary encoders and limit switches
trapdoor	1	limit switches

Mechanisms were divided into groups depending on how many positions they had and how they were encoded. By identifying these common operating modes only 4 software functions were required to handle the moves of all stages.

5. User Interface

The user interface has two alternate modes. The first is a standard command line. The second brings up a number of Motif X window displays which offer buttons for choosing or changing selected actions. At the present time only the most frequently used keywords are implemented in the two LRIS X window displays, written by Al Conrad of the Keck Observatory staff based on an initial design by J. Cohen and J.B.Oke.

The current list of implemented keywords for the LRIS has about 55 items (nouns). The command line (and internal software) allows only simple manipulations of keywords, along the lines of "show redfilt" (this results in a status display for the red filter changer), "modify redfilt = visual" (changes the red filter changer so that the visual filter is in the optical path), and "configure slitmask 16hr_field = 1 Abell_104 = 2" (associates names with positions of the mechanism). The number of verbs ("show" etc.) supported is under 5, making the command line interface very easy to learn to use.

The system for internally processing these keywords, passing the appropriate messages around, etc., was written by the HIRES group based in part on a previously existing package in use at the time at Lick Observatory. They also wrote the software for passing messages to/from the telescope drive and control system.

S. Southard, Jr, wrote an X windows display server currently used by the PGPLOT and Figaro pacakges. The same software is used to provide real-time image display of the data being transferred from the CCD. This software's functionality is described in this year's software report for Figaro in the Bulletin of the AAS.

The astrometric software to assign objects to multi-slits was written by Caltech undergraduate student D. Clowe with help from J.Mould based on software written for the Norris spectrograph by J.Cromer. J.Cohen has modified large parts of this code and written the other necessary pieces. This pre-observing planning software is the only part of the code written in Fortran; all the CCD, instrument, and motor control software is in C.

6. Current Status

The current status is that the preliminary design review for this system was held (together with the identical HIRES review) on Jan. 29, 1990. The instrument is now complete. It was shipped to Hawaii in May 1993. There have been 3 instrument commissioning runs at the Keck telescope on Mauna Kea. The overall impression of the software (and not just by us) is that we have met our goals. It is robust, easy to use, and powerful enough to perform the necessary functions. We look forward to routine observing by people not associated with the LRIS project as the final test of our success. We expect routine observing to start early in 1994.

Acknowledgments. The W. M. Keck Observatory is operated as a scientific partnership between the California Institute of Technology and the University of California. It was made possible by the generous gift of the W. M. Keck Foundation, and the support of its president, Howard Keck.

Astronomical Data Analysis Software and Systems III
ASP Conference Series, Vol. 61, 1994
D. R. Crabtree, R. J. Hanisch, and J. Barnes, eds.

Distributing Functionality in the Drift Scan Camera System[1]

Tom Nicinski, Penelope Constanta-Fanourakis, Bryan MacKinnon, Don Petravick, Catherine Pluquet, Ron Rechenmacher, and Gary Sergey

Fermi National Accelerator Laboratory, PO Box 500, Batavia, IL 60510

Abstract. The Drift Scan Camera (DSC) System acquires image data from a CCD camera. The DSC is divided physically into two subsystems which are tightly coupled to each other. Functionality is split between these two subsystems: the front-end performs data acquisition while the host subsystem performs near real-time data analysis and control. Yet, through the use of backplane-based Remote Procedure Calls, the feel of one coherent system is preserved. Observers can control data acquisition, archiving to tape, and other functions from the host, but, the front-end can accept these same commands and operate independently. The DSC meets the needs for such robustness and cost-effective computing.

1. Introduction

The Drift Scan Camera System (DSC) was developed at Fermilab for two major reasons: to be used as a science instrument on the ARC 3.5m telescope at Apache Point Observatory, New Mexico, and to provide experience with using large CCDs (Charge Coupled Devices) for the Sloan Digital Sky Survey (SDSS). Because Fermilab is a High Energy Physics laboratory, its experience in producing high speed data acquisition systems was beneficial in the development of the DSC.

The DSC is split into two major functional units:

- Front-end data acquisition (DA). Its duties include:
  - acquiring data from the CCD Camera
  - displaying acquired data in real-time
  - storing acquired data on disk in the Frame Pool
  - archiving acquired data to 8mm tape.
- Host processing. Its duties include:
  - controlling front-end operations
  - analyzing Frame Pool data in near real-time.

[1]Sponsored by DOE Contract number DE-AC02-76CHO3000.

Some quality goals of the DSC include:

- Responsiveness. The front-end must acquire data at high rates without loss. In addition, it must process host requests for data during acquisition. The host must be able to retrieve and analyze acquired data in near real-time. This is necessary to allow the host to determine, from acquired data, that problems need to be fixed (such as a misaligned instrument rotator).
- Robustness. The front-end data acquisition must be capable of operating without the presence of the host, including an unexpected loss of the host.
- User-friendliness. Graphical User Interfaces (GUIs) may be used. But, DSC must have a command language allowing operation without GUIs.

These goals also apply to the SDSS System being developed. Since SDSS data rates are considerably higher, off-loading operations to the host is even more vital.

2. Isolating Functionality within DSC Components

The two major DSC functional units are isolated physically in two subsystems (see Figure 1).The choice of machines was dictated by cost, computing power, availability of commercial hardware, and the ability to use in-house built hardware. VMEbus is an ideal backplane with the necessary bandwidth. One machine to handle all DSC needs could not be found, necessitating a separate host.

The Instrument Control Computer (ICC) contains the front-end DA while host analysis is done on the Online Analysis Computer (OAC). The ICC is a 20 MIP Motorola MVME167b Single Board Computer. The OAC is a 30 MIP Silicon Graphics 4D/35. These machines' VMEbuses are tightly coupled with an HVE Engineering VME Repeater (link). ICC and OAC processes communicate with each other via shared memory.

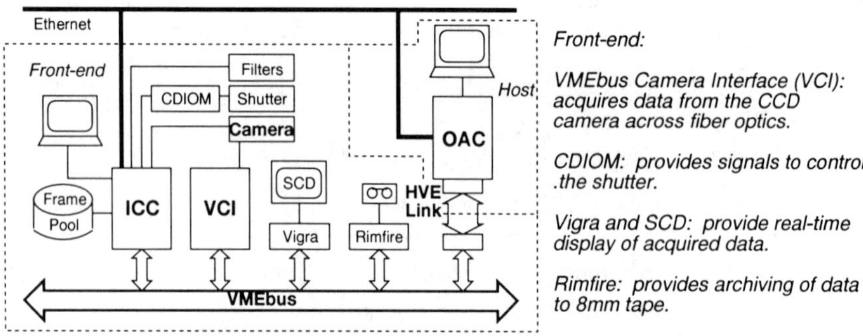

Figure 1. DSC Block Diagram.

3. The Need for Backplane Communications

The data rates of the DSC are not excessive, but the need for the OAC to analyze large data sets in near real-time necessitated backplane communications across shared memory. The DSC operates in one of two modes: drift scanning or staring. With a CCD size of 2048 × 2048 pixels (2 bytes each), the DA acquires data at 461 KBytes/sec. at the maximum drift rate. This data is packaged into Frames and stored on disk in the Frame Pool. The nominal Frame size is the CCD size, approximately 8 MBytes. At the maximum drift rate, an 8 MByte Frame is generated every 18.1 seconds. Buffering Frames to disk allows clients (including the archiver) to access data asynchronous to data acquisition.

For the host to analyze Frames in near real-time, Frames must be moved from the Frame Pool to host tasks quickly. Using the backplane interconnect, Frame transfer times are comparable to the time required to retrieve Frames from disk, about 10 seconds for an 8 MByte Frame; minimal time is spent on the actual transfer. However, coaxial Ethernet involves additional overhead along with transfer times an order of magnitude slower. Thus, using shared memory across the backplane provides the host considerably more analysis time between Frames (without resorting to double buffering in the science code). Besides moving Frames quickly from the ICC Frame Pool to the OAC, with tight coupling, other data can be shared transparently:

- Frame Pool directory. Maintaining, in shared memory, the list of all existing Frames permits OAC tasks to search the directory.
- Status entries. They provide information about the states, capacities, etc., of different DSC components. Host user interfaces can effectively inform users without repeatedly querying the ICC. Control parameter entries are also used to affect many system operations.

The shared data resides in ICC memory. This permits the ICC to continue operations in case the OAC is unavailable. The sharing is relatively efficient, compared to socket-based protocols, since only one indirect memory access is needed. If tasks outside the ICC and OAC need access to the shared data, a server on the OAC can be written to provide that access.

Sharing data across memory also improves system efficiency and reliability:

- Communication overheads are greatly reduced by using DMAs rather than, for example, socket-based protocols.
- Some processing tasks are localized to the "requester." For example, as mentioned above, OAC tasks perform Frame Pool directory searches, freeing up additional cycles for the ICC.
- Fewer points of failure exist (backplanes are more reliable than networks).

4. Distributing Control

Besides distributing functionality, control of the ICC is also distributed. A client/server model is used, where the clients can be on any machine, including the ICC, and the servers are on the ICC. ICC servers are theoretically not service-specific; they can handle all requests from all clients. In practice, servers handle specific jobs. Thus, by using varying server task priorities, client access to the ICC is prioritized. For example, a lower priority server task handles client requests for Frames; these requests cannot then block data acquisition.

Through more shared memory, the ICC can be controlled by the OAC. Remote Procedure Calls (RPC) across the backplane were chosen over a socket-based RPC because of better robustness. Although the hardware independent backplane RPC is somewhat complex, ICC servers and OAC clients can "connect" and "disconnect" from it without affecting their respective peers. The verification of safe backplane critical sections was considered simpler than handling connection requests and abnormal disconnects in a socket-based protocol.

4.1. Modularizing Commands

Having ICC servers handle all client requests is greatly simplified by using one common interpretive language throughout the DSC. Tcl (Ousterhout 1993) is a C and LISP-like extensible command interpreter. Commands can be added as Tcl procedures or by easily interfacing C routines to the interpreter. DSC uses low-level C routines with Tcl routines layered above to provide a user interface. As Tcl behaves the same on all platforms, users see no distinctions between issuing commands from the OAC or the ICC.

OAC clients issue Tcl commands to ICC servers via the backplane RPC while ICC clients execute Tcl commands directly. The GUIs on the OAC, acting as clients, emit Tcl commands. Not only is a user-friendly environment provided, but the system can be run with only line oriented commands, as the GUIs use the same Tcl interface to ICC and OAC functions. GUIs can be brought up quickly by using Tk (Ousterhout 1993), the Tcl toolkit for X Windows.

4.2. Using Existing Apache Point Facilities

DSC also interfaces to existing Apache Point machines on the local network:

- Telescope Control Computer. It controls the 3.5m telescope and enclosure. It broadcasts UDP packets containing telescope position/time pairs.
- Master Computer. Through a TCP/IP socket-based protocol, the DSC system issues telescope and enclosure movement commands and obtains information about the telescope, weather, etc.

5. Results

The DSC System has met its goals successfully. It's quite responsive and robust. Users have quickly written considerable code (much in Tcl) to perform analysis on the OAC and to control the ICC.

The DSC DA and communications architecture will be used as the base for the more ambitious SDSS System. Due to modularity, many DSC software components can be reused with little or no changes. With transparent access to shared data, such as the Frame Pool directory, the OAC off-loads work from the ICC. With SDSS' higher data rates (7 MBytes/sec. to 9 ICCs), such off-loading is necessary.

References

Ousterhout, John K. 1993, "An Introduction to Tcl and Tk", Addison-Wesley

Astronomical Data Analysis Software and Systems III
ASP Conference Series, Vol. 61, 1994
D. R. Crabtree, R. J. Hanisch, and J. Barnes, eds.

A Software State Machine for Computing Astronomical Coordinates

Jeffrey W. Percival

Space Astronomy Laboratory, 1150 University Avenue, Madison, WI 53706

Abstract. We consider the common problem of computing apparent and topocentric places of stars for the purpose of pointing a telescope. Detailed algorithmic descriptions exist (see, for example, Kaplan et al. 1989). In addition, several software packages such as NOVAS and Starlink's SLALIB by Patrick Wallace considerably ease the burden in building specific application programs. A few problems remain, however. Portability can be a problem, in that some real-time platforms have grudging or non-existent support for Fortran, which is the language of implementation for NOVAS and SLALIB (SLALIB in C is now available). Also, efficiency can be a problem if the subroutines try to do too much, not allowing the programmer to fragment the calculation as needed. SLALIB offers many convenient entry points, which avoids this problem, but the programmer is still left to weave the subroutines together to achieve a desired result.

We have designed a portable software state machine, written in C, for use in the WIYN Telescope Control System. The state machine is in the form of a graph, with the nodes representing coordinate states (heliocentric FK4, topocentric apparent, or galactic, for example) and the edges representing the calculations required to move between states. The programmer provides a starting state and coordinate state vector, and a desired ending state. Using the current state and desired end state, the machine marches through the graph, performing the transitions in the proper order.

This approach has several advantages. First, not only are the calculations well-defined, as they are in existing subroutine libraries, but their order of execution is embedded in the machine, rather than merely specified in documentation, removing a source of programming error. Second, each transition can be implemented exactly once, in exactly one place, while the state machine dynamically changes the order of events according to the state transition table. Third, transitions can be implemented in both directions, allowing both production and reduction of coordinates.

We present a brief overview of the state machine, including the state diagram and transition table.

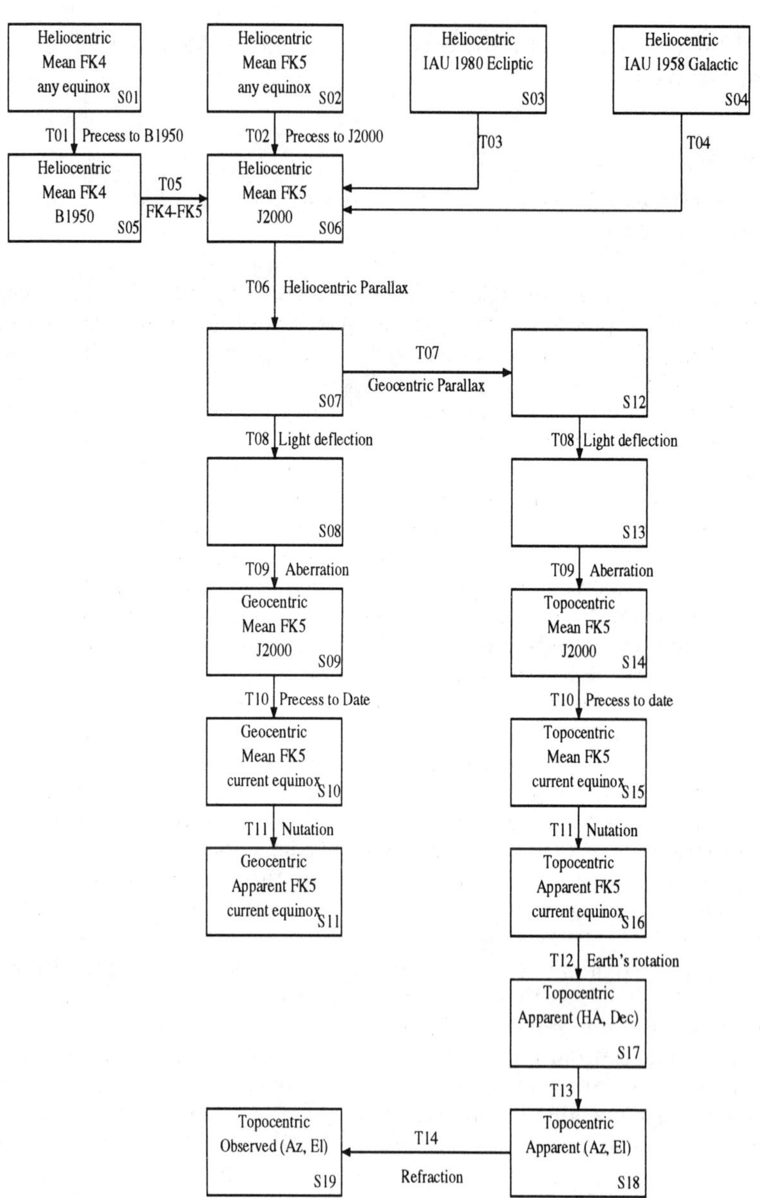

Figure 1. Telescope Pointing Machine (based on the Keck pointing flow by P. T. Wallace).

es

esigned to give application programmers a basic set of positional-
s which were accurate and easy to use. The library is:

vailable — from the author and from the Starlink Project, in-
ource code and documentation (Wallace 1993).

l and maintained.

— coded in standard languages and available for multiple com-
d operating systems.

ly commented, both for maintainability and to assist those wish-
nibalize the code.

hy — some care has gone into testing SLALIB, both by compar-
published data and by checks for internal consistency.

— corners are not cut, even where the practical consequences
most cases be negligible.

ensive, without including too many esoteric features required only
ists.

— almost all the routines have been written to satisfy real needs
ed during the development of real-life applications.

ent-independent — the package is completely free of pauses,
), etc.

ined — SLALIB calls no other libraries.

loes not pretend to be canonical. It is in essence an anthology,
dopted algorithms are liable to change as more up-to-date ones
vailable.

ions aren't orthogonal — there are several cases of different rou-
g similar things, and many examples where sequences of SLALIB
simply been packaged, all to make applications less trouble to

holes — there is no support for orbital calculations, for example,
gh-precision Solar-System predictions.

s not homogeneous, though important subsets (for example the
routines) are.

y is not foolproof. You have to know what you are trying to do
eading textbooks on positional astronomy), and it is the caller's
lity to supply sensible arguments (although enough internal val-
done to avoid arithmetic errors).

Table 1. State table (part 1) for the Telescope Pointing Machine.

start state	destination state									
	S01	S02	S03	S04	S05	S06	S07	S08	S09	S10
S01	+T00 S01	+T01 S05	+T01 S05	+T01 S05	+T01 S05	+T01 S05	+T01 S05	+T01 S05	+T01 S05	+T01 S05
S02	+T01 S06	+T00 S02	+T01 S06	+T01 S06	+T01 S06	+T01 S06	+T01 S06	+T01 S06	+T01 S06	+T01 S06
S03	+T03 S06	+T03 S06	+T00 S03	+T03 S06	+T03 S06	+T03 S06	+T03 S06	+T03 S06	+T03 S06	+T03 S06
S04	+T04 S06	+T04 S06	+T04 S06	+T00 S04	+T04 S06	+T04 S06	+T04 S06	+T04 S06	+T04 S06	+T04 S06
S05	-T01 S01	+T02 S06	+T02 S06	+T02 S06	+T00 S05	+T02 S06	+T02 S06	+T02 S06	+T02 S06	+T02 S06
S06	-T02 S05	-T01 S02	-T03 S03	-T04 S04	-T02 S05	+T00 S06	+T05 S07	+T05 S07	+T05 S07	+T05 S07
S07	-T05 S06	-T05 S06	-T05 S06	-T05 S06	-T05 S06	-T05 S06	+T00 S07	+T07 S08	+T07 S08	+T07 S08
S08	-T07 S07	-T07 S07	-T07 S07	-T07 S07	-T07 S07	-T07 S07	-T07 S07	+T00 S08	+T08 S09	+T08 S09
S09	-T08 S08	-T08 S08	-T08 S08	-T08 S08	-T08 S08	-T08 S08	-T08 S08	-T08 S08	+T00 S09	+T01 S10
S10	-T01 S09	-T01 S09	-T01 S09	-T01 S09	-T01 S09	-T01 S09	-T01 S09	-T01 S09	-T01 S09	+T00 S10
S11	-T09 S10	-T09 S10	-T09 S10	-T09 S10	-T09 S10	-T09 S10	-T09 S10	-T09 S10	-T09 S10	-T09 S10
S12	-T06 S07	-T06 S07	-T06 S07	-T06 S07	-T06 S07	-T06 S07	-T06 S07	-T06 S07	-T06 S07	-T06 S07
S13	-T07 S12	-T07 S12	-T07 S12	-T07 S12	-T07 S12	-T07 S12	-T07 S12	-T07 S12	-T07 S12	-T07 S12
S14	-T08 S13	-T08 S13	-T08 S13	-T08 S13	-T08 S13	-T08 S13	-T08 S13	-T08 S13	-T08 S13	-T08 S13
S15	-T01 S14	-T01 S14	-T01 S14	-T01 S14	-T01 S14	-T01 S14	-T01 S14	-T01 S14	-T01 S14	-T01 S14
S16	-T09 S15	-T09 S15	-T09 S15	-T09 S15	-T09 S15	-T09 S15	-T09 S15	-T09 S15	-T09 S15	-T09 S15
S17	-T10 S16	-T10 S16	-T10 S16	-T10 S16	-T10 S16	-T10 S16	-T10 S16	-T10 S16	-T10 S16	-T10 S16
S18	-T11 S17	-T11 S17	-T11 S17	-T11 S17	-T11 S17	-T11 S17	-T11 S17	-T11 S17	-T11 S17	-T11 S17
S19	-T12 S18	-T12 S18	-T12 S18	-T12 S18	-T12 S18	-T12 S18	-T12 S18	-T12 S18	-T12 S18	-T12 S18

1. Overview

The Telescope Pointing Machine is a software state machine that produces or reduces astronomical coordinates.

- It is coded entirely in C.
- It can perform a single calculation (i.e., precession) or it can perform a whole coordinate flow (i.e., galactic to refracted Az/El).
- It performs its calculations entirely in a 6-dimensional rectangular coordinate space (position and velocity). Spherical projections do not enter in at any point, nor is spherical trigonometry used at any point. No CPU cycles are used going back and forth between rectangular and spherical coordinate representations.

The state transition table is a two-dimensional matrix of entries, indexed by the current and destination state. Each entry specifies two data: the next transition to execute, and the state that will result. In this way, for example, precession can be specified at many points in the table, each one with a different beginning and ending state.

Note that a state table entry does *not* specify the whole calculation between the start state (row) and the destination state (column). This would require common calculations, such as precession, to appear many times in the state table and in the state machine. Rather, each state table entry specifies only the *next* calculation, as well as where to go in the table to find the next step. In this way, the state machine uses each basic subroutine only once, in one place.

References

Kaplan et al. 1989, AJ, 97, 1197

Astronomical Data Analysis Software and Sy
ASP Conference Series, Vol. 61, 1994
D. R. Crabtree, R. J. Hanisch, and J. Barne

The SLALIB Library

Patrick Wallace
Starlink Project, Rutherford Ap
Chilton, Didcot, Oxon OX11 0(

Abstract. The Starlink SLA]
mainly concerned with astronoı
orous but easy to use impleme
transformations, such as FK4 t
rical and computational faciliti
positional-astronomy applicatio

The SLALIB versions curr
and run on VAX/VMS, several
C version has recently been cor

1. Introduction

The Starlink Project (Lawden 1990)
provide UK astronomers with comput
analysis. Included in Starlink's sizable
number of application programs to do
transformations, astrometry, radial vel
ysis and so on. These applications all
called SLALIB[1]. The library is also w
tions by individual user-programmers,
and internationally, and is in service
example UKIRT, JCMT, WHT, ARC.

SLALIB is descended from routir
puters in the mid-1970s. The arrival
more comprehensive package to be desi
a time when the adoption of the IAU
would have to cope with a mixture of
clature. Development was essentially c

Despite its wide use, SLALIB h
literature; the present paper serves bo
to announce the availability of a new ir

[1]The name isn't an acronym; it just stands foı

2. Objectiv

SLALIB was d
astronomy too

- Readily
  cluding s
- Supporte
- Portable
  puters a
- Thoroug
  ing to ca
- Stable.
- Trustwoı
  ison with
- Rigorous
  would in
- Comprel
  by specia
- Practical
  encounte
- Environ
  stops, I/
- Self-cont

A few *caveats*:

- SLALIB
  and the
  become a
- The func
  tines doi
  calls hav
  write.
- There ar
  and no h
- SLALIB
  FK4/FK
- The libra
  (e.g., by
  responsib
  idation is

3. Contents

SLALIB contains 140 routines covering the following topics:

- String Decoding, Sexagesimal Conversions
- Angles, Vectors & Rotation Matrices
- Calendars, Timescales
- Precession & Nutation
- Proper Motion
- FK4/5, Elliptic Aberration
- Geocentric Coordinates
- Apparent & Observed Place
- Refraction & Air Mass
- Ecliptic, Galactic, Supergalactic Coordinates
- Ephemerides
- Astrometry
- Numerical Methods

The Fortran version weighs in at roughly 14000 lines. Individual modules are from 30 to 1000 lines long. A typical SLALIB routine is about 100 lines long, begins with 40 lines of comments describing purpose, arguments, etc., and has 40 lines of code intermingled with 20 comment lines (all inclusive of blank lines).

4. Portability

The Fortran version of SLALIB uses ANSI Fortran 77 with a few commonplace extensions. Three out of the 140 routines require platform-specific techniques and exist in different versions. SLALIB has been implemented on the following platforms: VAX/VMS, PC (MS-DOS and Microsoft Fortran), DECstation (Ultrix), DEC Alpha (OSF/1), Sun/SunOS, Sun/Solaris, CONVEX, Perkin-Elmer and Fujitsu. All but the last three are supported by Starlink and/or the author.

5. The C Version

A C version of SLALIB has recently been completed. It began as a hand-coded transcription by Steve Eaton (University of Leeds), was enhanced by John Straede (AAO) and Martin Shepherd (Caltech), and was then revised, completed and validated by the present author.

The revision process involved a number of key design decisions. The result was an implementation in the ANSI dialect, rather than K&R or C++. The functionality matches that of the Fortran SLALIB; a redesign to correct deficiencies in the latter and to exploit features of the C language was rejected, mainly for expedience but also for the convenience of existing users of the Fortran version, some of whom have implemented C "wrappers". The function names could not be left the same as the Fortran versions because of potential linking problems when both forms of the library are present; the format chosen was such that `SLA_REFRO` becomes `slaRefro`. The types of arguments follow the Fortran version, except that `int` was chosen for integers rather than `long`. Argument passing is by value (except for arrays and strings of course) for given arguments and reference for returned arguments. The code is laid out with all the interface information at the top (`#include`s and the function declaration), followed by the prologue comments, followed by the code.

6. The Future

The C version is functionally the same as the Fortran SLALIB. Despite this the two libraries will be kept intact and independent; there is no intention at present to implement one version as a set of wrappers around the other, something that would hinder code re-use. Each version will continue to be supported.

The homogeneity and ease of use of SLALIB could perhaps be improved by using C++ and object-oriented techniques. For example "celestial position" could be a class and many of the transformations could happen automatically. This requires further study and would almost certainly result in a complete redesign.

Similarly, the impact of Fortran 90 has yet to be assessed. Once compilers become widely available, some internal recoding may be worthwhile in order to simplify parts of the code. However, as with C++, a redesign of the application interfaces will be needed if the capabilities of the new language are to be exploited to the full.

References

Lawden, M.D. 1990, Starlink General Paper 31

Lawden, M.D. 1991, Starlink User Note 1

Wallace, P.T. 1993, Starlink User Note 67

Astronomical Data Analysis Software and Systems III
ASP Conference Series, Vol. 61, 1994
D. R. Crabtree, R. J. Hanisch, and J. Barnes, eds.

Interactive Dynamic Mission Scheduling for ASCA

A. J. Antunes and F. Nagase

Institute of Space and Astronomical Science, 3-1-1 Yoshinodai, Sagamihara-shi, Kanagawa-ken, Japan 229

T. Isobe

Center for Space Research, Massachusetts Institute of Technology, 77 Massachusetts Ave, Cambridge, MA 02139

Abstract. The Japanese X-ray astronomy satellite ASCA (Advanced Satellite for Cosmology and Astrophysics) mission requires scheduling for each 6-month observation phase, further broken down into weekly schedules at a few minutes resolution. Two tools, SPIKE and NEEDLE, written in Lisp and C, use artificial intelligence (AI) techniques combined with a graphic user interface for fast creation and alteration of mission schedules. These programs consider viewing and satellite attitude constraints as well as observer-requested criteria and present an optimized set of solutions for review by the planner. Six-month schedules at 1 day resolution are created for an oversubscribed set of targets by the SPIKE software, originally written for HST and presently being adapted for EUVE, XTE and AXAF. The NEEDLE code creates weekly schedules at 1 min resolution using in-house orbital routines and creates output for processing by the command generation software. Schedule creation on both the long- and short-term scale is rapid — less than 1 day for long-term, and one hour for short-term.

1. Introduction

The Japanese X-ray astronomy satellite ASCA (Advanced Satellite for Cosmology and Astrophysics) was launched in February 1993 and underwent an initial 2 month check-out period and a 6-month Performance Verification (PV) period prior to initiating 6 month Guest Observer (GO) periods. ASCA has four modules of multilayer thin-foil mirrors with four focal plane detectors — two Gas Imaging Spectrometers (GIS) and two Solid State Imaging Spectrometers (SIS) — and is sensitive over about 0.4 keV to 10 keV.

As ASCA is capable of both imaging and spectroscopic study, with two different types of instrument, observations of all classes of targets are feasible, from stars to the cosmic X-ray background. ASCA observes most targets for the nominal period of 1/2 or 1 day, with an expected average yield of 40 ksec of good time during each day. Thus ASCA will observe about two hundred targets during each 6 month observation phase.

The long term scheduling tool SPIKE was developed by programmers at the Space Telescope Science Institute (Johnston 1990) for use with the Hubble Space Telescope. SPIKE exploits artificial intelligence (AI) techniques to handle constraints and search strategies and allows interaction through a graphical display. The short term scheduling tool NEEDLE was developed at ISAS to complement SPIKE for daily scheduling to minute accuracy. DP10 is an ISAS in-house set of orbital routines.

All scheduling tasks are done on a Sun SPARCStation 2 at ISAS (provided by M.I.T.) in an X-windows environment. SPIKE is written in ANSI standard Common Lisp with embedded C code, and NEEDLE and DP10 are written entirely in C.

2. Suitability

The operation of ASCA includes constraints based on satellite limitations as well as observational requirements for individual targets. Typical constraints involved include:

- Target Visibility
- Solar Angle (between Solar Paddles and Sun)
- Moon Contamination in FOV
- Angle to Bright Earth Limb
- SAA Passages
- Star Tracker limits with respect to the Sun, Moon, Bright Earth
- Contact Pass Availability and Telemetry Requirements
- Time Critical and Phase Critical Observation Windows
- Telemetry Limits, Contact Passes and Data Downlinks

These multiple levels of constraint are handled by the SPIKE long term scheduler. The constraints are combined via a function called "Suitability", representing the overall preference of an observation for any given time. The automatic scheduling strategies for SPIKE work directly with this Suitability Function to create efficient schedules. As the process involves both rule-based and neural network approaches, multiple runs should be done, and a final solution chosen by the operator. Editing and customizing the final schedule is simple due to the Graphical User Interface (GUI).

3. Benchmarks and TOO

Scheduling involves two problems — the need to make an efficient schedule, and the ability to do so within a reasonable amount of time. In addition, schedule flexibility is required to handle unexpected events (such as entering Safe Hold mode) or sudden new possibilities — Targets Of Opportunity (TOO). The SPIKE/NEEDLE software set enables fast schedule generation which is highly optimized within the limit of target constraints.

A complete long-term schedule of 6-7 months, including 258 targets with 16 time-critical targets (as was typical for the first GO), takes approximately 6 hours to prepare detailed orbit files, and one day to schedule. Computer time to make a schedule is on order of minutes, but time must be taken by the planning staff to choose and adjust the details of the schedules created.

To create a typical one-week schedule (each target is generally 1 or 1/2 day in duration) takes approximately one hour, again dominated by human evaluation time, with computer calculation time taking less than 5 minutes.

ASCA is able to schedule a TOO, such as SN1993J, very quickly. Schedule alteration time is approximately one hour, and subsequent command load preparation requires less than one day.

4. ASCA Operations

4.1. Planning Procedure

Planning begins with a list of accepted proposals, generally oversubscribing the amount of time available by a factor of 25%. This is loaded from the Observation Data Base (ODB) into SPIKE and run until a viable schedule is found. This is reviewed and adjustments are made as needed. The resulting Long Term Schedule (LTS) is loaded into the ODB and also made publically available. Each week, a Short Term Schedule (STS) is made using NEEDLE (which gets its input from SPIKE), and this information is loaded into the ODB and also sent directly to the attitude and command planning team. After the decision to observe a TOO is made by the project manager, the TOO observation is typically allocated for the next upcoming week. When TOO or schedule variations occur, the LTS is also updated while keeping time-critical targets as they are, and reloaded into the ODB. At any time ASCA users can access the ODB to check and confirm their observation times.

4.2. Other Tools — DP10

DP10 is a set of routines which include a graphic front end and PostScript output. It is used to plot the conditions of a single target for 15 orbits (just over 1 day), including occultation of the source, day and night side of the satellite, SAA, contact pass, DSN downlink and altitude information, geomagnetic rigidity, and star tracker information. The "touban" (Duty Scientist) uses DP10 plots as a reference when adjustments to the plan are required.

5. ASCA PV Phase

During the PV phase, this scheduling system (SPIKE-NEEDLE) has worked satisfactorily. The system has shown good flexibility when faced with unexpected events such as transfer to safe hold mode, inserting of TOO observations, and the cancellation of observations due to large typhoons damaging the Kagoshima Space Center. We expect similar success with the first GO phase, already in progress.

Acknowledgments. We are grateful to G. Clark of M.I.T. for SPIKE support and to M. Johnston of ST ScI for creating SPIKE. The authors thank all the ASCA team members.

References

Johnston, M.D. 1990, in Proc. Sixth IEEE Conf. on Artificial Intelligence Applications (Santa Barbara, March 5-9, 1990), Los Alamitos, CA: IEEE Computer Society Press 1990, 184

Part 6. Systems Software, Software Development Methods, and Data Structures

Astronomical Data Analysis Software and Systems III
ASP Conference Series, Vol. 61, 1994
D. R. Crabtree, R. J. Hanisch, and J. Barnes, eds.

The Client/Server Software Model and its Application to Remote Data Access

Gareth J. Bevan

Center for EUV Astrophysics, 2150 Kittredge St., Berkeley, CA 94530

Abstract. The results of the Extreme Ultraviolet Explorer (EUVE) spectroscopic data server and a prototype client implemented at the Center for EUV Astrophysics (CEA) in the Image Reduction and Analysis Facility (IRAF) are presented. Current work on extending the data server to work with interactive client programs will also be presented. The client/server approach will enable scientist to selectively retrieve their own proprietary data, in addition to public data, across the Internet. Data selection will prove extremely useful for EUVE spectroscopic data given the large volumes of data accompanying an observation. Preliminary results of investigations into implementing clients in standard astronomical software including the Astrophysical Data System (ADS) and the IRAF as well as new innovative approaches are discussed.

1. The Remote Procedure Call Server

The RPC server can be envisioned as calling functions in one program (the server) from another program (the client). The data returned by these function calls must be passed back by value and not by reference as the programs may (and usually) run on different host machines. The Server and Client program must 'know' about each other at compile time. For example, the server may have a program and version number by which it is referenced by the client in order to make the connection.

The advantages of this protocol are that if the server has a large start up cost then one gains by only having to invoke the program once. However the program is always stored in memory and each function can not take too long to run or other client programs may not be able to access the server.

Figure 1 shows the RPC protocol. Note that the functions in the server need not be called in any special order i.e., there are many entry points into the Server program.

2. The Invoked Server

The other type of Client/Server programs communicate with each other by means of an agreed upon set of commands. These commands are passed between the programs via sockets. Many programs exist, inetd in Unix for example, that will map standard in and standard out to a socket. This enables programmers

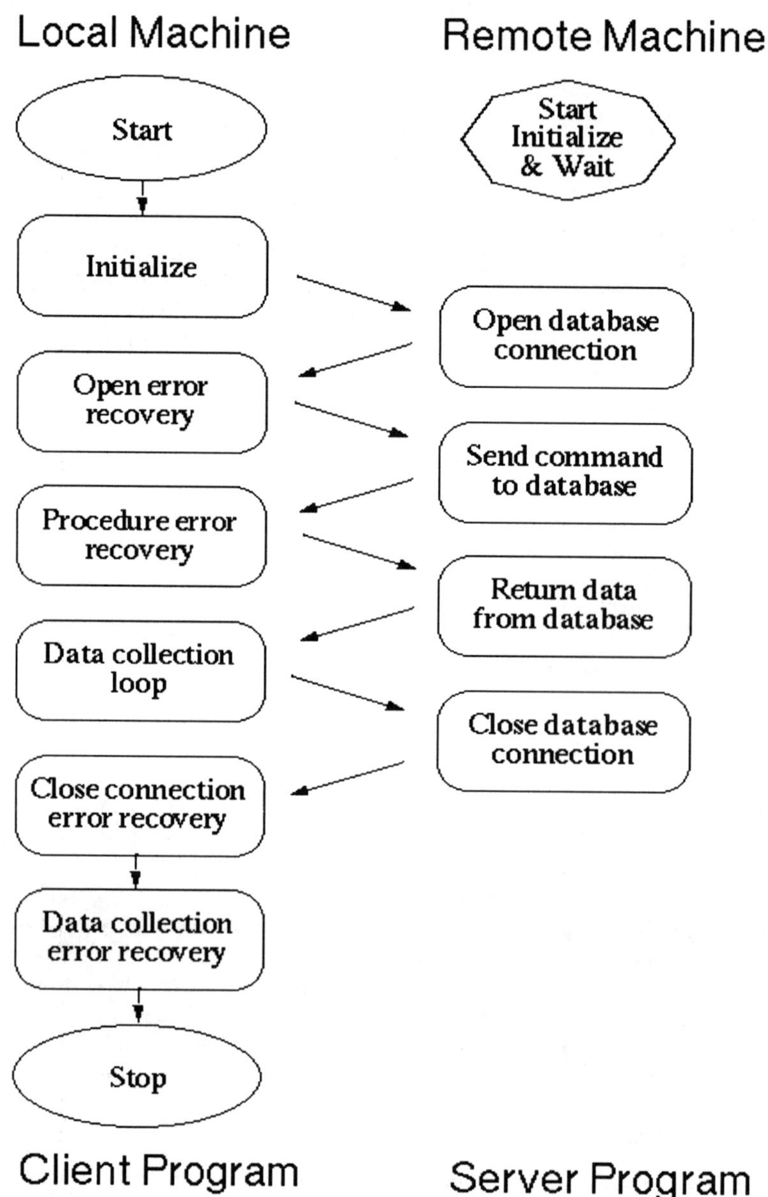

Figure 1. The Remote Procedure Call Communication Protocol.

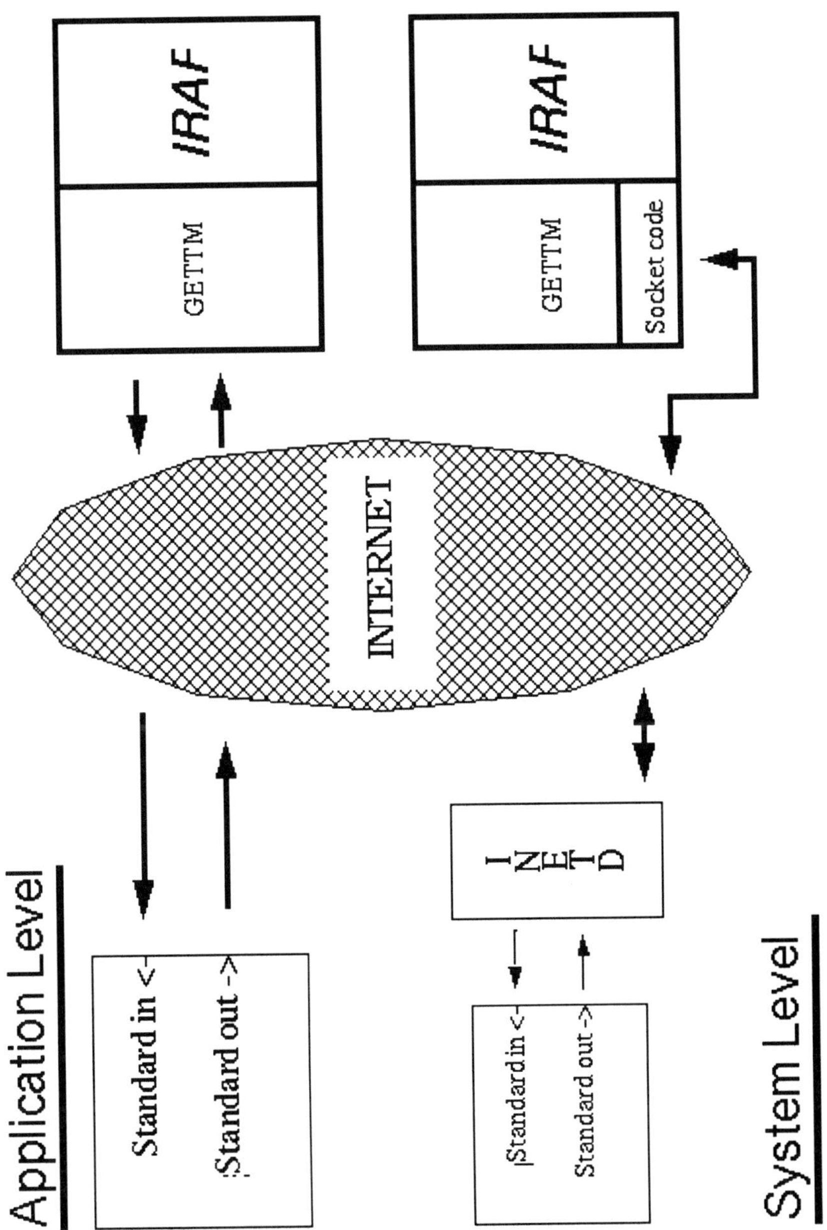

Figure 2. The CEA EUVE Telemetry Server.

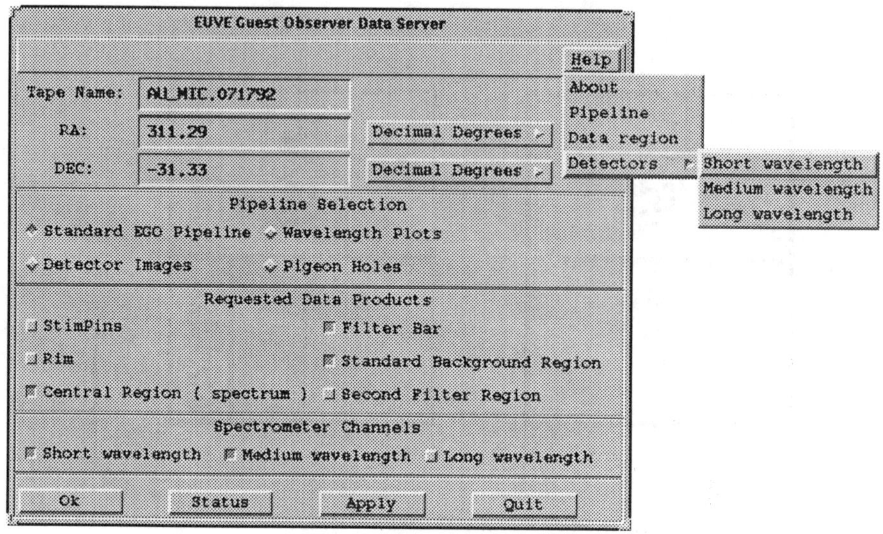

Figure 3. The CEA EUVE Telemetry Server.

to write code that reads from standard in and writes to standard out, that can communicate across the network.

The advantages of this protocol are that the Server program is only invoked when it is needed and does not use up memory when it is not in use.

3. The IRAF task GETTM

The Center for EUV Astrophysics Guest Observer Program has developed a task in IRAF to retrieve Guest Observer data across the Internet from the optical juke-box at CEA. The IRAF client task calls C library routines that send messages, via sockets, to the inet daemon at CEA. This then invokes the CEA Telemetry sever which writes the telemetry on standard out.

Using the Client/Server model in this way meant that the complicated Telemetry Server could be written in C, while the less complicated formatting code could be written in SPP, the native language in IRAF.

Figure 2 shows how the different parts of the telemetry server look to the application and what happens from a system point of view.

4. The ADS

Figure 3 shows a proposed GUI for the ADS. Guest Observers would be able to request data sets from the Guest Observer Center. Data reduction could be carried out using the computer resources at CEA and then the final data products could be delivered either by ftp or by mail depending on the size of the data products.

Astronomical Data Analysis Software and Systems III
ASP Conference Series, Vol. 61, 1994
D. R. Crabtree, R. J. Hanisch, and J. Barnes, eds.

The SAO-IIS Communication Package

J. R. Wright

Canada-France-Hawaii Telescope Corporation, P. O. Box 1597, Kamuela, HI 96743

A. R. Conrad

W. M. Keck Observatory, P. O. Box 220, Kamuela, HI 96743

Abstract. Cooperating processes send image data to `SAOimage` via the `IIS` transmission protocol. Due to the lack of an easy-to-use subroutine library, relatively few applications have made use of this feature. With `SAO-IIS`, we provide a package that programmers can use to send commands and send and receive image data to `SAOimage` via the `IIS` protocol.

We describe three applications of `SAO-IIS`: two data taking systems and an image analysis tool for segment stacking. We discuss options for coordinating multiple clients of `SAOimage`. Finally, we briefly discuss the future of display server communications.

1. Introduction

The `SAO-IIS`[1] library fills a short term need for easy access to `SAOimage` from autonomous applications. Admittedly, an application written with `SAO-IIS` falls into the trap of protocol-dependence, but it is small, simple and easy to understand. These worthy attributes are often lost in the long term quest for device independence.

2. `SAOimage` and `IRAF`

An `IRAF` tool, such as **`imexam`**, communicates with `SAOimage` (VanHilst 1991) via a named pipe. The data exchange format is a descendant of a protocol designed during the early seventies for `IIS`[2] frame buffers (see Table A). Because `IRAF` programmers use the device independent subroutines in `IRAF/GIO` (Tody 1984), they are spared the tedious and error prone task of formatting low level `IIS` commands.

---

[1] `SAO-IIS` is pronounced S-A-O-I-squared-S

[2] International Imaging Systems.

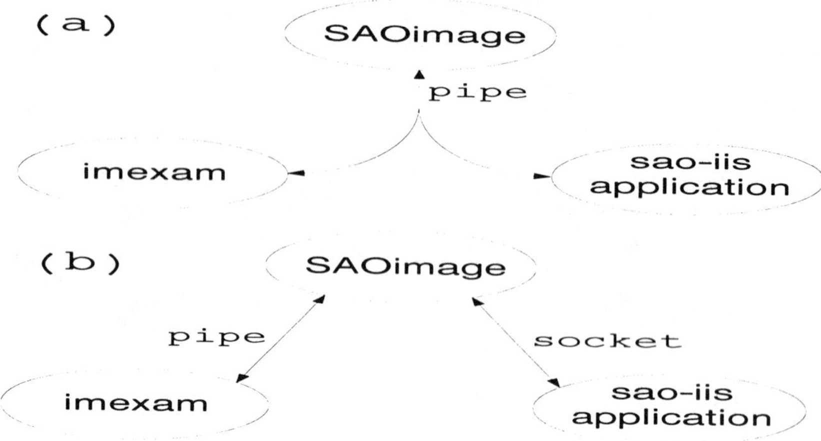

Figure 1. Communication Paths to SAOimage

3. The SAO-IIS Library

Although IRAF/GIO facilitates writing an IRAF application using the IIS protocol, to date writing a non-IRAF IIS application has been more difficult. Not only has there been no subroutine library, but IRAF's IIS protocol has evolved, is undocumented and must be reverse engineered from existing code. (The module $iraf/unix/sun/fifo.c is the best source for learning the protocol.)

The SAO-IIS library provides a callable subroutine package for use by autonomous, non-IRAF applications. The SAO-IIS functions are summarized in Table B.

4. Pipes versus Sockets

The named pipe mechanism constrains an SAOimage client (e.g., IRAF) to run on the same machine as SAOimage. Moreover, the named pipe restricts SAOimage to a single connection. Although several clients can communicate with SAOimage via the single connection (see Figure 1(a)), no mechanism is provided for arbitration and the result is crosstalk which will typically wedge one of the clients.

imexam adds considerable value to SAOimage. Without imexam, an SAOimage client must either reimplement or do without image quality estimates, surface plots, contour plots, radial profiles, and, perhaps most essential, the one dimensional "cut" provided by the imexam 'v' command.

We plan to allow SAOimage clients to become SAOimage/imexam clients by providing access to the SAOimage socket interface (see Figure 1(b)).

5. Sample Applications

5.1. gstack

The Shack-Hartmann camera used to align the Keck primary mirror segments is not available when we observe at the Cassegrain focus. When we observe at Cassegrain, the `gstack` program provides an alternative method for measuring the segment stack. Because it is familiar to our users and provides all of the needed features, we selected `SAOimage` for `gstack` image display. The `gstack` user manages several image frames in memory, so we decided to send frames directly to `SAOimage` via the named pipe. The `gstack` program motivated the first version of `SAO-IIS`.

5.2. GECKO

The CFHT coudé spectrograph GECKO (Wright 1993) is completely under computer control. All aspects of the instrument can be controlled from the observing account, opening the possibility of remote observing. Because of the complicated nature of the instrument (at last count: 32 motors, 130 indicator switches, 5 lamps, exposure meter in addition to the usual facilities of our observing environment) it is necessary to provide the observer with a sophisticated user interface. One of the design philosophies has been that the user is more interested in the acquired data rather than the instrument details. Using `SAO-IIS` we will be able to offset, tip, tilt and focus using interactions with `SAOimage`. This offers a much more intuitive interface than a command line based program, or even a form based program.

5.3. Autoguider

The Keck autoguider system (Lupton 1993) displays frames via a special purpose display server developed in house. Before converting to `SAOimage`, we must be sure we can efficiently overlay line graphics and text. We must also be able to display a guide box with a binning factor that differs from that of the surrounding field. Work in these areas is in progress.

6. Future

Now that the dust has settled after the transition from special purpose frame buffers to `X` window display servers, we all need to prepare ourselves for the next big shake up: the second generation image display program. Several new display programs will emerge soon, including `SaoTNG` (Mandel, 1993) and `ximtool`.

The communication path is key in our choice of which display server to use. It seems that every possibility is being tried. To name a few: `RPC`, `X` properties, `sockets`, `named pipes`, System V shared memory, and linkable widgets.

But what about `IIS`? Admittedly, `IIS` is just some of the frame buffer dust that hasn't settled since the transition to `X` windows. On the other hand, the serious contenders for the universal image display program will likely support `IIS` for backward compatibility. Should we just stick with `IIS`? Probably not. Huge (billion bit) images are looming on the horizon. Pure volume suggests the choice will be shared memory for applications running on the same machine.

A IIS Command Format

Size	Name	Description
16 bits	`tid`	transfer id – identifies direction and specific conditions of each Data or Control transfer
16 bits	`thingct`	thing count – two's complement of number of data elements to be acted upon
16 bits	`subunit`	subunit select – specifies which subunit within the Image Processor is to be accessed, and specifies the function of that access
16 bits	`checksum`	checksum – calculated so that the sum of all eight header words is 177777 octal
16 bits	`x`	x register – not used consistently
16 bits	`y`	y register – used only by Refresh Memory subunit
16 bits	`z`	z register – bit encoded word used to configure the Image Processor hardware, but not used consistently among the subunits
16 bits	`t`	t register – not used consistently
variable	data	data transferred

B SAO-IIS Function Descriptions

Function Name	Description
`opensao()`	open the connection to `SAOimage`
`closesao()`	close the connection to `SAOimage`
`selectframe()`	select which frame of memory to operate on
`send_image()`	send image data to be displayed
`send_erase()`	erase the image display
`send_wcs()`	send World Coordinate System data
`read_cursor()`	read cursor X and Y position and key pressed
`read_image()`	read image data from display
`read_wcs()`	read World Coordinate System data
`draw_circle()`	overlay a circle on the image

References

VanHilst, M. 1991, "User Manual for SAOimage"

Tody, D. 1987, "'Graphics I/O (IRAF Help Page)"

Lupton, W. F. 1993, "Keck Autoguider and Camera Server Architecture", this volume

Mandel, E. 1993, "SAOimage: The Next Generation Overview and Requirements"

Wright, J. R. 1993, "Software System Definition for the CFHT Coudé F/4 Spectrograph"

International Imaging Systems, Model 75 Digital Image Processor Theory of Operation and Maintenance Guide

Astronomical Data Analysis Software and Systems III
ASP Conference Series, Vol. 61, 1994
D. R. Crabtree, R. J. Hanisch, and J. Barnes, eds.

Codes for the Modelling of Stellar Structures: Parallel Implementations on a Workstation LAN

M. Pucillo, G. Bono, P. A. Mazzali, F. Pasian, R. Smareglia

Osservatorio Astronomico, Via G.B.Tiepolo 11, 34131 — Trieste, Italy

Abstract. In the past couple of years, modeling of stellar structures has become increasingly important at our institute. The codes used for this kind of computations are very demanding in terms of CPU power, and most of them are suited to parallel processing. In this paper, we discuss practical examples of porting codes for stellar structures modeling, developed on sequential computers, into a parallel environment, obtained with the workstation LAN and a software for distributed processing.

1. Status and Developments of Computing Infrastructure

In the past decade our institute has developed an extensive and heterogeneous network of workstations, which are mostly used for the interactive reduction and analysis of observational data. The power of such computing equipment has considerably increased over the past couple of years; in the meantime, modeling of stellar structures has become an increasingly important topic at the institute.

The codes used for stellar structures modeling were developed on sequential computers. They are very demanding in terms of CPU power, and most of them seem suited to parallel processing. It seemed logical to try and optimize the use of the available computing power to run modeling codes when the workstations did not perform any interactive task. This could be accomplished by using the available CPUs in parallel, managing them with a software for distributed processing. The alternate solution, i.e., porting the codes to parallel hardware such as the Parastation or Meiko machines, is really too expensive for our medium-sized institute in terms of both purchase and maintenance costs. Furthermore, the development of a parallel computing facility based on networked workstations allows a great amount of flexibility in coping with the dramatic technological developments (the trend is towards doubling the computing power every 2 years while cutting prices by 50% — Pucillo, 1993). The decision was made to buy the `ParaSoft` software `Express`, which allows program parallelization using an etherogeneous workstation network.

The Unix LAN at our institute is composed of a total 15 workstations, but only 6 of them have been used for the initial tests reported in this paper: 2 `Sun Sparc10/30` (28 Mips, 11 Mflops each), 2 `HP 710` (26 Mips, 12 Mflops each), and 2 `HP 715/50` (24 Mips, 14 Mflops each), for a total power of 74 Mflops. The workstations are connected via a 10 Mbps Ethernet. Future developments, foreseen in early 1994, include 2 new `HP 715` workstations and the use of a network based on `FDDI` (100 Mbps).

2. Modelling Codes

A Monte Carlo model is used to compute synthetic spectra of supernovae in the photospheric epoch; the code, hereafter referred to as SNMC, is thoroughly discussed in Mazzali & Lucy (1993) and is based on the Schuster-Schwarzschild approximation. A large number of energy packets, whose frequencies sample that of the assumed photospheric black body spectrum, are emitted in random directions at the photosphere and followed as they scatter their way through the envelope. The final wavelength of the packet depends on the initial wavelength and on the packet's scattering history.

The Smooth Particle Hydrocode (Bono & Capuzzo Dolcetta, 1993), hereafter referred to as SPH, has been developed together with an N-body code to simulate physical mechanisms that govern the star formation processes in giant molecular clouds and the mutual interaction between an open cluster and its placental cloud.

A non-linear, non-local and time-dependent code (HYCN) has been developed (Stellingwerf 1982, Bono & Stellingwerf 1993, Bono et al. 1993) to study the pulsation characteristics of variable stars (Cepheid, RR Lyrae, Delta Scuti). Local conservation equations plus a transport equation (diffusion approximation) have been used to derive the convective hydrodynamic equations. Along a pulsation cycle non-local effects such as convective overshooting, super-adiabatic gradients and the coupling between convective structure and pulsation can be easily evaluated. An eddy viscosity pressure term has been introduced for taking into account turbulence effects on small scale lengths.

3. Methods and Results

`Express`, the software for distributed processing we have used in these tests, does not allow an immediate parallelization of the codes, but it rather needs an *ad-hoc* analysis for every single code.

A `host-node` structure has been chosen to perform parallelization of the programs. In this type of approach the computational-intensive aspects of the application are extracted and executed on the parallel computer nodes. The interface or control portions of the code remain on the host computer. The interface between these two programs is provided by `Express` function calls which allow data to be transferred between host and nodes. In this way an existing piece of code can be maintained almost completely intact, only a small portion being actually extracted and parallelized.

The general structure for the parallelized codes is shown in Figure 1. Under such a configuration, the total time of execution T_{p_tot} of a parallelized program is given by:

$$T_{p\_tot} = T_s + T_c + T_p$$

where T_p is the time used for the computation of the parallelized sections of the program, T_c is the time "lost" in interprocess communications, T_s is the time used for the computation of the intrinsically sequential sections of the program. T_s also contains the overhead of `Express` synchronizing functionalities: this is mostly due to a different data organization before and after the parallelized loop with respect to the sequential program, and to the consequent need for

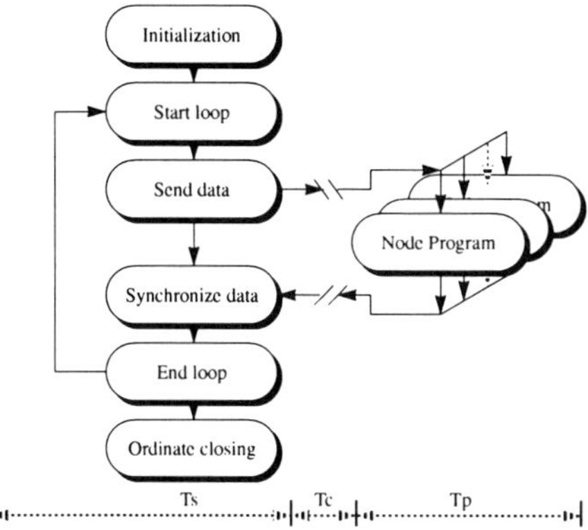

Figure 1. The general structure for parallelized codes.

re-organization. Part of the overhead time is also accounted under T_p. T_c may be critical, especially if plenty of data need to be exchanged among the various sections of the program. This value can become critical in the case of massive data exchange on a 10 Mbps Ethernet network. We define the efficiency of parallelization E as:

$$E = \frac{T_{seq}/N_{nodes}}{T_{p\_tot}}$$

where T_{seq} is the time used for the computation on a sequential machine, and N_{nodes} is the number of nodes the parallelized program is run on. E_{FDDI}, the efficiency using FDDI, can be estimated.

The results of the parallelization tests are given in Table 1; they will be briefly discussed in the following. All the tests on the 3 chosen algorithms have been performed on 4 CPUs (the HP 7xx computers, for homogeneity).

Table 1. Performance of analyzed codes in both sequential and parallel environments. Refer to text for meaning of symbols.

Code	Sequential	Parallel					
	T_{seq}	T_s	T_c	T_p	T_{p_tot}	E	E_{FDDI}
SNMC	10m 53s	0m 15s	0m 35s	2m 49s	3m 39s	75%	88%
SPH	35m 00s	7m 20s	7m 10s	10m 49s	25m 19s	35%	50%
HYCN	5m 44s	3m 09s	5m 02s	0m 58s	9m 09s	16%	32%

In terms of total elapsed time, the parallelized SNMC program proved to be faster by a factor of 3 with respect to its sequential version. There is no external loop in the program. Therefore, except for an initial broadcast, the computation is performed entirely within the nodes, and only at the end the data are fed back to the host, which performs data aggregation and the final sequential steps of the program. This configuration leads to a 75% efficiency using Ethernet. It was estimated that T_c could drop from the current value of 35s down to 2s if `FDDI` were to be used, thus raising the efficiency to 88%.

In the SPH program, where the parallelized version is faster than the sequential one only by a factor of 1.4, there is an external loop which is called a relatively small number of times, and thus the sequential section of the program is still an acceptable fraction (T_s) of the total computation time. Efficiency is fairly low, but still acceptable and could be significantly improved with `FDDI`.

In the HYCN program, the external loop, and consequently the sequential section of the program, is predominant. Therefore the parallelized version is slower than the sequential one, since additional burden is introduced by inter-process communication (T_c). This code is currently being analyzed in order to determine whether structural modifications could lead to an acceptable parallel performance.

4. Conclusions

The main problems in porting already-developed and "production-type" modeling codes (at least the ones the authors are familiar with) to a parallel environment lie on the fact that they have been designed and developed (usually in `FORTRAN`) when only sequential computers were commonly available, and their logic is therefore intrinsically sequential; they have been upgraded by several different persons; the physics underneath may not allow parallelization.

This work has shown that porting computational software in a parallel environment given by a LAN of workstations and by a software for distributed processing is possible, and may reach a high degree of efficiency provided that the physics underneath the scientific program is intrinsically "parallel"; the code is properly structured to allow parallelization; the LAN connecting the various nodes of this parallel computing structure has an adequate speed to support the amount of data exchanged. The approach may of course be different in the case of parallel or vectorial machines, where a proper compiler is usually available.

Acknowledgments. This work has been partially supported by CNR. We are grateful to our colleagues and friends Andrea Balestra and Claudio Vuerli for their dedication in taking care of OAT Unix computers, network, and users, while still performing research work.

References

Bono, G. & Capuzzo Dolcetta, R., 1993, ApJ, submitted

Bono, G. & Stellingwerf, R.F., 1993, ApJ, in press

Bono, G., Smareglia, R., Pulone, L., Balestra, A., Vuerli, C., Cumani, C., Marcucci, P., Pucillo, M., Santin, P., Pasian, F., Rusconi, L., & Sedmak, G., 1993, in: Astronet 1993, Memorie S.A.It., in press

`Express` User's Guides, 1992, ParaSoft Corp., Pasadena, CA

Mazzali P.A., & Lucy L.B., 1993, A&A, in press

Pucillo, M., 1993, in: *Astronet 1993*, Memorie S.A.It., in press

Stellingwerf, R.F., 1982, ApJ, 262, 230

Astronomical Data Analysis Software and Systems III
ASP Conference Series, Vol. 61, 1994
D. R. Crabtree, R. J. Hanisch, and J. Barnes, eds.

IRAF/STSDAS in OSF/1

Nelson Zarate
Space Telescope Science Institute, Baltimore, MD 21218
zarate@stsci.edu

1. Introduction

The IRAF system has been ported to the new operating system OSF/1 running on a Digital 3000/500 Alpha machine. We used version 1.3 of the system, which has the following characteristics:

- 64-bit kernel architecture based on the Carnegie Mellon University's Mach V2.5 kernel design with components from BSD 4.3 and 4.4, Unix System V, and other sources.
- OSF/Motif Application environment for Motif V1.3.3. Motif 1.3.3 is the default graphical user interface for OSF/1.
- POSIX compliance with the interfaces required by the IEEE 1003.1-1900 standard and FIPS 151-1.
- Realtime supports.
- Full complement of dynamic shared libraries, based on System V semantics, giving increased system performance, reduced minimum hardware requirements, and easier system management.
- Compatible with BSD 4.3 and System V programming interfaces to ensure compatibility with Ultrix applications.
- The system loaned to us by Digital has the following characteristics: DEC 3000 Model 50 with 150 Mhz clock speed, 252 megabytes of main memory, 1.2 gigabytes of disk space, a CD-ROM, and a DAT drive. We thank Digital for the use of this machine.

The IRAF software system can be divided in four major components:

- The Command Language (CL). Provides the IRAF user interface. The CL is used to run tasks, set user parameters, edit files, edit task parameters, interact with graphics, and perform other similar functions.
- Application Programs. Consists of IRAF utilities, such as the `dataio`, `language`, `system`, `plot`, and `images` packages as well as the various scientific applications. These applications are mostly written in the SPP language, but Fortran programs are easily incorporated in the IRAF viroment by using the IMFORT or the STSDAS IRAF77 Fortran interfaces.

- The Virtual Operating System (VOS). Consists of all libraries used by IRAF applications, such as the I/O routines for handling files, graphics, and images, as well as device routines such as those for network communication, memory management, and other tasks. VOS routines are completely independent of the host operating system since they are written in SPP.

- The Host System Interface (HSI). Consist of bootstrap utilities like mkpkg, rtar, wtar, xc, generic, etc., as well as the kernel interface. The kernel interface is a set of about 50 routines, written in C which are Fortran callable, that handle all of the possible host operating system calls needed by the VOS; this makes the kernel the only major part of IRAF that must be changed when porting the software to another operating system.

2. The IRAF Port

2.1. Portability Problems

The 64-bit OSF/1 OS provides support for greater addressibility and larger ranges for scalar arithmetic operations. Providing these functions for user applications involves changing one or more of the scalar datatypes. Unfortunaly, this change causes interoperability problems between 32- and 64-bit systems and were the main cause of portability problems for IRAF.

- `Sizeof(*) != Sizeof(int)`
- `Sizeof(long) = 64-bit`
- `Sizeof(*)    = 64-bit`

The major hurdle in porting IRAF to OSF/1 is handling the 64-bit memory addresses and pointers. This becomes a problem when dealing with the VOS routines, which are Fortran-callable, because these routines use 32-bit pointer references. This conflict is resolved by making all of the addresses and memory references that are passed to or from the kernel stay within the 32-bit range.

2.2. Kernel Routines

We began porting IRAF to OSF/1 using the IRAF version 2.10 for the IBM/AIX system as a starting point. The AIX system is also POSIX compliant and relies heavily on Unix System V.

The approach that we used was to handle the kernel interface (Z-routines) under DEC's C compiler and then look carefully at any changes needed to routines that returned errors or warnings.

During the first compilation, only four of the Z-routines needed to be modified to correct compilation errors. After the decision was made to use only 32-bit pointers and memory addresses, all arguments that had been `long` were changed to `int`.

2.3. Linking Applications

Any system routines that return a memory address during run-time are likely to use values in the top 32 bits of the 64-bit address. This requires that the linker arrange for addresses to fit into 32 bits, which can be done without affecting the performance of the application. The OSF/1 linker options allow the programmer to control the base addresses for the text and data segments in the executable.

- `-T nnnnnnnnnnnnnnnn` specifies that the text segment origin should be at hexadecimal address nnnnnnnnnnnnnnnn. The stack grows downwards below the text segment origin.
- `-D nnnnnnnnnnnnnnnn` specifies that the data segment origin should be at hexadecimal address nnnnnnnnnnnnnnnn. The bss segment and then the heap are placed immediately above the data segment.

The setting chosen to link the IRAF program was:

- `-T 30000000 -D 65000000`

3. IRAF Shareable Image LibS.so

Under OSF/1 the size of object modules are about 30% larger than their counterparts under Ultrix. This heavy penalty in disk space and linking time would make OSF/1 unattractive to programmers. Ultrix has already had problems because of its lack of support for shared libraries.

OSF/1 has excellent support for shared libraries; shared libraries are used by default when developing applications and the linking times are faster than for any machine of this class from any competitor.

For IRAF development the default static libraries are: Libos.a (the IRAF kernel routines), Libsys.a (the VOS core system routines), Libex.a (the VOS high level system routines), and Libvops.a (the vector operator routines). All four of these libraries are put together in one shared image to be used by all IRAF applications, saving an enormous amount of disk space.

The following procedure is used to produce the IRAF shared image: In one line using the `ld` Unix command, all object modules are extracted from the static libraries and a shared image is created with a default extension of `so`. The interesting aspect of this is that the data segment of the share image and the data segment of the executable program should not overlap. This is the value for the `LibS.so`:

- `-T 30000000 -D 50000000`

With regards to common areas that need to be protected in OSF/1, the default is no share, unlike the VMS counterpart where special linking command qualifiers are necessary to secure common areas from being accessed by other processes.

It is now possible to change the content of some routines that are part of the shared image without needing to relink the applications that use them. It is important to conserve the order in which the routines are declared the very first time that the shared image is built.

4. Size of Object Files and Executables.

A simpler instruction set goes directly with the size of the resulting object code from a compiler output. Here is a random list that compares the sizes of one OSF/1 and Ultrix object library and one executable file.

Filename	OSF/1	Ultrix (Mbytes)
Libsys.a	3.1	1.9
x_dataio.e	2.15	1.43 (No share image)
x_dataio.e	0.27	n/a (With share image)

5. Performance

Our favorite program to use when benchmarking a machine is the Lucy deconvolution task. This program makes heavy demands on memory usage and floating point operations since it does mostly Fast Fourier Transform (FFT) computations.

LUCY Deconvolution Benchmark (256x256 32bits/pixel) (10 iterations)

Machine	system-time (seconds)	user-time (seconds)
DEC AXP OSF/1	7.9	12
Sun SPARC 2	64.1	156

Astronomical Data Analysis Software and Systems III
ASP Conference Series, Vol. 61, 1994
D. R. Crabtree, R. J. Hanisch, and J. Barnes, eds.

Public Access Programming: Opening The Black Box That Hides Internal Data

E. Mandel

Harvard-Smithsonian Center for Astrophysics, Cambridge, MA 02138

R. Swick

Digital Equipment Corporation, Nashua, NH 03062

Abstract. We describe a simple and effective use of the *X* Toolkit (*Xt*) selection mechanism by which data in an *X* program can be tagged with string identifiers and then accessed externally by other programs. We will discuss our design goals, the technical challenges we faced — including extensions to the *Xt* selection implementation — and the user interface and application programming interface that we developed to meet these challenges. We also will discuss the important implications that our scheme has for programs such as *SAOimage*.

1. Introduction

Data analysis and display programs typically are written as "black boxes" with only a limited specification of input and output data. Such programs are linear and transient: they are activated in response to a single event (such as a "run" command) and they exit after transforming their input data into output data. Because these programs do not persist in memory after having completed their task, their internal data also do not persist and are not available to external processes.

There are analysis programs that *do* persist over time and are not linear in their operation. Rather than performing a single task once, they perform an action (or different actions) repeatedly in response to user events. The internal data of these event-driven programs generally are not available to external processes either, but unlike the transient programs mentioned above, there is no compelling reason why this should be so. Indeed, there are important cases where it is desirable to read or even modify a program's "internal" data through external means. The *SAOimage* image display program is a good example: over the years there have been numerous requests by users and developers to be able to read and/or write *SAOimage*'s image data, region information, colormaps, etc., externally.

It even can be argued that external access to "internal" data is critical for the proper integration of programs such as *SAOimage* into the heterogeneous data analysis environments used by astronomers today. A good image display program should be able to interface with the many analysis tools now available. To do this, it must be able to send data and information to analysis programs for further transformation, and also be able to receive the results of these transformations for further display.

These considerations have guided efforts at Smithsonian Astrophysical Observatory to develop a successor to the popular *SAOimage* image display program. From the beginning, we have recognized that the central problem we face is not the display of data, but how the image display program will interact with other programs and systems. Indeed, our discussions with users and developers have centered on the need for the image display to be capable of being controlled by external processes (which send it data and information) and also to be capable of controlling processes (by sending data and information to those processes). For example, many users have requested that *SAOimage* support quick-look analysis of the displayed data, i.e., that *SAOimage* be able to launch arbitrary external processes to retrieve the image data, transform it, and display results, perhaps even by writing data back to the image display. This kind of functionality will help make maximal use of existing analysis software.

2. XPA Design Goals

We therefore set about to develop a general mechanism by which arbitrary programs could communicate data and information to and from *X* programs such as *SAOimage*. From the outset, our design goals included the following:

- The mechanism should be driven by the external process, which either sends "named" data or requests "named" data to or from an *X* program.
- The mechanism should allow for the simultaneous sending and receiving of data associated with the same "name".
- *X* programs should send or receive data using the standard event-driven callback paradigm already familiar to *X* application programmers.
- External processes should be able to access data in *X* programs using a simple application programming interface (*API*) that does not require knowledge of *X* or linking against the *X* libraries.
- In addition to a low-level *API*, there should be a set of high-level programs that send or receive data at the command line.
- The mechanism should not add substantial overhead to the application, so that it may be liberally applied and remain dormant until invoked.

3. Layered Implementation of XPA

The result of our efforts is the "*X* Public Access" mechanism (*XPA*), a layered interface using *X* Toolkit (*Xt*) selections, by which data in an *X* program can be tagged with string identifiers and then accessed externally by other programs. We chose to implement *XPA* using the *Xt* selection mechanism because of the universal availability of the *Xt* Toolkit and also because the selection mechanism hides the platform-dependent implementation of the interprocess communications. Layering *XPA* on *Xt* selections also allows an analysis program to be executed on a remote network host with no additional effort on the part of either the programmer or the user.

The *XPA* interface consists of three main parts:

- The *NewXPA* subroutine is used by *X* programs to tag data with a string identifier and to register send and receive callbacks for these data.
- The *OpenXPA* subroutine is used by external applications to exchange data with an *X* application.
- Two high-level programs, *xpaset* and *xpaget*, allow data to be sent or received from the command line.

The *XPA* mechanism is conceptually rather simple. An *X* application such as *SAOimage* calls the *NewXPA* routine to associate a string name with a particular piece of data. The routine allows one to specify send and receive callback procedures which will be executed by the program when an external process either asks for this data or sends new data. Either of the callbacks can be omitted, so that a particular data item can be specified as being read-only, read-write, or write-only. Having tagged one or more data items in this way, the *X* program creates its graphical interface and enters its usual *X* event loop.

The tagged data are retrieved using either *xpaget* (at the command line) or *OpenXPA* for reading (inside a program). Both take the name of the tag as input. The *xpaget* program returns the data from the *X* application through its standard output. The *OpenXPA* routine (called for reading) returns a Unix file handle which can be used to retrieve the data using standard I/O calls. The *xpaset* program will send data from its standard input to an *X* application, while *OpenXPA* (called for writing) will return a Unix file handle through which data can be written to the *X* application using standard I/O calls. Note that *OpenXPA* simply opens a pipe to *xpaset* or *xpaget*. The *XPA* code is not added directly to the application and thus the *X* libraries need not be linked into the external program.

Arbitrarily large amounts of data can be transferred to and from *X* programs using *XPA* (subject, of course, to limitations on available memory). The system is designed so that the data associated with a given tag can be read and written simultaneously. This is accomplished by having the application manage its own "current" copy of the data, which data are sent to external requesting processes. Newly received data are maintained internally by *XPA* until the user-specified receive callback can be executed safely to replace the old data with the new. Because of this double buffering, a program can retrieve data from *SAOimage*, transform it in some way, and then send the transformed data back to *SAOimage* for display with one filter command, such as: *xpaget* data | transform | *xpaset* data.

Other features of the *XPA* interface include the ability to retrieve the list of tagged data items in an application and the ability of an external program to include an arbitrary "parameter" string in communications with the *X* application, so that the latter can use this information when transferring data.

4. Extensions to the Xt Selection Mechanism

The *Xt* selection mechanism, which allows *X* applications to exchange data, required extensions in order to implement *XPA*. The selection mechanism, as supplied by MIT, relies on timeouts to determine whether a transfer has been interrupted. For *XPA*, we had to disable the *Xt* timeouts in order to allow the computation and sending of data over an arbitrarily long period of time. This would be the case, for example, if a convolution program was sending image data to an *X* application one line at a time.

Instead of timeouts, therefore, *XPA* uses two mechanisms to ensure that both sides of a transfer are active. The external receiver code (for *xpaget* or *OpenXPA*) sets a special *X* property on an *X* window. The sending process monitors this property. If the receiver terminates, the property will disappear and the sender can take appropriate action to abort the transfer and clean up. On the other hand, to monitor a sending process, the receiver code checks to make sure that the owner of the selection is the same window as the original owner when the receive request was issued. If the selection owner changes, this means that the original sender is no longer sending data. The receiver then can take appropriate action.

5. Use of XPA in SAOimage

As previously mentioned, the *XPA* interface will be used by *SAOimage* to allow any external program to retrieve information or to control the display directly, by accessing the tagged data items in the program. Programs will be able to load images, modify the colormap scheme, get or set regions of interest, etc. Note that *SAOimage* itself will use the *XPA* functionality to "externalize" algorithms that usually are considered to be "internal". One very important application of the externalization of data transformation algorithms is the use of separate file formatting programs to convert different data formats into a standard *FITS* format that will be understood by *SAOimage*. Under this scheme, new file formats will be supported simply by writing a *FITS* converter that interprets the data and sends it directly to *SAOimage*, without having to change the *SAOimage* program itself. This will allow different projects to display their own private file formats easily.

6. Conclusion

XPA is a simple interface that allows any program to communicate arbitrary data to and from an *X* program. By facilitating the sharing of data between programs, *XPA* takes us one step closer to an era in which our different analysis tools and systems can work together as an integrated whole.

Acknowledgments. This work was supported under NASA contracts to the *IRAF* Technical Working Group (NAGW-1921), the *AXAF* High Resolution Camera (NAS8-38248), the *AXAF* Support Team (NAS8-36123) and the *AXAF* Science Center (NAS8-39073).

Astronomical Data Analysis Software and Systems III
ASP Conference Series, Vol. 61, 1994
D. R. Crabtree, R. J. Hanisch, and J. Barnes, eds.

Off-the-Shelf Control of Data Analysis Software

Stephen Wampler

Gemini Project, 950 North Cherry Ave, Tucson, AZ 85726

Abstract. The Gemini Project has proposed using low-cost, off-the-shelf software to abstract out both the control and distribution of data analysis from the functionality of the data analysis software. The goal is to allow users to select analysis routines from both ADAM and IRAF as appropriate, distributing these routines across a network of machines.

1. Problem Statement

The international collaboration behind the Gemini Project makes the issue of data reduction somewhat more interesting than with many telescope projects. In particular, there are parties that prefer ADAM as well as parties that prefer IRAF for data reduction and analysis. The Gemini Project believes that the telescope system itself should not impose a particular data reduction package system upon users, but should allow convenient access to the tools available in a variety of packages. Some of the requirements for this common interface are:

A. *Accessible* - Users of ADAM or IRAF should feel comfortable using the data reduction interface.

B. *Flexible* - It should be possible to interface to other data analysis tools than just ADAM and IRAF, when appropriate.

C. *Distributable* - It should be easy to distribute processes across a heterogeneous network. Parallel execution should be possible.

D. *Simple* - The user interface must be simple to learn and use, regardless of the user's background.

E. *Connectivity* - There should be a mechanism for handling the different data formats required of the underlying reduction packages.

F. *Cheap* - Since the Gemini Project is not prepared with a massive infrastructure for developing and maintaining this interface, it must be low-cost, both in development and in maintenance.

2. Off-the-Shelf Software

Given the large cost of developing and subsequently maintaining *in-house* software, it makes sense to look for available tools. If the source code for these tools is available, then they may be useful even if they don't completely cover the required functionality.

In general, adapting a generic solution to a specific problem is less costly than developing a specific solution from the ground up. Part of the reason for this is that most projects contain a great deal of software that duplicates functionality found in other projects (the "infrastructure"). A general solution typically provides the majority of this infrastructure.

3. Khoros

The Gemini Project has looked at several tools for this task. One that shows particular promise is the *Khoros* scientific visualization system, originally developed by the University of New Mexico and now provided by Khoral Research, Incorporated. The salient features of Khoros that are needed here are:

A. the *Cantata* visual programming language provided as part of the Khoros system, and

B. the support in Cantata for distributed processing.

4. Matching the Problem Requirements to Khoros

It is useful to examine the earlier requirements in the context of using Khoros:

A. *Accessible* - Since the functional units would be those currently present in ADAM and IRAF, the behavior of these programs remains consistent with prior use.

B. *Flexible* - Khoros can easily incorporate existing software from a variety of sources. This is simpler if the software was developed under the Unix paradigm of small functional units, but not a requirement.

C. *Distributable* - Cantata provides the user with a "point-and-click" selection method for distributing programs across a network. Khoros runs on most Unix platforms, from Linux and 386BSD (free Unix systems for 386/486 computers) to Crays. In addition, the user may easily select a preferred transport mechanism (shared memory, pipes, files, sockets, etc.) for each connection between programs. Finally, Cantata manages control flow automatically and is capable of scheduling parallel execution along separate control flow lines.

D. *Simple* - The Cantata visual language is extremely easy to learn while still providing the features commonly found in shell languages - selection and looping constructs, variables, and procedures.

E. *Connectivity* - Khoros version 2.0 provides a *data transport abstraction* that can be extended to handle file format conversions as appropriate between packages.

F. *Cheap* - For the use envisioned here, Khoros is free. (There are restrictions on *redistributing* Khoros, but not on *acquiring* it.)

5. Sample Cantata Programs

To illustrate the flavor of using Cantata, the figures show a few programs written in Cantata. Figure 1 is a "canned" program developed outside the astronomical community that provides a simplistic edge detection operation. Figures 2 shows the input and output to this program. Figure 3 shows a considerably more complex Cantata program, designed to reduce motion-blur effects in images. Sample input and output for this program are given in Figure 4.

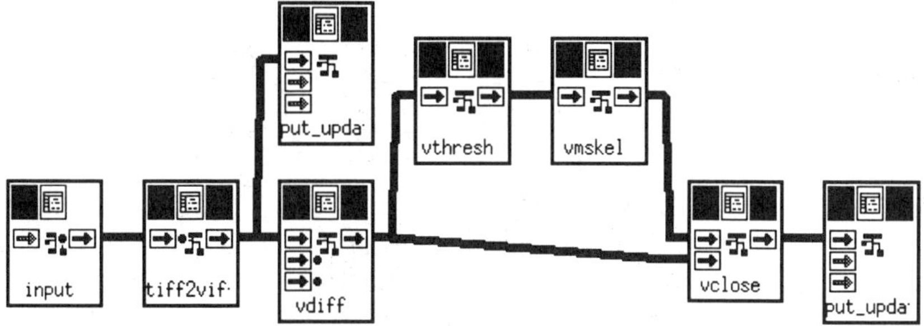

Figure 1. A simple edge detection program.

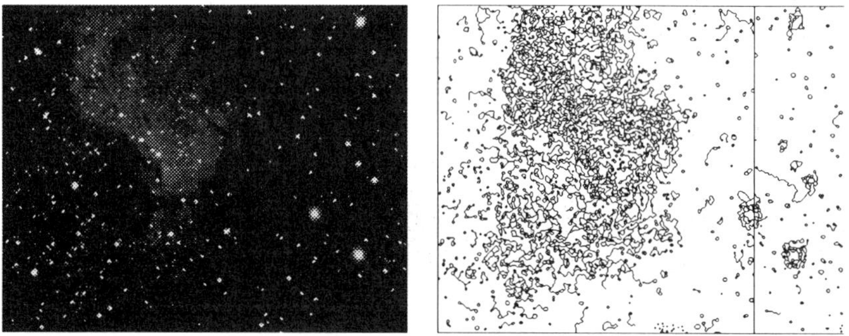

Figure 2. Input and output of edge detection program.

6. What Needs to be Done

While Khoros provides a great deal of the needed infrastructure, there are still missing pieces:

A. The "wrappers" for individual modules in ADAM and IRAF need to be written. It should be possible to automate this task, in a manner similar to the way that Dave Terrett has for adding TCL/Tk interfaces to ADAM(Terrett 1993).

B. The automatic format conversions between modules requires the most work in order to avoid unneeded conversions. The current version of Khoros (1.5) is missing the data transport abstraction that is part of Khoros 2.

References

Terrett, D. 1993, ADAM Workshop presentation

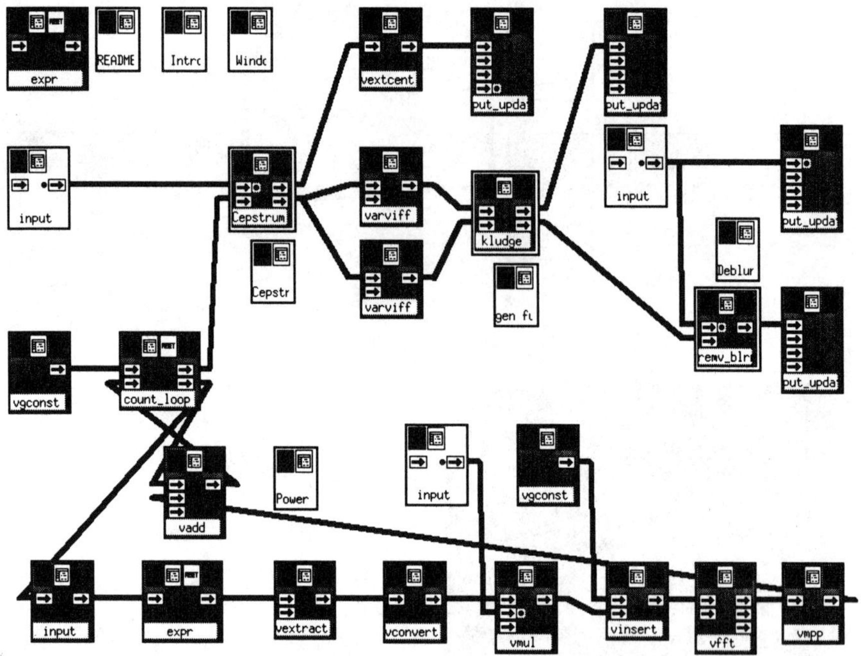

Figure 3. A motion-blur reducing program.

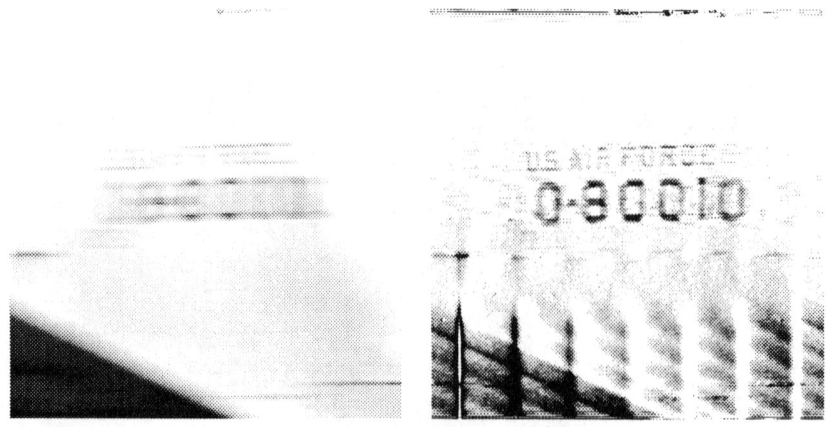

Figure 4. Input and output of motion blur program.

Astronomical Data Analysis Software and Systems III
ASP Conference Series, Vol. 61, 1994
D. R. Crabtree, R. J. Hanisch, and J. Barnes, eds.

Retrospective of the Software Metrics and Staffing Profile for COBE Ground System Software

Karin Loya Babst

Computer Sciences Corporation, Goddard Space Flight Center, Greenbelt, MD 20771

Richard J. Hollenhorst and James E. Stephens

Code 733, Goddard Space Flight Center, Greenbelt, MD 20771

Abstract. A large amount of software (S/W) for all ground systems elements of the COsmic Background Explorer (COBE) was considered to be "launch-critical", including S/W to support the scientific aspects of the mission. The latter S/W was comprised of the data structures, the data management S/W required to access them, mission planning S/W for establishing the mission timeline and for the preparation and validation of instrument command loads, ingest S/W for the acquisition of data and its migration into the data structures and science data processing S/W for attitude determination and instrument data refinement.

This paper presents, for the period which began with requirements analysis and culminated in the launch capability, the S/W metrics employed in estimating/tracking the work, a correlation with the detailed staffing profile which achieved COBE scientific S/W's launch-readiness, and a retrospective correlation of the latter with a profile generated by a COCOMO model-based tool.

1. Introduction

Estimating S/W size, especially for a large endeavor, is a non-trivial task. It is undertaken early in the S/W lifecycle to achieve a rough order of magnitude cost estimate which feeds into the project planning and scheduling process. Sizing estimates allow staff phasing trade-offs to be considered and controlled. A baseline is established by which quantitative progress can be monitored. Quantized changes to the baseline may be incorporated as requirements are developed in more detail. Monitoring progress against the plan through metrics is mandated for any project which has critical time constraints (as well as a cost ceiling) for the delivered system, to assure that the work will complete on schedule. The metrics provide management with a mechanism for early identification of problem areas Regular maintenance and reporting of metrics allow project managers to objectively monitor the project progress and effectively maximize the team productivity. Ultimately, the metrics for a project facilitate estimating and planning for future projects. This paper documents the COBE experience.

2. COBE

The COBE satellite, designed and built by the GSFC, housed three instruments, the Differential Microwave Radiometer (DMR), the Far InfraRed Absolute Spectrophotometer (FIRAS), and the Diffuse InfraRed Background Experiment (DIRBE), the latter two cryogenically cooled. COBE's objective to measure the cosmic microwave background radiation over the entire sky was achieved in six months; with all measurements continuing until the cryogen expired 10 months after launch. The computer hardware (h/w) and S/W systems for data acquisition and subsequent ground systems data processing were managed by the GSFC. The h/w and S/W environment that supported the science data processing was known as the COBE Science Data Room (CSDR).

The development of the CSDR S/W, comprised of VAX/VMS Fortran code, followed a modified waterfall model with separate S/W builds or releases, each offering increased functionality. This incremental build approach was chosen for its obvious advantages: the operational use (from Integration and Test on) of the core subset of the S/W is maximized, ensuring stability and robustness of critical elements of the system. The incremental build approach allows evaluation and refinement of the user interface, establishes system scaffolding that makes the development process more efficient; and allows early test of external interfaces which, if problematic, have the potential for a higher cost/schedule impact than internal interfaces. The build approach also establishes and exercises configuration management procedures early in the development period.

3. Metrics, Results

GSFC's chosen metric for estimating and monitoring the development was the functional module. The rationale was that the specified functionality could be adequately identified from prime/derived requirements, and broken down into small representative units. For planning and scheduling the work, this seemed to be a reasonable metric because each functional unit has an associated overhead due to the common work performed on it as a unit: compilation, configuration, and testing. The functional module was assumed to average approximately 50 Lines of Code (LOC) for estimation purposes. An automated module tracking utility program was developed to produce graphs that showed the actual number of modules configured/tested versus the estimated functional modules over the period from 1986 through 1989. Figure 1 depicts the progress as of Build 4 and shows that the functional modules completed according to schedule.

At COBE's launch, 8331 modules (488,197 LOC) were configured, averaging 58.6 LOC each. The planned number of functional modules (2525) to be completed by launch was at least matched by an equal number of non-functional modules. As well, some transported code was counted at that time, although it had been configured earlier. The S/W consisted of approximately 62% new code, 23% transported, and 15% adapted.

The reconstructed staff profile for the CSDR S/W development reflects the period from the System Concept Review through launch. This profile represents the directly- charged staff, but it does not include additional Science Working Group resources who supported independent functional validation of

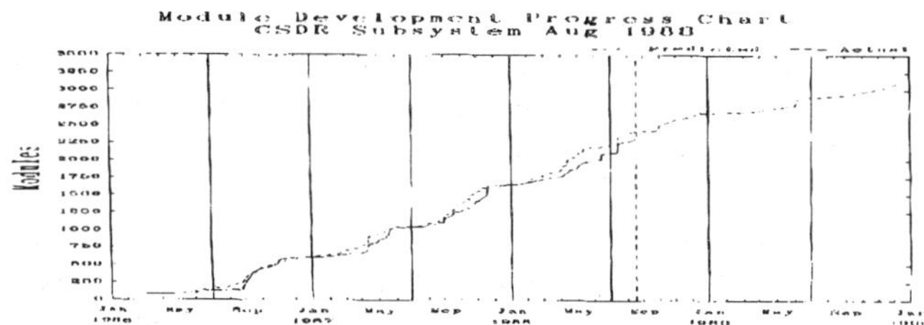

Figure 1. August 1988 CSDR Development Progress Chart

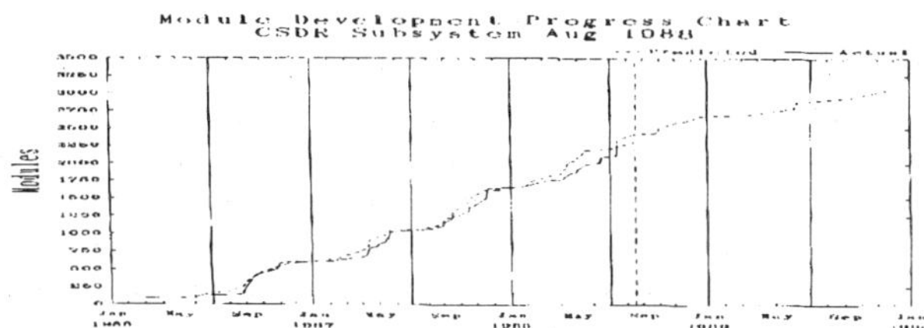

Figure 2. COBE Science Data Room Staff Profile

the algorithms. The total cumulative staff- months for the CSDR S/W was approximately 2013 as shown in Figure 2.

4. COCOMO and COCOMO-based SOFTE

Several forms of the COnstructive COst MOdel (COCOMO) for S/W estimation, from basic to detailed, are described by Barry W. Boehm in Software Engineering Economics, published by Prentice-Hall, Inc., in 1981. COCOMO-based estimation has been applied and evaluated over a sample range of applications, with results accurate within a margin of 20% of actual costs for 68% to 70% of the time, according to Boehm.

Use of COCOMO requires that certain underlying assumptions are met, abbreviated as follows: 1) Software Life-Cycle is Waterfall (or a refinement); 2) Life-cycle phases are Feasibility, Requirements, Product Design, Detailed Design, Integration, Operational Implementation, and Maintenance; 3) Primary cost driver is Delivered Source Instructions (Lines of Code or LOC in this paper); 4) COCOMO's estimate is from Design to Delivery; 5) Non-productive slack time is small (this sweeps up the assumption that the management is good: organized, plans well, staffs appropriately, keeps team motivated and committed); 6) Requirements specification does not change substantially; 7) Peripheral activities such as user training are not included; 8) All direct-charge labor (including project management, administration, librarian, configuration management and test staff, systems engineers) is included in the cost estimate. Facility, institutional, and corporate staff costs are not included.

The CSDR S/W development met these assumptions. To answer the question of whether COCOMO could have been used early to predict the total staff months required for the quantity of work represented by the total LOC, a tool was used. One of many COCOMO- based tools which have been developed and utilized in the S/W community is CSC's SOFTware Estimator (SOFTE). SOFTE implements the basic COCOMO model with an enhancement which adds requirements definition phase (5% of the total labor). With specific inputs, SOFTE generates a "shortest possible" project schedule, an associated staff profile, and statistics such as total project staff months, expected months to delivery, etc. The project management may then tune the plan; i.e., level the staff and extend the schedule because of other objectives (such as keeping expertise available after delivery of a system to support first real operational usage of the S/W) or fiscal year budget constraints.

SOFTE requires a breakdown of the total S/W size (Delivered Source Instructions or LOC) in the following categories, and allows a separate productivity rate to be established for each: New, Adapted (developed for another application), Converted (to another language), Transported (to a different platform). The SOFTE default productivity rates (in LOC per 8 hour staff day) which were accepted as "reasonable" by the authors and used for the results presented in this paper are: New Code, 7.5 LOC/Staff Day; Adapted Code, 15 LOC/Staff Day; Converted Code 30 LOC/Staff Day, and Transported Code, 60 LOC/Staff Day. SOFTE assumes 157 productive hours per month per staff member.

The productivity rates shown above were used with the following CSDR measured LOC data: 302,200 New LOC, 76,000 Adapted LOC, and 110,000 Transported LOC. This resulted in an estimate of 2403 staff-months for the project. Compared with the measured number of 2013, it is 19% greater (falls within the 20% margin predicted for COCOMO). Since in fact SOFTE adds 5% for requirements definition, the 2403 staff-months, reduced by 5%, brings the estimate within 13% of the actual. Considering that the tool-based result included validation resources that were not included in the CSDR's staffing profile, the gap between the estimated and actual narrows further.

5. Summary/Conclusions

COBE's scientific mission has been highly successful. The CSDR h/w and S/W environment provided an exceptionally stable and robust basis for the necessary data processing: there were no computer hardware or software failures during the L&EO period. The CSDR development was characterized by an effective management structure, good tools, policies, procedures and decisions that resulted in a high-quality software system. In monitoring the progress, the functional module proved to be highly effective as a metric. With the CSDR's early estimates a COCOMO-based tool could have been meaningfully employed to predict staff-months.

Acknowledgments. The COBE Science Data Room S/W development was managed by the Goddard Space Flight Center under Contract NAS5-30726 and sponsored by NASA's Astrophysics Division. The authors express appreciation to Mr. Glenn Harris and Mr. Feng-Yueh Huang of CSC for their S/W engineering consultation which was invaluable.

Astronomical Data Analysis Software and Systems III
ASP Conference Series, Vol. 61, 1994
D. R. Crabtree, R. J. Hanisch, and J. Barnes, eds.

Compression Software for Astronomical Images

J. P. Véran[1], J. R. Wright

Canada-France-Hawaii Telescope Corporation[2], P.O. Box 1597, Kamuela, HI 96743

Abstract. Astronomical images are usually very noisy and don't shrink well with the typical lossless compression algorithms. COMPFITS is a utility program which analyses and pre-processes the images before sending them to the lossless compression program of your choice. The compression is still lossless (i.e., the reverse process gives a file identical to the original) but the compression ratio achieved is usually much better. This allows fast, simple compression programs to achieve almost the same performance as time consuming, more sophisticated ones. Therefore COMPFITS is very efficient whenever computing time (for compression and/or decompression) is a critical parameter. Test charts are given to show how COMPFITS compares with other programs compressing unprocessed data. Tests were also carried out on medical images, which are also noisy, but the results are different. This stresses the fact that compression of astronomical images is a specific problem. Options are available to perform lossy compression as well. COMPFITS is freely available and can be retrieved via anonymous ftp.

Warning: Because of space constraints, we couldn't publish the paper initially written. This is only a summary of it. The full paper is much more detailed and contains more results. It can be obtained by anonymous ftp from *ftp.cfht.hawaii.edu*, file *full_paper.ps* in the *pub/compfits* directory.

1. Introduction

Currently, large format CCDs are widely used in ground based telescopes. At the moment, CFHT operates several 2048 × 2048 pixel (2K by 2K) CCDs with 16 bits/pixel, leading to 8 megabyte images. With about 100 images taken each night, up to 800 megabytes per night needs to be stored as a permanent record and later made available to the astronomical community. Soon, the new 4K by 4K CCDs will be available, and even larger formats are already in sight. To store such a tremendous quantity of data has become a real problem, as even optical disk technology can hardly keep up.

---

[1]Formerly with Ecole Nationale Supérieure des Télécommunications, Paris, France.

[2]Operated by the National Research Council of Canada, le Centre National de Recherche Scientifique de France, and the University of Hawaii.

Compression is meant to address this storage problem. As a requirement for a permanent archive database for astronomical images, it is necessary that the uncompressed image is bit-for-bit identical to the original image; that is, the compression has to be lossless. It is also necessary that the compression process does not require too much computation time so as to keep up with the data flow in the telescope archive pipeline (i.e., the process that transfers any frame taken at the summit observatory down to the corporation headquarters and stores it permanently, currently on optical disk).

Many lossless compression programs are widely and publicly available. However, except one called FITSPRESS, none of them is specific to astronomical images and therefore none of them achieves an acceptable performance on them. FITSPRESS (Press 1992) achieves much better results than the usual compression programs on astronomical images. However, it requires a much longer computation time, which makes it not suitable for the CFHT archive pipeline.

2. Theory of Operation

Because of imperfections in the acquisition process as well as intrinsic effects, an astronomical image usually contains a lot of noise. This noise mostly impacts low order (least significant) bitplanes in which, at least for compression purposes, it can be considered as a white uncorrelated random signal. It is a well known fact that such a signal can't possibly be compressed and is even likely to grow upon any attempt of compression. Our guess is that this is exactly what happens with the usual compression programs: even if the high order bitplanes shrink very well, the low order ones actually grow, making the overall performance poor.

Our program COMPFITS is designed to prevent this undesirable effect. First the image is translated from the usual binary code to the Gray code, in an attempt to decorrelate the different bitplanes and increase the coherency of each of them. Then, the first order entropy of each bitplane is computed, starting with the least significant one. The first order entropy, quantity commonly used in Information Theory, is, for the bitplane of order n, defined as:

$$H(n) = p_0(n)log_2 p_0(n) + p_1(n)log_2 p_1(n) \tag{1}$$

where $p_0(n)$ (resp. $p_1(n)$) is the probability of occurance of the symbol 0 (resp. 1) in this bitplane. The low order bitplanes contain mostly random noise, therefore their entropy is very close to 1 bit/pixel. The high order bitplanes should contain almost no noise, therefore their entropy is normally much less than 1 bit/pixel. The bitplane for which the entropy becomes slightly different from 1 bit/pixel defines the limit between a low order partition (LOP) of bitplanes, deemed too noisy to be compressible and a high order partition (HOP) of bitplanes, deemed worth to be compressed. COMPFITS then proceeds to send the HOP to an external compression program. The concatenation of the LOP (NOT compressed) and the compressed HOP then makes up the lossless compressed image. This image is usually much smaller than the compressed image obtained by running the same external compression program directly on the original image. The value of $1 - H(n)$ beyond which the corresponding bitplane is considered to belong to the HOP is a parameter that has been determined experimentally so as it gives the optimal partition most of the time, optimal meaning that the same process with any other partition yields a bigger compressed image.

3. Implementation

COMPFITS needs an external program to do the compression work on the HOP and therefore is not a compression program per se. Communication with the external compression program is made through pipes so that the overhead required by COMPFITS is minimal. Any lossless compression program able to read from its standard input and write to its standard output is able to work with COMPFITS. The user can give the name of the one he wants on the command line of COMPFITS, along with the options to be passed to it. However a program that assumes a 16 bit input stream is obviously the most suitable.

An option is available to turn COMPFITS into a lossy compression program. When enabled, the lossy mode saves only the compressed high order partition and discards the low order partition. The compression ratio achieved is of course much better but the reconstruction will not be exact as the least significant bitplanes are lost. However, the quality of the reconstructed image will be good as the discarded bitplanes are presumed to contain white random noise. No particular efforts have been made to quantify this quality nor to enhance this feature.

In a general archive prospect, it seems useful to keep the FITS header (if any) uncompressed and easily available at the beginning of the compressed file. Two special cards are added to it so a decompression process may find out the compression parameters it needs (especially the partition used).

The routines are written in C. The full package is now publicly available by anonymous ftp from *ftp.cfht.hawaii.edu* in the *pub/compfits* directory.

4. Results and Performance

The external compression program found to work best with COMPFITS is the 16 bit Lempel-Ziv based COMPACT 1.0, written by Gene H. Olson. The plot in Figure 1 shows how COMPFITS + COMPACT 1.0 compare with other compression programs in term of the size of the compressed image. The set of astronomical frames chosen includes object images (actual image of the sky, most of the frames taken), bias images (read out of the CCD with no exposure time allowed) and flat images (image of an area of uniform illumination). The graph shows that there is not much difference between the performance of the different algorithms for highly compressible images. However, for slightly compressible images (= very noisy), COMPFITS allows one to gain up to a factor of two in the compression compared to Unix COMPRESS and can be more efficient than FITSPRESS, our reference in term of compression ratio. Typical computing time for the compression of an eight megabyte image on a Sun SPARC10 is about 60 seconds for Unix COMPRESS, 34 seconds for COMPACT 1.0, 35 seconds for COMPFITS + COMPACT 1.0 and 210 seconds for FITSPRESS.

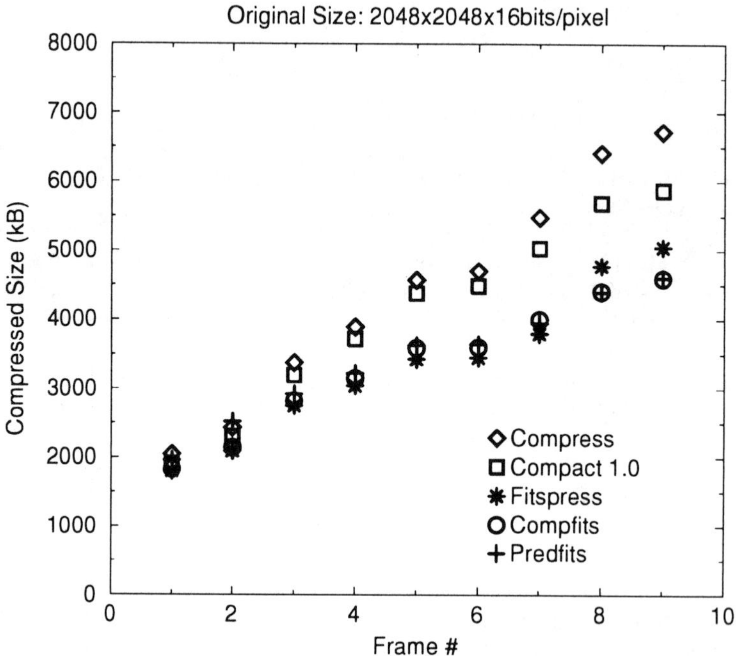

Figure 1. Compression of astronomical images.

Acknowledgments. We would like to thanks all the members of the CFHT software group for their help and support, especially Bernt Grundseth, head of the group, for initiating and supervising this research project. We are also grateful to the following persons for their valuable suggestions and comments: L. Berman (National Library of Medicine, MD), B. Lucier (Perdue University), J. Glaspey (CFHT), N. Farvardin (University of Maryland), H. Maitre (Ecole Nationale Supérieure des Télécommunications, Paris, France), W. Press (Harvard-Smithsonian Center for Astrophysics, MA) and P. Tischer (Monash University, Victoria, Australia).

References

Press, W. H. 1992, "Wavelet-Based Compression Software for FITS Images", in Astronomical Data Analysis Software and Systems I, A.S.P. Conf. Ser., Vol. 25, eds. D.M. Worrall, C. Biemesderfer & J. Barnes

Astronomical Data Analysis Software and Systems III
ASP Conference Series, Vol. 61, 1994
D. R. Crabtree, R. J. Hanisch, and J. Barnes, eds.

FITS Image Compression Using Variable Length Binary Tables

W. D. Pence

NASA/GSFC, Greenbelt, MD 20771

Abstract. This paper describes a technique in which the compressed byte stream from each row of a FITS image is stored in the corresponding row of a FITS binary table with a variable length vector column. The advantage of this technique is that the compressed image is itself also a valid FITS file and thus is completely self-documented via the header keywords and is naturally machine-independent. Also, since each row of the image is stored independently it is easy to randomly access any section of the image.

A prototype implementation of this compression scheme has been written as an add-on set of subroutines for the FITSIO subroutine library for reading and writing FITS files. Some of the features of this prototype and suggestions for future work are discussed.

1. Introduction

The FITS format is now routinely used to store and distribute astronomical data files, and it is often desirable to compress the files to reduce the disk storage space requirements and decrease network transmission times. This paper describes a particular compression technique for FITS images in which each row of the original FITS image is compressed and then stored in a FITS binary table with a variable length vector column. Each row of the binary table then corresponds to the same row in the original FITS image. The main advantages of this technique are (1) the compressed image is itself a valid FITS file and thus, is completely self-documented via the header keywords and is naturally machine-independent, and (2) it provides easy random access to any part of the image since each row is stored separately. This second feature can also be a disadvantage however because a single row may not contain enough pixels to be efficiently compressed. Some ideas on overcoming this disadvantage are discussed in the "Future Work" section below.

2. Implementation

A prototype implementation of this compression scheme has been written as an add-on set of subroutines for the FITSIO subroutine library (Pence 1992) for reading and writing FITS files. Application programs can call this subroutine interface to simply read or write a specified row of an image without knowing anything about the underlying compression algorithm; the subroutine interface transparently determines whether the image is compressed, and if so, does the required compression or decompression on the fly as the data is transferred to or from the FITS file. In the current prototype a single simple compression algorithm has been implemented which works just on integer data. This is only suitable for certain types of images, but it has achieved compression ratios greater than 10 on various sparse X-ray images. In principle, support for any number of additional compression algorithms could be added to this interface to deal with a other kinds of images. A keyword in the header of each compressed FITS file would identify which particular compression algorithm had been used on the file, and would be used to determine which decompression routine to use to read the data. One advantage of this type of scheme is that application software can directly access the compressed data, which means the user requires much less diskspace on which to hold the data files. This is unlike other data compression schemes (e.g., using the Unix compress program) which require that the user uncompress the entire data file before it can be analyzed by other software.

This prototype also demonstrated another advantage of using compressed FITS files: it was much faster to read the compressed image than it was to read the original image because the smaller number of disk I/O operations more than compensated for the additional processing time needed to uncompress the file.

3. Future Work

This project has raised a number of issues that would need to be addressed before this prototype would be more generally useful:

1) More powerful and robust compression algorithms need to be supported. Fortunately there are many compression algorithms freely available in the public domain, so it should not be a problem incorporating them into this interface.

2) Having to compress each row of the image separately is too restrictive, so a more general way of compressing different subsections of an image at a time are required. On one hand one would like to compress large sections of the image since some algorithms only achieve maximum efficiency when dealing with long sequences of bytes. On the other hand, compressing smaller chunks of the data improves the ability to randomly access different parts of the image without having to uncompress many unneeded image pixels. One solution to this problem would be to allow an image to be subdivided into an an arbitrary number of rectangular subimages, and then compress each subimage separately. The FITS file designer would then have the flexibility of choosing the optimal number and size of subimages to use on any given image. At one limit this could be used to compress the entire image in one big subimage. Or it could be used to compress each row of the image separately as was done in the current prototype. Under this scheme additional columns would need to be added to the table to identify the boundaries of each compressed subsection.

3) Similar compression techniques need to be developed to handle FITS table data, as well as FITS images. Increasingly, large data sets are being produced in the FITS BINTABLE format so it is becoming more critical to find effective ways to compress these types of files. This is a more difficult challenge because of the heterogeneous types of data that can be represented in different

columns of a FITS table, so that the same compression algorithm may not work effectively on all columns. One may need to devise methods in which each column of the table is compressed separately with it's own optimal compression algorithm.

References

Pence, W. D. 1992, in Astronomical Data Analysis Software and Systems I, A.S.P. Conf. Ser., Vol. 25, eds D. Worrall, C. Biemesderfer & J. Barnes, 22

Astronomical Data Analysis Software and Systems III
ASP Conference Series, Vol. 61, 1994
D. R. Crabtree, R. J. Hanisch, and J. Barnes, eds.

Investigating HDF as an Astronomical Transport and Archival Format

D. G. Jennings, T. A. McGlynn and J. M. Jordan

Compton Observatory Science Support Center, CSC/GSFC, Greenbelt, MD 20771

Abstract. This paper investigates the viability of using the *Hierarchical Data Format* (HDF), developed at the National Center for Supercomputing Applications, for astronomical data transport and archiving. First we provide an overview of the HDF structure and software interface. Next, we list advantages and disadvantages of using HDF as a transport and archival format. Finally, we postulate several possible enhancements to HDF based upon the attributes of FITS, as well as several enhancements to FITS based upon the attributes of HDF, that could increase the long term usefulness of both formats.

1. Introduction

The Flexible Image Transport System (FITS) has become a widely accepted and very successful standard for data transport and data archiving in the astronomical community. While FITS performs its function as a data standard admirably (especially since the addition of the BINTABLE binary table extension), it does contain several design limitations that might impede its long-term usefulness.

In an effort to stimulate discussion and debate on the evolution of formats in astronomy, we present some initial thoughts on using the Hierarchical Data Format (HDF) for the transport and archiving of astronomical data. HDF possesses several desirable attributes such as a common software interface, data object grouping structures and applicability to other scientific disciplines that makes it a potentially useful format for the astronomical community. Section 2 gives a brief overview of the HDF project and its supported data types. Sections 3 and 4 describe the HDF common software interface and byte structures. Sections 5 and 6 discuss the advantages and disadvantages of using HDF as a transport and archival format. Finally, section 7 investigates the possibility of augmenting FITS with HDF-like characteristics, as well as augmenting HDF with FITS-like characteristics, in order to enhance the general versatility and inter-convertibility of each.

2. HDF Overview

The Hierarchical Data Format (HDF) project began in 1988 at the National Center for Supercomputing Applications (NCSA) in response to their need for a way to store and transport many differing and machine dependent scientific data types in a standard format. Since its conception, HDF has grown in use far beyond its home institution. It is used by many diverse institutions (private industry, national laboratories, universities, NASA) and its appeal crosses the boundaries of many scientific disciplines. HDF is in the public domain, although many proprietary software tools make use of HDF. Unlike FITS, HDF is defined in terms of its application interfaces. Thus, **HDF is both a data format and a Common Software Interface (CSI).**

HDF currently supports six data structure types, or data models: 8-bit raster images, 24-bit raster images, color palettes, scientific data sets (similar to FITS primary arrays), annotations (text entries, similar to FITS ASCII tables), and Vdata (similar to FITS binary tables).

3. The HDF Common Software Interface (CSI)

The HDF common software interface consists of three layers: a low level interface library, data model application interfaces and command-line utilities.

The *low level interface library* is a tool kit for software developers who wish to define new data models within HDF. This level of the CSI performs file I/O, error handling, memory management and physical formatting of data.

The *data model application interfaces* consist of six independent modules, each specifically built to handle one of the supported HDF data models. They are designed to hide the details of data formatting from the user and can be called from either C or Fortran programs.

The *command-line utilities* are a collection of programs distributed with the HDF CSI. The functionality of the command-line utilities ranges from general purpose, such as listing the contents of an HDF file, to special purpose, such as converting data between different HDF data types (e.g., raster images to scientific data sets).

In addition to the HDF CSI, many HDF-based data analysis tools exist such as the NCSA Visualization Tools (NCSA 1993), SpyGlass (SpyGlass 1993) and IDL (RSI 1993).

4. HDF Structure

HDF data files are collections of *data descriptors* and *data elements*. A data element, along with its associated data descriptor, constitutes the fundamental unit of an HDF file known as a *data object*. Data objects contain data (data element) and a 12 byte header (data descriptor) that identifies the structure of the data contained by the data object.

Data descriptors consist of 4 fields that determine their associated data element's type, sequence and position within the file. Data elements are the raw data part of a data object, stored as a set of contiguous bytes starting at the byte offset specified in the data descriptor.

The data descriptors of an HDF file are grouped into structures of linked lists known as *data descriptor blocks* or DD blocks. Each DD block contains a header that tells how many data descriptors belong to it and where the next data descriptor block within the file can be found. Storing the data descriptors separately from the actual data in linked lists allows for quick random access of data elements within an HDF file.

Instead of containing data elements, data objects may contain pointers to other data objects. Such data objects are known as *data groups*. Data groups may themselves be assigned to other data groups, allowing the building of complex structures within an HDF file. Members of a data group are accessible either individually or as part of their group.

5. HDF as a Transport Format

HDF possess several attributes that makes it a good astronomical data transport candidate. First, the HDF common software interface actually enforces a practical definition of the data format. The capabilities of the CSI confine the data format structures, and the encoded format syntax acts as a *defacto* standard for the format. In essence, the HDF CSI provides the most effective form of configuration control possible. Second, lots of commercial and NCSA supplied visualization and analysis tools that use HDF are available. Third, HDF development is centrally controlled at NCSA, and thus can adapt quickly to the needs of the scientific community. Last, the hierarchical grouping capability of HDF allows for the encapsulation of very complex data structures.

HDF also possess some disadvantages as an astronomical data transport format. With the notable exception of IDL, most astronomical data analysis packages do not (yet) support HDF; therefore, it is not easy to make quick use of HDF data files. Futhermore, a simple byte dump of a HDF file will not reveal the file's contents. This makes it difficult to determine the content, or even the data format, of HDF data files without special software and/or *a priori* knowledge of the data file origination.

6. HDF as an Archival Format

HDF has several properties that make it a potentially useful archival format. First, the HDF CSI assures that every HDF file written adheres to a strict, universally agreed upon and implemented rule set; this is very important if future researchers are to understand the contents of large archival data sets. Second, the hierarchical grouping capability of HDF allows very complex data structures to be represented with a minimum of data "massaging". Also of note, NASA's Earth Observing System Data and Information System (EOSDIS) will use HDF is its archival and transport format. This alone should assure that HDF will be a supported format for decades to come.

Unfortunately, HDF has several problems as an archival format. HDF is tied to a single organization (NCSA); therefore, it is subject to the fortunes of that organization. Another important drawback to HDF is that its data format specifications are not themselves archived in any permanent manner (ie., published in journals). This endangers their accessibility to future generations of researchers. HDF also lacks a human readable metadata description, such as the FITS header, that both accompanies the contents and acts as part of the data. This means that all data archived in HDF relies upon external software tools and documentation to describe the meaning of its data objects.

7. Combining the Best of Both (Data Standard) Worlds

Evaluating the usefulness of data formats "foreign" to astronomy has the associated benefit of pointing out some of the strengths and weaknesses of the FITS format, as well as those of the format under scrutiny. We believe that both FITS and HDF could gain by adopting some of the attributes possessed by the other.

For example, FITS lacks a set of common tools for reading and writing data files. By adopting a universally used common software interface such as the HDF CSI, the astronomical community could assure that its membership produces syntactically correct and universally readable FITS files that make use of the latest data structures (ie., image extensions). Tools such as FITSIO and FTOOLS (Pence and Jennings 1993) are good candidates for a FITS CSI. FITS also lacks an effective way to allow hierarchical groupings of data objects. By defining a new FITS extension to mimic the hierarchical grouping capabilities of HDF data objects, FITS files could accommodate data structures of much greater complexity than they are now capable of doing.

HDF would benefit by requiring human-readable metadata descriptions of its data models, built according to rules similar to those used for FITS headers, and by adopting the "primary header" convention of a human-readable manifest as the first data object in a HDF file. Future users of HDF would also benefit from the publishing of HDF format specifications in recognized journals.

8. Summary

Investigating the use of other data formats such as HDF is a worth-while exercise because it allows the astronomical community to stay abreast of the data standards used by other scientific disciplines and points out possible deficiencies in our own data transport and archiving schemes.

While HDF is by no means an ideal replacement for FITS, it could be reasonably utilized as an astronomical data transport and archival format. Since many other scientific disciplines use HDF, and its use is growing, we believe the astronomical community should be considering ways to enhance FITS (and requesting enhancements to HDF by NCSA) so that at least conversion between the two formats is possible.

For more information on HDF see the documents *Getting Started with NCSA HDF* and *NCSA HDF Specifications*. Both are available via anonymous ftp on **ftp.ncsa.uiuc.edu**.

References

NCSA, 1993. "PolyView Users Manual, ImageTool Users Manual, IsoVis Users Manual", NCSA, University of Illinois at Urbana-Champaign

SpyGlass, 1993. "SpyGlass Transform", SpyGlass, Urbana-Champaign, Il

RSI, 1993. "IDL Users Guide", Research Systems, Inc., Boulder, Co

Pence & Jennings, 1993. "FITS Software Update", Legacy, number 3, Goddard Space Flight Center, Greenbelt, Md

Author Index

Subject Index

Colophon

These Proceedings were prepared with LaTeX using a style file customized for the PASP series. Authors submitted manuscripts electronically, either by electronic mail or by depositing files in an designated anonymous FTP area. All of the 118 papers in this volume were submitted with usable LaTeX markup generated by the authors themselves.

Final proof pages for the entire book (from page v on) were produced at DAO on an HP Laserjet IIIci. Tomas Rokicki's *dvips* program was used to translate the device-independent output from LaTeX. This combination of PostScript printer and driver program enabled us to take advantage of the PostScript language to merge figures into the pages as they were printed.

64 of the papers submitted were accompanied by 158 figures. Of these papers, 57 authors submitted their figures as Encapsulated PostScript (EPS) files. There were 143 figures submitted as EPS files, and we were able to incorporate 142 of them directly into the manuscripts as part of the printing process. 56 papers with figures were generated completely electronically requiring no cutting and pasting. 1 paper included EPS files for some figures but not all. The automatic inclusion of EPS graphics is advantageous and produces a superior product, and it is our hope that graphics-producing software will evolve toward a common implementation of the EPS standard. Already, we see a vast improvement in the production of EPS figures over the past year when we compare the results of this Proceedings with that of last year.

Most of the front and back matter was generated mechanically from the material submitted by the authors. A database of pertinent information about each paper was maintained, and the table of contents, author index, etc. were derived from it. The overall pagination of the volume was determined by software after the papers were edited and ordered, and the running heads and folio numbers were applied by LaTeX using standard markup commands. The software that was developed for these chores is generalized so that it can be used for other projects, and it is available to other editors through the PASP Conference Series Office.